TURBULENT PREMIXED FLAMES

A work on turbulent premixed combustion is timely because of increased concern about the environmental impact of combustion and the search for new combustion concepts and technologies. An improved understanding of fuel-lean turbulent premixed flames must play a central role in the fundamental science of these new concepts. Lean premixed flames have the potential to offer ultra-low-emission levels, but they are notoriously susceptible to combustion oscillations. Thus sophisticated control measures are inevitably required. The editors' intent is to set out the modelling aspects in the field of turbulent premixed combustion. Good progress has been made recently on this topic. Thus it is timely to edit a cohesive volume containing contributions from international experts on various subtopics of the lean premixed flame problem.

Dr. Nedunchezhian Swaminathan is a Lecturer in the Department of Engineering at the University of Cambridge and Director of Studies at Robinson College. He has published more than 90 research articles on turbulent flames and combustion and on the numerical simulation of turbulence and combustion.

Professor K. N. C. Bray is Professor Emeritus in the Department of Engineering at the University of Cambridge. He is the author of numerous refereed research publications. Among his many honors, he was elected a Fellow of the Royal Society.

Turbulent Premixed Flames

Edited by

Nedunchezhian Swaminathan
University of Cambridge

K. N. C. Bray
University of Cambridge

CAMBRIDGE UNIVERSITY PRESS
Cambridge, New York, Melbourne, Madrid, Cape Town,
Singapore, São Paulo, Delhi, Mexico City

Cambridge University Press
32 Avenue of the Americas, New York NY 10013-2473, USA

Published in the United States of America by Cambridge University Press, New York

www.cambridge.org
Information on this title: www.cambridge.org/9781107679061

© Nedunchezhian Swaminathan and K. N. C. Bray 2011

First published 2011
First paperback edition 2013

A catalogue record for this publication is available from the British Library

Library of Congress Cataloguing in Publication Data

Turbulent premixed flames / [edited by] Nedunchezhian Swaminathan,
Kenneth Bray.
   p. cm.
Includes bibliographical references and index.
ISBN 978-0-521-76961-7 (hardback)
1. Combustion engineering. 2. Flame. 3. Turbulence. I. Swaminathan,
Nedunchezhian. II. Bray, K. N. C. (Kenneth Noel Corbett), 1929– III. Title.
QD516.T84 2011
621.402'3 – dc   222010041887

ISBN 978-0-521-76961-7 Hardback
ISBN 978-1-107-67906-1 Paperback

Contents

Preface

Turbulent combustion is rich in physics and is also a socioeconomically important topic. Much research has been carried out in the past half century, resulting in many important advances. It is simply impractical to review and discuss all of these studies and to explain their results in a single book. Many volumes published in the past reviewed such material, and we draw attention in particular to the two volumes edited by Paul Libby and Forman Williams, published in 1980 by Springer-Verlag and in 1994 by Academic Press, covering both non-premixed and premixed turbulent flames and the book by Norbert Peters published by Cambridge University Press in 2000.

The recent outcry on environmental impacts of anthropogenic sources has resulted in a strong emphasis on alternative means for power generation by use of wind and solar energies. However, for high-energy density applications such as transportation, combustion of fossil or alternative fuels will remain indispensable for many decades to come. Thus more stringent emissions legislation will be introduced and constantly revised in order to curtail the growing impact of combustion equipment on the environment. Fuel-lean combustion in a premixed or partially premixed mode is known to be a potential route to control emissions and to improve efficiency in both automobile engines and gas turbines. These two modes of combustion involve a strong coupling among various physical processes; to name a few of these processes, turbulence, chemical reactions, molecular diffusion, and large- and small-scale mixing are involved in turbulent lean combustion. Recognising these interplays, Chapter 1 surveys the current modelling approaches for turbulent premixed flames from a physical perspective. Chapter 2 describes these modelling methods in some detail. Lean premixed flames are notoriously unstable, and thus any analysis of them would be incomplete without addressing their instabilities. This topic is covered in Chapter 3. Current practice and challenges of fuel-lean combustion in practical devices such as internal combustion engines and gas turbines are described in Chapter 4, and future technological and scientific directions are discussed in Chapter 5.

We aim to provide a simple and clear discussion in terms of physical processes that can be followed by a graduate student with a good background in fluid mechanics, and combustion and some knowledge of turbulence modelling, together with some analytical skills. Many recent references are provided for further reading. We also attempted to make the discussion interesting for readers who are already familiar with turbulent premixed combustion, to help them to appreciate the complexity of this subject and the seriousness of the challenge involved in successfully introducing fuel-lean combustion to practical systems.

We would like to thank all the contributors, without whose dedication and cooperation it would not have been possible to meet our objectives in this volume.

N. Swaminathan
K. N. C. Bray
Cambridge, UK
March 2011

Contributors

D. Bradley, FRS, FREng, FIMechE, FInstP, is a Research Professor in the School of Mechanical Engineering at the University of Leeds, with interests in most aspects of combustion and energy. He was Editor of *Combustion and Flame* for 14 years and served as a consultant to several energy and engine companies. He has been involved in investigations of aircraft fires and large explosions. **e-mail:** d.Bradley@leeds.ac.uk.

K. N. C. Bray has degrees from Cambridge, Princeton, and Southampton Universities. He was Professor of Gas Dynamics at Southampton and then Hopkinson and Imperial Chemical Industries Professor of Applied Thermodynamics at Cambridge, where he is now an Emeritus Professor. His current research interests in combustion include the various ways in which turbulence interacts with combustion processes. He was Editor of *Combustion and Flame* from 1981 to 1986 and was elected to a Fellowship of The Royal Society in 1991, **e-mail:** knc.bray@btinternet.com.

N. Chakraborty is a Senior Lecturer in the Department of Engineering, University of Liverpool. He obtained a Ph.D. in 2004 from Cambridge University, Engineering Department, supported by a Gates Cambridge Scholarship. His research involves modelling of turbulent premixed flames and localised ignition of turbulent gaseous and droplet-laden inhomogeneous mixtures by use of direct numerical simulation. He also works on turbulent convective heat transfer with phase change in applications such as casting, welding, and laser melting. **e-mail:** n.chakraborty@liverpool.ac.uk.

M. Champion is CNRS scientist and works at the Laboratoire de Combustion et Déetonique, Poitiers. He obtained
his Ph.D. in 1980. His research interests are all related
to the fields of fluid mechanics, combustion and turbulent
flows, with a special emphasis on the modelling of turbulent flames. Since 1981 he has been regularly invited as
a Visiting Professor by the Department of Mechanical and
Aerospace Engineering of the University of California, San
Diego. **e-mail:** michel.champion@lcd.ensma.fr.

P. Domingo is CNRS scientist at CORIA. She obtained her
Ph.D. from the University of Rouen in 1991 and then joined
Stanford University Aerospace Department as a post-doc.
She has developed fully coupled solvers for plasma flows
and flames to analyze reacting flow physics by using direct numerical simulation and large-eddy simulation. She
also interacts with industry to help in the optimisation of
reliable, fuel-efficient, and environmentally friendly combustion systems. **e-mail:** domingo@coria.fr.

A. P. Dowling is the Head of Department of Engineering,
University of Cambridge, where she is Professor of Me-
chanical Engineering and Chairman of the University Gas
Turbine Partnership with Rolls-Royce. Her research is pri-
marily in the fields of combustion, acoustics, and vibration
and is aimed at low-emission combustion and quiet vehi-
cles. She is a Fellow of the Royal Society, Royal Academy of
Engineering, and is a Foreign Member of the U.S. National
Academy of Engineering and of the French Academy of
Sciences. **e-mail:** apd1@cam.ac.uk.

J. F. Driscoll is a Professor of Aerospace Engineering at the
University of Michigan, where he conducts fundamental
research in the area of premixed turbulent flames as well as
applied research in lean premixed gas turbine combustors
and scramjet devices. He has developed specialised diagnostics to image the physics of flame–eddy interactions. He
is a Fellow of the AIAA and is a former editor of *Combustion and Flame*. **e-mail:** jamesfd@umich.edu.

L. Y. M. Gicquel received his Ph.D. from the State University of New York at Buffalo in 2001. He then joined the CERFACS computational fluid dynamics team at Toulouse, France, to become a Senior Researcher, and he contributes to the development of massively parallel large-eddy simulations and acoustic solvers for industrial applications. His areas of expertise cover turbulent reacting flows, large-eddy simulations, direct numerical simulations, two-phase flows, pollutant emissions, stochastic processes, and combustion instabilities. **e-mail:** lgicquel@cerfacs.fr.

S. Hayashi is a Professor in the Science Engineering Department, Hosei University, Japan. He obtained his Ph.D. in engineering in 1971 from the University of Tokyo and was involved in research and project management in The National Aerospace Laboratory and Japan Aerospace Exploration Agency from 1972 to 2009. His research includes low-emission gas turbine combustion and laser diagnostics of sprays. **e-mail:** hayashi@hosei.ac.jp.

B. Jones is a consultant, specialising in gas turbine combustion, and is a Fellow of the Institute of Mechanical Engineering. He worked in this field from 1968 to 2003 for Rolls-Royce plc. From 1990 he was responsible for management of the design of new aeroengine combustion systems and their demonstration, and was responsible for the rapid expansion of participation by Rolls-Royce in collaborative combustion research within the EC and in the Far East. **e-mail:** bryn.jones@kausis.adsl24.co.uk.

P. Lindstedt received his M. Eng. degree from Chalmers University in 1981 and his Ph.D. from Imperial College in 1984 where he was appointed Professor in 1999. He served as Deputy Editor of *Combustion and Flame* (2000–2010) and as Colloquium Co-Chair for the 30th International Combustion Symposium. His research interests are focussed on the interactions of chemistry with flow, and he has received the Sugden Award and the Gaydon Prize. **e-mail:** p.lindstedt@imperial.ac.uk.

Y. Mizobuchi is a Senior Researcher in Aerospace Research and Development Directorate, Japan Aerospace Exploration Agency. He obtained his Ph.D. in 1995 from the University of Tokyo. His research interest includes numerical elucidation of turbulent combustion using supercomputer systems. He also engages in research and development activities on prediction and control of combustion instability and environmental acceptability of aerospace propulsion systems. **e-mail:** mizo@chofu.jaxa.jp.

A. S. Morgans is a Lecturer in the Department of Aeronautics, Imperial College London. She obtained a Ph.D. in Acoustics from Cambridge University Engineering Department in 2003. She then held a Royal Academy of Engineering/EPSRC Fellowship for research into the active control of combustion instabilities (2004 and 2009) at Cambridge and then at Imperial College. Her research interests include acoustics, combustion instabilities, and active control of flows. She was awarded the SET for Britain's 'Top Younger Engineer's Award' and the Gold Medal in 2004 for her work on controlling combustion instabilities. **e-mail:** a.morgans@imperial.ac.uk.

V. Moureau is a CNRS scientist at CORIA. He obtained a Ph.D. in 2004 from Institut Français du Pétrole and Ecole Centrale Paris. After a two-year post-doctoral fellowship at Stanford University in the Center for Turbulence Research, he joined Turbomeca, SAFRAN group, as a combustion engineer from 2006 to 2008. His research is focussed on turbulent combustion and spray modelling and on the development of YALES2 solver for large-eddy simulations and direct numerical simulation of turbulent flows in complex geometries. **e-mail:** vincent.moureau@coria.fr.

A. Mura is a senior scientist working for CNRS, Poitiers, France. After an academic cursus at the University of Rouen (INSA), he obtained a Ph.D. from ESM2 (Ecole Centrale Marseille). His teaching activities include postgraduate level courses and international short lectures. His research is devoted to the analysis of multiphase and reactive media with a broad range of technical issues encountered in practical systems, spanning from combustion in engines to rocket propulsion. **e-mail:** arnaud.mura@lcd.ensma.fr.

F. Nicoud has been a Professor in the Mathematics Department at the University of Montpellier since 2001. He teaches applied mathematics and scientific computing at the School of Engineering. He gained his Ph.D. in 1993 from Institute National Polytechnique in Toulouse and then was appointed a Senior Researcher at CERFACS in 1995. He held a fellowship at Center for Turbulence Research, Stanford University. His research interests span from analytical and numerical studies of thermoacoustic instabilities and combustion noise and wall-modelling issues in complex turbulent flows to computation of blood flow under physiological conditions. **e-mail:** franck.nicoud@univ-montp2.fr.

T. J. Poinsot received a Ph.D. in heat transfer from Ecole Centrale Paris in 1983 and his These d'Etat in combustion in 1987. He is a Research Director at CNRS, Head of the Computational Fluid Dynamics Group at CERFACS, Senior Research Fellow at Stanford University, and consultant at Institut Français du Pétrole, Air Liquide, Daimler. He teaches numerical methods and combustion in Ecole Centrale Paris, ENSEEIHT, ENSICA, Supaero, UPS, Stanford, and VKI. He has authored more than 120 papers in journals and 200 communications. He has co-authored a textbook *Theoretical and Numerical Combustion* with D. Veynante. He is an associate editor of *Combustion and Flame*. **e-mail:** Thierry.Poinsot@cerfacs.fr.

N. Swaminathan is a University Lecturer in the Cambridge University Engineering Department and a Fellow and Director of Studies at Robinson College. He gained a Ph.D. in 1994 from the University of Colorado, Boulder, and his research interest includes modelling and simulations of turbulent combustion. His professional experiences span from academia to industries. He has served on the editorial board of the *Open Fuels and Energy Science* journal since 2009 and he has been a member of EPSRC (Engineering and Physical Sciences Research Council) College, UK, since 2006; **e-mail:** ns341@cam.ac.uk.

A. M. K. P. Taylor graduated in 1975 and obtained his Ph.D. in 1981 from Imperial College. After brief visits to the University of Karlsruhe (as a DAAD scholar) and the NASA Lewis Research Center, he was a Royal Society University Research Fellow from 1985 to 1990, subsequently being appointed Lecturer in the Department of Mechanical Engineering. He was promoted to a Professor of Fluid Mechanics in the same department in 1999. His interest in basic research is in the fields of two-phase flow and combustion, combined with the development of optical instruments for these two areas. His applied research includes spray drying and combustion in gas turbines and, of course, internal combustion engines. **e-mail:** a.m.taylor@imperial.ac.uk.

Y. Urata graduated in 1984 from the Department of Mechanical Engineering, Tohoku University, Japan. Currently, he is a Chief Engineer at the Automobile R&D Center of Honda Research and Development Co., Ltd. His research interests include internal combustion engines (gasoline), combustion analysis with visualization technique, and variable valve trains. **e-mail:** Yasuhiro_ Urata@n.t.rd.honda.co.jp.

L. Vervisch is a Professor at the National Institute of Applied Science of Normandy and Researcher at the CNRS laboratory CORIA. He completed a Ph.D. in 1991 at Laboratoire National d'Hydraulique in Chatou, followed by a post-doc at the Center for Turbulence Research, Stanford University, and then became a Junior Member of Institute Universitaire de France in 2003. He uses direct numerical simulations to understand laminar and turbulent flames. He also contributes to the development of subgrid-scale closures for large-eddy simulations and Reynolds-averaged Navier–Stokes computations. Burners and aeronautical and internal combustion engines are the main targets of those studies. Vervisch is a co-editor of the journal *Flow Turbulence and Combustion*. **e-mail:** vervisch@coria.fr.

D. Veynante got his Ph.D. from Ecole Centrale Paris in 1985 and joined the CNRS. His research is devoted to turbulent combustion covering theoretical analysis, modelling, numerical simulations, experiments, and corresponding data processing. He has published more than 50 papers in international journals. He teaches combustion at Ecole Centrale de Paris, Ecole Centrale de Nantes, and in various spring or summer schools. He has co-authored a book titled *Theoretical and Numerical Combustion* (Edwards, 2001, second edition in 2005) with T. Poinsot and got the Grand Prix Institut Français du Pétrole from the French Sciences Academy in 2003. Since 2007, he has also been Deputy Scientific Director of CNRS Institute for Engineering and Systems Sciences (INSIS), in charge of about 50 labs working on fluid mechanics, combustion, heat transfer, plasma, and chemical engineering. **e-mail:** denis.veynante@em2c.ecp.fr.

1 Fundamentals and Challenges

By N. Swaminathan and K. N. C. Bray

1.1 Aims and Coverage

Currently the energy required for domestic and industrial use and for transportation is predominantly met by burning fossil fuels. Although alternative sources are evolving, for example, by harvesting wind and solar energies, energy production by means of combustion is expected to remain dominant for many decades to come, especially for high-power density applications. Thus pollutant emission regulations for power-producing devices are set with the aim of reducing the impact of combustion on the environment, both by curtailing pollutant emissions and by increasing the efficiency of combustion equipment. Conventional combustion technologies are unable to achieve these two demands simultaneously but *fuel-lean* combustion, in which the fuel–air mixture contains a controlled excess of air, has the potential to fulfil both requirements.

The aim of this book is to bring together a review of the physics of lean combustion and its current modelling practices, together with a description of scientific challenges to be faced and ways to achieve stable lean combustion in practical devices. Additional material on non-reacting turbulent flows may be found in [1–5], and basic theories of combustion and turbulent reacting flows and their governing equations are presented in [6–9].

The present chapter sets the scene for the remainder of this volume. It has three main aims: (1) to provide a brief review of the governing equations and auxiliary relations that are required for the description of turbulent premixed combustion, (2) to review the present status of the analysis of turbulent premixed combustion, and (3) to identify flame data that can be employed to verify modelling assumptions and to propose experiments that could usefully add to such data. Different levels of detail in numerical simulations are identified, and various regimes of combustion are defined. The important concept of a *turbulent flame speed* is also introduced. Although this quantity is central to some modelling methods, it can be defined in several different ways, and consequently comparisons between measurements and theoretical predictions are often a difficult task. The review provided here is not intended to be exhaustive but to be sufficiently detailed in providing background for later chapters. Earlier comprehensive treatments of turbulent premixed flames are provided in a number of references [e.g., 9–11].

Various modelling approaches are described in Chapter 2. As a consequence of averaging, these models are required to provide statistical information related to the unresolved small-scale structure of a turbulent flame and the two-way interaction between heat release and turbulence. It is intended that this presentation will help the reader to appreciate the physics of lean premixed and partially premixed flames, its links to modelling, and the inter-relationship among the various modelling approaches.

As we shall see, lean premixed flames are inherently unstable, and thus the discussion would be incomplete without the description of various instability processes, presented in Chapter 3. As explained there, the instabilities can be broadly classified as thermodiffusive, hydrodynamic, and thermoacoustic, based on the physical processes involved. Thermodiffusive instabilities are related to differences in the diffusion rates of mass and heat to and from the flame front, respectively. If the mass diffusion rate of reactant to the flame front is larger than the rate of heat diffusion away from the front, then the flame becomes unstable. The strong density jump across a perturbed flame front and the corresponding induced velocity changes lead to an inherent thermal instability, called the Darrieus–Landau instability. When this is coupled with buoyancy or an imposed pressure gradient, another hydrodynamic instability, called Rayleigh–Taylor instability, results. Thermoacoustic oscillations result when heat release fluctuations are in phase with fluctuations in pressure. The thermodiffusive and Darrieus–Landau instabilities are important in premixed laminar flames but are usually overwhelmed by sufficiently intense turbulence. However, the Rayleigh–Taylor and thermoacoustic instabilities can play significant roles in turbulent flames. The physics of these instabilities and methods to capture their effects on turbulent premixed flames are described in Chapter 3. The effects of thermoacoustic instabilities are of vital importance in gas turbine engines, so the science behind them and various strategies adopted to control them are also fully discussed in Chapter 3.

In appropriate circumstances, lean flames emit a very low level of pollutants and thus provide an ideal candidate for environmentally friendly engines and power-generation devices. However, premixing of fuel and oxidiser must occur inside the combustion chamber for safety reasons, creating only *partially* premixed reactants because of the limited space and time available for mixing. Nevertheless, depending on the level of partial premixing, it is still possible for a significant proportion of the combustion to occur in the premixed mode [12]. The scientific challenges involved in achieving stable lean combustion in practical devices, the physical processes involved, their interactions, and their modelling are all discussed in Chapter 4, in three different perspectives. The first section of Chapter 4 deals with the internal combustion (IC) engines employing intermittent combustion along with a detailed review of emissions legislation for automotive engines; the second and third sections consider continuous combustion systems, but differentiate the requirements for and challenges in aero gas turbines and their counterparts for power generation. This chapter also identifies some major challenges to be faced in future developments together with the factors driving them.

Chapter 5 discusses possible methods and technologies to meet the demands of the next and future generations of combustion devices. Scientific and technological challenges are also identified, and these challenges are discussed in three different perspectives: The concepts, specifically of combustion with high-temperature air and exhaust recirculation, are discussed in a broad sense in the first section of this chapter. The second section discusses the scientific aspects of partially premixed flames that will form the central element of future combustors and also identifies challenges for experimental investigation of lean combustion. The future modelling challenges are discussed in the third section.

Section 1.2 sets out to explain more fully what is meant by lean combustion, why it can be advantageous, what problems it introduces, and how these various topics are addressed in this book.

1.2 Background

It is convenient to identify two different modes for the combustion of a gaseous fuel: If the fuel and air are fully premixed before they enter the combustion zone, the flame is said to be *premixed*. On the other hand, when the fuel is kept separate until it burns, so that the reactants must diffuse towards each other before they react, a *diffusion* flame results. Combustion in a diffusion flame is centred on the stoichiometric or chemically balanced mixture of fuel and air, resulting in high temperatures and pollutant concentrations in combustion products. A difference between these two types of burning, leading to a burning-mode criterion known as the flame index [13, 14], is that fuel and air enter a premixed flame from the same side, whereas they go in from opposite sides of a diffusion flame. Fuel–air mixing is incomplete in many practical systems, leading to *partially* premixed combustion.

The term *lean* implies that the fuel–air mixture contains air in excess of that required by stoichiometry for a given amount of fuel, an amount that is determined by the overall energy output of the system and its thermal efficiency. The equivalence ratio, often denoted by ϕ, is defined as the ratio of the actual fuel-to-air mass proportion to its stoichiometric value and is typically small in most practical systems other than spark-ignited IC engines. At present, these are usually required to operate under stoichiometric conditions because of the requirements of the catalytic convertor. Future spark-ignition engines are expected to burn fuel lean, leading to a significant improvement in thermal efficiency, together with a significant reduction in emissions of oxides of nitrogen and carbon. However, careful design is required because of the inherent difficulty of achieving stable lean combustion, as explained in Chapters 4 and 5. Even when the *overall* equivalence ratio of the combustion system is very lean, some arrangements, such as exhaust gas recirculation (EGR for diesel engines), or flue gas recirculation (FGR for furnaces and boilers), or rich burn–quench–lean burn (RQL for gas turbines), are required for controlling nitric oxide emissions. This is because *local* combustion occurs mostly at the stoichiometric condition, resulting in a high flame temperature, which increases nitric oxide formation. The special arrangements such as the EGR just noted dilute the local combustible mixture with cooled combustion products and limit the peak flame temperature.

Figure 1.1 shows typical variations of flame temperature T_f and concentrations of oxides of nitrogen and carbon with equivalence ratio. The peak flame temperature and concentrations of pollutants decrease sharply as the equivalence ratio is reduced below unity, irrespective of operating pressure and reactant temperature.

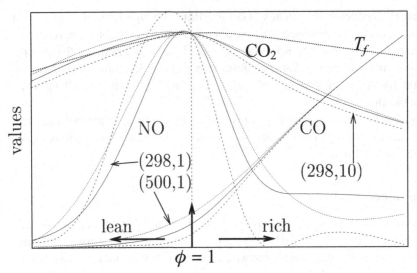

Figure 1.1. Typical variation of NO, CO, CO_2, and T_f with equivalence ratio. Effects of reactant temperature and operating pressure are also shown. (298.1) implies 298 K and 1 atm.

It is clear that locally lean combustion, which requires some degree of premixing, offers low levels of emission. Although this is very desirable, Fig. 1.2 shows that the laminar flame speed varies sharply with the equivalence ratio on the lean side. Thus a small change in local equivalence ratio, which may be caused by variations in local fluid dynamics induced by a number of factors in a practical situation, can lead to a substantial change in the local heat release rate. The sensitivity of the flame speed to the equivalence ratio ϕ is expressed by $d \ln \hat{S}_L / d\phi = m\phi / \hat{S}_L$, where $\hat{S}_L$ is the normalised flame speed and $m = d\hat{S}_L / d\phi$, as shown in Fig. 1.2. The following observations can be made from this figure, which is constructed with the available experimental data [15–18]: (1) The sensitivity does not depend on the fuel, but it depends on the reactant temperature, and (2) the hydrocarbons are more sensitive to local-equivalence-ratio oscillation than hydrogen and acetylene (compare the values for $T = 298$ K). These are purely thermochemical effects, which are compounded by effects of fluid dynamics and its interaction with thermochemistry in practical systems.

If the change in the local heat release rate is highly intermittent or if it is in phase with pressure oscillations, then the combustion process can become very rough, which is highly undesirable. Combustion instabilities are discussed in Chapter 3. Achieving smooth, stable lean combustion in practical devices is difficult because of the strong coupling between heat release from chemical reactions, diffusion, and fluid dynamics; see Chapter 4. The requirement to ensure thorough mixing between the fuel and air, often at significantly raised values of temperature and pressure, poses additional problems, because of the danger of the mixture auto-igniting upstream of the location where a flame is to be stabilised. In the purely chemical ignition or auto-ignition process [6], the temperature initially rises only slowly during an *ignition delay* stage in which a pool of reaction intermediates is formed before increasing more steeply in a second, heat release stage. Auto-ignitive burning and ignition delay times are discussed in Section 3.1.

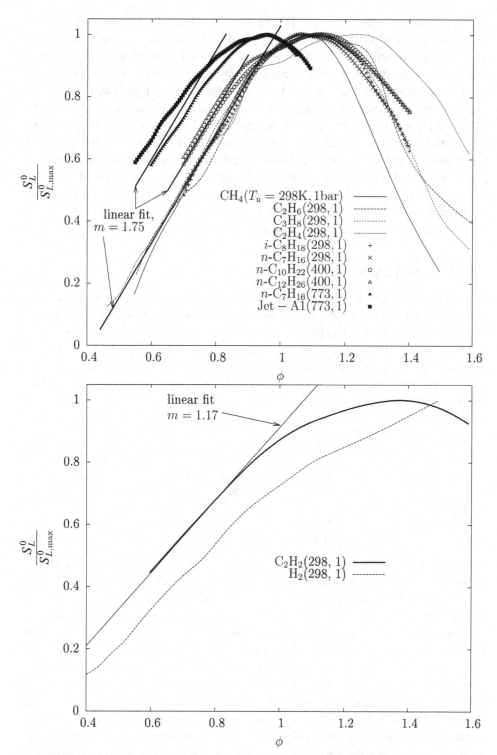

Figure 1.2. Variation of laminar flame speed with equivalence ratio for commonly used hydrocarbon–air mixtures. The flame speed is normalised by its maximum value $S_{L,\mathrm{max}}^0$, which is 0.416, 0.433, 0.449, and 0.722 m/s [15], 0.352 and 0.405 m/s [16], 0.658 and 0.682 m/s [17], and 2.312 and 2.023 m/s [18] in the same order as in the legend. For acetylene – and hydrogen–air mixtures [15] it is 1.559 and 2.856 m/s, respectively.

To avoid auto-ignition, many practical systems feed the fuel and air into the combustion zone while it is only *partially* premixed, and then the potential benefits of premixing are only partially achieved. All of these processes are invariably turbulent, which greatly compounds the complexity level. As explained in Section 1.4, turbulent flow and combustion involve a very wide range of length and time scales that cannot all be resolved in numerical simulations of practical systems. This difficulty is overcome by the introduction of some form of *averaging* into the governing differential equations. Averaging involves a loss of information, with the result that the number of unknowns always exceeds the number of equations, and an appropriate number of *model* expressions must be devised from an understanding of the controlling physics; see Chapter 2. Thus designing and constructing successful lean combustion systems are major challenges [19] that require a close and thorough understanding of all the relevant physical processes and their interactions.

1.3 Governing Equations

This section provides a brief summary of the equations governing the flow of a chemically reacting gas mixture; their derivation is described in detail elsewhere, for example by Williams [7] and Law [6]. These equations consist of conservation laws of mass, momentum, and energy and concentrations of reacting chemical species, together with equations specifying molecular transport and thermodynamic properties of the mixture and the rates of chemical conversion. Mass, momentum, and species conservation equations are

$$\frac{D\rho}{Dt} + \rho\frac{\partial u_k}{\partial x_k} = 0,$$

$$\rho\frac{Du_\ell}{Dt} = -\frac{\partial p}{\partial x_\ell} + \frac{\partial \tau_{\ell k}}{\partial x_k} + \rho\sum_{i=1}^{N} Y_i f_{i\ell},$$

$$\rho\frac{DY_i}{Dt} = -\frac{\partial J_{ik}}{\partial x_k} + \dot{\omega}_i. \tag{1.1}$$

Here and in the following chapters, Cartesian tensor notations are used with the indices k, ℓ, and m. These indices imply summation when they are repeated in the same term. In Eqs. (1.1), the substantial derivative is $D/Dt \equiv \partial/\partial t + u_k \partial/\partial x_k$ and the symbols x_k, u_k, ρ, p, and Y_i represent, respectively, the spatial coordinate in direction k, velocity component in that direction, gas density, pressure, and the mass fraction of a chemical species i, where $i = 1 \ldots N$. The quantity $f_{i\ell}$ is a body force per unit volume acting on species i in direction ℓ; if this body force is due to a gravitational acceleration g, the body-force term becomes $\overline{\rho} g_\ell$. Subscripts i and j are consistently used for chemical species; when their summation is required, it is shown explicitly. Also, $\dot{\omega}_i$ is the mass rate of production or destruction of species i that is due to chemical reactions. Finally, $\tau_{\ell k}$ is a shear-stress component in the ℓth direction on a surface whose outward normal is in the kth direction, which may be represented by

$$\tau_{\ell k} = \mu\left(\frac{\partial u_\ell}{\partial x_k} + \frac{\partial u_k}{\partial x_\ell} - \frac{2}{3}\frac{\partial u_m}{\partial x_m}\delta_{\ell k}\right), \tag{1.2}$$

where μ is viscosity coefficient of the mixture and $\delta_{\ell k}$ is the Kronecker delta. For present purposes the molecular diffusion flux J_{ik}, which is the mass molecular flux of species i in direction k, can be approximated by Fick's law:

$$J_{ik} = -\rho \mathcal{D}_i \frac{\partial Y_i}{\partial x_k}, \tag{1.3}$$

where $\mathcal{D}_i$ is a diffusion coefficient.

The thermodynamic properties of a mixture of ideal gases may be described by the equation of state,

$$p = \rho \mathcal{R} T \sum_{i=1}^{N} \frac{Y_i}{W_i}, \tag{1.4}$$

where W_i is the molecular weight of species i and $\mathcal{R}$ is the universal gas constant, 8.314 kJ kmol^{-1} K^{-1}. The specific enthalpy of the mixture is

$$h = \sum_{i=1}^{N} Y_i h_i, \tag{1.5}$$

and the specific enthalpy of a species i is

$$h_i = h_i^0 + h_i^s(T; T^0), \tag{1.6}$$

where h_i^0 is the standard specific heat of formation of species i at temperature T^0 and

$$h_i^s(T; T^0) = \int_{T^0}^{T} c_{p,i} \, dT, \tag{1.7}$$

is the specific sensible enthalpy of the species i relative to T^0. This can sometimes be approximated by treating $c_{p,i}$ as an appropriately chosen constant.

The energy equation can be written in several alternative forms [6, 9], one of which is

$$\rho \frac{Dh}{Dt} = \frac{Dp}{Dt} + \tau_{\ell k} \frac{\partial u_\ell}{\partial x_k} - \frac{\partial q_k}{\partial x_k} - \rho \sum_{i=1}^{N} \mathcal{D}_i f_{ik} \frac{\partial Y_i}{\partial x_k} + \dot{Q}_{\mathrm{rad}}. \tag{1.8}$$

In this equation q_k is the molecular flux of enthalpy, given by

$$q_k = -\hat{\lambda} \frac{\partial T}{\partial x_k} - \rho \sum_{i=1}^{N} h_i \mathcal{D}_i \frac{\partial Y_i}{\partial x_k}, \tag{1.9}$$

where $\hat{\lambda}$ is the thermal conductivity of the mixture and $\dot{Q}_{\mathrm{rad}}$ is the energy exchange per unit volume due to radiation.

Three dimensionless parameters are conventionally introduced to characterise molecular transport. The *Prandtl* number of the mixture is defined as $\mathrm{Pr} = \mu c_p / \hat{\lambda}$, where $c_p = \sum_{i=1}^{N} Y_i c_{p,i}$ is the specific heat at constant pressure of the mixture. The *Schmidt* number of species i is $\mathrm{Sc}_i = \mu / \rho \mathcal{D}_i$, and the *Lewis* number of species i is $\mathrm{Le}_i = \hat{\lambda} / c_p \rho \mathcal{D}_i = \mathrm{Sc}_i / \mathrm{Pr}$.

Equation (1.8) may be greatly simplified if certain restrictive conditions are met; see for example Libby and Williams [10]: (1) It is assumed that the flow Mach number is sufficiently low for compressibility effects to be neglected, allowing the first two terms on the right-hand side of Eq. (1.8) to be set to zero. (2) Radiative heat transfer and energy changes arising from body forces are assumed to be negligible. (3) It is also assumed that $Sc_i = Pr$ for all species; this requires both that $\mathcal{D}_i$ can be approximated by a common value $\mathcal{D}$ for all species and also that $Sc = Pr$, implying that the Lewis number is unity. Then the molecular-diffusion terms arising when Eq. (1.9) is substituted into Eq. (1.8) cancel, with molecular heat flux terms generated if Eqs. (1.5) and (1.6) are used to replace $\partial h_i / \partial x_k$ with $c_{p,i} \partial T / \partial x_k$. The molecular flux of enthalpy becomes $q_k = -\rho \mathcal{D} \partial h / \partial x_k$, and Eq. (1.8) can be written as

$$\rho \frac{Dh}{Dt} = \frac{\partial}{\partial x_k} \left(\rho \mathcal{D} \frac{\partial h}{\partial x_k} \right). \qquad (1.10)$$

Subject to appropriate initial and boundary conditions, Eq. (1.10) is satisfied by $h = $ constant, i.e., an *isenthalpic* flow. This simplified form of the energy equation is often used in turbulent combustion models, and it is applicable for different situations of lean premixed combustion noted in the previous section as long as restrictive conditions (2) and (3) are met. However, the prediction of lean premixed combustion and flow in IC engines will involve $\partial p / \partial t$ because of a cyclic variation of pressure inside the cylinder. Situations in which the preceding restrictive conditions are to be relaxed are noted appropriately in the following chapters.

1.3.1 Chemical Reaction Rate

Two different types of information are required for specifying the chemical-reaction-rate term $\dot{\omega}_i$ appearing in the last of Eqs. (1.1): a reaction *mechanism* and a set of reaction *rates*. The first of these consists of a set of n elementary reactions, the rth of which is written as

$$\sum_{i=1}^{N} v'_{ri} M_i \underset{k_{br}}{\overset{k_{fr}}{\rightleftharpoons}} \sum_{i=1}^{N} v''_{ri} M_i, \qquad (1.11)$$

where M_i is the chemical symbol for species i involved in the rth reaction and v' and v'' are the *stoichiometric coefficients* for species i in the forward and backward steps of the reaction, respectively. If reaction (1.11) describes the actual elementary process by which these species are created and destroyed, then the stoichiometric coefficients will be integers. If the reaction is a purely phenomenological description of the effects of several elementary reactions, then it is called a *global* reaction, which can have non-integer stoichiometric coefficients. As shown, the reaction is *reversible*, i.e., it proceeds in both directions. A reaction mechanism can be of any length, and detailed mechanisms consisting of some hundreds of elementary reactions are not uncommon (see for example [20, 21]).

The quantity $\dot{\omega}_i$ is the net result of all n chemical reactions, and thus

$$\dot{\omega}_i = \sum_{r=1}^{n} (v''_{ir} - v'_{ir}) \varpi_r W_i, \qquad (1.12)$$

where ϖ_r is the molar rate of reaction r, which can be expressed in the form

$$\varpi_r = k_{\mathrm{fr}} \prod_{j=1}^{N} \left(\frac{\rho Y_j}{W_j}\right)^{v'_{jr}} - k_{\mathrm{br}} \prod_{j=1}^{N} \left(\frac{\rho Y_j}{W_j}\right)^{v''_{jr}}, \tag{1.13}$$

where k_{fr} and k_{br} are the forward- and backward-rate coefficients, respectively, of reaction r, which are conventionally written in the form

$$k_{\mathrm{fr}}(T) = A_{\mathrm{fr}} T^{\alpha_{\mathrm{fr}}} \exp\left(\frac{-E_{\mathrm{fr}}^a}{\mathcal{R}T}\right). \tag{1.14}$$

A similar expression can be written for the specific rate constant k_{br} of the backward step of the elementary reaction r. The ratio of the forward-rate to the backward-rate constants is given by the equilibrium constant. The pre-exponential factor A_{fr}, the temperature exponent α_{fr}, and the activation energy E_{fr}^a are constants, which are often obtained from shock-tube experiments [22–24] and are of an empirical nature.

1.3.2 Mixture Fraction

If partially premixed combustion occurs, that is, if fuel and air are not fully mixed when they enter the combustion zone, then some means must be found to track variations in their ratio. This may be done by the introduction of so-called *conserved scalar* variables, which remain unaffected by the progress of chemical reactions. The mass fractions of chemical elements, the pth of which is denoted by ξ_p, where $p = 1, \ldots, q$, are conserved scalars and are related to the species mass fractions by

$$\xi_p = \sum_{i=1}^{N} \mu_{pi} Y_i,$$

in which μ_{pi} represents the number of kilograms of element p in 1 kg of species i. Because elements are conserved in chemical reactions, we have

$$\sum_{i=1}^{N} \mu_{pi} \dot{\omega}_i = 0.$$

It follows [10] from this expression, together with the third of Eqs. (1.1) and the simple Fick's law of diffusion expression of Eq. (1.3), that the element mass fraction ξ_p is governed by

$$\rho \frac{\mathrm{D}\xi_p}{\mathrm{D}t} = \frac{\partial}{\partial x_k} \sum_{i=1}^{N} \rho \mu_{pi} \mathcal{D}_i \frac{\partial Y_i}{\partial x_k}. \tag{1.15}$$

If the approximation $\mathcal{D}_i \approx \mathcal{D}$ can be justified, where $\mathcal{D}$ is a single molecular-diffusion coefficient applicable to all chemical species, then Eq. (1.15) may be seen to have the same form as simplified energy equation (1.10), namely,

$$\rho \frac{\mathrm{D}\xi_p}{\mathrm{D}t} = \frac{\partial}{\partial x_k} \rho \mathcal{D} \frac{\partial \xi_p}{\partial x_k}, \tag{1.16}$$

from which it follows that ξ_p and the specific enthalpy h are linearly related, provided they have the same boundary conditions. This can be achieved by the introduction of a normalised conserved scalar Z, the *mixture fraction*, based either on h or on an element mass fraction so

$$Z = \frac{h - h_0}{h_1 - h_0} = \frac{\xi_p - \xi_{p,0}}{\xi_{p,1} - \xi_{p,0}},$$

where the subscripts 0 and 1 respectively represent conditions in pure air and pure fuel, and

$$\rho \frac{DZ}{Dt} = \frac{\partial}{\partial x_k} \rho D \frac{\partial Z}{\partial x_k}. \tag{1.17}$$

However, it must be kept in mind that this formulation involves significant simplifications of molecular-transport processes. Such simplifications are known to lead to large errors in predicting laminar flame properties in some circumstances and, as we shall see, structures resembling laminar flames are often found to occur in premixed turbulent combustion.

1.3.3 Spray Combustion

Liquid fuel must be *atomised* into a fine spray that is either mixed with air before it enters the combustion chamber or is fed directly into the combustion zone. The physics of spray formation, evaporation, and combustion is described in detail elsewhere [25–32]. Our aim here is simply to identify the most important processes controlling this type of combustion. An isolated liquid drop, at rest in a stationary medium, evaporates in a time proportional to the square of its initial diameter. However, the process is more complex in the presence of relative motion, in which details of the flow fields and heat and mass transfer rates both inside and outside the drop must be taken into account [32]. The vapour evaporating from an isolated drop in a stagnant environment can burn in a laminar diffusion flame, forming an envelope surrounding the drop, and, once again, relative motion can complicate the picture, with burning sometime occurring in the wake behind the drop. Liquid sprays can burn in a variety of arrangements [33, 34]: in envelopes diffusion flames surrounding either single drops or groups of drops, in wakes behind drops, in flames that are distinct from the evaporating liquid drops, or in a combination of ways. Turbulence adds further complexity, but this categorisation can still be applied.

If fuel-lean combustion is to be achieved, the diffusion-flame mode of burning, which takes place under stoichiometric conditions, must always be minimised, if it cannot be eliminated. This means that burning in envelope or wake flames must be avoided to the extent possible; droplets need to evaporate as quickly as possible and their vapour thoroughly mixed with air before burning takes place. Efficient atomisation into a fine mist of droplets aids this process. Nevertheless, the need to avoid the possibility of auto-ignition or flashback limits the time available for evaporation and mixing, and spray combustion will generally be only partially premixed. The twin annular premixed system (TAPS), discussed in Sections 4.2 and 5.2, is a good example for a practical arrangment to minimise, if not to completely avoid, stoichiometric combustion.

1.4 Levels of Simulation

Turbulent combustion simulations can be categorised into three broad topics, viz., direct numerical simulation (DNS), Large-Eddy Simulation (LES), and Reynolds-averaged Navier–Stokes (RANS) simulation. Each of these has its own advantages and disadvantages, and the choice is dictated by the level of details required from their results. The essentials of these three types of simulation are described briefly in the next three subsections.

1.4.1 DNS

The accurate numerical simulation of turbulent flows is difficult because of the nonlinearity in the governing equations and the wide range of length and time scales involved. If we introduce a turbulence Reynolds number $\mathrm{Re}_T = u_{\mathrm{rms}}\Lambda/\nu$, where u_{rms}, Λ, and ν are characteristic values of root-mean-square turbulence velocity, turbulence length scale, and kinematic viscosity, respectively, then the range of length scales increases as $\mathrm{Re}_T^{3/4}$ and the range of time scales as $\mathrm{Re}_T^{1/2}$. A consequence is that the complete solution or *direct numerical simulation* of even non-reactive turbulent flows, at realistic Reynolds numbers, which are of the order of 10^6, is still prohibitively expensive. The situation is even more challenging in the presence of combustion: Chemical reactions introduce their own length and time scales, which may sometimes be smaller than the smallest scales of turbulence, thus further extending the range of scales to be resolved, whereas the chemical mechanism can greatly increase both the number of differential equations to be solved and their nonlinearity. Nevertheless, DNS of turbulent combustion in simplified geometries, with either a single global reaction or a more realistic but relatively small set of elementary reactions, has proved to be an invaluable research tool for exploring the basic physics and testing the validity of modelling hypotheses. DNS of *spray* combustion may be found in [35–37]. However, despite the continuing dramatic growth in computing power, a DNS of, say, a complete gas turbine combustor is not to be expected in the near future.

1.4.2 RANS

The classical solution to the problem of too wide a range of length and time scales in turbulent flows is to solve equations for *averaged* variables whose minimum scales are much larger than the smallest scales of the turbulent fluctuations. However, because the original equations are nonlinear, their averaged versions contain additional terms, for example Reynolds stresses, involving co-variances of fluctuations about the mean; and, to evaluate these co-variances, it is necessary to make assumptions about the small-scale structure that was lost because of averaging. It is possible to use the original equations to derive transport equations for the co-variances, but these in turn always involve higher-order co-variances – the so-called *closure* problem of this approach. In combustion, this problem is exacerbated by the need to model the average of highly nonlinear reaction-rate expressions. The averaging approach, referred to as *Reynolds-averaged Navier–Stokes* simulation, or URANS for unsteady RANS, is highly developed for both non-reacting and reacting flows; closure models are introduced in Section 1.7 and reviewed in detail in Chapter 2.

Various different types of average can be used in RANS. In flows that are stationary, a *time* average is formed over a time period T that is longer than the largest time scale of turbulent fluctuations so, for a variable $\varphi(\mathbf{x}, t)$,

$$\overline{\varphi}(\mathbf{x}) = \frac{1}{T} \int_0^T \varphi(\mathbf{x}, t) \mathrm{d}t,$$

whereas in flows with time-dependent mean values, an *ensemble* average of a sufficiently large number N_r of realisations of the flow must be used, and

$$\overline{\varphi}(\mathbf{x}, t) = \frac{1}{N_r} \sum_{s=1}^{N_r} \varphi_s(\mathbf{x}, t).$$

When predictions are compared with DNS data, which are often unsteady in the mean, a spatial average is sometimes used. If the DNS is two-dimensional (2D) in the mean, one can write

$$\overline{\varphi}(\mathbf{x}) = \frac{1}{L_Y} \int_0^{L_Y} \varphi(\mathbf{x}, t) \, \mathrm{d}y,$$

where y is distance in the direction in which the flow is homogeneous and L_Y is the width of the domain in this direction. It is usually assumed that these three types of average are equivalent to each other according to the *ergodic hypothesis*, although this may not necessarily be the case.

The fluctuation about the mean is denoted by $\varphi'(\mathbf{x}, t)$, where $\varphi(\mathbf{x}, t) = \overline{\varphi}(\mathbf{x}) + \varphi'(\mathbf{x}, t)$ and $\overline{\varphi'} = 0$. In fluid mechanics this type of average is referred to as a *Reynolds* average. It is usually more convenient in combustion problems to define mass-weighted or *Favre* mean variables as

$$\tilde{\varphi} = \frac{\overline{\rho \varphi}}{\overline{\rho}}.$$

In this case the fluctuation is written as $\varphi''(\mathbf{x}, t)$, where $\varphi(\mathbf{x}, t) = \tilde{\varphi}(\mathbf{x}) + \varphi''(\mathbf{x}, t)$. Note that $\overline{\rho \varphi''} = 0$ but $\overline{\varphi''}$ is not necessarily zero. Both the Reynolds and Favre means can be evaluated by time, ensemble, or spatial averaging.

It is common to use *phase* averaging when stationary or travelling waves are involved, as in the case of themoacoustic instablilities inside the combustor. The phase average is defined as [38]

$$\langle \varphi \rangle (\mathbf{x}, t) = \frac{1}{N} \sum_{i=1}^N \varphi(\mathbf{x}, t + i\tau), \tag{1.18}$$

where τ is the wave period. Using this average, one can write the turbulent variable as $\varphi(\mathbf{x}, t) = \langle \varphi \rangle + \varphi' = \overline{\varphi}(\mathbf{x}) + \varphi_p(\mathbf{x}, t) + \varphi'$, where φ_p represents the coherent content in φ. The phase and time averaging commute [38], and thus $\langle \overline{\varphi} \rangle = \overline{\langle \varphi \rangle}$ and $\overline{\varphi_p} = 0$.

1.4.3 LES

A weakness of RANS and URANS predictions is that, whereas the small-scale fluctuations, which are strongly influenced by molecular transport, are almost universal in character, large-scale fluctuations are more strongly dependent on details of the flow and its boundary conditions. Because most of the energy of the turbulence is contained in these larger eddying motions, it is difficult to find universal model expressions to close the system of equations. The *large-eddy simulation* technique was developed to overcome this difficulty. In LES, the large-scale motions are resolved in both space and time, but the small scales are not resolved, so the requirement for spatial resolution is related to the larger eddying motions. Small scales are filtered out by use of a localised spatial averaging (weighted by F) of the form

$$\overline{\varphi}(\mathbf{x}, t) = \int \varphi(\mathbf{x}', t) F[(\mathbf{x} - \mathbf{x}'); \Delta] \, d\mathbf{x}', \qquad (1.19)$$

where F is a *filter function* whose shape is chosen so that it approaches zero when $\mathbf{x} - \mathbf{x}'$ exceeds the chosen *filter size* Δ and also $\int F(\mathbf{x}') \, d\mathbf{x}' = 1$. The commonly used shapes for F and their characteristics can be found in [4, 9]. A couple of points worth noting at this stage are, that (1) the filtering operation does not commute with derivative operators, and (2) $\overline{\overline{\varphi}} \neq \overline{\varphi}$ and $\overline{\varphi'} \neq 0$, unlike in RANS averaging.

The preceding averaging or filtering process results in loss of information about small-scale motions, and consequently the filtered equations are unclosed because they contain co-variance terms analogous to the Reynolds stresses in RANS. The *subgrid scale* (SGS) model expressions with which the system of equations is closed are often developed from the corresponding RANS models. A discussion of such modelling for non-reacting flows can be found in [4]. For flows with density variation, such as combustion, it is convenient to solve the Favre mean version of the LES-filtered transport equations, which can be obtained by density-weighted filtering given by

$$\overline{\rho}\widetilde{\varphi}(\mathbf{x}, t) = \int \rho \varphi(\mathbf{x}', t) F[(\mathbf{x} - \mathbf{x}'); \Delta] \, d\mathbf{x}'. \qquad (1.20)$$

Equations (1.19) and (1.20) suggest that the LES solution is expected to have some dependence on the artificial (scale) parameter Δ. This raises some fundamental conceptual questions, as discussed by Pope [39]. Although it is not unexpected that this filter scale will influence the degree of resolution in the LES solution, the averaged solution obtained from the LES should be independent of the choice of Δ.

1.5 Equations of Turbulent Flow

The mass-weighted, Favre mean transport equations of turbulent reacting flow are obtained by the averaging of the equations of Section 1.3. Equations (1.1) give

$$\frac{\partial \overline{\rho}}{\partial t} + \frac{\partial \overline{\rho}\widetilde{u}_k}{\partial x_k} = 0,$$

$$\frac{\partial \overline{\rho}\widetilde{u}_\ell}{\partial t} + \frac{\partial \overline{\rho}\widetilde{u}_k\widetilde{u}_\ell}{\partial x_k} = -\frac{\partial \overline{\rho}\widetilde{u_\ell'' u_k''}}{\partial x_k} - \frac{\partial \overline{p}}{\partial x_\ell} + \frac{\partial \overline{\tau}_{\ell k}}{\partial x_k} + \overline{\rho}\sum_{i=1}^{N} \widetilde{Y_i f_{i\ell}},$$

$$\frac{\partial \overline{\rho}\widetilde{Y}_i}{\partial t} + \frac{\partial \overline{\rho}\widetilde{u}_k\widetilde{Y}_i}{\partial x_k} = -\frac{\partial \overline{\rho}\widetilde{u_k'' Y_i''}}{\partial x_k} - \frac{\partial \overline{J}_{ik}}{\partial x_k} + \overline{\dot{\omega}}_i. \qquad (1.21)$$

Similarly the simplified energy conservation expression of Eq. (1.10) becomes

$$\frac{\partial \overline{\rho} \widetilde{h}}{\partial t} + \frac{\partial \overline{\rho} \widetilde{u}_k \widetilde{h}}{\partial x_k} = -\frac{\partial \overline{\rho} \widetilde{u_k'' h''}}{\partial x_k} - \frac{\partial}{\partial x_k} \overline{\left(\rho D \frac{\partial h}{\partial x_k} \right)}, \qquad (1.22)$$

and the equation for the mean mixture fraction is

$$\frac{\partial \overline{\rho} \widetilde{Z}}{\partial t} + \frac{\partial \overline{\rho} \widetilde{u}_k \widetilde{Z}}{\partial x_k} = -\frac{\partial \overline{\rho} \widetilde{u_k'' Z''}}{\partial x_k} - \frac{\partial}{\partial x_k} \overline{\left(\rho D \frac{\partial Z}{\partial x_k} \right)}, \qquad (1.23)$$

from Eq. (1.17). These equations are applicable to both RANS and LES, where $\widetilde{\varphi'' \psi''} \equiv \widetilde{\varphi \psi} - \widetilde{\varphi} \widetilde{\psi}$. The important assumptions introduced in the derivation of Eqs. (1.10) and (1.17) should not be forgotten.

Several types of additional information are required before these equations can be solved: (1) The equations of state, Eqs. (1.4) and (1.5), must be averaged to provide relationships among $\overline{p}$, $\overline{\rho}$, $\widetilde{h}$, and $\widetilde{Y}_i$. (2) Models must be provided for the Reynolds stress components $\widetilde{u_\ell'' u_k''}$ and the corresponding scalar flux components $\widetilde{u_k'' Y_i''}$, $\widetilde{u_k'' h''}$, and $\widetilde{u_k'' Z''}$. (3) The highly nonlinear reaction-rate expression of Eqs. (1.12) and (1.13) must be averaged; the nature of the closure model for these mean reaction rates will depend on assumptions about the unresolved small-scale structure of the flame. (4) In many cases the Reynolds number is sufficiently large for all the mean molecular-transport terms in these first-moment transport equations to be ignored; otherwise simplifying assumptions are required. (5) Last, it may be noted that, if the final term in the second of Eqs. (1.21) represents buoyancy that is due to gravity, then $f_{i\ell} = g_\ell$ and no additional modelling assumption is needed. The strategies required to provide the additional information are discussed in Section 1.7.

1.6 Combustion Regimes

The quantities to be modelled to obtain a closed set of averaged equations involve statistical relationships between the fluctuating variables, and these relationships depend on the small-scale structure lost because of averaging. Damköhler [40] identified two limiting situations: If all length and time scales of the chemical reactions are small in comparison with the smallest scales of the turbulent flow, then combustion should be restricted to thin, laminar-like reaction zones. The turbulent flow stretches and distorts these so-called *laminar flamelet* combustion zones and so increases the mean rate of heat release. At the other extreme, if the chemical length and time scales are all large in comparison with the biggest scales of turbulence, the structure of the reaction zone would be expected to be more random.

To test and utilise this insight, one must first introduce characteristic scales of the turbulent flow and chemical reactions. Those of the turbulence can be expressed in terms of the mean turbulence kinetic energy $\overline{k} = \widetilde{u_\ell' u_\ell'}/2$ and its mean rate of dissipation

$$\overline{\varepsilon} = 0.5 \nu \overline{\left(\frac{\partial u_\ell}{\partial x_k} + \frac{\partial u_k}{\partial x_\ell} \right)^2}.$$

Then characteristic turbulence length and time scales are $\Lambda = (\overline{k}^{3/2}/\overline{\varepsilon})$ and $\tau_T = (\overline{k}/\overline{\varepsilon})$, respectively. A typical rms turbulence velocity $u_{\mathrm{rms}} = (2\overline{k}/3)^{1/2}$ is often used in

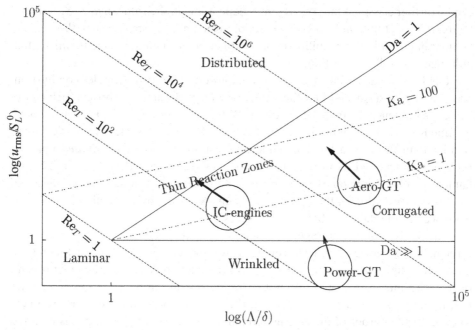

Figure 1.3. Turbulent combustion regime diagram. Typical combustion conditions in three main categories of practical engines are shown, and the arrows indicate the likely direction of change that is due to lean-burn technologies. GT, gas turbine.

place of $\bar{k}$. The smallest length scale, limited by viscous dissipation, is the Kolmogorov length $\eta_K = (\nu^3/\bar{\varepsilon})^{1/4}$, and the corresponding time scale is $\tau_K = (\nu/\varepsilon)^{1/2}$. It should be noted that, in combustion systems, the predicted magnitudes of these characteristic quantities can be significantly influenced by the choice of whether to evaluate the kinematic viscosity $\nu(T)$ in cold reactants or hot products. It is convenient to use the properties of an unstretched laminar flame to provide chemical scales. If the laminar burning velocity of the premixed reactants is S_L^0, then a characteristic diffusive thickness is $\delta = \mathcal{D}/S_L^0$, where $\mathcal{D}$ is a molecular-diffusion coefficient and $\tau_c = \delta/S_L^0$ is the corresponding time. This diffusive thickness δ is often called the Zeldovich thickness. One can also define a slope thickness, known as the thermal thickness δ_L^0, based on the maximum temperature gradient in and the temperature rise across an unstrained laminar flame.

These characteristic scales may now be combined to form dimensionless parameters. The ratio of turbulence time to laminar flame time, $\mathrm{Da} \equiv \tau_T/\tau_c$, is known as a Damköhler number. The quantity $\mathrm{Re}_T \equiv u_{\mathrm{rms}}\Lambda/\nu$ is a turbulence Reynolds number. If the Schmidt number is 1 then $\mathrm{Re}_T = u_{\mathrm{rms}}\Lambda/(S_L^0\delta)$. A third important parameter, known as the Karlovitz number, is the ratio of chemical to Kolmogorov time scales, $\mathrm{Ka} \equiv \tau_c/\tau_K = \delta^2/\eta_K^2$. The relationship between these parameters can be illustrated in the form of a *regime diagram* [7, 9, 41], in which either $(\mathrm{Re}_T, \mathrm{Da})$ or $(\Lambda/\delta, u_{\mathrm{rms}}/S_L^0)$ are used as coordinates. Such a diagram is shown in Fig. 1.3, depicting various features. The line identifying $\mathrm{Ka} = 1$ is called *Klimov–Williams line*. The classical argument is that, when $\mathrm{Ka} < 1$, laminar flame scales are smaller than all the relevant scales of the turbulence; turbulent eddies can then only wrinkle a laminar flame but cannot extinguish it. This is the *laminar flamelets* combustion regime, which can be further classified into two regimes depending on the value of u_{rms}/S_L^0 as in the figure. If Ka > 1 but Da > 1, some scales of turbulence are smaller than those of the laminar flame and the possibility of local extinction of a laminar flame is anticipated, although τ_c is still smaller than τ_T.

DNS data analysis has been used to explore the limits of flamelet combustion, and it has been suggested that the Klimov–Williams boundary, above which turbulence can influence the internal structure of a reaction zone, is shifted upwards above the line Ka = 1. The details are in [9]. Finally, if Da < 1, all scales of turbulence are smaller than those of a laminar flame and a more randomly *distributed* reaction regime is anticipated. However, as discussed at greater length in Section 2.1, thin flamelet-like reaction zones have been found to be more robust than is suggested by this interpretation of the regimes diagram. A review by Driscoll [42] finds that evidence of 'non-flamelet' behaviour is sparse.

The likely combustion regimes based on the data available in the literature for three major categories of practical engines are shown in Fig. 1.3. These estimates are made after the laminar flame scales are corrected for the temperature and pressure of the reactants. Aeroengines do not currently operate in purely premixed modes but, if one presumes a premixed mode, then the combustion conditions are likely to occur at the border of the corrugated flamelets and thin-reaction-zones regimes. The drive towards the lean-burn technologies that is due to emissions regulations is likely to push these conditions towards lower values of global or overall Da, because reaction rates are reduced in leaner mixtures. However, for gas turbines, it is quite unlikely to involve local Da < 1, unless high-temperature air or product dilution concepts are employed. These concepts and their possible modelling are discussed in Section 5.1.

1.7 Modelling Strategies

Two main types of model are required for converting the first-moment equations of Section 1.5 into a closed set: (1) a fluid mechanical model to describe the Reynolds stress and Reynolds flux terms by means of equations involving only known mean properties of the flow, and (2) a closure model for the mean values of highly nonlinear chemical-reaction-rate terms. Although we describe these models separately, it must be emphasised that, in reality, they interact strongly with each other. It is necessary to account properly for, on the one hand, the influence of combustion and heat release on the turbulent flow and, on the other, the effects of eddying turbulent motion on the heat and mass transfer processes that accompany chemical reaction. Combustion introduces very large variations in fluid density, and the corresponding changes in the volume of fluid elements induce significant local velocity changes. Very steep property gradients can result in regions of strong reaction, in which molecular-transport processes play a vital role. Turbulent motion distorts and stretches these thin regions and may further increase their gradients. A proper understanding of these physical processes and their interaction is essential if realistic models are to be developed. All models, together with their constituent submodels, must yield unique solutions, and the variables that they predict must always lie within the physical bounds of the quantities they represent. RANS models are considered first, and then their adaption for LES is discussed.

1.7.1 Turbulent Transport

In non-reactive turbulent flows, closure of the mean-flow equations is generally obtained either through the so-called k–ε model, in which additional transport equations are closed and solved for the mean turbulence kinetic energy $\overline{k}$ and its dissipation rate $\overline{\varepsilon}$, or in a more detailed second-moment closure in which the transport equation for $\overline{k}$ is replaced with closed transport equations for all Reynolds stress and flux components. In the presence of heat release it is convenient to replace $\overline{k}$ and $\overline{\varepsilon}$ with their Favre mean equivalents $\widetilde{k}$ and $\widetilde{\varepsilon}$. Their modelled transport equations may be written [10] as

$$
\overline{\rho}\frac{\partial \widetilde{k}}{\partial t} + \overline{\rho}\,\widetilde{u}_\ell \frac{\partial \widetilde{k}}{\partial x_\ell} = \frac{\partial}{\partial x_\ell}\left[\left(\mu + \frac{\mu_T}{\sigma_k}\right)\frac{\partial \widetilde{k}}{\partial x_\ell}\right] - \overline{\rho}\widetilde{\varepsilon}
$$

$$
\underbrace{-\overline{\rho u_\ell'' u_m''}\frac{\partial \widetilde{u}_m}{\partial x_\ell} - \overline{u_\ell''}\frac{\partial \overline{p}}{\partial x_\ell} + \overline{p'\frac{\partial u_\ell''}{\partial x_\ell}}}_{\mathcal{P}},
$$

$$
\overline{\rho}\frac{\partial \widetilde{\varepsilon}}{\partial t} + \overline{\rho}\,\widetilde{u}_\ell \frac{\partial \widetilde{\varepsilon}}{\partial x_\ell} = \frac{\partial}{\partial x_\ell}\left[\left(\mu + \frac{\mu_T}{\sigma_\varepsilon}\right)\frac{\partial \widetilde{\varepsilon}}{\partial x_\ell}\right] - c_{\varepsilon 2}\overline{\rho}\frac{\widetilde{\varepsilon}^2}{\widetilde{k}} - c_{\varepsilon 1}\frac{\widetilde{\varepsilon}}{\widetilde{k}}\mathcal{P}. \qquad (1.24)
$$

Additional terms are sometimes introduced to allow for other effects such as buoyancy [43] or compressibility in high-speed flows [10]. The equation for $\widetilde{\varepsilon}$ is essentially ad hoc and contains empirical coefficients whose usual values are $\sigma_\varepsilon = 1.3$, $c_{\varepsilon 1} = 1.44$, and $c_{\varepsilon 2} = 1.92$. The terms marked as $\mathcal{P}$ need modelling.

An important feature of this modelling methodology is its incorporation of a *gradient transport* assumption in which turbulent shear stresses are described, by analogy with the molecular process, through an *eddy viscosity*

$$
\mu_T = C_\mu \overline{\rho}\frac{\widetilde{k}^2}{\widetilde{\varepsilon}},
$$

where the usual value of the coefficient is $C_\mu = 0.09$. The Reynolds stress components are related to μ_T by

$$
\overline{\rho u_\ell'' u_m''} = -\mu_T\left(\frac{\partial \widetilde{u}_\ell}{\partial x_m} + \frac{\partial \widetilde{u}_m}{\partial x_\ell} - \frac{2}{3}\delta_{\ell m}\frac{\partial \widetilde{u}_k}{\partial x_k}\right) + \frac{2}{3}\overline{\rho}\widetilde{k}\delta_{\ell m}. \qquad (1.25)
$$

Using the decompositions $\rho = \overline{\rho} + \rho'$ in $\overline{\rho u_\ell''} = 0$ gives $\overline{u_\ell''} = -\overline{\rho' u_\ell''}/\overline{\rho}$. From the state equation $\rho T = \text{const.}$ and its averaged form $\overline{\rho}\widetilde{T} = \text{const.}$, when combustion occurs at constant pressure, one can show that $\overline{\rho' u_\ell''} = -\overline{\rho}\widetilde{u_\ell'' T''}/\widetilde{T}$. This readily gives a model

$$
\overline{u_\ell''} = \frac{\widetilde{u_\ell'' T''}}{\widetilde{T}}. \qquad (1.26)
$$

in terms of the scalar flux.

A scalar flux component may be determined from a gradient transport expression of the form

$$
\overline{\rho}\,\widetilde{u_\ell'' Y_i''} = -\frac{\mu_T}{\sigma_i}\frac{\partial \widetilde{Y}_i}{\partial x_\ell}, \qquad (1.27)
$$

where σ_i is an empirically determined turbulent Schmidt number that is usually taken to be about 0.7. However, it is found that the assumption of gradient scalar transport can fail in premixed turbulent combustion, and this has led some researchers [44–46] to develop more detailed second-moment models. Models of this kind can be illustrated from the work of Bray, Moss, and Libby (BML) [47, 48], described in Section 2.1.

It is convenient to introduce a *reaction progress variable* $c(\mathbf{x}, t)$, defined as a reduced temperature or reaction product mass fraction, having values of zero in reactants and unity in fully burned products. The transport equation for this instantaneous quantity is

$$\frac{\partial \rho c}{\partial t} + \frac{\partial \rho u_k c}{\partial x_k} = \frac{\partial}{\partial x_k} \left(\rho D \frac{\partial c}{\partial x_k} \right) + \dot{\omega}_c. \tag{1.28}$$

By Favre averaging this equation, one obtains the following balance equation for $\tilde{c}$:

$$\frac{\partial \bar{\rho} \tilde{c}}{\partial t} + \frac{\partial \bar{\rho} \tilde{u}_k \tilde{c}}{\partial x_k} = \frac{\partial}{\partial x_k} \left(\overline{\rho D \frac{\partial c}{\partial x_k}} - \overline{\rho u_k'' c''} \right) + \overline{\dot{\omega}}_c, \tag{1.29}$$

where the molecular transport is usually neglected compared with the turbulent transport.

Its *probability density function* or PDF is denoted by $\overline{P}(\zeta; \mathbf{x})$, where ζ is a stochastic variable corresponding to $c(\mathbf{x}, t)$. The PDF has the properties that its integral over all possible states is unity and the integral of $\overline{P}$ times a function of ζ gives the mean value of the function so, for example, the Favre mean of $c(\mathbf{x}, t)$ is

$$\tilde{c}(\mathbf{x}) = \frac{1}{\bar{\rho}} \int_0^1 \rho(\zeta) \, \zeta \, \overline{P}(\zeta; \mathbf{x}) \, d\zeta.$$

If the burning zone is made up of pockets of reactants, with probability α, and pockets of products, with probability β, separated from each other by reaction zones, having probability γ, then the PDF can be written as

$$\overline{P}(\zeta; \mathbf{x}) = \alpha \, \delta(\zeta) + \beta \, \delta(1 - \zeta) + \gamma f(\zeta; \mathbf{x}), \tag{1.30}$$

with $\alpha + \beta + \gamma = 1$. These weights depend on $\mathbf{x}$ and $f(\zeta; \mathbf{x})$ is the PDF of the *internal* part of the distribution. If it is also the case that burning is restricted to thin reacting zones or *flamelets*, whose characteristic dimension is small compared with all other scales of the flow, then $\gamma \sim \mathcal{O}(1/\mathrm{Da}) \ll 1$ and the Favre mean of any property $\varphi(c)$ can be written as

$$\bar{\rho} \tilde{\varphi}(\mathbf{x}) = \alpha \, \rho_u \varphi_u + (1 - \alpha) \rho_p \varphi_p + \mathcal{O}(\gamma), \tag{1.31}$$

where $\rho_u, \rho_p, \varphi_u$, and φ_p represent densities in reactants and products and values of φ in reactants and products, respectively.

These thin-flamlets arguments can be extended to evaluate *conditional mean* velocities in reactants and products, that is, mean values formed from velocity measurements taken only when $c = 0$ or $c = 1$, respectively, and written $\bar{u}_{k,u} = \langle u_k | c = 0 \rangle$ for reactants or $\bar{u}_{k,p} = \langle u_k | c = 1 \rangle$ for products, where the angle brackets indicate averaging. It may then be shown (see Section 2.1) that

$$\widetilde{u_k'' c''} = \tilde{c}(1 - \tilde{c})(\bar{u}_{k,p} - \bar{u}_{k,u}) + \mathcal{O}(\gamma),$$

so the direction of the scalar flux $\widetilde{u_k''c''}$ is the same as that of the conditional mean velocity difference $(\bar{u}_{k,p} - \bar{u}_{k,u})$. An explanation for *counter-gradient* scalar transport, that is, transport in the direction opposite to that indicated by Eq. (1.27), is that force fields that are due to pressure gradient [45] or Reynolds stress [49] act differentially on low-density burned-gas pockets within the flame brush.

Second-moment models, in which transport equations are used to determine second moments, variances, and co-variances, can be used to address these issues. Reynolds stress components must satisfy [50–52]

$$
\frac{\partial}{\partial t}\left(\bar{\rho}\widetilde{u_\ell''u_m''}\right) + \frac{\partial}{\partial x_k}\left(\bar{\rho}\tilde{u}_k\widetilde{u_\ell''u_m''}\right) = -\frac{\partial}{\partial x_k}\left(\bar{\rho}\widetilde{u_k''u_\ell''u_m''}\right) + \overline{u_\ell''}g_m + \overline{u_m''}g_\ell - \bar{\rho}\tilde{\varepsilon}_{\ell m}
$$

$$
- \left[\bar{\rho}\widetilde{u_k''u_\ell''}\frac{\partial\tilde{u}_m}{\partial x_k} + \bar{\rho}\widetilde{u_k''u_m''}\frac{\partial\tilde{u}_\ell}{\partial x_k}\right]
$$

$$
- \left[\overline{u_\ell''\frac{\partial p}{\partial x_m}} + \overline{u_m''\frac{\partial p}{\partial x_\ell}}\right], \tag{1.32}
$$

and the scalar flux component $\widetilde{u_\ell''Y_i''}$ obeys [10, 52, 53]

$$
\frac{\partial}{\partial t}\left(\bar{\rho}\widetilde{u_\ell''Y_i''}\right) + \frac{\partial}{\partial x_k}\left(\bar{\rho}\tilde{u}_k\widetilde{u_\ell''Y_i''}\right) = -\frac{\partial}{\partial x_k}\left(\bar{\rho}\widetilde{u_k''u_\ell''Y_i''}\right) + \overline{Y_i''}g_\ell + \overline{u_\ell''\dot{\omega}_i} - \bar{\rho}\tilde{\varepsilon}_{\ell i}
$$

$$
- \left[\bar{\rho}\widetilde{u_k''Y_i''}\frac{\partial\tilde{u}_\ell}{\partial x_k} + \bar{\rho}\widetilde{u_k''u_\ell''}\frac{\partial\tilde{Y}_i}{\partial x_k}\right]
$$

$$
- \overline{Y_i''\frac{\partial p}{\partial x_\ell}}. \tag{1.33}
$$

Closure models are required for all the terms on the right-hand sides of the preceding two equations. The viscous term in Eq. (1.32) is given by

$$
\bar{\rho}\tilde{\varepsilon}_{\ell m} = \overline{u_\ell''\frac{\partial\tau_{mk}}{\partial x_k}} + \overline{u_m''\frac{\partial\tau_{\ell k}}{\partial x_k}}
$$

$$
= \left[\frac{\partial\overline{\tau_{mk}u_\ell''}}{\partial x_k} + \frac{\partial\overline{\tau_{\ell k}u_m''}}{\partial x_k}\right] - \left[\overline{\tau_{mk}\frac{\partial u_\ell''}{\partial x_k}} + \overline{\tau_{\ell k}\frac{\partial u_m''}{\partial x_k}}\right].
$$

The parts in the first set of square brackets are usually included with the third-order correlation [54] because it represents diffusion, and those in the second set of square brackets are known as viscous dissipation, which is obtained from the modelled transport equation in Eq. (1.24) after the assumption of local isotropy [52, 54] is introduced; $\tilde{\varepsilon}_{\ell m} = 2\delta_{\ell m}\tilde{\varepsilon}/3$. Modelling of other terms in Eqs. (1.32), and (1.33) are discussed by Jones [52], and an example of their application specifically to premixed flames can be found in Section 2.4. The dissipation rate $\tilde{\varepsilon}$ is commonly modelled with the second of Eq. (1.24). It is worth noting that the pressure-related terms in the equations for $\tilde{k}, \tilde{\varepsilon}$, and $\widetilde{u_\ell''u_m''}$ can usually be neglected for open flames, but may not be so for flames inside combustors. Possible modelling of the pressure–dilatation term is discussed in Section 2.4.

The variance of $Y_i(\mathbf{x}, t)$ is also needed, and it obeys

$$
\bar{\rho} \frac{D\widetilde{Y_i''^2}}{Dt} = 2\,\overline{Y_i''\dot{\omega}_i''} - 2\bar{\rho}\,\tilde{\epsilon}_i - 2\overline{\widetilde{\rho u_k'' Y_i''}} \frac{\partial \widetilde{Y}_i}{\partial x_k} - \frac{\partial \overline{\rho u_k'' Y_i''^2}}{\partial x_k}
$$

$$
+ 2\frac{\partial}{\partial x_k}\left(\overline{\rho\hat{\alpha}\frac{\partial Y_i''^2}{\partial x_k}}\right) + 2\,\overline{Y_i''\frac{\partial}{\partial x_k}\left(\rho\hat{\alpha}\frac{\partial \widetilde{Y}_i}{\partial x_k}\right)}, \qquad (1.34)
$$

in which closures are needed for third-moment, chemical reaction terms and the mean scalar dissipation $\bar{\rho}\,\tilde{\epsilon}_i = \overline{\rho D(\partial Y_i/\partial x_k)^2}$. The last two terms are small and are usually neglected. The scalar dissipation rate has a dimension of $(\text{time})^{-1}$, and represents the reciprocal of a local time scale for molecular mixing. We will see later that the scalar dissipation plays an essential role in most mean reaction rate models for premixed combustion.

1.7.2 Reaction-Rate Closures

The PDF formalism also provides a means of evaluating the mean values of chemical reaction rates. For example, if the chemical source term in the transport equation for $c(\mathbf{x}, t)$ is $\dot{\omega}_c(c)$[1], its Reynolds mean is

$$
\overline{\dot{\omega}_c} = \int_0^1 \dot{\omega}_c\,\overline{P}\,\mathrm{d}\zeta = \gamma\int_0^1 \dot{\omega}_c f\,\mathrm{d}\zeta, \qquad (1.35)
$$

where the second expression follows from the PDF in Eq. (1.31) and the observation that $\dot{\omega}_c = 0$ when $c = 0$ or $c = 1$. By a similar argument, *any* property $\varphi(c)$ that is zero in both reactants and products, so $\varphi(c = 0) = \varphi(c = 1) = 0$, has a mean

$$
\bar{\varphi} = \int_0^1 \varphi\overline{P}\,\mathrm{d}\zeta = \gamma\int_0^1 \varphi f\,\mathrm{d}\zeta.
$$

Eliminating γ between these two equations, one sees that $\overline{\dot{\omega}_c}$ is proportional to $\bar{\varphi}$. The scalar dissipation $\rho\,\epsilon_c = \rho D(\partial c/\partial x_k)^2$, whose mean appears as the dissipation term in the transport equation for the variance $\widetilde{c''^2}$, is zero when $c = 0$ or 1 and is therefore proportional to the mean reaction rate. A physical interpretation [55] (see Section 2.1) is that, with $\gamma \ll 1$, reaction zones resemble unstretched laminar flames, for which both $\dot{\omega}_c$ and ϵ_c are unique functions of c. It can be shown [55] that

$$
\overline{\dot{\omega}_c} = \frac{2\bar{\rho}\,\tilde{\epsilon}_c}{(2C_m - 1)} + \mathcal{O}(\gamma), \qquad (1.36)
$$

where $C_m \equiv \overline{c\,\dot{\omega}_c}/\overline{\dot{\omega}_c}$ can usually be treated as a constant; typically it varies between 0.7 to 0.8 for lean hydrocarbon – and ultra-lean ($\phi \leq 0.4$) hydrogen–air flames. A typical value of this parameter for hydrogen–air mixtures with $0.4 < \phi \leq 1$ is about 0.6 [56]. In situations in which $\gamma \ll 1$, this equation allows the mean reaction rate to be calculated in terms of the mean scalar dissipation, which is found from a closed version of its transport equation. This approach is explored in Section 2.3.

---

[1] For a single reaction with a heat release parameter of $\tau = (T_b - T_u)/T_u$ and a Zeldovich number of $\hat{\beta} \equiv \alpha\,T_a/T_b$, the reaction rate is given by $\dot{\omega}_c = B\rho(1-c)\exp\left[\dfrac{-\hat{\beta}(1-c)}{1-\alpha(1-c)}\right]$, where $\alpha = \tau/(1+\tau)$ and $B = A_f\,T^f\exp(-\hat{\beta}/\alpha)$ when the Lewis number is unity.

The progress-variable approach previously outlined can be adapted to include the effects of fluid dynamics on the reaction surface, as proposed by Bradley [57], which is discussed in Section 2.3. The previous progress-variable approach can also be adapted for *partially premixed* combustion [58] in which mass fractions depend on mixture fraction Z as well as on c, so $Y_i(\mathbf{x}, t) = Y_i[c(\mathbf{x}, t), Z(\mathbf{x}, t)]$. The progress variable can be defined in terms of fuel mass fraction $Y_f(\mathbf{x}, t)$ as

$$c(\mathbf{x}, t) = \frac{Y_{f,r}[Z(\mathbf{x}, t)] - Y_f(\mathbf{x}, t)}{Y_{f,r}[Z(\mathbf{x}, t)] - Y_{f,p}[Z(\mathbf{x}, t)]},$$

where the notation indicates that the fuel mass fraction in unburned reactants, $Y_{f,r}$, depends on Z, which is a function of $\mathbf{x}$ and t. Similarly, $Y_{f,p}$, which is the mass fraction of fuel in equilibrium products, is also a function of Z. These dependencies introduce additional terms in the transport equation for $c(\mathbf{x}, t)$, Eq. (1.28), which can be accommodated [58] by replacing the true reaction rate $\dot\omega_c$ with an effective rate

$$\dot\omega_c^* = \dot\omega_c + A(\mathbf{x}, t) - B(\mathbf{x}, t), \tag{1.37}$$

where

$$A = \frac{2(Y_{f,r} - dY_{f,p}/dZ)}{Y_{f,r} - Y_{f,p}} \rho\, N_{zc},$$

$$B = \frac{1}{Y_{f,r} - Y_{f,p}} \frac{d^2 Y_{f,p}}{dZ^2} \rho\, N_z.$$

These expressions introduce two additional scalar dissipation rates: $N_Z = \mathcal{D}(\nabla Z \cdot \nabla Z)$ is the dissipation of Z, and $N_{Zc} = \mathcal{D}(\nabla Z \cdot \nabla c)$ is known as the *cross dissipation*. If Eq. (1.37) is averaged, then the dissipation rate $\tilde\epsilon_z$ of the variance of Z and $\tilde\epsilon_{zc}$ of the co-variance $\widetilde{z''c''}$ appear. It is clear that additional modelling is required.

Returning to fully premixed combustion, there are two ways in which PDFs can be evaluated to determine mean reaction rates from Eq. (1.35). The first, more empirical *presumed* PDF method assumes the PDF to be described by a specified algebraic expression, containing parameters that are related to moments of the argument(s) of the PDF. This approach becomes impracticable if the PDF depends on more than one or at most two stochastic variables. An appropriate form for a presumed PDF is provided by the beta function [10]; in the case of a *monovariate* PDF $\overline{P}(\zeta; \mathbf{x})$, where the stochastic variable ζ, which is related to a flow variable $\varphi(\mathbf{x}, t)$, lies between zero and unity, this leads to

$$\overline{P}(\zeta; \mathbf{x}) = \frac{\zeta^{(a-1)}(1 - \zeta)^{(b-1)}}{\hat\beta(a, b)},$$

where a and b are functions of position $\mathbf{x}$ as they are related to the first and second moments of φ by

$$a = \frac{\overline{\varphi}^2(1 - \overline{\varphi})}{\overline{\varphi'^2}} - \overline{\varphi}, \quad \text{and} \quad b = \frac{a(1 - \overline{\varphi})}{\overline{\varphi}}.$$

The normalising factor $\hat{\beta}$ is the beta function [59], which is related to the gamma function, and it is given by

$$\hat{\beta}(a, b) = \int_0^1 \zeta^{(a-1)}(1 - \zeta)^{(b-1)} \, \mathrm{d}\zeta = \frac{\Gamma(a)\Gamma(b)}{\Gamma(a + b)};$$

the gamma function can be calculated with a fifth- or eighth-order polynomial approximation in [59]. This presumed form provides an appropriate range of shapes: If a and b approach zero (in the limit of large variance), the PDF resembles a bimodal shape of Eq. (1.30) whereas, if a and b are large (in the limit of small variance), it develops a monomodal form with an internal peak. It has also been shown by Girimaji [60] that this PDF behaves likes a Gaussian when $\overline{\varphi'^2}$ is very small.

The alternative *transported* PDF method [61] provides a more rigorous means of computing PDF shapes and evaluating the influence of finite-rate chemical reactions. Take the PDF $\overline{P}(\zeta; \mathbf{x})$ as an example, where ζ is the stochastic variable corresponding to $\varphi(\mathbf{x}, t)$. If we had a fully resolved time-dependent solution, then $\varphi(\mathbf{x}, t)$ would be known, so $\zeta \equiv \varphi(\mathbf{x}, t)$, and the instantaneous or 'fine-grained' probability would be represented by the Dirac delta function $\delta(\varphi - \zeta)$. It can then be seen that the PDF is obtainable as the average of this delta function. Manipulation of the transport equation for $\varphi(\mathbf{x}, t)$, together with the properties of the Dirac delta function, then leads to the differential equation for $\overline{P}$ (see [62] and Section 2.4). However, as noted earlier, it is more convenient to work with mass-weighted, i.e., Favre mean, quantities in flows with large variations in density, leading to the introduction of a mass-weighted PDF $\widetilde{P}(\zeta; \mathbf{x}) = \overline{P}(\zeta; \mathbf{x})\rho/\overline{\rho}$, from which the Favre mean of $\varphi(\mathbf{x}, t)$ is obtained as

$$\widetilde{\varphi}(\mathbf{x}, t) = \int_{\varphi_{\min}}^{\varphi_{\max}} \zeta(\mathbf{x}, t) \, \widetilde{P}(\zeta; \mathbf{x}) \, \mathrm{d}\zeta.$$

The mass-weighted PDF obeys the equation

$$\overline{\rho}\frac{\mathrm{D}\widetilde{P}}{\mathrm{D}t} = -\frac{\partial}{\partial x_k}\left[\langle u_k''|\varphi = \zeta\rangle\overline{\rho}\,\widetilde{P}\right] - \frac{\partial}{\partial \zeta}\left[\dot{\omega}(\zeta)\overline{\rho}\,\widetilde{P}\right]$$
$$- \frac{\partial^2}{\partial \zeta^2}\left[\left\langle \mathcal{D}\frac{\partial \varphi}{\partial x_k}\frac{\partial \varphi}{\partial x_k}\middle|\varphi = \zeta\right\rangle\overline{\rho}\,\widetilde{P}\right]. \tag{1.38}$$

The terms on the left arise from time variations and convection. In the first term on the right, the notation $\langle u_k''|\varphi = \zeta\rangle$ represents the mean of $(u_k - \widetilde{u}_k)$, subject to the condition $\varphi = \zeta$, and this term describes the influence of turbulent transport on $\widetilde{P}$. It is usually represented by assuming a gradient transport expression. However, the transported PDF formulation can be extended [61] to encompass the *joint* PDF (JPDF) of a scalar variable and the flow velocity, and the turbulent transport term is then closed; an additional unclosed term then appears on the right-hand side of the equation, involving fluctuations in pressure. With this addition, both gradient and counter-gradient scalar transport can be predicted. Significantly, the second term on the right in Eq. (1.38), representing effects of chemical reaction, is *closed*, because $\dot{\omega}(\zeta)$ is a function of independent variable ζ and $\widetilde{P}$ is determined as part of the solution. Also, as explained in Section 2.4, the transported PDF formulation can again be extended to describe the *joint* PDF of several scalar variables – mass fractions and temperature – so detailed and realistic chemical kinetic mechanisms can be incorporated in closed form. These are very attractive features of the method, which are described in Section 2.4.

The final, so-called *mixing*, term in Eq. (1.38) containing the molecular-diffusion coefficient $\mathcal{D}$ describes the effect of molecular transport on the PDF; the quantity $\langle \mathcal{D}(\partial\varphi/\partial x_k)(\partial\varphi/\partial x_k)|\varphi = \zeta\rangle$ is the conditional mean scalar dissipation rate. If the Damköhler number Da is small, that is, if all turbulence scales are smaller than chemical scales, then the mixing term can be modelled as a function of turbulence quantities alone, and the equation can be closed. On the other hand, when Da $\gg$ 1, the smallest scales are chemical, and the gradient $\partial\varphi/\partial x_k$ approaches that of a laminar flame, so the mixing term is strongly influenced by chemical reaction and cannot be described simply as a function of turbulence quantities; that is, a mean-reaction-rate closure is again required. This problem is discussed in Section 2.4. Note that, in the most general case, with three spatial coordinates, three velocity components, N species, and temperature, the JPDF equation involves $7 + N$ independent variables.

Conditional moment closure (CMC) models [63–65], which can be derived from the JPDF transport equation, provide an alternative means of incorporating detailed chemical reaction mechanisms. The basic dependent variables of CMC are the *conditional* means of the species mass fractions $Y_i(\mathbf{x}, t)$ and temperature $T(\mathbf{x}, t)$, conditional on the value of a chosen scalar. In the case of non-premixed combustion, this is the mixture fraction $Z(\mathbf{x}, t)$, whereas the progress variable $c(\mathbf{x}, t)$ is selected [63, 66] for premixed combustion. The conditional mean of $Y_i(\mathbf{x}, t)$ is then

$$Q_i(\zeta; \mathbf{x}, t) = \langle Y_i(\mathbf{x}, t)|c = \zeta\rangle,$$

which obeys the transport equation [63, 66, 67]

$$\langle \rho|\zeta\rangle \frac{\partial Q_i}{\partial t} + \langle \rho u_k|\zeta\rangle \frac{\partial Q_i}{\partial x_k} = \frac{\mathrm{Le}_c}{\mathrm{Le}_i} \left\langle \rho\mathcal{D} \frac{\partial c}{\partial x_k} \frac{\partial c}{\partial x_k} \Big| \zeta \right\rangle \frac{\partial^2 Q_i}{\partial \zeta^2} + \langle \dot{\omega}_i|\zeta\rangle - \langle \dot{\omega}_c|\zeta\rangle \frac{\partial Q_i}{\partial \zeta} + e_{Q_i} + e_{y_i}, \tag{1.39}$$

where e_{Q_i} represents other molecular-diffusion terms and e_{y_i} involves the conditional fluctuation $y_i(\mathbf{x}, t) = Y_i(\mathbf{x}, t) - Q_i(\zeta; \mathbf{x}, t)$. Additional source or sink terms will arise depending on the precise definition of c, as noted in [66, 67]. The first term on the right-hand side of the equation contains the conditional scalar dissipation $\langle \rho\mathcal{D}(\partial c/\partial x_k)(\partial c/\partial x_k)|\zeta\rangle$, and the same modelling difficulties as in the transported PDF methods occur when Da $\gg$ 1. A key assumption of CMC is that fluctuations about the conditional mean are small. In *first-order* CMC, these conditional mean fluctuations are ignored and the conditional mean reaction rate in Eq. (1.39) is taken to have the same functional dependence on Q_i as that of the instantaneous reaction rate, i.e., $\langle \dot{\omega}_i|\zeta\rangle = \dot{\omega}(\mathbf{Q})$. An allowance for the conditional fluctuations is included in *second-order* CMC by including conditional variances and co-variances, which require further modelling. The CMC method is well advanced for non-premixed flames, but it is in its early stage for premixed flames, primarily because of the issues sourrounding the modelling of the conditional scalar dissipation rate. A preliminary application [68] of this methodology to lean premixed flames is encouraging. However, in a problem with three spatial coordinates and $N + 1$ scalar variables, there are $N + 1$ CMC equations, each with four independent variables.

In many situations of practical interest, Da > 1, and regions of chemical reaction form thin interfaces separating unburned reactants from fully burned combustion products. The mean burning rate can then be specified as the flame area per unit volume – the *flame surface density* (FSD) – multiplied by the rate of conversion of reactants to products per unit flame area. The *surface density function* $\Sigma(\zeta; \mathbf{x}, t)$, defined as the mean area per unit volume of the isosurface on which $c(\mathbf{x}, t) = \zeta$, is a distribution function, analogous to the PDF $\overline{P}(\zeta; \mathbf{x}, t)$. As shown in Section 2.2, the two are related by [69, 70]

$$\Sigma(\zeta; \mathbf{x}, t) = \left\langle \left(\frac{\partial c}{\partial x_k} \frac{\partial c}{\partial x_k} \right)^{1/2} \middle| c = \zeta \right\rangle \overline{P}(\zeta; \mathbf{x}, t), \qquad (1.40)$$

and $\Sigma(\zeta; \mathbf{x}, t)$ obeys the equation [71, 72]

$$\frac{\partial \Sigma}{\partial t} + \frac{\partial}{\partial x_k} \left[\langle u_k + S_{d,c} n_k \rangle_s \Sigma \right] = \left\langle (\delta_{km} - n_k n_m) \frac{\partial u_k}{\partial x_m} \right\rangle_s \Sigma + \left\langle S_{d,c} \frac{\partial n_k}{\partial x_k} \right\rangle_s \Sigma. \quad (1.41)$$

Here $S_{d,c}$ is the displacement speed of the isosurface $c(\mathbf{x}, t) = \zeta$ and n_k is the component of the unit normal vector $\mathbf{n}$, pointing towards small values of c in the direction of x_k. *Surface averages* are denoted by $\langle Q \rangle_s$ and are defined as

$$\langle Q \rangle_s = \frac{\langle Q \mathcal{G} | \zeta \rangle}{\langle \mathcal{G} | \zeta \rangle},$$

where $\mathcal{G} \equiv [(\partial c / \partial x_k)(\partial c / \partial x_k)]^{1/2}$ is the magnitude of the progress-variable gradient. Terms on the left-hand side of Eq. (1.41) represent unsteady effects and the influences of convection and isosurface propagation, respectively. The first term on the right represents effects of tangential strain that are due to the local velocity field, and the final term arises from combined effects of surface curvature and propagation. Note that the velocity appearing in this equation is the local velocity in the interior of the thin flame, at locations where $c(\mathbf{x}, t) = \zeta$, so extensive modelling is required to derive a closed version of Eq. (1.41). The FSD may be defined as the value of the surface density function at a chosen isosurface, for example, the value of c at which $\dot{\omega}_c(c)$ is maximum. Alternatively, a *generalised* FSD is [73]

$$\Sigma_g(\mathbf{x}) = \int_0^1 \Sigma(\zeta; \mathbf{x}) \, \mathrm{d}\zeta = \int_0^1 \langle \mathcal{G} | \zeta \rangle \overline{P}(\zeta; \mathbf{x}) \, \mathrm{d}\zeta = \langle \mathcal{G} \rangle. \qquad (1.42)$$

The local displacement speed $S_{d,c}$ is often estimated in terms of the burning velocity S_L^0 of an unstretched laminar flame, so $\rho(\zeta) S_{d,c} = \rho_u S_L^0$.

An alternative starting point for thin-flamelet combustion models, i.e., for $\mathrm{Da} > 1$, is provided by the *level set* or *G-equation* formalism [41, 74], which is also introduced in Section 2.2. A function $G(\mathbf{x}, t)$ is introduced, which satisfies $G(\mathbf{x}, t) = G_f$ on the thin-flame isosurface. The isosurface is assumed to be an interface between reactants and products, propagating with a velocity S_G relative to reactants, which is modelled in terms of the strain and curvature of the surface $G(\mathbf{x}, t) = G_f$. The Huygens-type equation, which is a kinematic equation, describes the evolution of G by

$$\frac{\partial G}{\partial t} + u_k \frac{\partial G}{\partial x_k} = S_G \left(\frac{\partial G}{\partial x_m} \frac{\partial G}{\partial x_m} \right)^{1/2}. \qquad (1.43)$$

The methodology is developed by Peters [41] and Peters et al. [74] by averaging Eq. (1.43). Although the function $G(\mathbf{x}, t)$ lacks physical significance everywhere except on the surface $G(\mathbf{x}, t) = G_f$, averaging involves values for which $G(\mathbf{x}, t) \neq G_f$, so a functional form must be specified. It can be shown [75] that the flame surface density is related to the G field by

$$\Sigma(\mathbf{x}) = \left\langle \left(\frac{\partial G}{\partial x_k} \frac{\partial G}{\partial x_k} \right)^{1/2} \middle| G = G_f \right\rangle \overline{P}(G_f; \mathbf{x}).$$

An approximate expression has been derived for $\overline{P}(G_f; \mathbf{x})$ [75], where $\overline{P}(G; \mathbf{x})$ is the PDF of G.

A striking and important feature of all the schemes for determining mean rates of chemical reaction introduced here is that a mean or conditional mean scalar dissipation function plays a central role in each case. In a presumed PDF model for the progress variable c, the mean dissipation $\widetilde{\epsilon}_c$ strongly influences the variance $\widetilde{c''^2}$, which, in turn, decides the width of the PDF $\widetilde{P}(\varsigma; \mathbf{x})$. As may be seen from Eq. (1.35), this PDF determines the mean rate. And, in the high Damköhler limit that approximates many practical situations, the mean rate is directly proportional to $\widetilde{\epsilon}_c$, as shown by Eq. (1.36). This simple expression reminds us that, in a thin flame, the local rate of reaction must be directly balanced by the rate at which molecular-transport processes carry heat and reactive species through the preheat zone at the cold side of the flame. In the transported PDF and CMC models, this physics is represented in the conditional scalar dissipation or micromixing model. The surface density function, on which FSD models are based, is defined in terms of a local gradient, as seen in Eq. (1.40), so the generalised FSD of Eq. (1.42) can be rewritten as

$$\Sigma_g = \frac{\overline{(\rho \epsilon_c)^{1/2}}}{\rho_u \mathcal{D}_u},$$

where it is assumed that $\rho \mathcal{D} = \rho_u \mathcal{D}_u$. Finally, the G-equation model of Peters [41] and Peters et al. [74] involves the scalar dissipation of G. In each of these models, the accepted practice has generally been to represent the scalar dissipation in terms of a relatively simple algebraic model, and this has often been borrowed from studies of non-reactive turbulent flows, namely

$$\widetilde{\epsilon}_c = C_D \frac{\widetilde{\varepsilon}}{\widetilde{k}} \widetilde{c''^2},$$

where C_D is a constant. In circumstances characterised by Da > 1, one must expect to find that $\widetilde{\epsilon}_c$ is influenced by the laminar flame time δ_L^0/S_L^0 in addition to the turbulence time $\widetilde{k}/\widetilde{\epsilon}$. The *scalar-dissipation-rate transport* approach, described in Section 2.3, recognises the key role played by the scalar dissipation and determines it from a closed version of its transport equation containing both characteristic scales. It is also possible to obtain an algebraic model for the FSD Σ_g as

$$\Sigma_g \delta_L^0 = \frac{2C_{D_c}}{(2C_m - 1)} \frac{\overline{\rho}}{\rho_u} \left(1 + \frac{2}{3} \frac{C_{\epsilon_c} S_L^0}{\sqrt{\widetilde{k}}} \right) \left(1 + \frac{C_D \widetilde{\varepsilon} \delta_L^0}{C_{D_c} \widetilde{k} S_L^0} \right) \widetilde{c''^2},$$

using the scalar-dissipation-rate approach [76]. The model parameters defined in Section 2.3 suggest that C_D/C_{D_c} is of the order of unity, and thus it follows that $C_D \widetilde{\varepsilon} \delta_L^0/(C_{D_c} \widetilde{k} S_L^0) \ll 1$ for large-Da flames. Hence the FSD scales with δ_L^0 rather than

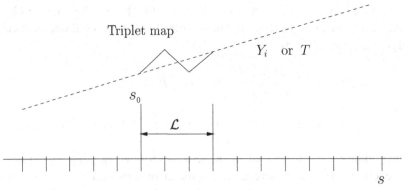

Figure 1.4. Illustration of triplet mapping in a LEM.

with Λ. Earlier models for the FSD scaling with Λ are discussed in the book edited by Libby and Williams [10].

From a physical point of view, the evolution of an instantaneous scalar value at a point inside a turbulent flame is influenced by the molecular diffusion, reaction, and convection as conveyed by the last of Eqs. (1.1). Thus, the mean reaction rate is also expected to be influenced by these processes, and the closure models previously discussed attempt to capture these effects by using a simplified description. In another approach, known as the *linear eddy model* (LEM), it was suggested in [77] that these effects can be accounted for by solving a one-dimensional (1D) unsteady reaction–diffusion equation along with a stochastic term representing the turbulent mixing or stirring. This approach was developed, elaborately discussed, and tested by Kerstein [77–83] for a variety of flows with scalar mixing and chemical reactions. Thus details can be found in those references. Here, the main features of this approach and its similarity, where there are any, to other approaches are discussed briefly.

The 1D unsteady equation for the mass fraction of species i is written as

$$\frac{\partial Y_i}{\partial t} = -\frac{1}{\rho}\frac{\partial J_i}{\partial s} + \frac{\dot{\omega}_i}{\rho} + \mathcal{F}_{\text{stir}}, \tag{1.44}$$

with the stirring term $\mathcal{F}_{\text{stir}}$ included. The direction s is taken to be aligned with the local normal specified by the gradient of Y_i. The size of this 1D domain is typically taken to be about 6Λ in numerical calculations. Because the reaction rate depends on temperature, Eq. (1.44) is to be supplemented with a similar equation for temperature T. In these two equations, the diffusion and reaction are described in a deterministic fashion and the stirring part is modelled with a stochastic approach. The turbulence is usually expected to increase the scalar gradient and isoscalar surface area by the action of the compressive principal strain rate on the scalar field (for an elaborate discussion of this physics see Section 2.3). This physics is simulated by a triplet mapping. A clear exposition of this mapping is given by Kerstein [82] and is schematically shown in Fig. 1.4, identifying two important parameters, viz., (1) the location s_0 for the mapping and (2) its size $\mathcal{L}$. The location s_0 is chosen randomly with equal probability for all discrete locations inside the 1D domain. Because $\mathcal{F}_{\text{stir}}$ represents the stirring by turbulent eddies ranging from Λ to η_K, the 1D domain is discretised according to DNS requirements and the Kolmogorov length scale is obtained with $\eta_K = 4\,\mathrm{Re}_T^{3/4}\,(\tilde{k}^{3/2}/\tilde{\varepsilon})$ in the context of RANS. The mapping size $\mathcal{L}$ is chosen randomly in the range $\Lambda \geq \mathcal{L} \geq \eta_K$ by use of a PDF given by

$$P(\mathcal{L}) = \frac{5}{3} \frac{\mathcal{L}^{-8/3}}{\eta_K^{-5/3} - \Lambda^{-5/3}}. \tag{1.45}$$

This PDF is obtained by drawing an analogy of the stirring event with the random walk of a marker particle and using the inertial range scaling of Kolmogorov turbulence [82, 84]. This analysis is also used to obtain a frequency, λ_f with a dimension of $L^{-1} T^{-1}$, of this stirring event as

$$\lambda_f = \frac{54}{5} \frac{\nu \, \mathrm{Re}_T}{C_\lambda \Lambda^3} \frac{(\Lambda/\eta_K)^{5/3} - 1}{1 - (\eta_K/\Lambda)^{4/3}}, \tag{1.46}$$

where C_λ is a model parameter. Now, one can see that Eq. (1.44) can be solved using any standard numerical technique, noting that it is an instantaneous representation, although an operator-splitting method is useful because of the stiffness associated with $\dot{\omega}_i$. These solutions are then ensemble or time averaged to obtain the mean reaction rate and the density or temperature required for the computational fluid dynamics (CFD) solution. Application of this methodology to turbulent premixed combustion is discussed in [84–87].

An important application of simulation codes for turbulent combustion is to estimate the influence of design changes on emissions of pollutant species and greenhouse gases, and this necessitates the incorporation of realistic chemical kinetic mechanisms. The growing use of non–fossil fuels increases the importance of this requirement. At present there are two types of approach to this problem. Detailed kinetic mechanisms can be directly incorporated into transported PDF, CMC, and LEM models, although, to reduce computing costs, it is usual to use one of several available methods (intrinsic low-dimensional manifold, in situ adaptive tabulation, computational singulor perturbation, etc.) to limit the size of the mechanism [20, 21, 88–90]. Also, as pointed out earlier, our current understanding of possible chemical kinetic effects on the mixing or scalar dissipation processes in these models is still incomplete. The alternative approach (FGM, FPI) [89, 91, 92] (also see Section 2.2), applicable to the various thin-flamelet models, is to assume combustion to occur under conditions similar to those to be found in an unstretched laminar flame. Then data from a laminar flame calculation at appropriate values of mixture fraction, pressure, and initial temperature can be tabulated as a function of the progress variable, and mean species mass fractions can be estimated from

$$\overline{\rho}(\mathbf{x}) \widetilde{Y}_i(\mathbf{x}) = \int_0^1 \rho(\zeta) Y_i(\zeta) \overline{P}(\zeta; \mathbf{x}, t) \, \mathrm{d}\zeta,$$

where $\overline{P}(\zeta; \mathbf{x}, t)$ is a presumed PDF, calculated as described earlier.

1.7.3 Models for LES

In LESs, mean fluid mechanical and reaction-rate closures are required for replacing the SGS information that is lost as a result of the volume-averaging process of Eq. (1.19). Turbulent flow models are based on closures developed for non-reactive flow applications. In the *Smagorinsky* model [93] for constant-density flows, unresolved Reynolds stresses are represented by

$$\overline{u_k u_m} - \overline{u}_k \overline{u}_m - \frac{\delta_{km}}{3} \left[\overline{u_\ell u_\ell} - \overline{u}_\ell \overline{u}_\ell \right] = -\hat{\nu}_T \left[\frac{\partial \overline{u}_k}{\partial x_m} + \frac{\partial \overline{u}_m}{\partial x_k} \right], \qquad (1.47)$$

where $\hat{\nu}_T$ is a SGS eddy viscosity given by $\hat{\nu}_T = (C_S \Delta)^2 |\overline{S}|$. Here Δ is size of the LES filter and $\overline{S}$ is the resolved part of the shear stress with its component $\overline{S}_{km}$ appearing inside square brackets in Eq. (1.47). However, the Smagorinsky model is found to be too dissipative and the coefficient C_S depends on the flow configuration. In *dynamic* or *scale-similarity* models [94, 95], the coefficient $C_S(\mathbf{x}, t)$ and the dissipation are determined locally as part of the solution, by introducing two filter scales, the scale Δ of the LES and a second larger scale $\widehat{\Delta}$, refiltering the LES solution at this second scale, and using the identity $\overline{u_k u_m} = \widehat{\overline{u_k u_m}}$, which assumes that the smallest of resolved scales and the largest of unresolved scales behave similarly. The presence of heat release does not seem to affect this [96]. Further details and application to variable-density flows can be found in many references cited in the following discussion. Unresolved scalar fluxes in combustion simulations are often described in terms of a gradient transport assumption,

$$\widetilde{u_k Y_i} - \tilde{u}_k \tilde{Y}_i = -\frac{\hat{\nu}_T}{\sigma_i} \frac{\partial \tilde{Y}_i}{\partial x_k}, \qquad (1.48)$$

where σ_i is a Schmidt number and $\hat{\nu}_T$ is determined from the model for unresolved Reynolds stresses. However, by filtering DNS data, Tullis and Cant [97] showed that counter-gradient scalar fluxes can occur. These fluxes are considered to come from laminar flamelets, and thus they may be included by modelling the SGS flux as $\overline{\rho u c} - \overline{\rho} \tilde{u} \tilde{c} = \rho_u s_L (\overline{c} - \tilde{c})$. Also, an alternative view of treating this subgrid flux as a source under some restrictive assumptions was also expressed [98].

Heat release occurs at scales that are unresolved in combustion LES, so models are required. These are often developed from models that have previously been used in RANS calculations. The current LES approaches for premixed flames can be broadly categorised as (1) eddy break-up (EBU) and presumed PDF models, (2) FSD models, (3) thickened-flame models, (4) G-equation models, and (5) linear eddy models (LEM-LES). The transported PDF [99] and CMC [100, 101] methods have been used as SGS models in the LES of non-premixed flames. Although these methods can be extended to premixed combustion, no attempts to do so have yet been made. The methods just noted are subsequently briefly reviewed and discussed below to identify their essential features. This discussion is not intented to be exhaustive, and interested readers are referred to the cited references for a comprehensive discussion.

(1) EBU AND PRESUMED PDF MODELS. In EBU modelling for RANS, the averaged reaction rate is expressed as a sum of two inverse time scales: One is related to the chemical kinetics and the other one is related to turbulent mixing, as noted by Spalding, Magnussen, and their co-workers. This modelling is extended [102–104] to LES by replacing the turbulent mixing time scale with the corresponding one for the SGS. This mixing time scale is typically expressed [103] as $\tau_m \approx k_{sgs}/\varepsilon_{sgs}$ with $\varepsilon_{sgs} = c_\varepsilon k_{sgs}^{3/2}/\Delta$, where c_ε is a model parameter. The subgrid kinetic energy k_{sgs}

is obtained by the solution of a modelled balance equation [103] or the dynamic procedure. In another approach [103, 105] the filtered reaction rate is obtained with $\bar{\omega} = \int P_{sgs}(\mathbf{Y})\dot{\omega}(\mathbf{Y}) \, d\mathbf{Y}$, where P_{sgs} is the subgrid PDF, which is taken to be a multidimensional Guassian function. The calculations [103] of a turbulent lean premixed propane flame stabilised behind a 'V' gutter shows that the predicted mean quantities of engineering interest are insensitive to the SGS combustion model. However, the choice of numerical solver and grid resolution can affect the details of predictions in LES of reacting [106] and non-reacting [107] flows, which is a somewhat less attractive feature.

(2) FSD MODELS. In the FSD and thickened-flame approaches, the general philosophy to model the filtered reaction rate is to write $\bar{\omega} = \rho_u S_L^0 \overline{\Sigma}$, where $\overline{\Sigma} = \overline{|\nabla c|}$ is the FSD per unit volume in a given computational cell, which is usually referred to as a filtered FSD or sometimes it is called subgrid FSD. An algebraic model for this quantity was proposed [73] as

$$\overline{\Sigma} = 4\Xi_\Delta \sqrt{\frac{6}{\pi} \frac{\bar{c}(1 - \bar{c})}{\Delta}}, \tag{1.49}$$

where $\Xi_\Delta = \overline{|\nabla c|}/|\nabla \bar{c}|$ is the subgrid flame-wrinkling factor, which is modelled by two methods. Although the simplest one is to treat this factor as unity [73], an algebraic expression

$$\Xi_\Delta = \left(1 + \frac{\Delta}{\eta_c}\right)^\beta$$

was proposed [108, 109] using fractal analysis in which the exponent β is related to the fractal dimenison $\hat{D}$ by $\beta = \hat{D} - 2$. The typical value of $\hat{D}$ varies between 2 and 3 [110, 111] and it scale dependence was also investigated using dynamic procedure [109]. The inner cut-off length scale η_c can be the Kolmogorov or Gibson length scales, or δ_L^0 [110], or the inverse of mean flame curvature [108, 109], $|\langle \partial n_k / \partial x_k \rangle_s|^{-1}$. It seems that a good prediction of $\bar{\omega}$ when these models are used strongly depends on the choices for the exponent and the inner cut-off length scale. These two quantities can also depend on the Karlovitz number [112] as the fractal nature of the flame surface depends on the combustion regime.

In the second method, a transport equation for Ξ_Δ is derived and solved [113–115] along with additional modelling for generation and removal rates of Ξ_Δ. These modellings introduce further uncertainies, although the calculations were shown [113–115] to compare well with measured fluid dynamic quantities.

The algebraic model given in Eq. (1.49) is used in [116] to calculate turbulent premixed flames propagating in a channel with a square obstruction. The computed results showed an underprediction of about 20%–30% in peak pressure and also a time lag in the pressure history compared with experimental measurements. The reason for this is unclear.

The FSD required for obtaining the filtered reaction rate can be written [110] as

$$\overline{\Sigma} = \underbrace{|\nabla \bar{c}|}_{\text{resolved}} + \underbrace{\left(\overline{|\nabla c|} - |\nabla \bar{c}|\right)}_{\text{unresolved}} = |\nabla \bar{c}| + \mathcal{U}. \tag{1.50}$$

By taking the resolved part to be from Eq. (1.49) with $\Xi_\Delta = 1$, the unresolved part is obtained [110] as

$$\mathcal{U} \approx C_s \left[\frac{\overline{\tilde{c}(1-\tilde{c})}}{\Delta} - \frac{\widehat{\tilde{c}}(1-\widehat{\tilde{c}})}{\widehat{\Delta}} \right], \tag{1.51}$$

using the dynamic procedure with a test filter of size $\widehat{\Delta}$. The model parameter C_s, obtained [110] by fractal analysis with δ_L^0 as the inner cut-off scale, tends to 1.8 when the filter size Δ becomes small (<2 mm). The SGS contribution $\mathcal{U}$ can become significant, irrespective of the combustion regime, which can influence the LES prediction when algebraic models are used.

In an alternative approach to modelling Ξ_Δ, a transport equation for the filtered FSD is derived and solved [73, 111, 117–119]. This equation is written as follows after generalised FSD equation (1.41) is filtered:

$$\frac{\partial \overline{\Sigma}}{\partial t} + \frac{\partial}{\partial x_k} \left[\langle u_k + S_{d,c} n_k \rangle_s \Sigma \right] = \overline{\left\langle (\delta_{km} - n_k n_m) \frac{\partial u_k}{\partial x_m} \right\rangle_s \Sigma} + \overline{\left\langle S_{d,c} \frac{\partial n_k}{\partial x_k} \right\rangle_s \Sigma}. \tag{1.52}$$

The filtered convective term and the two terms on the right-hand side of Eq. (1.52) need to be modelled, and they are discussed in [111, 117–119]. An elaborate discussion on this topic can be found in Section 2.2. It was also noted [120] that the physical realisablity requirement (i.e., $\overline{\Sigma} \geq 0$) may not always be satisfied by the preceding algebraic and transported FSD models and care needs to be exercised.

(3) THICKENED-FLAME MODEL. The preceding subgrid modelling approaches for the filtered reaction rate assume the flame front to be a laminar flame, which is typically smaller than the LES mesh size. Thus the flame-front structure is not captured, and so alternative modelling methods are needed to predict minor species, pollutants, and their formation rates. A simple approach developed specifically to resolve the flame-front structure in LES involves *artificially thickened flames* [111]. An unstretched laminar flame propagates at a speed $S_L^0 \sim (\hat{\alpha} A)^{1/2}$, where A is the pre-exponential factor of the heat release rate, $\hat{\alpha} \sim S_L^0 \delta_\ell$ is the thermal diffusivity, and δ_ℓ is the flame thickness, from which it follows that $\delta_\ell \sim (\hat{\alpha}/A)^{1/2}$. Consequently, if the thermal diffusivity is artificially increased by a factor F and the pre-exponential factor is decreased by the same amount, the flame speed remains unchanged, whereas the flame thickness is increased by the factor F. This quantity is chosen so that the reaction layer thickness $F\delta_\ell$ can be resolved on the LES computational mesh. Also, the flame evolution time is artificially increased by the factor F. An allowance is then made for SGS flame wrinkling by use of a stretch efficiency function [111] to compensate for the decrease in flame wrinkling because of artificial thickening. Other approaches, such as filtered tabulated chemistry [98] or multidimensional flamelet-generated manifold [121] methods with a presumed subgrid PDF [91] can be used in conjunction with the thickened-flame approach. Details are discussed in Section 2.2.

(4) G-EQUATION MODELS. This method is also known as flame-tracking method [122–125], in which the flame front is treated as a thin interface and thus the standard filter kernel cannot be used because the filtering has to be done along this interface where

G is meaningful [126–128]. This introduces an additional quantity, the conditionally filtered flow velocity, requiring modelling that is typically represented [127] as the sum of the Favre filtered velocity $\tilde{v}$ and a correction obtained by use of the velocity jump across a laminar front. A form of this filtered equation is written as Eq. (2.64), and the related modelling is given in Section 2.2. Here we just note that the the important modelling part of this approach is in the turbulent burning speed S_T, which is usually written using [129, 130]

$$\frac{S_T}{S_L^0} = 1 + \mathcal{A} \left(\frac{u'}{S_L^0} \right)^n ,$$

where $\mathcal{A}$ and n are model parameters taken to be constants or estimated using the dynamic procedure as explained in [125]. Other possible modelling of S_T is discussed in [131]. It is widely accepted that other effects, such chemical kinetics, non–unity Lewis number, etc., can be incorporated into the laminar flame-speed S_L^0 part of the model.

(5) LEM–LES. The LEM approach discussed earlier within the RANS context can be extended to LES, as was discussed in [132–134]. The essence of this extension is to limit the stochastic stirring event to scales below the filter scale Δ so that the triplet mapping is limited to the range $\Delta \geq \mathcal{L} \geq \eta_K$. This is achieved by replacing Λ in Eqs. (1.45) and (1.46) with Δ. Other subtleties in this approach, such as the influence of volumetric expansion on the flow field or LEM domain size, need to be considered carefully while this model is implemented. These are explained in [132–134], and an application of the methodology to simulate soot formation can be found in [135].

1.8 Data for Model Validation

Empiricism is unavoidable in the development of submodels to replace unknown terms in the averaged transport equations of both RANS and LES, and comparisons with data from experiments and DNS provide a vital link in the development process. Until relatively recently, experimental comparisons were mainly restricted to a global level, in which a RANS or averaged LES prediction of a combustion flow field is compared with averaged data from experiments. However, the RANS or LES calculation necessarily involves many closure assumptions, so it is difficult to isolate the performance of a particular submodel in this way. Only in the past few years has it become possible to make more detailed measurements that are directly relevant to a specified unknown term in an averaged transport equation.

An example of the first type of comparison is provided by turbulent flame speeds. An example of the use of turbulent flame-speed data to test a turbulent combustion model is provided in Section 2.3. The idea that the propagation of a premixed turbulent flame can be characterised as a burning velocity, by analogy with laminar flames, has been explored for many years, and a large amount of information is available[2] from theoretical analysis [136–140], experiments (see [42] for further references), RANS calculations (see [140, 141] for further references), and DNS (see for example [72]), so it provides a way to test the overall performance of a

---

[2] It is not possible to cite all the work on this topic here. However, the relatively recent works cited here provide a good start for an interested reader.

RANS model against published experimental data. A burning velocity – laminar or turbulent – can be defined as the mass flow per second of reactants converted to combustion products, within a specified streamtube, divided by the density of unburned reactants and streamtube area, namely

$$S_c = \frac{1}{\bar{\rho} A_m} \int \bar{\dot{\omega}}_c A(n) \, \mathrm{d}n,$$

where n is a coordinate normal to the local mean flame brush, $A(n)$ is the streamtube area at location n, A_m is the streamtube area at a chosen location $n = m$, and integration is extended through the whole streamtube volume within the thickness of the brush. The propagation speed defined in this way is known as the *consumption speed*. It is important to note that the thickness of an averaged turbulent flame brush is often significant in comparison with other scales of the averaged field, and the streamtube area is generally not constant, so care is required in the choice of an appropriate normalising flame area A_m.

Alternatively, the *displacement speed* of the flame S_T is the propagation speed relative to the mean flow of a chosen isosurface of the mean flame, for example, the surface on which $\tilde{c} = \tilde{c}_m$. It is usual to choose an isosurface close to the cold side of the flame, and then $\tilde{c}_m \ll 1$. The displacement and consumption speeds of a given combustion flow can be significantly different from each other [42, 142]. In addition, Driscoll [42] pointed out that the turbulent burning velocity should not be regarded as a universal property, because it is influenced by the type of burner used to stabilise the flame. This sensitivity arises because the process of flame wrinkling develops in different ways, depending on the choice of burner geometry. For example, the wrinkles on a Bunsen flame grow in amplitude as they are convected downstream, whereas those on an expanding spherical flame are continually stretched and grow in wavelength.

Experimental data can also be used to make a more direct test of a specific submodel within a complete RANS or LES model. For example, simultaneous, local time-resolved measurements of velocity $\mathbf{u}(\mathbf{x}, t)$ and the mass fraction $Y_i(\mathbf{x}, t)$ provide information from which both the scalar flux $\widetilde{u_k'' Y_i''}(\mathbf{x})$ and the gradient $\partial \tilde{Y}_i / \partial x_k(\mathbf{x})$ can be calculated. The validity of the *gradient transport* submodel of Eq. (1.27) can then be directly tested. Clear evidence of *counter-gradient* scalar transport in premixed turbulent combustion was found in this way, as discussed in Section 2.1. A second example is provided by the FSD Σ. A reasonable approximation to the mean of this quantity, which, as we have seen, is central to some mean-reaction-rate models, can be found (see for example [143, 144]) from the length of isocomposition lines in 2D laser-sheet images. Alternatively, the area of the isocomposition surface can be determined from two parallel laser sheets. Recent advances in laser diagnostics for combustion now offer the prospect of direct experimental validation of a wider range of important submodels under conditions that cannot be achieved by DNS.

There is a continuing need for well-planned experiments to provide data for comparison with turbulent combustion models, and this need is particularly apparent in the case of premixed and partially premixed combustion at high turbulence intensities. Experimental configurations should be sufficiently simple to ease problems of numerical simulation, and boundary conditions should be accurately determined.

The value of an experiment will be greatly increased if several different flow properties, preferably including velocity, can be measured simultaneously, with good space and time resolution. A sufficiently large database of this kind will allow co-variances to be estimated and related to flow conditions.

DNS aims to provide a fully resolved numerical solution to the complete system of equations of a chemically reactive flow, largely avoiding the need for empirical models. The output from such a simulation can be used to evaluate the unknown statistical quantity that is the subject of a specific submodel, together with the combination of mean variables forming the submodel, allowing the validity of the submodel to be tested. Many comparisons of this kind can be found in the literature, from which it is clear that DNS comparisons provide a very powerful tool for model development and testing. Examples of the use of DNS data in this way may be found in various sections of Chapter 2. Nevertheless, the limitations of DNS must be kept in mind when it is used for model testing: The Reynolds number is usually very low in comparison with those of practical interest, the flow geometry is greatly simplified, and the chemical kinetic mechanism is often simplified to make the calculation feasible.

Users of RANS or LES simulation codes to predict flow fields involving turbulent combustion frequently need to be reminded that their codes contain numerous empirical submodels that have, at best, been tested in a few corners of their multi-dimensional parameter space. Model developers must do more to test their models over a wider range of conditions. They must work towards a goal to quantify the range of validity of their models with greater accuracy and minimise, if not eliminate, the empiricism in current modelling.

REFERENCES

[1] A. S. Monin and A. M. Yoglom, *Statistical Fluid Mechanics* (MIT Press, Cambridge, MA, 1975), two volumes.

[2] J. O. Hinze, *Turbulence* (McGraw-Hill, New York, 1975).

[3] P. A. Libby, *An Introduction To Turbulence* (Taylor & Francis, Washington, D.C., 1996).

[4] S. B. Pope, *Turbulent Flows* (Cambridge University Press, Cambridge, 2000).

[5] P. A. Davidson, *Turbulence: An Introduction for Scientists and Engineers* (Oxford University Press, Oxford, 2004).

[6] C. K. Law, *Combustion Physics* (Cambridge University Press, Cambridge, 2006).

[7] F. A. Williams, *Combustion Theory* (Addison-Wesley, Redwood City, CA, 1985).

[8] S. R. Turns, *An Introduction to Combustion: Concepts and Application*, 2nd ed. (McGraw-Hill, New York, 2000).

[9] T. J. Poinsot and D. Veynante, *Theoretical and Numerical Combustion* (Edwards, Philadelphia, PA, 2001).

[10] P. A. Libby and F. A. Williams, Fundamental aspects and a review, in P. A. Libby and F. A. Williams (eds.), *Turbulent Reacting Flows* (Academic, New York, 1994), pp. 1–62.

[11] D. Veynante and L. Vervisch, Turbulent combustion modeling, *Prog. Energy Combust. Sci.* **28**, 193–266 (2002).

[12] C. K. Westbrook, Y. Mizobuchi, T. J. Poinsot, P. J. Smith, and J. Warnatz, Computational combustion, *Proc. Combust. Inst.* **30**, 125–157 (2005).

[13] H. Yamashita, M. Shimada, and T. Takeno, A numerical study on flame stability at the transition point of jet diffusion flames, *Proc. Combust. Inst.* **26**, 27–34 (1996).

[14] Y. Mizobuchi, S. Tachibana, J. Shinio, and S. O. T. Takeno, A numerical analysis of the structure of a turbulent hydrogen jet lifted flame, *Proc. Combust. Inst.* **29**, 2009–2015 (2002).

[15] C. K. Law, A compilation of experimental data on laminar burning velocities, in N. Peters and B. Rogg (eds.), *Reduced Kinetic Mechanisms for Applications in Combustion Systems*, Lecture Notes in Physics Series, **384** (Springer-Verlag, Berlin, 1993), pp. 15–26.

[16] K. Kumar, J. E. Freeh, C. J. Sung, and Y. Huang, Laminar flame speeds of pre-heated iso-octance/O_2/N_2 and n-heptane/O_2/N_2 mixtures, *J. Propulsion Power* **23**, 428–436 (2007).

[17] K. Kumar and C. J. Sung, Laminar flame speeds and extinction limits of pre-heated n-decane/O_2/N_2 and n-dodecane/O_2/N_2 mixtures, *Combust. Flame* **151**, 209–224 (2007).

[18] O. Schafer and S. Wittig, The laminar flame speed of aviation fuel–air mixtures at elevated temperatures, Tech. Rep. (Institut für Thermische Strömungsmaschinen, University of Karlsruhe, Karlsruhe Germany, 2003).

[19] D. Dunn-Rankin (ed.), *Lean Combustion Technology and Control* (Academic/Elsevier, San Diego, CA, 2008).

[20] N. Peters and B. Rogg (eds.), *Reduced Kinetic Mechanisms for Applications in Combustion Systems*, Lecture Notes in Physics Series, **384** (Springer-Verlag, Berlin, 1993).

[21] M. D. Smooke (ed.), *Reduced Kinetic Mechanisms and Asymptotic Approximations for Methane–Air Flames*, Lecture Notes in Physics Series, **384** (Springer-Verlag, Berlin, 1991).

[22] C. K. Westbrook and F. L. Dryer, Chemical kinetic modeling of hydrocarbon combustion, *Prog. Energy Combust. Sci.* **10**, 1–57 (1984).

[23] R. D. Kern and K. Xie, Shock tube studies of gas phase reactions preceding the soot formation process, *Prog. Energy Combust. Sci.* **17**, 191–210 (1991).

[24] K. A. Bhaskaran and P. Roth, The shock tube as wave reactor for kinetic studies and material systems, *Prog. Energy Combust. Sci.* **28**, 151–192 (2002).

[25] W. A. Sirignano, Fuel droplet vaporization and spray combustion theory, *Prog. Energy Combust. Sci.* **9**, 291–322 (1983).

[26] G. M. Faeth, Evaporation and combustion of sprays, *Prog. Energy Combust. Sci.* **9**, 1–76 (1983).

[27] G. M. Faeth, Mixing, transport and combustion in sprays, *Prog. Energy Combust. Sci.* **13**, 293–345 (1987).

[28] K. K. Kuo, *Recent Advances in Spray Combustion: Spray Atomization and Drop Burning Phenomena*, Vol. 166 of Progress in Astronautics and Aeronautics Series (AIAA, Reston, VA, 1996).

[29] K. K. Kuo, *Recent Advances in Spray Combustion: Spray Combustion Measurements and Model Simulation*, Vol. 171 of Progress in Astronautics and Aeronautics Series (AIAA, Reston, VA, 1996).

[30] S. S. Sazhin, Advanced models of fuel droplet heating and evaporation, *Prog. Energy Combust. Sci.* **32**, 162–214 (2006).

[31] M. Birouk and I. Gökalp, Current status of droplet evaporation in turbulent flows, *Prog. Energy and Combust. Sci.* **32**, 408–423 (2006).

[32] X. Jiang, G. A. Siamas, K. Jagus, and T. G. Karayiannis, Physical modelling and advanced simulations of gas-liquid two-phase jet flows in atomization and sprays, *Prog. Energy Combust. Sci.* **36**, 131–167 (2010).

[33] H. H. Chiu and T. M. Liu, Group combustion of liquid droplets, *Combust. Sci. Technol.* **17**, 127–142 (1977).

[34] H. H. Chiu, H. Y. Kim, and E. J. Croke, Internal group combustion of liquid droplets, *Proc. Combust. Inst.* **19**, 971–980 (1982).

[35] J. Réveillon and L. Vervisch, Accounting for spray vaporization in turbulent combustion modeling, in *Proceedings of the Summer Program 1998, CTR* (Stanford University, Stanford, CA, 1998), pp. 25–38.

[36] J. Réveillon and L. Vervisch, Analysis of weakly turbulent dilute-spray flames and spray combustion regimes, *J. Fluid Mech.* **537**, 317–347 (2005).

[37] Y. Baba and R. Kurose, Analysis and flamelet modelling for spray combustion, *J. Fluid Mech.* **612**, 45–79 (2008).

[38] A. K. M. F. Hussain and W. C. Reynolds, The mechanics of an organized wave in turbulent shear flow, *J. Fluid Mech.* **41**, 241–258 (1970).

[39] S. B. Pope, Ten questions concerning the large-eddy simulation of turbulent flows, *New J. Physics* **6**, 1–24 (2004).

[40] G. Damköhler, The effect of turbulence on the flame velocity in gas mixtures, Tech. Rep. TM 1112. National Advisory Committee for Aeronautics, USA (NACA, 1947) (also *Z. Elektrochem* **46**, 601, 1940).

[41] N. Peters, *Turbulent Combustion* (Cambridge University Press, Cambridge, 2000).

[42] J. F. Driscoll, Turbulent premixed combustion: Flamelet structure and its effect on turbulent burning velocities, *Prog. Energy Combust. Sci.* **34**, 91–134 (2008).

[43] M. Vachon and M. Champion, Integral model of a flame with large buoyancy effects, *Combust. Flame* **63**, 269–278 (1986).

[44] P. A. Libby and K. N. C. Bray, Variable density effects in premixed turbulent flames, *AIAA J.* **15**, 1186–1193 (1977).

[45] P. A. Libby and K. N. C. Bray, Countergradient diffusion in premixed turbulent flames, *AIAA J.* **19**, 205–213 (1981).

[46] J. B. Moss, Simultaneous measurements of concentration and velocity in an open premixed turbulent flame, *Combust. Sci. Technol.* **22**, 119–129 (1980).

[47] K. N. C. Bray and J. B. Moss, A unified statistical model of the premixed turbulent flame, *Acta Astron.* **4**, 291–320 (1977).

[48] K. N. C. Bray, P. A. Libby, G. Masuya, and J. B. Moss, Turbulence production in premixed tubulent flames, *Combust. Sci. Technol.* **25**, 127–140 (1981).

[49] G. Masuya and P. A. Libby, Nongradient theory for oblique turbulent flames with premixed reactants, *AIAA J.* **19**, 1590–1599 (1981).

[50] M. Hallback, A. V. Johansson, and A. D. Burden, The basics of turbulence modelling, in M. Hallback, D. S. Henningson, A. V. Johansson, and P. H. Alfredsson (eds.), *Turbulence and Transistion Modelling*, Kluwer Academic, Dordrecht, The Netherlands, 1996), pp. 81–154.

[51] B. W. Launder, Advanced turbulence models for industrial applications, in M. Hallback, D. S. Henningson, A. V. Johansson, and P. H. Alfredsson (eds.), *Turbulence and Transistion Modelling* (Kluwer Academic, Dordrecht, The Netherlands, 1996), pp. 193–231.

[52] W. P. Jones, Turbulence modelling and numerical solution methods for variable density and combusting flows, in P. A. Libby and F. A. Williams (eds.), *Turbulent Reacting Flows* (Academic, New York, 1994), pp. 309–374.

[53] D. Veynante, A. Trouve, K. N. C. Bray, and T. Mantel, Gradient and counter-gradient scalar transport in turbulent premixed flames, *J. Fluid Mech.* **332**, 263–293 (1997).

[54] B. E. Launder, G. J. Reece, and W. Rodi. Progress in the development of a Reynolds-stress turbulence closure *J. Fluid Mech.* **68**, 537–566 (1975).

[55] K. N. C. Bray, The interaction between turbulence and combustion, *Proc. Combust. Inst.* **17**, 223–233 (1979).

[56] J. W. Rogerson and N. Swaminathan, Correlation between dilatation and scalar dissipation in turbulent premixed flames, in *Proceedings of the European Combustion Meeting*, Mediterranean Argonomic Institute of Chania, Crete, Greece. 2007.

[57] D. Bradley, How fast can we burn?, *Proc. Combust. Inst.* **24**, 247–262 (1992).

[58] K. N. C. Bray, P. Domingo, and L. Vervisch, Role of the progress variable in models for partially premixed turbulent combustion, *Combust. Flame* **141**, 431–437 (2005).

[59] P. J. Davis, Gamma functions and related functions, in M. Abramowitz and I. A. Stegun (eds.), *Handbook of Mathematical Functions* (Dover, New York, 1970).

[60] S. S. Girimaji, Assumed β-PDF model for turbulent mixing: Validation and extension to multiple scalar mixing, *Combust. Sci. Technol.* **78**, 177–196 (1991).

[61] S. B. Pope, PDF methods for turbulent reactive flows, *Prog. Energy Combust. Sci.* **11**, 119–192 (1985).

[62] E. E. O'Brien, The probability density function (pdf) approach to reacting turbulent flows, in P. A. Libby and F. A. Williams (eds.), *Turbulent Reacting Flows* (Springer-Verlag, Berlin/Heidelberg, 1980), pp. 185–218.

[63] A. Y. Klimenko and R. W. Bilger, Conditional moment closure for turbulent combustion, *Prog. Energy Combust. Sci.* **25**, 595–687 (1999).

[64] A. Y. Klimenko, Multicomponent diffusion of various admixtures in turbulent flow, *Fluid Dyn.* **25**, 327–334 (1990).

[65] R. W. Bilger, Conditional moment closure for turbulent reacting flow, *Phys. Fluids* **5**, 436–444 (1993).

[66] R. W. Bilger, Conditional moment closure modelling and advanced laser measurements, in T. Takeno (ed.), *Turbulence and Molecular Processes in Combustion* (Elsevier Science, New York, 1993), pp. 267–285.

[67] N. Swaminathan and R. W. Bilger, Analyses of conditional moment closure for turbulent premixed flames, *Combust. Theory Model.* **5**, 241–260 (2001).

[68] S. M. Martin, J. C. Kramlich, G. Kosaly, and J. J. Riley, The premixed conditional moment closure method applied to idealised lean premixed gas turbine combustors, *J. Eng. Gas Turbines Power* **125**, 895–900 (2003).

[69] S. B. Pope, Computations of turbulent combustion: Progress and challenges, *Proc. Combust. Inst.* **23**, 591–612 (1990).

[70] L. Vervisch, E. Bidaux, K. N. C. Bray, and W. Kollmann, Surface density function in premixed turbulent combustion modeling, similarities between probability density function and flame surface approaches, *Phys. Fluids* **7**, 2496–2503 (1995).

[71] S. M. Candel and T. J. Poinsot, Flame stretch and the balance equation for the flame area, *Combust. Sci. Technol.* **70**, 1–15 (1990).

[72] A. Trouve and T. Poinsot, The evolution equation for the flame surface density in turbulent premixed combustion, *J. Fluid Mech.* **278**, 1–31 (1994).

[73] M. Boger, D. Veynante, H. Boughanem, and A. Trouvé, Direct numerical simulation analysis of flame surface density concept for large eddy simulation of turbulent premixed combustion, *Proc. Combust. Inst.* **27**, 917–925 (1998).

[74] N. Peters, P. Terhoeven, J. H. Chen, and T. Echekki, Statistics of flame displacement speeds from computations of 2-D unsteady methane–air flames, *Proc. Combust. Inst.* **27**, 833–839 (1998).

[75] K. N. C. Bray and N. Peters, Laminar flamelets in turbulent flames, in P. A. Libby and F. A. Williams (eds.), *Turbulent Reacting Flows* (Academic, New York, 1994), pp. 63–113.

[76] K. N. C. Bray and N. Swaminathan, Scalar dissipation and flame surface density in premixed turbulent combustion, *C. R. Mec.* **334**, 466–473 (2006).

[77] A. R. Kerstein, A linear-eddy model of turbulent scalar transport and mixing, *Combust. Sci. Technol.* **60**, 391–421 (1988).

[78] A. R. Kerstein, Linear-eddy modeling of turbulent transport. II: Application to shear layer mixing, *Combust. Flame* **75**, 397–413 (1989).

[79] A. R. Kerstein, Linear-eddy modelling of turbulent transport. Part 3. mixing and differential molecular diffusion in round jets, *J. Fluid Mech.* **216**, 411–435 (1990).

[80] A. R. Kerstein, A linear-eddy modeling of turbulent scalar transport. Part 4. Structure of diffusion flames, *Combust. Sci. Technol.* **81**, 75–96 (1992).

[81] A. R. Kerstein, Linear-eddy modeling of turbulent scalar transport. Part 5. Geometry of scalar interfaces, *Phys. Fluids A* **3**, 1110–1114 (1991).

[82] A. R. Kerstein, Linear-eddy modelling of turbulent scalar transport. Part 6. Microstructure of diffusive scalar mixing fields, *J. Fluid Mech.* **231**, 361–394 (1991).

[83] A. R. Kerstein, Linear-eddy modelling of turbulent scalar transport. Part 7. Finite rate chemistry and multi-stream mixing, *J. Fluid Mech.* **240**, 289–313 (1992).

[84] P. A. McMurtry, S. Menon, and A. R. Kerstein, Linear eddy modeling of turbulent combustion, *Energy Fuels* **7**, 817–826 (1993).

[85] S. Menon, P. A. McMurtry, and A. R. Kerstein, A liner eddy mixing model for turbulent combustion: Application to premixed flames, in *Proceedings of the AIAA 31st Aerospace Sciences Meeting*, AIAA paper 93-0107 (AIAA, Reston, VA, 1993).

[86] T. Smith and S. Menon, Model simulations of freely propagating turbulent premixed flames, *Proc. Combust. Inst.* **26**, 299–306 (1996).

[87] V. Sankaran and S. Menon, Structure of premixed turbulent flames in the thin-reaction-zones regime, *Proc. Combust. Inst.* **28**, 203–209 (2000).

[88] S. B. Pope, Computationally efficient implementation of combustion chemistry using in situ adaptive tabulation, *Combust. Theory Model.* **1**, 44–63 (1997).

[89] U. Maas and S. B. Pope, Simplifying chemical kinetics: Intrinsic low dimensional manifolds in composition space, *Combust. Flame* **88**, 239–264 (1992).

[90] S. H. Lam and D. A. Goussis, The CSP method for simplifying kinetics, *Int. J. Chem. Kinet.* **26**, 461–486 (1994).

[91] P. Domingo, L. Vervisch, S. Payet, and R. Hauguel, DNS of a premixed turbulent V flame and LES of a ducted flame using a FSD-PDF subgrid scale closure with FPI-tabulated chemistry, *Combust. Flame* **143**, 566–586 (2005).

[92] B. Fiorina, O. Gicquel, L. Vervisch, N. Datrabiha, and S. Carpentier, Approximating the chemical structure of partially premixed and diffusion counterflow flames using FPI flamelet tabulation, *Combust. Flame* **140**, 147–160 (2005).

[93] J. Smagorinsky, General circulation experiment with the primitive equations. I: The basic experiment, *Mon. Weather Rev.* **91**, 99–164 (1963).

[94] M. Germano, U. Piomelli, P. Moin, and W. H. Cabot, A dynamic subgrid-scale eddy viscosity model, *Phys. Fluids* **238**, 1760–1765 (1991).

[95] M. Germano, Turbulence: The filtering approach, *J. Fluid Mech.* **238**, 325–336 (1992).

[96] J. Réveillon and L. Vervisch, Response of the dynamic LES model to heat release induced effects, *Phys. Fluids* **8**, 2248–2250 (1996).

[97] S. Tullis and R. S. Cant, Scalar transport modelling in large eddy simulation of turbulent premixed flames, *Proc. Combust. Inst.* **29**, 2097–2104 (2002).

[98] B. Fiorina, R. Vicquelin, P. Auzillon, N. Darabiha, O. Gicquel, and D. Veynante, A filtered tabulated chemistry model for LES of premixed combustion, *Combust. Flame* **157**, 465–475 (2010).

[99] F. A. Jaberi, P. J. Colucci, S. James, P. Givi, and S. B. Pope, Filtered mass density function for large-eddy simulation of turbulent reacting flows, *J. Fluid Mech.* **401**, 85–121 (1999).

[100] S. Navarro-Martinez, A. Kronenburg, and F. D. Mare, Conditional moment closure for large eddy simulations, *Flow Turbulence Combust.* **75**, 245–274 (2005).

[101] A. Triantafyllidis, E. Mastorakos, and R. L. G. M. Eggels, Large eddy simulations of forced ignition of a non-premixed bluff-body methane flame with conditional moment closure, *Combust. Flame* **156**, 2328–2345 (2009).

[102] C. Fureby and S. I. Möller, Large eddy simulation of reacting flows applied to bluff body stabilized flames, *AIAA J.* **33**, 2339–2347 (1995).

[103] S. I. Möller, E. Lundgren, and C. Fureby, Large eddy simulation of unsteady combustion, *Proc. Combust. Inst.* **26**, 241–248 (1996).

[104] E. Giacomazzi, V. Battaglia, and C. Bruno, The coupling of turbulence and chemistry in a premixed bluff-body flame as studied by LES, *Combust. Flame* **138**, 320–335 (2004).

[105] F. Gao, *Annual Research Briefs* (Center for Turbulence Research, Stanford University, Stanford, CA, 1993), pp. 187–197.

[106] K. J. Nogenmyr, C. Fureby, X. S. Bai, P. Petersson, R. Collin, and M. Linne, Large eddy simulation and laser diagnostic studies on a low swirl stratified premixed flame, *Combust. Flame* **156**, 25–36 (2009).

[107] S. E. Gant, Reliability issues of LES–related approaches in an industrial context, *Flow Turbulence Combust.* **84**, 325–335 (2010).

[108] F. Charlette, C. Meneveau, and D. Veynante, A power-law flame wrinkling model for LES of premixed turbulent combustion part I: Non-dynamic formulation and initial tests, *Combust. Flame* **131**, 159–180 (2002).

[109] F. Charlette, C. Meneveau, and D. Veynante, A power-law flame wrinkling model for LES of premixed turbulent combustion part II: dynamic formulation, *Combust. Flame* **131**, 181–197 (2002).

[110] R. Knikker, D. Veynante, C. Meneveau, A priori testing of a similarity model for large eddy simulations of turbulent premixed combustion, *Proc. Combust. Inst.* **29**, 2105–2111 (2002).

[111] O. Colin, F. Ducros, D. Veynante, and T. Poinsot, A thickened flame model for large eddy simulations of turbulent premixed combustion, *Phys. Fluids* **12**, 1843–1863 (2000).

[112] N. Chakraborty and M. Klein, A priori direct numerical simulation assessment of algebraic flame surface density models for turbulent premixed flames in the context of large eddy simulation, *Phys. Fluids* **20**, 085108, 1–14 (2008).

[113] H. G. Weller, G. Tabor, A. D. Gosman, and C. Fureby, Application of a flame-wrinkling LES combustion model to a turbulent mixing layer, *Proc. Combust. Inst.* **27**, 899–907 (1998).

[114] C. Fureby, A computational study of combustion instabilities due to vortex shedding, *Proc. Combust. Inst.* **28**, 783–791 (2000).

[115] C. Fureby, Large eddy simulation of combustion instabilities in a jet engine afterburner model, *Combust. Sci. Technol.* **161**, 213–243 (2000).

[116] M. P. Kirkpatrick, S. W. Armfield, A. R. Masri, and S. S. Ibrahim, Large eddy simulation of a propagating turbulent premixed flame, *Flow Turbulence Combust.* **70**, 1–19 (2003).

[117] E. R. Hawkes and R. S. Cant, Implications of a flame surface density approach to large eddy simulation of premixed turbulent combustion, *Combust. Flame* **126**, 1617–1629 (2001).

[118] N. Chakraborty and R. S. Cant, A priori analysis of the curvature and propagation terms of the flame surface density transport equation for large eddy simulation, *Phys. Fluids* **19**, 105101, 1–22 (2007).

[119] N. Chakraborty and R. S. Cant, Direct numerical simulation analysis of the flame surface density transport equation in the context of large eddy simulation, *Proc. Combust. Inst.* **32**, 1445–1453 (2009).

[120] E. R. Hawkes and R. S. Cant, Physical and numerical realizability requirements for flame surface density approaches, *Combust. Theory Model.* **5**, 699–720 (2001).

[121] P. Nguyen, L. Vervisch, V. Subramanian, and P. Domingo, Multidimensional flamelet-generated manifolds for partially premixed combustion, *Combust. Flame* **157**, 43–61 (2010).

[122] A. Kerstein, W. Ashurst, and F. A. Williams, Field equation for interface propagation in an unsteady homogeneous flow field, *Phys. Rev. A* **33**, 2728–2731 (1988).

[123] S. Menon and W. H. Jou, Large-eddy simulations of combustion instability in an axisymmetric ramjet combustor, *Combust. Sci. Technol.* **75**, 53–72 (1991).

[124] V. Smiljanovski, V. Moser, and R. Klein, A capturing-tracking hybrid scheme for deflagration discontinuities, *Combust. Theory Model.* **1**, 183–215 (1997).

[125] H. G. Im, T. S. Lund, and J. H. Ferziger, Large eddy simulation of turbulent front propagation with dynamic subgrid models, *Phys. Fluids* **38**, 3826–3833 (1997).

[126] H. Pitsch, L. Duchamp, and D. Lageneste, Large-eddy simulation of premixed turbulent combustion using a level-set approach, *Proc. Combust. Inst.* **29**, 2001–2008 (2002).

[127] H. Pitsch, A consistent level set formulation for large-eddy simulation of premixed turbulent combustion, *Combust. Flame* **143**, 587–598 (2005).

[128] H. Pitsch, Large-eddy simulation of turbulent combustion, *Annu. Rev. Fluid Mech.* **38**, 453–482 (2006).

[129] J. Shinjo, Y. Mizobuchi, and S. Ogawa, LES of unstable combustion in a gas turbine combustor, in A. Veidenbaum, K. Joe, H. Amano, and H. Aiso (eds.), *High Performance Computing: Proceedings of 5th International Symposium on High Performance Computing* (Springer-Verlag, Berlin/Heidelberg/New York, 2003).

[130] G. Eggenspieler and S. Menon, Large-eddy simulation of pollutant emission in a DOE-HAT combustor, *J. Propul. Power* **20**, 1076–1085 (2004).

[131] W.-W. Kim and S. Menon, Numerical modeling of turbulent premixed flames in the thin-reaction-zones regime, *Combust. Sci. Technol.* **160**, 119–150 (2000).

[132] V. K. Chakravarthy and S. Menon, Large-eddy simulation of turbulent premixed flames in the flamelet regime, *Combust. Sci. Technol.* **162**, 175–222 (2001).

[133] V. Sankaran and S. Menon, Subgrid combustion modeling of 3-D premixed flames in the thin-reaction-zone regime, *Proc. Combust. Inst.* **30**, 575–582 (2005).

[134] S. Undapalli, S. Srinivasan, and S. Menon, LES of premixed and non-premixed combustion in a stagnation point reverse flow combustor, *Proc. Combust. Inst.* **32**, 1537–1544 (2009).

[135] H. E. Asrag, T. Lu, C. K. Law, and S. Menon, Simulation of soot formation in turbulent premixed flames, *Combust. Flame* **150**, 108–126 (2007).

[136] A. N. Lipatnikov and J. Chomiak, Turbulent flame speed and thickness: phenomenology, evaluation, and application in multi-dimensional simulations, *Prog. Energy Combust. Sci.* **28**, 1–74 (2002).

[137] A. N. Lipatnikov and J. Chomiak, Molecular transport effects on turbulent flame propagation and structure, *Prog. Energy Combust. Sci.* **31**, 1–73 (2005).

[138] K. N. C. Bray, Studies of the turbulent burning velocity, *Proc. R. Soc. London A* **431**, 315–335 (1990).

[139] H. Kolla, J. W. Rogerson, N. Chakraborty, and N. Swaminathan, Scalar dissipation rate modelling and its validation, *Combust. Sci. Technol.* **181**, 518–535 (2009).

[140] H. Kolla, J. W. Rogerson, and N. Swaminathan, Validation of a turbulent flame speed model across combustion regimes, *Combust. Sci. Technol.* **182**, 284–308 (2010).

[141] H. Kolla and N. Swaminathan, Strained flamelets for turbulent premixed flames, I: Formulation and planar flame results, *Combust. Flame* **157**, 943–954 (2010).

[142] R. W. Bilger, S. B. Pope, K. N. C. Bray, and J. F. Driscoll, Paradigms in turbulent combustion research, *Proc. Combust. Inst.* **30**, 21–42 (2005).

[143] Y.-C. Chen and R. W. Bilger, Experimental investigation of three-dimensional flame front structure in premixed turbulent combustion-I: hydrocarbon/air bunsen flames, *Combust. Flame* **131**, 400–435 (2002).

[144] Y.-C. Chen and R. W. Bilger, Detailed measurements of local scalar-front structures in stagnation-type turbulent premixed flames, *Proc. Combust. Inst.* **30**, 801–808 (2005).

2 Modelling Methods

2.1 Laminar Flamelets and the Bray, Moss, and Libby Model
By K. N. C. Bray

This section is concerned with premixed turbulent combustion in circumstances in which the heat release reactions can be considered 'fast', a situation that is common in practical combustion systems. Several dimensionless parameters can help to identify the meaning of fast chemistry in this context. A Damköhler number, $\mathrm{Da} = \tau_T/\tau_c$, compares a characteristic turbulent flow time scale $\tau_T = \tilde{k}/\tilde{\epsilon}$ with a laminar flame time $\tau_c = \delta/S_L^0$. As explained in Chapter 1, when $\mathrm{Da} \gg 1$, combustion occurs in thin wrinkled interfaces separating unburned reactants from fully burned products. If also a Karlovitz number $\mathrm{Ka} = \tau_L/\tau_K$ is sufficiently small, where τ_K is the Kolmogorov time characterising the smallest eddies in a turbulent flow, these reacting interfaces will resemble unstretched laminar flames. It is then possible to divide a Reynolds-averaged Navier–Stokes (RANS) simulation or large-eddy simulation (LES) into two separate parts: a chemical kinetic analysis of a laminar flame and a subsequent fluid flow calculation. A third dimensionless quantity is $\hat{N} = \tau S_L^0/u_{\mathrm{rms}}$, where $\tau = (T_b/T_u) - 1$ is a heat release parameter and T_u and T_b are the temperatures in unburned reactants and fully burned combustion products, respectively. As explained in Chapter 1, gradient scalar transport expressions are found [1] to fail unless $\hat{N} \leq O(1)$. Mura and Champion [2] showed that the velocity gradient in a laminar flame is greater than the typical velocity gradient that is due to turbulence if $\hat{N} > \mathrm{Da}^{-1/2}$; dissipative effects that are due to gradients in laminar flames can then become significant and should be taken into account.

Second-moment or Reynolds stress turbulence models [3, 4] were first developed in order to avoid the gradient transport and near-isotropic turbulence assumptions of k–ϵ models in non-reactive flow calculations and were subsequently applied to turbulent combustion. The first such combustion model to fully exploit the limit $\mathrm{Da} \gg 1$ in premixed turbulent combustion was proposed by Bray, Moss, and Libby (BML) [5–8], and is reviewed here. The BML model has been adapted to predict combustion processes in complex engineering flows [9], in which its performance is similar to that of other second-moment models. However, by partitioning the probability density function (PDF) into separate contributions from reactants, products, and burning zones, it also provides a mechanism for taking account of the fact that burning occurs in thin flamelets, leading to very sudden changes in gas density and other flow properties.

This model, which is discussed in Subsection 2.1.2, provides an intuitive framework for describing such systems and for identifying interactions between flow and combustion. 'Intermittency' effects that are due to the passage of thin-flame reaction zones past a fixed point are found to be important: A measurement at such a location within the flame brush would reveal the alternate appearance of unburned and fully burned gases separated by brief periods of chemical change that are due to the passage of a thin flamelet. Application of the BML model to combustion in a flow approaching a stagnation point is presented in Subsection 2.1.3 as an example. This is followed in Subsection 2.1.4 by a review of information concerning gradient and counter-gradient scalar transport, and Subsection 2.1.5 discusses laminar flamelet models. A simple laminar flamelet mean-burning-rate model is developed in Subsection 2.1.6. The aim is to illustrate relevant physical processes rather than to develop a detailed and accurate method of prediction, which can be found in [11] and other references subsequently cited. This section is concerned with RANS calculations. However, it should be noted that some of the issues identified here also arise in LES closures; see, for example, the LES flame surface density (FSD) analysis of Chakraborty and Cant [10].

2.1.1 The BML Model

Several reviews of the basic BML model are available [8, 11], and only a brief summary is presented here, followed by a review of some more recent developments. The essential features are as follows.

(1) THERMOCHEMISTRY. An adiabatic, low-Mach-number flow is considered, characterised by a Lewis number of unity. A *progress variable* $c(\mathbf{x}, t)$ is introduced and is defined here as a normalised temperature, so $c(\mathbf{x}, t) = [T(\mathbf{x}, t) - T_u]/(T_b - T_u)$. Alternatively, c can be defined in terms of the total mass fraction of combustion products. An essential assumption is that all thermochemical variables, including the rate of heat release, can be expressed as functions of c alone, and this requires changes in pressure to be small enough to be thermodynamically negligible. Additionally the BML model assumes constancy of mean molecular weight and specific heat, and the thermodynamic state of the mixture can then be reduced to

$$\rho(\mathbf{x}, t) = \frac{\rho_u}{1 + \tau c(\mathbf{x}, t)}. \tag{2.1}$$

However, more realistic equations of state can be accommodated without difficulty, so long as the state remains as a function of c alone. It should also be noted that non-unity Lewis number effects, which are not considered here, can be important in premixed combustion: see, for example, Chakraborty and Cant [12].

(2) MODELLED EQUATIONS. The Favre mean first-moment transport equations (see Sections 1.5 and 1.7) for an assumed adiabatic low Mach number flow, at a high

Reynolds number, with a Lewis number Le $= 1$, are

$$\frac{\partial \overline{\rho}}{\partial t} + \frac{\partial \overline{\rho} \, \tilde{u}_k}{\partial x_k} = 0,$$

$$\frac{\partial \overline{\rho} \, \tilde{u}_\ell}{\partial t} + \frac{\partial \overline{\rho} \, \tilde{u}_k \, \tilde{u}_\ell}{\partial x_k} = -\frac{\partial \overline{p}}{\partial x_\ell} + \frac{\partial (\overline{\tau}_{\ell k} - \overline{\rho \, u_k'' u_\ell''})}{\partial x_k},$$

$$\frac{\partial \overline{\rho} \, \tilde{c}}{\partial t} + \frac{\partial \overline{\rho} \, \tilde{u}_k \, \tilde{c}}{\partial x_k} = \frac{\partial}{\partial x_k}\left(\overline{\rho v \frac{\partial c}{\partial x_k}} - \overline{\rho \, u_k'' c''} \right) + \overline{\dot{\omega}}_c, \qquad (2.2)$$

where the molecular transport terms $\overline{\tau}_{\ell k}$ and $\rho v \frac{\partial c}{\partial x_k}$ are negligible at high Reynolds numbers, except near solid walls [13]. Buoyancy effects are assumed negligible here, but were treated in a BML analysis by Libby [14], whose predictions were subsequently confirmed by direct numerical simulation (DNS) [15]. Closure assumptions are required for the mean chemical source term $\overline{\dot{\omega}}_c$ together with the Reynolds stress components $\widetilde{u_\ell'' u_k''}$ and Reynolds flux components $\widetilde{u_k'' c''}$, which are obtained from the second-moment equations:

$$\frac{\partial \overline{\rho} \, \widetilde{c''^2}}{\partial t} + \frac{\partial \overline{\rho} \, \tilde{u}_k \, \widetilde{c''^2}}{\partial x_k} = -2 \overline{\rho} \widetilde{u_k'' c''} \frac{\partial \tilde{c}}{\partial x_k} - \frac{\partial}{\partial x_k}(\overline{\rho} \widetilde{u_k'' c''^2}) + 2\overline{c'' \dot{\omega}_c} - \overline{\rho} \tilde{\epsilon}_c,$$

$$\frac{\partial \overline{\rho} \widetilde{u_\ell'' u_m''}}{\partial t} + \frac{\partial \overline{\rho} \tilde{u}_k \widetilde{u_\ell'' u_m''}}{\partial x_k} = -\overline{\rho}\left[\widetilde{u_k'' u_\ell''} \frac{\partial \tilde{u}_m}{\partial x_k} + \widetilde{u_k'' u_m''} \frac{\partial \tilde{u}_\ell}{\partial x_k} \right]$$

$$- \frac{\partial \overline{\rho} \widetilde{u_k'' u_\ell'' u_m''}}{\partial x_k} - \left[\overline{u_\ell'' \frac{\partial p}{\partial x_m}} + \overline{u_m'' \frac{\partial p}{\partial x_\ell}} \right] - \overline{\rho} \tilde{\epsilon}_{\ell m},$$

$$\frac{\partial \overline{\rho} \widetilde{u_\ell'' c''}}{\partial t} + \frac{\partial \overline{\rho} \tilde{u}_k \widetilde{u_\ell'' c''}}{\partial x_k} = -\left[\overline{\rho} \widetilde{u_k'' c''} \frac{\partial \tilde{u}_\ell}{\partial x_k} + \overline{\rho} \widetilde{u_k'' u_\ell''} \frac{\partial \tilde{c}}{\partial x_k} \right]$$

$$- \frac{\partial \overline{\rho} \widetilde{u_k'' u_\ell'' c''}}{\partial x_k} - \overline{c'' \frac{\partial p}{\partial x_\ell}} + \overline{u_\ell'' \dot{\omega}_c} - \overline{\rho} \tilde{\epsilon}_{\ell c}. \qquad (2.3)$$

Here the closure requirements include third-moment turbulent transport terms, which can be approximated by gradient transport expressions of varying degrees of complexity [11], the simplest of which is [9]

$$\widetilde{u_k'' u_\ell'' u_m''} = -C_s \frac{\tilde{k}^2}{\tilde{\epsilon}} \frac{\partial \widetilde{u_\ell'' u_m''}}{\partial x_k},$$

where C_s is a constant. Alternatively [16],

$$\widetilde{u_k'' u_\ell'' u_m''} = -C_s \frac{\tilde{k}}{\tilde{\epsilon}} \widetilde{u_m'' u_n''} \frac{\partial \widetilde{u_k'' u_l''}}{\partial x_n},$$

where $C_s \simeq 0.25$. The velocity–reaction-rate co-variance is modelled as

$$\overline{u_\ell'' \dot{\omega}_c} = (\mathcal{A} - \tilde{c}) \frac{\overline{u_\ell'' c''}}{\tilde{c}(1 - \tilde{c})} \overline{\dot{\omega}}_c,$$

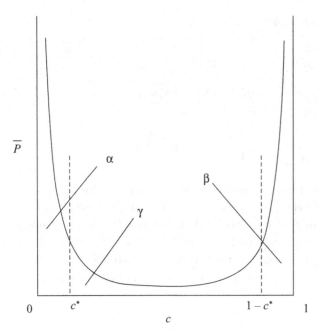

Figure 2.1. Sketch of the PDF $\overline{P}(c; \mathbf{x})$ of Eq. (2.4).

where $\mathcal{A}$ is a constant, normally taken to be 0.83. The dissipation terms in the Reynolds stress and Reynolds flux equations are assumed [11] to be

$$\overline{\rho}\tilde{\varepsilon}_{\ell m} = \frac{2}{3}\overline{\rho}\tilde{\varepsilon}\delta_{\ell m} + \kappa_2 \frac{\widetilde{u_\ell'' c''}\,\widetilde{u_m'' c''}}{[\tilde{c}(1-\tilde{c})]^2}\overline{\dot{\omega}}_c; \quad \overline{\rho}\tilde{\varepsilon}_{\ell c} = \kappa_1 \frac{\widetilde{u_\ell'' c''}}{\tilde{c}(1-\tilde{c})}\overline{\dot{\omega}}_c,$$

where [7] $\kappa_1 \simeq 0.85$ and $\kappa_2 \simeq 0.25$ are constants (but see [17], in which a more general model is proposed). The important scalar dissipation term $\overline{\rho}\tilde{\varepsilon}_c = \overline{\mathcal{D}(\partial c/\partial x_k)^2}$ in the variance equation and the pressure gradient co-variances $\overline{c''\partial p'/\partial x_k}$ and $\overline{u_\ell''\partial p'/\partial x_k}$, which are known to play an essential role in models for non-reactive flows, are discussed later. Finally, an essentially empirical transport equation is required for the viscous dissipation $\tilde{\varepsilon}$; see Eq. (1.24) in Chapter 1.

(3) PROBABILITY DENSITY FUNCTION. The reacting flow field is partitioned [18] into three categories: *reactants*, for which $0 \le c \le c^*$, *reaction zones*, for which $c^* \le c(\mathbf{x}, t) \le (1 - c^*)$, and *combustion products*, for which $(1 - c^*) \le c \le 1$, whose probabilities are denoted by $\alpha(\mathbf{x})$, $\gamma(\mathbf{x})$, and $\beta(\mathbf{x})$, respectively, with $\alpha + \beta + \gamma = 1$, see Fig. 2.1. In the limit $c^* \to 0$, the Reynolds-averaging PDF of c can be written as

$$\overline{P}(c; \mathbf{x}) = \alpha\,\delta(c) + \beta\,\delta(1-c) + \gamma\,[H(0) - H(1)]f(c; \mathbf{x}), \qquad (2.4)$$

where the Dirac delta functions $\delta(c)$ and $\delta(1 - c)$ represent reactants and products, respectively. Also, H is the Heaviside function, $f(c; \mathbf{x})$ is the *internal* part of the PDF $\overline{P}(c; \mathbf{x})$, and

$$\int_0^1 f(c; \mathbf{x})\,\mathrm{d}c = 1.$$

In the thin-flame limit with Da $\gg 1$ we have $\gamma \ll 1$. No assumption is required at this point as to whether or not the reaction zone is to be approximated as a laminar flame.

It is sometimes convenient to define a Favre or mass-weighted PDF, equivalent to Eq. (2.4) and related to $\overline{P}$ by $\tilde{P}(c; \mathbf{x}) = \rho(c)/\overline{\rho}\overline{P}(c; \mathbf{x})$. Then

$$\tilde{P}(c; \mathbf{x}) = \tilde{\alpha}\,\delta(c) + \tilde{\beta}\,\delta(1-c) + \tilde{\gamma}\,[H(0) - H(1)]\tilde{f}(c; \mathbf{x}), \qquad (2.5)$$

so that the Favre mean of any thermochemical variable $\varphi(\mathbf{x}, t)$ is

$$\tilde{\varphi}(\mathbf{x}) = \int_0^1 \varphi(\mathbf{x}, t)\tilde{P}(c; \mathbf{x})\,\mathrm{d}c.$$

It can be shown, using the relationship between $\tilde{P}$ and $\overline{P}$ together with the normalisations of $f(c)$ and $\tilde{f}(c)$, that the two sets of variables are related by

$$\alpha = \tilde{\alpha}\frac{\overline{\rho}}{\rho_u}, \qquad \beta = \tilde{\beta}\frac{\overline{\rho}}{\rho_P},$$

$$\gamma = \tilde{\gamma} \int_{c^*}^{1-c^*} \frac{\overline{\rho}}{\rho(c)}\tilde{f}(c)\,\mathrm{d}c, \qquad f(c) = \frac{\overline{\rho}}{\rho}\frac{\tilde{f}(c)}{\int_{c^*}^{1-c^*}(\overline{\rho}/\rho)\tilde{f}(c)\,\mathrm{d}c}.$$

Continuing in terms of the Reynolds PDF $\overline{P}$ of Eq. (2.4), the Reynolds mean of φ is

$$\overline{\varphi}(\mathbf{x}, t) = \int_0^1 \varphi(\mathbf{x}, t)\overline{P}(c; \mathbf{x})\,\mathrm{d}c = (1-\overline{c})\varphi_R + \overline{c}\varphi_P + \mathcal{O}(\gamma), \qquad (2.6)$$

where φ_R and φ_P are the values of φ in reactants and products, respectively. These terms represent the effects of *intermittency* between reactants and products, as the thin-flame surface moves past the location $\mathbf{x}$. Because $\dot{\omega}_c(\mathbf{x}, t)$ is zero everywhere outside the reaction zones, its mean value,

$$\overline{\dot{\omega}}_c(\mathbf{x}) = \gamma(\mathbf{x}) \int_{c^*}^{1-c^*} \dot{\omega}_c f(c; \mathbf{x})\,\mathrm{d}c, \qquad (2.7)$$

is proportional to γ. If the turbulent transport terms in the $\tilde{c}$ transport equation, Eq. (2.2), are estimated to be $O(\rho_u/\tau_T)$, whereas $\overline{\dot{\omega}}_c$ is $O(\gamma\rho_u/\tau_L)$, then, if $\overline{\dot{\omega}}_c$ is to be of a magnitude similar to that of the transport terms, it follows that $\gamma = O(1/\mathrm{Da})$. Consequently, when Da $\gg 1$, $\gamma(\mathbf{x}, t) \ll 1$, and Eq. (2.6) shows that, for all properties other than those like $\dot{\omega}_c(c)$, which are zero in reactants and products, $\tilde{\varphi}$ can be very simply approximated in terms of φ_R and φ_P. We also have

$$\overline{\rho} = \frac{\rho_u}{1 + \tau\tilde{c}}; \; \alpha = \frac{1 - \tilde{c}}{1 + \tau\tilde{c}} + \mathcal{O}(\gamma); \; \beta = \frac{(1 + \tau)\tilde{c}}{1 + \tau\tilde{c}} + \mathcal{O}(\gamma).$$

The first of these equations is exact. The Favre variance of c is given by

$$\overline{\rho c''^2} = \alpha\rho_u\tilde{c}^2 + \beta\rho_P(1 - \tilde{c}^2) + \gamma \int_{c^*}^{1-c^*} \rho(c - \tilde{c})^2\tilde{f}\,\mathrm{d}c$$

$$= \overline{\rho}\tilde{c}(1 - \tilde{c}) + \mathcal{O}(\gamma), \qquad (2.8)$$

when $c^* \ll 1$. It follows that, when $\gamma \ll 1$, the variance transport equation is reduced to a second transport equation for the Favre mean $\tilde{c}$. The reconciliation of these

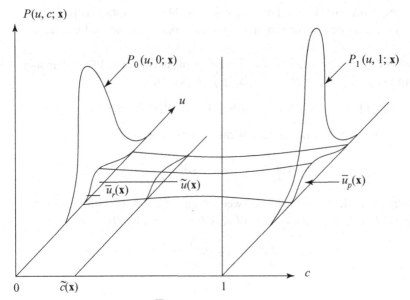

Figure 2.2. Sketch of JPDF $\overline{P}(c, u_k; \mathbf{x})$.

two transport equations leads [19] to the following important 'reaction–dissipation' relationship, which has been confirmed by analysis of DNS data [20], namely,

$$\overline{\dot{\omega}}_c = \frac{2}{2C_m - 1} \overline{\rho} \tilde{\epsilon}_c + \mathcal{O}(\gamma), \tag{2.9}$$

as has been noted in Eq. (1.35). The scalar dissipation $\tilde{\epsilon}_c$ can be interpreted as the reciprocal of a local time scale for molecular transport, and Eq. (2.9) shows the mean rate of heat release to be linked to the rate at which molecular processes transport heat and reactive species between reactants and products. With $\gamma \ll 1$, both of these processes occur in the thin reactive–diffusive interfaces separating reactants from products. It may be seen that, if $\overline{\dot{\omega}}_c$ is modelled, ϵ_c can be found from Eq. (2.9), and vice versa. The modelling of $\tilde{\epsilon}_c$ is explored in Section 2.3.

The partitioning illustrated in Eq. (2.6) can be extended [21] to the joint PDF (JPDF) of a velocity component $u_k(\mathbf{x}, t)$ and the progress variable $c(\mathbf{x}, t)$:

$$\overline{P}(c, u_k; \mathbf{x}) = \alpha \, \delta(c) f_R(u_k; \mathbf{x}) + \beta \, \delta(1 - c) f_P(u_k; \mathbf{x})$$

$$+ \gamma [H(0) - H(1)] f_F(c, u_k; \mathbf{x}), \tag{2.10}$$

as illustrated in Fig. 2.2. Conditional mean velocities in reactants and products can then be defined from

$$\alpha \overline{u}_{k,R}(\mathbf{x}) = \int_0^{c*} \int_{-\infty}^{\infty} u_k(\mathbf{x}, t) \overline{P}(c, u_k; \mathbf{x}) \, du_k \, dc,$$

$$\beta \overline{u}_{k,P}(\mathbf{x}) = \int_{1-c*}^{1} \int_{-\infty}^{\infty} u_k(\mathbf{x}, t) \overline{P}(c, u_k; \mathbf{x}) \, du_k \, dc, \tag{2.11}$$

where $c* \to 0$ and it follows that

$$\tilde{u}_k = (1 - \tilde{c}) \overline{u}_{k,R} + \tilde{c} \overline{u}_{k,P} + \mathcal{O}(\gamma). \tag{2.12}$$

Useful expressions following from these definitions include

$$\widetilde{u_k''c''} = \tilde{c}(1 - \tilde{c})(\overline{u}_{k,P} - \overline{u}_{k,R}) + \mathcal{O}(\gamma),$$

$$\overline{u}_{k,R} = \tilde{u}_k - \frac{\widetilde{u_k''c''}}{1 - \tilde{c}} + \mathcal{O}(\gamma),$$

$$\overline{u}_{k,P} = \tilde{u}_k + \frac{\widetilde{u_k''c''}}{\tilde{c}} + \mathcal{O}(\gamma),$$

$$\overline{u_k''} = \frac{\tau \widetilde{u_k''c''}}{1 + \tau\tilde{c}} + \mathcal{O}(\gamma). \tag{2.13}$$

It can be seen that, if $\tilde{u}_k$ and $\widetilde{u_k''c''}$ are determined from transport equations, values of $\overline{u}_{k,R}, \overline{u}_{k,P}$, and $\overline{u_k''}$ readily follow. The important implications of the first of Eqs. (2.13) for gradient or counter-gradient scalar transport will be discussed later. Here we simply note that $\widetilde{u_k''c''}$ arises because of intermittency between reactants and products; its sign is determined by the relative magnitudes of the conditional mean velocities $\overline{u}_{k,P}$ and $\overline{u}_{k,R}$.

A component of Favre mean square velocity is

$$\widetilde{u_k''^2} = (1 - \tilde{c})(\widetilde{u_k''^2})_R + \tilde{c}(\widetilde{u_k''^2})_P + \tilde{c}(1 - \tilde{c})(\overline{u}_{k,P} - \overline{u}_{k,R})^2 + \mathcal{O}(\gamma). \tag{2.14}$$

The first and second terms on the right-hand side of Eq. (2.14) represent contributions from turbulence in reactants and products, respectively, and the third term, which appears as a flame-generated contribution to the turbulence intensity, is in fact again a consequence of the intermittency between the mean velocities in reactants and products. Its properties do not resemble those of a quasi-random turbulent field.

An alternative to this BML partitioning of the JPDF $\overline{P}(c, u_k; \mathbf{x})$ into contributions from reactants, products, and burning zones is to formulate *conditional mean* transport equations for reactants and products [22–24]. These equations contain additional terms, representing fluxes at the interface with the thin reaction zone, which must be modelled. An advantage of such a formulation is that the flow in reactants and products may be treated as having constant density. A possible disadvantage is that the interface flux terms depend on the reaction-zone structure at a chosen isosurface location, and satisfactory general closure models have not yet been demonstrated.

It is worth noting here that *any* multivariate PDF $\overline{P}(c, \varphi_1, \varphi_2 \ldots ; \mathbf{x})$ involving $c(\mathbf{x}, t)$ along with other variables $\varphi_1(\mathbf{x}, t), \varphi_2(\mathbf{x}, t), \ldots$, can be partitioned in a manner similar to that shown in Eq. (2.10), allowing conditional mean quantities in reactants, products, and reaction zones to be defined and their relationships explored. For example, premixed turbulent combustion with variable enthalpy is analysed in a partitioned bivariate PDF of specific enthalpy and progress variable, [25] and in transient combustion in a spark-ignition engine in which the bivariate PDF is $\overline{P}(c, T; \mathbf{x}, t)$ [11]. A similar analysis is applied [26] to situations in which several chemical reactions occur sequentially. Finally, BML fluid mechanics is recovered in an analysis [27] incorporating systematically reduced reaction-rate mechanisms.

The application of the partitioning introduced in Eq. (2.6) can be further illustrated by considering the pressure gradient co-variances appearing in the transport equations for Reynolds stress and flux components. Taking the term in the scalar flux equation as an example, we have

$$\overline{c'' \frac{\partial p'}{\partial x_k}} = \alpha \left(\overline{c'' \frac{\partial p'}{\partial x_k}} \right)_R + \beta \left(\overline{c'' \frac{\partial p'}{\partial x_k}} \right)_P + \gamma \left(\overline{c'' \frac{\partial p'}{\partial x_k}} \right)_F. \qquad (2.15)$$

Note that, although $\gamma \ll 1$, gradients through reaction zones may be larger than those that are due to turbulence if $\hat{N} = \tau S_L^0 / u_{\mathrm{rms}} > \mathrm{Da}^{-1/2}$ [2], so the final term in Eq. (2.15) must be retained. Domingo and Bray [18] developed a model for these co-variances assuming that the reaction zones could be approximated by unstretched laminar flames. The reactants and products contributions are

$$\left(\overline{c'' \frac{\partial p'}{\partial x_k}} \right)_R = \left\langle c'' \frac{\partial p'}{\partial x_k} | c = 0 \right\rangle = -\tilde{c} \frac{\partial (\overline{p}_R - \overline{p})}{\partial x_k},$$

$$\left(\overline{c'' \frac{\partial p'}{\partial x_k}} \right)_P = \left\langle c'' \frac{\partial p'}{\partial x_k} | c = 1 \right\rangle = (1 - \tilde{c}) \frac{\partial (\overline{p}_P - \overline{p})}{\partial x_k}, \qquad (2.16)$$

and the conditional mean pressure gradients $\partial p_R / \partial x_k$ and $\partial p_P / \partial x_k$ are evaluated from Euler equations applied upstream and downstream of the thin flame, respectively, but neglecting additional terms representing fluxes at the interface with the reaction zone [24]. The flamelet contribution,

$$\gamma \left(\overline{c'' \frac{\partial p'}{\partial x_k}} \right)_F = \int_{c^*}^{1-c^*} \left\langle c'' \frac{\partial p'}{\partial x_k} | c \right\rangle \overline{P} \, dc, \qquad (2.17)$$

with $c^* \to 0$, is evaluated by means of laminar flamelet assumptions, leading to

$$\gamma \left(\overline{c'' \frac{\partial p'}{\partial x_k}} \right)_F = \frac{1}{2} \tau \langle \mathbf{n} \cdot \mathbf{N}_k \rangle \overline{\Sigma} \rho_u S_L^2 (0.7 - \tilde{c}), \qquad (2.18)$$

where $\overline{\Sigma}$ is the flame surface density, $\mathbf{n}$ and N_k are unit vectors normal to the flamelet, pointing towards reactant and in the direction of x_k, respectively, and $\langle \mathbf{n} \cdot N_k \rangle$ is treated as a constant of the order of unity. Models of these pressure gradient co-variances with a wider range of applicability may be found in [12, 17, 28].

2.1.2 Application to Stagnating Flows

Predictions from the BML model were systematically tested and compared with experimental data in a series of papers [29–32] addressing the problem of a planar turbulent flame brush stabilised in the vicinity of a stagnation point, which is established either between two opposed turbulent jets, as illustrated in Fig. 2.3(a), or on the surface of a plate arranged perpendicular to a single turbulent jet, as in Fig. 2.3(b). These are convenient configurations for experiments and a large amount of data were published [33–40].

Although the experiments of Kalt et al. [38] include the highest available turbulence levels, much of the available experimental data involve turbulence of low intensity, as measured by a parameter $\delta = k_1 / w_1^2$, where k_1 and w_1 represent the turbulence kinetic energy and mean axial flow velocity, respectively, at the nozzle exit.

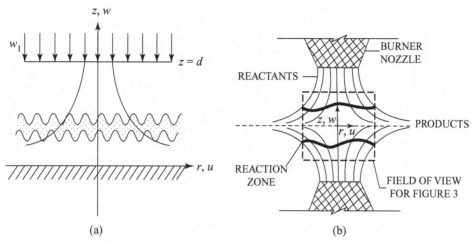

Figure 2.3. Sketches of stagnation point flow geometries: (a) stagnation plate (b) counterflow.

Figure 2.3 shows flow arrangements for stagnation point flow. The nozzle is placed at a distance d from the stagnation plane, the axial coordinate is z, measured from the stagnation point, and the radial coordinate is r. A dimensionless quantity $\tilde{z} = z/d$ is introduced for the analysis. With $\delta \ll 1$, this flow is dominated by pressure gradients, and turbulent transport terms in the RANS equations are shown [29] to be negligible. In the first stage of the analysis [29], the flame brush is shown to be planar in the vicinity of the axis of symmetry. The relative unimportance of Reynolds stresses and fluxes when $\delta \ll 1$ is tested by first *imposing* a mean progress-variable distribution $\tilde{c}(\tilde{z})$ from experiment and calculating $\overline{\rho}(\tilde{z})$ from Eq. (2.1) and then solving the mean-flow equations with these stresses and fluxes set to zero and showing that predictions agree well with published experimental data. A second stage [30] begins with $\tilde{c}(\tilde{z})$, $\overline{\rho}(\tilde{z})$, $\tilde{w}(\tilde{z})$, and $U(\tilde{z}) = \tilde{u}d/w_1 r$, all known from these calculations [29]. With these quantities provided, the *uncoupled* second-moment transport equations are closed and solved to predict the Reynolds stresses $\widetilde{u''^2}$ and $\widetilde{w''^2}$ and the Reynolds flux $\widetilde{w''c''}$. It is found that, in these circumstances, the pressure gradient co-variance expressions described earlier [see Eqs. (2.17) and (2.18)] play an essential role in these equations and give good agreement with available experimental data.

The turbulence intensity components, Fig. 2.4, are shown to be anisotropic and strongly peaked inside the flame brush as a result of flame-generated 'turbulence' that is due to intermittency.

A further stage in the analysis of flames in stagnating flows [31] uses the measured mean progress-variable distributions $\tilde{c}(\tilde{z})$ from five different experiments to calculate the corresponding experimental mean-reaction-rate distributions, $\overline{\dot{\omega}}_c(\tilde{z})$. It then makes use of earlier calculations [29, 30] to compare predictions from five different mean-reaction-rate models with the experimental data. Only qualitative agreement is found. Finally [32] the coupled model flow equations are solved directly, using a new laminar flamelet mean-reaction-rate model [41] [described in Subsection 2.1.5; see Eq. (2.36)] and without any assumed mean progress-variable distribution, and predictions are again found to be in satisfactory agreement with published experimental data. It is also shown that an approximately sixfold increase

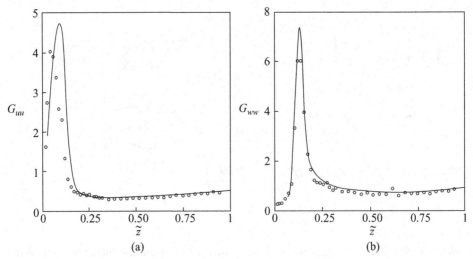

Figure 2.4. Turbulence intensity components from the model of Bray et al. [30] (lines), compared with experimental data of Li et al. [34] (data points): (a) radial intensity $G_{uu} = \widetilde{u''^2}/k_0$, (b) axial intensity $G_{ww} = \widetilde{w''^2}/k_0$, where k_0 is the turbulence kinetic energy at the nozzle exit. From Bray et al. [30].

in a mean-strain-rate function is required for displacing the flame brush from a location where flashback to the nozzle is about to occur to a position so close to the wall that extinction is imminent.

2.1.3 Gradient and Counter-Gradient Scalar Transport

An important result of this asymptotic analysis of flames near stagnation points [30] is that, to the lowest order in δ, the scalar flux transport equation reduces to an algebraic expression:

$$\widetilde{w''c''} = -\tau S_L \langle \mathbf{n} \cdot \mathbf{N}_z \rangle \tilde{c}(1 - \tilde{c}). \tag{2.19}$$

Counter-gradient scalar transport [6] therefore occurs because $\langle \mathbf{n} \cdot \mathbf{N}_z \rangle$ is positive here, so $\widetilde{w''c''}$ has the same sign as $d\tilde{c}/dz$, both being negative in the coordinate system shown in Fig. 2.3. The scalar flux is in the opposite direction to that predicted by the usual gradient transport expression of non-reactive turbulent flows, namely

$$\widetilde{w''c''} = -\nu_T \frac{d\tilde{c}}{dz} \tag{2.20}$$

where ν_T is a turbulent diffusivity.

As may be seen from Fig. 2.5, Eq. (2.19) gives excellent agreement with available experimental data. Other experiments [42] and DNS calculations [1, 20, 43] confirm the occurrence of counter-gradient transport.

Counter-gradient transport in premixed flames was first observed in experiments by Moss [44] and was predicted in RANS calculations by Libby and Bray [6]. It was also predicted in calculations from other second-moment turbulent combustion models [17, 45], and Zimont and Biagioli [46] proposed a model in which scalar transport is described by the sum of a k–ϵ gradient transport expression and 'gas dynamic' counter-gradient quantity. A theoretical and DNS analysis of planar turbulent flames by Veynante et al. [1] identifies the increased conditional mean velocity

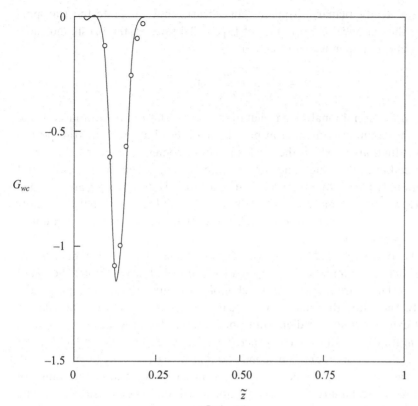

Figure 2.5. Axial scalar flux $G_{wc} = \widetilde{w''c''}/k_0^{1/2}$ from the model of Bray et al. [30] (lines), compared with experimental data of Li et al. [34] (data points). From Bray et al. [30].

in burned gases, resulting from the reduction in density, relative to the reactants, as a cause of counter-gradient transport. The analysis therefore introduces the velocity jump τS_L^0 across a laminar flame as the driver of counter-gradient transport, opposed by a turbulence quantity αu_{rms}, where u_{rms} is the turbulence rms velocity ahead of the flame. The quantity α is an 'efficiency factor' [47] measuring the influence of eddies of different scales on a laminar flame. The scalar flux in an assumed planar turbulent flame brush is then modelled as

$$\widetilde{u''c''} = \tilde{c}(1 - \tilde{c})(\tau S_L^0 - 2\alpha u_{rms}).\qquad(2.21)$$

This expression predicts that a transition from gradient to counter-gradient transport occurs when

$$\hat{N}^* \equiv \frac{\tau S_L^0}{2\alpha u_{rms}} \geq 1.\qquad(2.22)$$

Typical values of α are close to 0.5, so $\hat{N}^*$ is similar in magnitude to the parameter $\hat{N} = \tau S_L^0/u_{rms}$ introduced earlier. This criterion is supported by DNS data from [1] and elsewhere. It is also in reasonable agreement with the experimental data of Kalt et al. [38] in stagnation point flames in intense turbulence, although these authors also suggest a modified expression.

In contrast to the expression of Veynante et al. in [1], Eq. (2.22), which depends on the velocity jump τS_L across a flamelet, the alternative criterion proposed by Swaminathan et al. [48, 49] involves the *internal structure* of the thin reaction zone.

Considering a steady, planar, turbulent flame brush, they integrate the transport equation for the variance of c [the first of Eqs. (2.3)] through the brush and show that the sign of $\widetilde{u''c''}$ depends on the sign of

$$I = \int_{-\infty}^{\infty} \overline{c'' \left(\frac{\partial u_k}{\partial x_k} \right)} \, dx. \qquad (2.23)$$

If the peak in the conditional mean dilatation $\langle \partial u_k / \partial x_k | c \rangle$ occurs at a large value of c, $I > 0$ and counter-gradient transport is favoured, but gradient transport is indicated in situations in which the peak dilatation comes at smaller values of c. They conclude that, unlike methane–air flames, counter-gradient transport will not occur in stationary planar hydrogen–air flames with equivalence ratios greater than about 0.5. However, in a more recent work, Chen et al. [50] show that mean flow divergence can modify this picture in such a way that either gradient or counter-gradient transport can occur when $I > 0$.

Chakraborty and Cant [12] use DNS of planar turbulent flame brushes to explore the influence of Lewis number $Le = \hat{\lambda}/c_p \rho \mathcal{D}$ on scalar turbulent transport. The small-Lewis-number flames are found to be much more strongly wrinkled and propagate more quickly. They show that, under the same initial conditions of turbulence, flames with $Le < 1$ exhibit counter-gradient transport, whereas flames with $Le > 1$ tend to have gradient transport. The efficiency factor α in Eq. (2.22) is evaluated and found to increase as Le decreases, and they conclude that the criterion of Veynante et al. cannot correlate their data. Chakraborty and Cant [12] then test a wide range of models to close the scalar flux transport equation and develop new closures including Lewis number effects.

It is important to note that most of the experimental and DNS data that were used to investigate scalar transport and to assess the criterion of Veynante et al. [1] involve planar turbulent flame brushes in which Reynolds shear stresses are absent. A model calculation by Masuya and Libby [51], representing a V-flame highly constrained within a duct, indicates the presence of a mean scalar flux *along* the flame brush, a direction in which the mean progress-variable gradient is zero. The existence of this *non*-gradient transport is not predicted by inequality in Eq. (2.22). Also, experimental data [42, 52] and DNS calculations [20] in more complex flow fields sometimes show that the various components of the scalar flux at a given location involve a combination of gradient and counter-gradient transport. The transition between these two types of transport process in complex flows deserves further study.

Most LESs of premixed turbulent combustion assume a gradient expression for subgrid scalar transport. However, Tullis and Cant [53], using filtered DNS data, show that regions of both gradient and counter-gradient transport can frequently occur within the same simulation; LES closures were proposed [53] to address this issue.

2.1.4 Laminar Flamelets

Damköhler [54] proposed that a premixed flame in turbulence of sufficiently low intensity and large scales can be viewed as a wrinkled laminar flame of increased surface area per unit streamtube area, leading to a turbulent burning velocity

$S_T = S_L A_T / A$, where A_T is the area of wrinkled laminar flame in a streamtube of cross-sectional area A. The conditions under which an unstretched laminar flame can provide a valid model for localised zones of heat release in premixed turbulent combustion were reviewed by several authors, including Peters [55], Poinsot and Veynante [56], and Driscoll [57], and are often presented in the form of a regimes diagram such as that shown in Fig. 1.3. The classical perspective assumes that laminar flamelets will persist so long as a laminar flame is thinner than all scales of turbulence, that is, when the Karlovitz number $\mathrm{Ka} = \tau_c / \tau_K < 1$. Thickened flamelets are assumed to occur whenever $\mathrm{Ka} \geq 1$. However, much experimental and DNS evidence [57] now suggests that laminar-like reaction-zone structures can be much more robust than is implied by the $\mathrm{Ka} = 1$ criterion. An explanation may be found in studies [58, 59] of individual vortices interacting with laminar flames: Small eddies are found to be unable to disrupt the flame structure, as they are rapidly dissipated by the combined effects of viscosity and dilatation. Consequently laminar-like flame zones persist in much more intense turbulence than was previously thought possible. However, it must be kept in mind that, however small Ka may be, composition gradients in the outermost edges of these flame zones will always be influenced by turbulent fluctuations. Some models [46, 55] describe these interactions in terms of a preheat zone that is thickened by turbulence followed by a thin laminar reaction layer. Another perspective [60] argues that the range of validity of flamelet models may be extended by including the influence of stretch on a laminar flame.

Laminar flamelet models can be implemented in several different ways: One approach [61] is to represent the reaction zones in terms of a laminar flame speed (with or without allowance for the influence of strain and curvature) and a flame surface area or FSD $\overline{\Sigma}$. The mean chemical source term in the $\tilde{c}$ equation is then $\overline{\dot{\omega}}_c = \overline{\rho}_u S_L^0 I_0 \overline{\Sigma}$, where I_0 is a factor to allow for the influence of flame stretch. Alternatively, in a level-set or G-equation approach [55], the rate of propagation of a chosen isosurface is modelled. Another way to apply a laminar flamelet model is to assume (see Section 2.2 and [62, 63]) that all chemical species mass fractions $Y_i(\mathbf{x}, t)$ and their reaction rates $\omega_i(\mathbf{x}, t)$ at given temperature, pressure, and composition are unique functions of c, determined from the properties of an unstretched laminar flame. This assumption can be combined with several different alternative mean-reaction-rate models; for example, in a presumed PDF model (see Chapter 1, Subsection 1.7.2), the chemical kinetic mechanism allows $\dot{\omega}_c(c)$ to be expressed in terms of the flamelet reaction rates $\dot{\omega}_{i,F}(c)$. Then

$$\overline{\dot{\omega}}_c = \int_0^1 \frac{\overline{\rho}}{\rho(c)} \, \dot{\omega}_{c,F}(c) \, \tilde{P}_{\mathrm{pr}}(c; \mathbf{x}) \, \mathrm{d}c, \qquad (2.24)$$

where $\tilde{P}_{\mathrm{pr}}(c; \mathbf{x})$ is the presumed Favre PDF, and

$$\tilde{Y}_i = \int_0^1 Y_{i,F}(c) \tilde{P}_{\mathrm{pr}}(c; \mathbf{x}) \, \mathrm{d}c, \qquad (2.25)$$

where $Y_{i,F}(c)$ represents flamelet composition.

A more restrictive but potentially powerful laminar flamelets approach, explored in the next subsection, is to assume [21, 62] that reaction zones can be replaced with unstretched laminar flames not only in composition space but also in physical space.

Laminar flame stretch, which is included in the model of Bradley et al. [45], is briefly considered at the end of this section.

2.1.5 A Simple Laminar Flamelet Model

To develop this presumed PDF laminar flamelet analysis we recall from Chapter 1, Eq. (1.39), that the PDF $\overline{P}(c_*; \mathbf{x})$ on a chosen internal isosurface $c = c_*$ is related to the surface density function $\Sigma(c_*; \mathbf{x})$ by [64, 65]

$$\Sigma(c_*; \mathbf{x}) = \left\langle \left\| \frac{\partial c}{\partial x_k} \right\| \middle| c_* \right\rangle \overline{P}(c_*, \mathbf{x}). \tag{2.26}$$

In the limit of thin weakly wrinkled laminar flamelets, $\Sigma(c_*; \mathbf{x})$ is almost independent of the choice of the isosurface c_*, so long as c_* is not too close to zero or unity and is identified as the FSD $\overline{\Sigma}(\mathbf{x})$. The progress-variable gradient at a point $c = c_*$ in a steady, unstretched laminar flame is a function of c_*

$$\left\langle \left\| \frac{\partial c}{\partial x_k} \right\| \middle| c_* \right\rangle = \left(\frac{\mathrm{d}c}{\mathrm{d}\eta} \right)_F. \tag{2.27}$$

Although this assumption may be valid in the interior of the reaction zone, where the composition gradients are large, it must become less accurate at the edges of this zone. It follows from Eq. (2.27) that the interior part of the PDF $\overline{P}(c; \mathbf{x})$ of Eq. (2.4) as a function of c is [41, 62]

$$f_F(c) = C_1 \left(\frac{\mathrm{d}c}{\mathrm{d}\eta} \right)_F^{-1}, \tag{2.28}$$

where C_1 is a normalising factor given by

$$1 = C_1 \int_{0^+}^{1^-} \frac{\mathrm{d}c}{(\mathrm{d}c/\mathrm{d}\eta)_F} = C_1 \int_{\eta_{min}}^{\eta_{max}} \mathrm{d}\eta.$$

Here 0^+ and 1^- represent the limits within which $\Sigma(c; \mathbf{x})$ is to be assumed constant and $\eta = \eta_{min}$ and η_{max} are the corresponding values of η. It then follows that $C_1 = 1/\Delta\eta$, where $\Delta\eta = \eta_{max} - \eta_{min}$ is a measure of the laminar flame thickness.

The flamelets contribution to the Reynolds mean of any property φ can then be found as

$$\overline{\varphi}(\mathbf{x}) - \alpha(\mathbf{x})\varphi_R - \beta(\mathbf{x})\varphi_P = \gamma(\mathbf{x}) \int_{0^+}^{1^-} \varphi_F(c) f_F(c) \, \mathrm{d}c$$

$$= \gamma(\mathbf{x}) \int_{0^+}^{1^-} \varphi_F(c) \frac{C_1}{(\mathrm{d}c/\mathrm{d}\eta)_F (c)} \, \mathrm{d}c$$

$$= \frac{\gamma(\mathbf{x})}{\Delta\eta} \int_{\eta_{min}}^{\eta_{max}} \varphi_F[c(\eta)] \, \mathrm{d}\eta, \tag{2.29}$$

where the integral can be determined from a laminar flame calculation.

This analysis was used in Ref. [18] to model flamelet contributions to the pressure gradient co-variances in the second-moment transport equations; see Eq. (2.18). The co-variances were found to play an important role in the closure of these equations, so the successful prediction of second-moment quantities, illustrated in Figs. 2.4 and 2.5, provides support for the validity of the approach. We now illustrate the application of the analysis by using it to derive simple laminar flamelet expressions for the mean reaction rate and by recovering the relationship, Eq. (2.9), between this and the mean scalar dissipation. The mean reaction rate can be expressed in several different ways, the first of which is

$$\bar{\omega}_c = \gamma C_1 \int_{\eta_{min}}^{\eta_{max}} \dot{\omega}_c[c(\eta)] \, d\eta \quad = \frac{\gamma}{\Delta \eta} \rho_u S_L^0, \tag{2.30}$$

where the c equation for an unstretched laminar flame is used to evaluate the integral, assuming $\eta_{max} \to \infty$ and $\eta_{min} \to -\infty$.

To evaluate this equation, the burning-mode (BM) probability γ can be related to the variance $\widetilde{c''^2}$ by use of the PDF of Eq. (2.4) to find expressions for $\tilde{c}$ and $\widetilde{c''^2}$ in terms of α, β, and γ and recalling that $\alpha + \beta + \gamma = 1$. The result is

$$\gamma = \frac{\epsilon \, \Delta \eta \, \tilde{c}(1 - \tilde{c})}{(1 + \tau \tilde{c}) \delta^*}, \tag{2.31}$$

where $\epsilon(\mathbf{x})$ is a variance parameter defined as

$$\epsilon = 1 - \frac{\widetilde{c''^2}}{\tilde{c}(1 - \tilde{c})},$$

which may be predicted from a solution to the transport equation for $\widetilde{c''^2}$. We have $\epsilon \ll 1$ in the thin-flamelets limit in which $\tilde{\gamma} \ll 1$ and $\epsilon \to 1$ if the variance approaches zero. It may be seen that $\epsilon = \mathcal{O}(\gamma) = \mathcal{O}(1/Da)$ in the present analysis. Also $\delta^* = \int_{\eta_{min}}^{\eta_{max}} c(1 - c)/(1 + \tau c) \, d\eta$ is another measure of the laminar flame thickness, which reaches a constant value as η_{max} and η_{min} approach $\pm\infty$. Note that $\gamma \sim \Delta \eta$ because, in this simple model, the reaction-zone probability increases when the laminar flame limits η_{max} and η_{min} are extended.

It follows from Eqs. (2.30) and (2.31) that

$$\bar{\omega}_c = \frac{\epsilon \tilde{c}(1 - \tilde{c})}{(1 + \tau \tilde{c}) \delta^*} \rho_u S_L^0, \tag{2.32}$$

which removes the apparent sensitivity of $\bar{\omega}_c$ in Eq. (2.30) to the choice of η_{max} and η_{min}. This expression can be rewritten in terms of the generalised FSD Σ_G. The SDF of an isosurface $c = c^*$ is

$$\Sigma(c^*) = \left\langle |\frac{\partial c}{\partial x_k}| |c^* \right\rangle f_F(c^*) = \frac{\gamma}{\Delta \eta}, \tag{2.33}$$

and, as assumed earlier, this is independent of c^*. It follows that the generalised FSD is

$$\Sigma_g \equiv \int_0^1 \Sigma(c^*) \, dc^* \quad = \bar{\Sigma} = \frac{\gamma}{\Delta \eta} = \frac{\epsilon \tilde{c}(1 - \tilde{c})}{(1 + \tau \tilde{c}) \delta^*}. \tag{2.34}$$

An alternative way of presenting this laminar flamelet mean-reaction-rate model, in terms of chemical kinetic parameters rather than laminar flame properties, can be obtained from

$$\bar{\omega}_c = \gamma \int_{0^+}^{1^-} \dot{\omega}_c(c) f_F(c) \, dc. \tag{2.35}$$

If we now introduce a dimensionless reaction-rate function $\Omega_c(c) = \dot{\omega}_c/\rho B$, where B is a characteristic reaction-rate constant with dimension (time)$^{-1}$, and substitute f_F

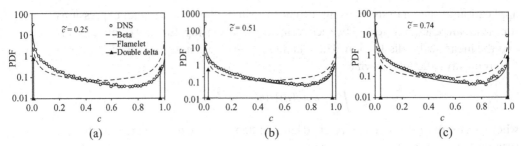

Figure 2.6. Laminar flamelet PDF (heavy line), beta function PDF (light line), and twin-delta PDF (triangles), compared with DNS data of Rutland and Cant [43] (data points); from Bray et al. [41].

from Eq. (2.28) and γ from Eq. (2.31), we find [41]

$$\bar{\dot{\omega}}_c = \epsilon \bar{\rho} \tilde{c}(1 - \tilde{c}) I_F B, \qquad (2.36)$$

where

$$I_F = \frac{1}{\delta^*} \int_{\eta_{min}}^{\eta_{max}} \frac{\Omega_c(\eta)}{1 + \tau c(\eta)} d\eta$$

is a constant, related to the assumed global reaction-rate mechanism, whose value can be calculated from a laminar flame solution. Comparing Eqs. (2.36) with the earlier mean-rate expression of Eq. (2.30) and using Eq. (2.31), we find that $I_F B = S_L^0 / \delta^*$. Equation (2.36) shows the mean reaction rate to be the product of a turbulent mixing factor $\bar{\rho} \tilde{c}(1 - \tilde{c})\epsilon$ and a reaction-rate factor $I_F B$. Note that, if the order of magnitude $\epsilon = \mathcal{O}(\gamma) = \mathcal{O}(1/\text{Da})$ were to be replaced with an ad hoc model of the form $\epsilon = \text{constant}/\text{Da}$, the chemical time scale in Eq. (2.36) would cancel, the turbulence time $\tilde{k}/\tilde{\epsilon}$ would control reaction, and an eddy break-up (EBu) mean-reaction-rate expression would be recovered. However, experience shows this to be incorrect, as the mean rate depends on both characteristic scales.

It was shown in [41] that, in the limit $\gamma \ll 1$, other presumed PDF shapes $\bar{P}_{pr}(c; \mathbf{x})$ lead to exactly the same mean-reaction-rate expression as Eq. (2.36), but with the constant I_F replaced with a constant I_{pr} whose magnitude depends on the reaction-rate expression and the shape of the PDF $\bar{P}_{pr}(c; \mathbf{x})$. It is also shown in [41] that I_F and I_{pr} approach constant values when the flamelet limits η_{min} and η_{max} approach $\pm\infty$. The flamelet PDF is shown to give the best agreement with the DNS data of Rutland and Cant [43], as shown in Fig. 2.6, and to provide an excellent match to the mean heat release rate from the DNS; see Fig. 2.7. By way of contrast, a beta function PDF gives a mean rate 2.4 times as large as that required to match the DNS, whereas a double delta leads to a rate that is 8.6 times too large.

We turn now to the scalar dissipation $\tilde{\epsilon}_c$, which must be modelled in order to determine ϵ from the $\widetilde{c''^2}$ equation. The instantaneous scalar dissipation function is $\rho N = \rho_u \mathcal{D}_u (\partial c/\partial x_k)(\partial c/\partial x_k)$ if it is assumed that $\rho\mathcal{D} = \rho_u \mathcal{D}_u$. We further assume that the gradient within a flamelet is large in comparison with mean property gradients, and $\overline{\rho N}$ can then be equated to $\overline{\rho\epsilon_c}$, which is the dissipation term in the transport equation for the variance $\widetilde{c''^2}$. Thus

$$\bar{\rho} \tilde{\epsilon}_c \simeq \overline{\rho N} = \rho_u \mathcal{D}_u \frac{\gamma}{\Delta\eta} \int_{0^+}^{1^-} \left(\frac{dc}{d\eta}\right)_F dc. \qquad (2.37)$$

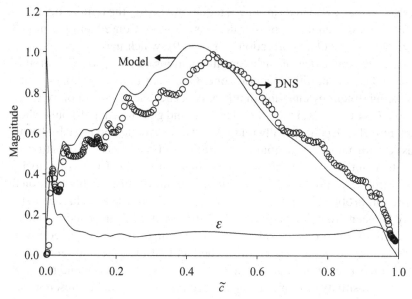

Figure 2.7. Mean heat release rate $\overline{\omega}_c/\epsilon$ and ϵ vs $\tilde{c}$: Predictions compared with DNS data of Rutland and Cant [43] (data points). From Bray et al. [41].

Equations (2.30) and (2.37) both contain the BM probability γ, which can be eliminated by forming the ratio of Eqs. (2.30) and (2.37). If we introduce the laminar flame thickness $\delta = \mathcal{D}_u/S_L^0$, we find

$$\frac{\overline{\rho}\tilde{\epsilon}_c}{\overline{\dot{\omega}}_c} = \delta \int_{0+}^{1^-} \left(\frac{dc}{d\eta}\right)_F dc, \tag{2.38}$$

which resembles the reaction–dissipation relationship of Eq. (2.9). To demonstrate that these two expressions are compatible, we may calculate $C_m \equiv \overline{c\dot{\omega}_c}/\overline{\dot{\omega}}_c$ from

$$\overline{c\,\dot{\omega}_c} = \frac{\gamma}{\Delta\eta} \int_{\eta_{\min}}^{\eta_{\max}} c\,\dot{\omega}_F\,d\eta$$

$$= \frac{\gamma}{\Delta\eta}\rho_u S_L^0 \int_{\eta_{\min}}^{\eta_{\max}} c\frac{dc}{d\eta}\,d\eta - \rho_u \mathcal{D}_u \frac{\gamma}{\Delta\eta} \int_{\eta_{\min}}^{\eta_{\max}} c\frac{d^2 c}{d\eta^2}\,d\eta,$$

where the c equation for an unstretched laminar flame has been used. Evaluation of the integrals then shows that C_m in this flamelet model is a constant, given by

$$C_m - \frac{1}{2} = \delta \int_{0+}^{1^-} \left(\frac{dc}{d\eta}\right)_F dc, \tag{2.39}$$

and substitution of this into Eq. (2.38) allows Eq. (2.9) to be recovered.

The flamelet mean-reaction-rate expressions of Eqs. (2.32) and (2.36) contain the chemical time scale $\delta^*/S_L^0 = 1/I_F B$, but characteristic turbulence scales can enter the present description only by means of $\epsilon(\mathbf{x})$. This quantity can be determined from a solution of the transport equation for the variance $\widetilde{c''^2}$, which determines the balance between the reactive part $\gamma(\mathbf{x})$ of the PDF $\overline{P}(c;\mathbf{x})$ and the non-reactive components $\alpha(\mathbf{x})$ and $\beta(\mathbf{x})$. Although an expression for the scalar dissipation $\tilde{\epsilon}_c$ in this equation may be obtained because of Eq. (2.36) and the reaction–dissipation expression,

Eq. (2.9), with $\tilde{\epsilon}_c(\mathbf{x})$ expressed in terms of $\widetilde{c''^2}$, it does not satisfy the requirement that all terms in the variance equation must vanish as $\widetilde{c''^2} \to 0$. Consequently a model must be provided for the $\mathcal{O}(\gamma)$ correction to Eq. (2.9), which may be viewed as a representation of the influence of turbulent fluctuations on the progress-variable gradient at the edges of the thin reaction zone. One such model, [32], meets this 'realisability' requirement by ensuring that $\tilde{\epsilon}_c$ approaches the classical non-reactive flow expression, $\tilde{\epsilon}_c = C_\epsilon \widetilde{c''^2} \tilde{\epsilon}/\tilde{k}$, in the limit Da $\ll 1$, and goes to Eq. (2.9) only when Da $\gg 1$. Such models are conceptually related to that of Anand and Pope [66], who added a classical turbulence contribution to the conditional scalar dissipation in their laminar flamelet closure of the transport equation for the J-PDF of velocity and progress variable in premixed combustion. A consequence of the additional term in the present problem is that $\tilde{\epsilon}_c(\mathbf{x})$ then depends on both sets of characteristic scales. It may be seen from Eqs. (2.32) and (2.36) that the mean reaction rate is extremely sensitive to the predicted value of $\widetilde{c''^2}$; for example, if $\epsilon = 0.01$, a reduction of 0.5% in $\widetilde{c''^2}$ can reduce $\tilde{\epsilon}_c$ and $\overline{\omega}_c$ by 50%. The fact that the $\widetilde{c''^2}$ equation at large Damköhler numbers is essentially a balance between scalar dissipation and reaction effects suggests a sensitivity of ϵ_c and the mean heat release rate to the chosen scalar dissipation model; see Section 2.3.

An aim of model development for premixed turbulent combustion must be to derive models that are applicable in a wider range of burning regimes, and laminar flamelet combustion should be included within this range. Although the present model is greatly simplified, it does capture important features of the flamelet heat release process. It provides a straightforward and tested route to the prediction of flamelet contributions to the closure of terms such as pressure gradient co-variances [18]. As shown in Fig. 2.8, combustion flow-field predictions [32] using the flamelet model agree well with experimental data. The separation of mean heat release rate expression (2.36) into the product of a turbulent mixing factor and a chemical kinetic factor permits the influence of the shape of the interior PDF $f(c;\mathbf{x})$ to be quantified [41] and has recently been extended [27] to include more complex chemical kinetic mechanisms.

Finally, the theoretical framework previously presented to describe burning in terms of assumed unstretched laminar flamelets will be adapted to make qualitative assessments of the influence on the mean burning rate of, first, laminar flame stretch, and then a thickening of the laminar flame preheat zone.

STRETCHED LAMINAR FLAMELETS. Bradley et al. [45, 67] describe a second-moment, stretched laminar flamelet model in which the heat release rate is obtained in the form $\dot{\omega}_c(c, K) \approx \dot{\omega}_c(c, K = 0)f(K)$ from laminar flame calculations, where $K = (u_{\mathrm{rms}}/\lambda)(\delta/S_L)$ is the Karlovitz stretch factor, λ is the Taylor microscale of turbulence, and $f(K)$ is a correction factor for the influence of stretch. The mean rate $\overline{\omega}_c$ is obtained by multiplying this rate expression by the JPDF $\overline{P}(c, K)$ and integrating twice. Taking $\overline{P}(c, K) = \overline{P}(c)\overline{P}(K)$, they find

$$\overline{\omega}_c = P_b \int_0^1 \dot{\omega}(c, K = 0)\overline{P}(c)\,dc,$$

where

$$P_b = \int_{K_{\min}}^{K_{\max}} f(K)\overline{P}(K)\,dK.$$

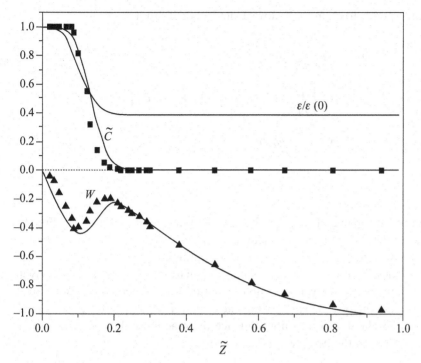

Figure 2.8. Flamelet predictions [32] (lines) compared with the experimental data of Li et al. [34] (data points), where $W = \tilde{w}(\tilde{z})$, $\tilde{z} = z/d$. From Bray et al. [32].

It may be seen that the mean heat release rate is modelled as the product of a probability P_b that the stretch rate can sustain a flamelet and a mean heat release rate, calculated in the absence of stretch effects. Bradley et al. [45] assume a beta function PDF $\overline{P}(c; \mathbf{x})$. If we repeat this analysis, replacing the beta function with our unstretched flamelet PDF, we obtain

$$\overline{\dot{\omega}}_c = P_b \epsilon \frac{\rho_u \tilde{c}(1 - \tilde{c})}{(1 + \tau \tilde{c})} \frac{S_L^0}{\delta^*}, \qquad (2.40)$$

where Eq. (2.32) has been used and S_L^0 and δ^* are to be evaluated with $K = 0$. Bradley et al. [67] calculate P_b as a function of K, with Markstein number Ma as a parameter; with Ma > 0, they find it to decrease monotonically with K but, when Ma < 0, P_b can increase with increasing K. This analysis suggests that our simple laminar flamelet model can be adapted to include the influence of stretched laminar flames with Le ≈ 1 simply by including the flamelet burning probability P_b, as illustrated in Eq. (2.40).

THICKENED PREHEAT ZONE. Here we demonstrate that, as expected on physical grounds, a thickening of the preheat zone in this model will not modify the predicted mean heat release rate so long as the thickening is confined to a non-reactive portion of the flamelet. The internal PDF of Eq. (2.28) is modified as follows: The progress-variable gradient in a *non-reactive* portion of the preheat zone where $0 \leq c \leq c^*$ is assumed to be $1/\vartheta$ times that in a laminar flame, where $\vartheta \geq 1$; for $c^* \leq c \leq 1$, the gradient is that of a laminar flame. Then the constant C_1 of Eq. (2.28) is

$$C_1 = \frac{1}{\Delta \eta_T} \equiv \frac{1}{\vartheta(\eta^* - \eta_{\min}) + (\eta_{\max} - \eta^*)}, \qquad (2.41)$$

where η^* corresponds to c^*, and it follows from Eq. (2.30) that

$$\bar{\dot{\omega}}_c = \gamma \int_0^1 \dot{\omega}_c f_F \, dc$$

$$= \gamma C_1 \left[\int_{\eta_{\min}}^{\eta_{\max}} \dot{\omega}_c \, d\eta + (\vartheta - 1) \int_{\eta_{\min}}^{\eta^*} \dot{\omega}_c \, d\eta \right]$$

$$= \gamma C_1 \int_{\eta_{\min}}^{\eta_{\max}} \dot{\omega}_c \, d\eta$$

$$= \frac{\gamma \rho_u S_L^0}{\Delta \eta_T}.$$

However, in the thin-flamelets regime, the flamelet burning probability γ is directly proportional to the flamelet thickness $\Delta \eta_T = \vartheta(\eta^* - \eta_{\min}) + (\eta_{\max} - \eta^*)$, so the influence of the thickening parameter ϑ cancels from this equation.

In classical descriptions of a laminar flame [68], in terms of an inert preheat layer followed by a thin reaction layer, the rate of reaction is governed by the molecular-transport flux from the preheat zone to the reaction layer. It follows that a thickening of a part of the preheat zone will not influence the reaction unless the gradient at the boundary between the two layers is modified.

2.1.6 Conclusions

The BML model considers a burning zone formed by one or more thin wrinkled flame surfaces, identifies 'turbulence' production and scalar flux changes that are due to intermittency between reactants and combustion products, and provides models to describe these processes. In this subsection we have assumed the wrinkled flame surface to be adequately represented by an unstretched laminar flame and shown that the U-shaped interior PDF $f(c)$ is then proportional to the reciprocal of the known laminar flame gradient $dc/d\eta$. Simple algebraic expressions for the mean reaction rate, proportional to the scalar variance quantity ϵ, then follow. To calculate ϵ, the variance $\widetilde{c''^2}$ must first be determined by closing and solving the first of Eqs. (2.3). Closure of this equation requires a model for the scalar dissipation $\tilde{\varepsilon}_c$, and such models are discussed at length in Section 2.3. Other models, involving a more detailed description of the interaction between combustion and turbulent flow than has been presented here, are described in the following pages.

2.2 Flame Surface Density and the G Equation
By L. Vervisch, V. Moureau, P. Domingo, and D. Veynante

FSD and the field equation (or G equation) are fundamental tools of premixed turbulent combustion modelling [55, 56, 69, 70]. They have been intensively studied in RANS [71–74], LES [75–81], and DNS [82–86] contexts and also in experiments [87, 88].

The FSD and G equation were introduced in combustion mainly for two reasons. The first comes from the direct observation of flames: The reactive fronts behave as more-or-less convoluted surfaces separating fresh and burnt gases. The second is related to the discussion in earlier sections of the molecular-diffusion-controlled character of combustion: According to Fick's law, molecular diffusion depends on species gradients developing between isoconcentration surfaces, and the knowledge of the properties of these surfaces is thus a basic ingredient of turbulent combustion modelling to which detailed chemical kinetic information can readily be added.

This section reports on the basis of FSD and field equation or G equation. The interlinks between these approaches and the PDF formalism are also discussed in the context of the introduction of tabulated detailed chemistry in the numerical simulation of flames.

2.2.1 Flame Surface Density

BASIS OF FSD. Scalar surfaces in turbulent flows are non-planar 2D objects evolving in time, which are not easy to comprehend directly: The density of isosurface per unit volume, Σ, was then introduced as an Eulerian quantity [89] defined at every point in the flow as

$$\Sigma(\varphi^*; \mathbf{x}, t) = \lim_{\delta V \to 0} \frac{\mathcal{A}_{\varphi^*}}{\delta V}, \tag{2.42}$$

to denote the limit, for a measuring volume δV that goes to zero, of the ratio between $\mathcal{A}_{\varphi^*}$, the area of the iso-φ^* scalar surface, and the volume δV centred at $\mathbf{x}$.

Considering a small volume $dV_{\varphi^*} = \mathcal{A}_{\varphi^*} d\xi$, built over the $\mathcal{A}_{\varphi^*}$ surface from a displacement $d\xi = \xi^+ - \xi^-$ in the direction normal to the surface and centred on the surface, with $\varphi(\xi^+, t) = \varphi^* + d\varphi/2$ and $\varphi(\xi^-, t) = \varphi^* - d\varphi/2$, the probability of finding $\varphi = \varphi^*$ at x may be written as

$$\overline{P}(\varphi^*; \mathbf{x}, t)d\varphi = \lim_{\delta V \to 0} \frac{dV_{\varphi^*}}{\delta V}, \tag{2.43}$$

where $\overline{P}(\varphi^*; \mathbf{x}, t)$ is the PDF of φ^*. This PDF can be defined here in both RANS [90–92] and LES [93] frameworks. The combination of relations (2.42) and (2.43) yields

$$\Sigma(\varphi^*; \mathbf{x}, t) = \overline{P}(\varphi^*; \mathbf{x}, t)\mathcal{G}(\varphi^*), \tag{2.44}$$

where $\mathcal{G}(\varphi^*) \equiv \left(\overline{d\varphi/d\xi \mid \varphi = \varphi^*}\right)$ is the conditional mean in RANS, or the conditional filtered value in LES, of the scalar gradient estimated for $\varphi = \varphi^*$. This relation may be derived in a more formal way by considering the integral over the volume V of the fine-grained PDF $\delta[\varphi^* - \varphi(\mathbf{x}, t)]$ [90], leading to the Σ definition [94, 95]:

$$\Sigma(\varphi^*; \mathbf{x}, t) = \overline{P}(\varphi^*; \mathbf{x}, t)\mathcal{G}(\varphi^*). \tag{2.45}$$

From here onwards, $\mathcal{G}(\varphi^*) \equiv \left(\overline{|d\varphi/dx_k| \mid \varphi = \varphi^*}\right)$. Relation (2.45) indicates that the FSD is a conditional moment of $\overline{P}(\varphi^*, \gamma^*; \mathbf{x}, t)$, the joint scalar, scalar-gradient PDF:

$$\Sigma(\varphi^*; \mathbf{x}, t) = \int_0^\infty \gamma^* \overline{P}(\varphi^*, \gamma^*; \mathbf{x}, t)d\gamma^*, \tag{2.46}$$

where γ^* is the magnitude of the scalar gradient. With this definition, $\Sigma(\varphi^*; \mathbf{x}, t)$ features similarities with conditional moment closure (CMC) [96, 97], also focussing on the behaviour of thermochemical variables (species concentrations, enthalpy) for given isoscalar values.

To avoid focussing on a single isoscalar surface, a generalised FSD was also introduced by integration through the flame [70]:

$$\overline{\Sigma}(\mathbf{x}, t) = \int_0^1 \Sigma(\varphi^*; \mathbf{x}, t) \mathrm{d}\varphi^* = \overline{|\nabla\varphi|}. \tag{2.47}$$

Surface densities are then all related to scalar gradients and provide direct information on the scalar field topology.

If φ denotes a premixed flame progress variable used to tabulate chemistry [98–101] then the averaged (RANS) or filtered (LES) burning rate $\overline{\dot{\omega}}_\varphi$ required to simulate premixed flame propagation may be written with Eq. (2.45) as

$$\overline{\dot{\omega}}_\varphi = \int_0^1 \dot{\omega}(\varphi^*)\overline{P}(\varphi^*)\mathrm{d}\varphi^* = \int_\epsilon^{1-\epsilon} \dot{\omega}(\varphi^*)\frac{\Sigma(\varphi^*; \mathbf{x}, t)}{\mathcal{G}(\varphi^*)} \mathrm{d}\varphi^*, \tag{2.48}$$

where $\dot{\omega}(\varphi^*)$ denotes the tabulated chemical source that vanishes for $\varphi \to 0$ and $\varphi \to 1$, so that the integral can be completed in the range $\varphi^* \in [\epsilon, 1-\epsilon]$, where ϵ is a small number avoiding the singularity appearing when $|\nabla\varphi| \to 0$. Relation (2.48) shows that the FSD and the conditional mean scalar gradient are fundamental components of burning rates in both RANS and LES.

The surface density evolves in accordance with the scalar equation controlling the iso-φ^*:

$$\frac{\partial\rho\varphi}{\partial t} + \frac{\partial\rho u_k \varphi}{\partial x_k} = \frac{\partial}{\partial x_\ell}\left(\rho\mathcal{D}_\varphi\frac{\partial\varphi}{\partial x_\ell}\right) + \dot{\omega}_\varphi, \tag{2.49}$$

where ρ is the mass density, u is the velocity vector, and $\mathcal{D}_\varphi$ is the molecular-diffusion coefficient of φ assuming Fick's law. In a reference frame attached to the thin-flame front, with ξ as the direction normal to the reaction zone, relation (2.49) becomes

$$\frac{\partial\rho\varphi}{\partial t} + \frac{\partial\rho u_\xi \varphi}{\partial \xi} = \frac{\partial}{\partial \xi}\left(\rho\mathcal{D}_\varphi\frac{\partial\varphi}{\partial \xi}\right) + \dot{\omega}_\varphi - \dot{K}_\varphi, \tag{2.50}$$

where $\dot{K}_\varphi$ is a stretch leakage term that cumulates budgets of fluxes occuring along the flame surface and is thus representative of transverse convection and diffusion resulting from straining and curvature of the flamelet surface. The case of a freely propagating 1-D unstrained premixed flame corresponds to $\dot{K}_\varphi = 0$. For such an unstrained and steadily propagating flame, the integration across the reaction zone of the scalar budget with boundary conditions $\varphi(-\infty) = 0$, $\varphi(+\infty) = 1$, and $\partial\varphi/\partial\xi = 0$ at $\pm\infty$ gives

$$\int_{-\infty}^{+\infty} \dot{\omega}_\varphi(\xi)\mathrm{d}\xi = \rho_u S_L^0, \tag{2.51}$$

where ρ_u is the fresh gas mass density and S_L^0 is the burning velocity. In thin reaction zones, the iso-φ^* surfaces stay almost parallel and Σ weakly depends on φ^*; relations (2.48) and (2.51) provide

$$\overline{\dot{\omega}}_\varphi = \int_\epsilon^{1-\epsilon} \dot{\omega}(\varphi^*)\frac{\Sigma}{\mathcal{G}(\varphi^*)} \mathrm{d}\varphi^* = \Sigma\int_\epsilon^{1-\epsilon} \frac{\dot{\omega}(\varphi^*)}{\mathcal{G}(\varphi^*)} \mathrm{d}\varphi^* = \Sigma\int_{-\infty}^{+\infty} \dot{\omega}_\varphi(\xi) \, \mathrm{d}\xi = \rho_u S_L^0 \Sigma,$$

$$\tag{2.52}$$

where it is assumed that $\mathcal{G}(\varphi) = |\mathrm{d}\varphi/\mathrm{d}\xi|$ because the gradient measured through the laminar flame is a good approximation to the conditional mean gradient in the thin-flame reaction-zone regime. Hence the joint knowledge of the FSD and of the laminar burning velocity S_L^0 is sufficient to get a first approximation of the heat release rate in premixed turbulent combustion, further motivating the use of FSD. Algebraic closures were proposed to express the FSD from resolved quantities [102] or a balance equation for Σ may be solved.

FSD BALANCE EQUATION. The FSD balance equation may be derived from kinematic considerations [94, 103], from surface analysis [82], or from scalar-gradient and PDF equations [65]; in all cases, the scalar budget is written in its propagative form after the introduction of the progression velocity of the isosurface relative to the flow (or displacement speed) $S_d(\mathbf{x}, t)$ to express the absolute velocity of the isoscalar surface as

$$u_k^{\text{Abs}_\varphi} = u_k + S_d n_k, \tag{2.53}$$

with $n_k = -(\partial\varphi/\partial x_k)/|\nabla\varphi|$, the normal to the isoscalar surface, pointing towards small scalar values. It is imperative that the displacement speed and the absolute velocity be for isoscalar surfaces. By use of the absolute velocity and the mass continuity equation, the scalar balance equation reads as

$$\rho\frac{\partial\varphi}{\partial t} + \rho u_k^{\text{Abs}_\varphi}\frac{\partial\varphi}{\partial x_k} = 0, \tag{2.54}$$

leading to

$$\frac{\partial\varphi}{\partial t} + u_k\frac{\partial\varphi}{\partial x_k} = -S_d n_k\frac{\partial\varphi}{\partial x_k} = S_d|\nabla\varphi|. \tag{2.55}$$

This form of the scalar balance equation is further discussed in the subsection on the G equation.

Equating Eqs. (2.49) and (2.55), we obtain

$$S_d = \frac{1}{\rho|\nabla\varphi|}\left[\frac{\partial}{\partial x_k}\left(\rho\mathcal{D}_\varphi\frac{\partial\varphi}{\partial x_k}\right) + \dot\omega_\varphi\right]. \tag{2.56}$$

The diffusive contribution may be decomposed exactly into a planar part to which a curvature correction is added [55, 65, 83, 104]:

$$S_d = \frac{1}{\rho|\nabla\varphi|}\left[n_k n_\ell\frac{\partial}{\partial x_k}\left(\rho\mathcal{D}_\varphi\frac{\partial\varphi}{\partial x_\ell}\right) + \dot\omega_\varphi\right] - \mathcal{D}_\varphi\frac{\partial n_\ell}{\partial x_\ell}. \tag{2.57}$$

Now the corresponding FSD equation reads [56, 70] as

$$\frac{1}{\Sigma}\left[\frac{\partial\Sigma}{\partial t} + \frac{\partial\langle u_k + S_d n_k\rangle_s\,\Sigma}{\partial x_k}\right] = \left\langle\frac{\partial u_k}{\partial x_k} - n_\ell n_k\frac{\partial u_\ell}{\partial x_k}\right\rangle_s + \left\langle S_d\frac{\partial n_k}{\partial x_k}\right\rangle_s. \tag{2.58}$$

In this equation, the surface average in RANS (or filtered surface average in LES) $\langle Q\rangle_s$ is defined as

$$\langle Q\rangle_s = \frac{\left(\overline{Q\mathcal{G}|\varphi^*}\right)}{\left(\overline{\mathcal{G}|\varphi^*}\right)}. \tag{2.59}$$

The right-hand side of Eq. (2.58) is the flame stretch rate, composed of an in-plane strain-rate term $\langle\partial u_k/\partial x_k - n_\ell n_k(\partial u_\ell/\partial x_k)\rangle_s$ and a curvature contribution $\langle S_d(\partial n_k/\partial x_k)\rangle_s$. Numerous closures exist in the literature to express the right-hand side of Eq. (2.58) (see [70] for a detailed review); it is usually decomposed into $\mathcal{T}_1$, a strain-rate term acting at the resolved scales of the simulation (for instance, induced by the mean flow field in RANS or space-filtered field in LES), to which are added $\mathcal{T}_2$, a strain-rate term that is due to the unresolved turbulent motions, and, P, a term representative of flame-area consumption by curvature or flame interaction. Various basic questions arise concerning the introduction of surface-based averaged (or filtered) quantities within usual flow solvers, which are based on Eulerian single-point quantities, a point that was further discussed in [105].

The most recent developments in FSD modelling applicable to premixed combustion are within a LES context [80]. The balance equation for Σ then reads as

$$
\frac{\partial \Sigma}{\partial t} + \frac{\partial u_k \Sigma}{\partial x_k} = \underbrace{\frac{\partial}{\partial x_k}\left(D_\Sigma \sigma_c \frac{\partial \Sigma}{\partial x_k}\right)}_{(i)} + \underbrace{\left(\frac{\partial \hat{\bar{u}}_k}{\partial x_k} - \overline{n_\ell n_k}\frac{\partial \hat{\bar{u}}_\ell}{\partial x_k}\right)\Sigma}_{(ii)} + \underbrace{\frac{\Gamma}{\sigma_c}\frac{u'_{\hat{\Delta}}}{\hat{\Delta}}\Sigma}_{(iii)}
$$

$$
+ \underbrace{S_d\left(\frac{\bar{n}_k}{\partial x_k}\right)\Sigma}_{(iv)} - \underbrace{\frac{\partial S_\varphi \bar{n}_k \Sigma}{\partial x_k}}_{(v)}
$$

$$
+ \underbrace{\beta_c S_L^0 \frac{\varphi^* - \bar{\varphi}}{\bar{\varphi}(1 - \bar{\varphi})}(\Sigma - \Sigma^{\text{lam}})\Sigma}_{(vi)}, \tag{2.60}
$$

where (i) is the turbulent transport of the FSD, i.e., transport by unresolved subgrid scale (SGS) velocity fluctuations, (ii) and (iii) are the resolved and SGS parts of the strain rate, (iv) is the resolved curvature, (v) is the resolved propagation in the direction normal to the flame front, and (vi) is a dissipative term, combining SGS contributions of propagation and curvature. $u'_{\hat{\Delta}}$ is the SGS velocity fluctuation taken at a specific filter scale $\hat{\Delta}$ that is different from the LES filter size Δ used for the momentum equations to account for the specific character of the response of the flame surface to the turbulent eddies cascade [106]. D_Σ, σ_c, Γ, β_c, and Σ^{lam} are model parameters, some of them are dynamically computed from the resolved fields; details may be found in [80]. One of the major advantages of Eq. (2.60) lies in the fact that it features a correct asymptotic behaviour if applied to laminar flames; when the turbulent velocity fluctuations vanish, laminar flame speed is properly reproduced by Eq. (2.60), a point that may be of importance in LES to describe the stabilisation of reaction zones through laminar flame propagation.

2.2.2 The G Equation for Laminar and Corrugated Turbulent Flames

The FSD and G-equation approaches both describe premixed flames as an interface between a fresh mixture and burned gases. This geometrical description is closely related to the flamelet assumption in the sense that, once the interface is defined as a certain isosurface of temperature or species concentration in the flame, all the flamelets may be parameterised by a distance to this interface. This concept can in fact be generalised as long as the scale of the flame is much smaller than the flow characteristic length scales, as for instance in large-scale fire. This allows a proper decoupling of the flame–turbulence interactions from the chemistry.

For laminar and corrugated turbulent flames with constant composition and constant thermodynamic state of the fresh mixture, the flame modelling is reduced to the transport of the local FSD, or equivalently to tracking the interface position and its evolution. In the context of the so-called G-equation model, the interface is defined as a surface $G = G_0$ of a level-set scalar G. This model was originally introduced in [69] to illustrate the relationship between the wrinkling of a premixed turbulent flame and its turbulent flame speed in the corrugated-flamelets regime.

It must be kept in mind that the level set scalar G has a significance only at the $G = G_0$ surface. Elsewhere the level-set scalar may take any value. Nevertheless, this implicit definition of the flame-front position imposes a requirement to solve the entire G field to achieve the transport of the interface. For numerical reasons, the G scalar is often chosen as the signed distance to the $G = G_0$ surface, which implies that $G_0 = 0$. Additionally, the usual convention is to set $G > 0$ in the burned gases and $G < 0$ in the fresh mixture. Then the transport equation for the level-set scalar G is simply derived from the temperature or any reaction progress variable with particular attention paid to the velocity of the G_0 surface.

As observed in relation (2.55), from a kinematic standpoint, the balance between the diffusion and the reaction terms may be interpreted as an intrinsic displacement speed of the interface, and relation (2.57) decomposes the diffusion contribution into its normal and tangential parts. If the flame is unstrained, the normal part of ρS_d, combined with the chemical source, is equal to the laminar burning velocity S_L^0 multiplied by the density of the fresh mixture ρ_u. The second contribution on the right-hand side of Eq. (2.57) may be positive or negative and represents the effect of the curvature of the flame front on the displacement speed. This contribution is noted as $\rho_u S_{\kappa,L} = -\rho \mathcal{D}_\varphi (\partial n_k / \partial x_k)$ or $\rho_u S_{\kappa,L} = -\rho \mathcal{D}_\varphi \kappa$, where $\kappa = \partial n_k / \partial x_k$ is the curvature of the flame front. From this analysis the governing balance equation becomes

$$\frac{\partial \rho \varphi}{\partial t} + \frac{\partial \rho u_k \varphi}{\partial x_k} = -\rho_u (S_L^0 + S_{\kappa,L}) n_k \frac{\partial \varphi}{\partial x_k} \qquad (2.61)$$

or, in non-conservative form,

$$\frac{\partial \varphi}{\partial t} + u_k \frac{\partial \varphi}{\partial x_k} = -\frac{\rho_u}{\rho} (S_L^0 + S_{\kappa,L}) n_k \frac{\partial \varphi}{\partial x_k}. \qquad (2.62)$$

The transport equation of the level-set G is then derived from progress-variable Eq. (2.62) by means of a change of variables $\partial G = \partial \varphi / |\nabla \varphi|$:

$$\frac{\partial G}{\partial t} + u_k \frac{\partial G}{\partial x_k} = -\frac{\rho_u}{\rho} (S_L^0 + S_{\kappa,L}) n_k \frac{\partial G}{\partial x_k}. \qquad (2.63)$$

This transport equation for the level-set G clearly exhibits the different contributions to the flame-front velocity, namely the flow velocity u, the laminar burning velocity S_L^0, and the curvature-induced velocity $S_{\kappa,L}$. This transport equation is valid for laminar flames and turbulent flames in the corrugated-flamelets regime [55]. For other regimes, or if the wrinkling of the flame cannot be resolved, some additional modelling is required.

In the wrinkled- and corrugated-flamelets regimes, the smallest turbulent scales are larger than the flame thickness and the turbulent eddies may wrinkle only the flame front [55]. In the thin-reaction-zones regime, the smallest eddies may enter the preheating zone and thicken the flame brush, but these eddies do not interact with the reaction zone, which is an order of magnitude smaller than the flame thickness. In the previous regimes, the fresh mixture and the burned gases are still separated by the inner layer, where the reactions occur. The flamelet assumption is therefore valid at the scale of the flame thickness and may be used for further modelling. This is not the case in the broken-reaction-zones regime, where the turbulent scales may quench the flame and lead to local extinctions. Then the flamelet description is not valid anymore because the gradients along the flame front are large. Fortunately this combustion regime is of limited interest for practical applications because the local extinctions produce many pollutants such as unburned hydrocarbons and soot, and they strongly affect the conversion rate of the burner.

In the other combustion regimes, a G-equation model may be derived from an averaged progress variable in a RANS simulation or from a filtered progress variable in a LES following the procedure just detailed. In both cases, the G-equation model for turbulent premixed combustion takes the form

$$\frac{\partial \check{G}}{\partial t} + \tilde{u}_k \frac{\partial \check{G}}{\partial x_k} = -\frac{\rho_u}{\overline{\rho}}(S_T + S_{\kappa,T})\check{n}_k \frac{\partial \check{G}}{\partial x_k}, \qquad (2.64)$$

where $\tilde{u}$ denotes the Favre-averaged or filtered velocity, $\overline{\rho}$ is the Reynolds-averaged or filtered density, $\check{n}$ is the unit normal of the mean or filtered flame front pointing towards the unburned gases, and S_T and $S_{\kappa,T}$ are the intrinsic burning velocity of the mean or filtered flame front and the increase or decrease of the burning velocity that is due to resolved curvature, respectively. The mean or filtered flame front is implicitly defined as the $\check{G} = \check{G}_0$ isosurface. In this equation, the unclosed terms that require a substantial modelling effort are the turbulent burning speeds S_T and $S_{\kappa,T}$.

The derivation of analytical closures for the turbulent burning speed S_T has been a very active field since the pioneering work of Damköhler [54]. Detailed reviews may be found in [55, 108]. One commonly used closure takes the following form:

$$\frac{S_T - S_L^0}{S_L^0} = C\left(\frac{u_{\mathrm{rms}}}{S_L^0}\right)^m, \qquad (2.65)$$

where u_{rms} is the turbulence rms velocity and C is a constant. The power m was found experimentally to be close to 0.7. Nevertheless, this expression fails to agree with experimental data in all of the combustion regimes. For this reason, similar expressions were derived for both RANS simulation [55] and LES [109], respectively, and take the form

$$\frac{S_T - S_L^0}{S_L^0} = -\frac{b_3^2}{2b_1}\frac{\mathcal{D}_T}{\mathcal{D}}\frac{S_L^0}{u'} + \sqrt{\left(\frac{b_3^2}{2b_1}\frac{\mathcal{D}_T}{\mathcal{D}}\frac{S_L^0}{u'}\right) + b_3^2\frac{\mathcal{D}_T}{\mathcal{D}}}, \qquad (2.66)$$

where the constants are $b_3 = 1.0$ and $b_1 = 2.0$. In a RANS simulation, $\mathcal{D}_T$ is the progress-variable turbulent diffusivity. In LES, $\mathcal{D}_T$ is also the progress-variable turbulent diffusivity used to represent unresolved SGS turbulent transport, but u' is equal to the turbulent rms velocity at the filter scale u'_Δ. This expression satisfies two important asymptotic behaviors: In the large-Damköhler-number limit, the turbulent burning velocity S_T scales linearly with the turbulence rms velocity u', and in the small-Damköhler-number limit, S_T scales with $u'\sqrt{\mathrm{Da}}$. Other S_T expressions have been discussed in slightly different contexts, with algebraic closures [110] or from an efficiency function used to fine-tune flame thickening in LES [106] or again from a power-law flame-wrinkling model [78, 79].

Concerning the resolved curvature contribution $S_{\kappa,T}$, a closure similar to the laminar case was proposed [55], simply accounting for the turbulent diffusivity:

$$S_{\kappa,T} = -\overline{\rho}\,(\mathcal{D} + \mathcal{D}_T)\,\frac{\partial n_k}{\partial x_k}.\tag{2.67}$$

This expression is also used in LES [109] if the Damköhler number $\mathrm{Da}_\Delta = S_L^0 \Delta/(u'_\Delta l_F)$ is lower than unity, where l_F is the reference laminar thermal flame thickness. It was then argued that the turbulent curvature contribution should vanish in the large-Damköhler-number limit and proposed for $\mathrm{Da}_\Delta > 1$:

$$S_{\kappa,T} = -\overline{\rho}\left(\mathcal{D} + \mathcal{D}_T \mathrm{Da}_\Delta^{-2}\right)\frac{\partial n_k}{\partial x_k}.\tag{2.68}$$

Other expressions for the turbulent burning speed may be derived from higher-order transport equations [55, 111] or from the FSD, for instance, alternative approaches for determining S_T are also discussed in Subsection 2.3.6. The G-equation and FSD models share the same geometrical description of the turbulent flame brush, and many models may be transposable from one framework to the other.

In a RANS simulation or a LES of turbulent premixed flames, the G equation provides only the position of the mean or filtered flame brush and its evolution. Nevertheless, additional terms in the Navier–Stokes equations also have to be closed. In the low-Mach-number limit, the density has to be prescribed everywhere in the computational domain. Otherwise, in the generic framework of compressible flows, the heat release in the energy or enthalpy equation and the source term of any progress variable or reactive species must be modelled. These necessary closures have to be derived from the $\check{G}$ field to ensure the consistency of the model.

Several approaches exist to evaluate the density or the source terms. The first one is a purely kinematic approach, in which the $\check{G} = \check{G}_0$ isosurface is considered to be a sharp interface between the fresh mixture and the burned gases [75, 76, 112]. Consequently the density or any variable involved in the combustion process is constant on each side of the level set and discontinuous at the interface $\check{G} = \check{G}_0$. Specific numerical methods may be used in this case to deal with the discontinuity, like the ghost-fluid method [113]. Alternatively, the temperature or density field may be filtered on the computational mesh to avoid any numerical difficulty, as is also done in the thickened-flame model [106]. In the G-equation context, this latter method usually requires many points in the flame front. Because the thickness of the flame-brush is of primary importance to capture the flame–turbulence interactions in LES, a model was designed based on the ghost-fluid method featuring an arbitrary thickness of the flame brush [114]. This thickness becomes an input to the closure and may be obtained from analytical expressions [55]. Following the same kinematic approach, the heat release and the progress-variable source term may be evaluated from the volume swept by the level set because of the burning speed [115]. However, none of these methods gives a consistent modelling of the flame-brush thickness. Moreover, the coupling with a compressible solver for the computation of combustion instabilities, for instance, is difficult to perform.

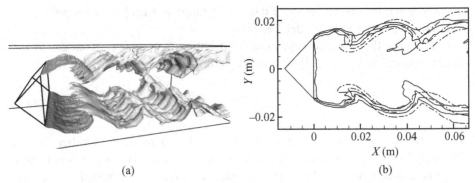

Figure 2.9. (a) LES: snapshot of the flame surface; (b) comparison of experimental (solid lines) [107] and LES (dot–dashed lines) progress-variable isocontours. The values of the isocontours are 0.1, 0.5, and 0.9. The 0.1 isocontours are close to the channel walls and the 0.9 isocontours are close to the midplane, from [81].

These two last issues were recently addressed in a LES context [81] with a coupled G-equation–progress-variable model. The G equation is transported to ensure the correct propagation speed of the flame brush, and the progress variable provides the species profiles in the flame. The progress-variable source term is modeled from the $\check{G}$ field so that the progress-variable propagation speed is always consistent with the $\check{G}$ field. Figure 2.9 shows the application of the coupled G-equation–progress-variable strategy to a turbulent premixed flame anchored by a triangular flame holder. The stoichiometric premixed propane–air flame under atmospheric pressure develops downstream of the triangular body [107], the flame surface interacts with the rolling up of the shear layers, and the progress-variable isocontours are well reproduced by the simulation [Fig. 2.9(b)].

2.2.3 Detailed Chemistry Modelling with FSD

CHEMISTRY TABULATION. Detailed chemistry tabulation is widely used for the simulation of turbulent flames; this tabulation may be completed before the burner is simulated to drastically reduce CPU (i.e., CPU, central processing unit) time or in situ during the simulation. In both cases, the detailed chemistry tabulation may be conducted from the generic chemical point of view [116, 117], thus neglecting molecular-diffusion effects or considering canonical flame configurations to directly include diffusion of species and heat, for instance, to mimic local flamelet behaviour [55, 99, 100, 118–123], a methodology that was in fact first introduced for RANS computations of diffusion [124] and premixed [98] flames. These flamelet methods have been successfully used in RANS simulations and LESs of turbulent flames with steady premixed or non-premixed (diffusion) generated manifolds (i.e., composition space trajectories tabulating chemistry from a few coordinates [63, 125–127]), along with unsteady diffusion flamelets [109, 128]. The LES of the forced ignition of an annular bluff-body burner was conducted using such tabulated chemistry [129]; the objective was to estimate the prediction capabilities of the LES in a fully transient phenomenon. Figure 2.10 shows snapshots of 1 of the 20 ignition sequences that were simulated by varying the spark position within the mixing zones, following a preceding experimental study [130]. The LES observation agrees with experiment,

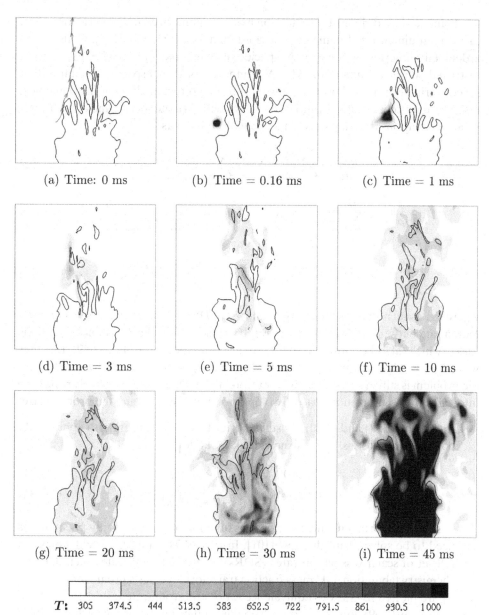

Figure 2.10. LES-resolved instantaneous snapshots of temperature after sparking in a bluff-body burner, from [129]. Solid black line: Isoline of stoichiometric mixture fraction. An image covers a domain dimension of 70×70 mm; the temperature scale saturates at 1000 K to favour low temperature levels. The flow streamline is shown at sparking position in (a).

and comparisons with available measurements show that strain-rate effects need to be included in the modelling of the filtered burning rate; a correction to the usual presumed PDF modelling associated with tabulated chemistry was then discussed and validated to address this point [129], and this correction makes use of the FSD concept suggesting that coupling PDF with FSD is an interesting option to improve SGS modeling.

Flamelet-based chemistry tabulation may be summarized in a generic formalism called multidimensional flamelet-generated manifolds (MFMs) [101]. Considering a detailed chemistry scheme of N species of mass fractions Y_i ($i = 1, \ldots, N$), a restricted subset of dimension $M < N$ composed of φ_j subspace (or composition space) variables is first chosen, for instance from a direct analysis of the dynamical response of the chemical system. The exact projection of the species and temperature balance equations into this subset may be written [101] as

$$\rho \frac{\partial Y_i}{\partial \tau} + \sum_{j=1}^{M} \frac{\partial Y_i}{\partial \varphi_j} \dot{\omega}_j = \sum_{j=1}^{M} \sum_{k=1}^{M} \frac{\rho \chi_{jk}}{\mathrm{Le}_i} \frac{\partial^2 Y_i}{\partial \varphi_j \partial \varphi_k} + \dot{\omega}_i, \tag{2.69}$$

$$\rho \frac{\partial T}{\partial \tau} + \sum_{j=1}^{M} \frac{\partial T}{\partial \varphi_j} \left[\dot{\omega}_j - \sum_{k=1}^{M} \frac{\rho \chi_{jk}}{c_p} \left(\frac{\partial c_p}{\partial \varphi_k} + \sum_{i=1}^{N} \frac{c_{P_i}}{\mathrm{Le}_i} \frac{\partial Y_i}{\partial \varphi_k} \right) \right]$$

$$= \sum_{j=1}^{M} \sum_{k=1}^{M} \rho \chi_{jk} \frac{\partial^2 T}{\partial \varphi_j \partial \varphi_k} - \frac{1}{c_p} \sum_{i=1}^{N} h_i \dot{\omega}_i, \tag{2.70}$$

where the usual notations are adopted and $\chi_{ij} = \mathcal{D} \nabla \varphi_i \cdot \nabla \varphi_j$ is the cross-scalar dissipation rate between the scalars φ_i and φ_j. If one solves for the M balance equations for φ_j, which take the form of the usual species equations as in Eq. (2.49), together with $N + 1 - M$ equations in φ_j space for Y_i and temperature [Eqs. (2.69) and (2.70)], the problem is still of size $N + 1$ and therefore not yet reduced, but exactly projected into a given φ_j composition space. Two options allow for reducing the problem size:

- Equations (2.69) and (2.70) are solved prior to the simulation, with appropriate boundary conditions, to construct a look-up table,

$$Y_i = Y_i^{\mathrm{MFM}}(\varphi_1, \ldots, \varphi_M, \tau; \chi), \tag{2.71}$$

$$T = T^{\mathrm{MFM}}(\varphi_1, \ldots, \varphi_M, \tau; \chi), \tag{2.72}$$

to store the thermochemical response as function of φ_j so that only M equations need to be solved with the flow [101]. In Eqs. (2.71) and (2.72), χ denotes the full set of scalar dissipation rates (SDRs). Figure 2.11 illustrates such detailed chemistry tabulation, OH mass fraction trajectories versus a progress of reaction Y_c are plotted for various equivalence ratios and strain-rate levels.
- Equations (2.69) and (2.70) are solved during the simulation, according to local flow conditions, but for a reduced number of dimensions M. These equations are usually easier to solve numerically than the primitive species equations defined in physical space, because the projection into the composition space φ_j naturally acts as a zoom into the thin-flame zone, hence reducing the resolution requirement [55].

In both options for $M = 1$, if φ_1 is a reactive species (passive scalar) usual premixed (diffusion) flamelet modelling is recovered. The situation with $M > 1$ allows for improving the description of chemistry in lean premixed combustion modelling or for addressing hybrid combustion regimes, as in stratified or partially premixed flames [101]. The first option is quite easy to couple with flow solvers, but may lack generic character because the boundary condition of the flamelets are fixed once and for all, as for instance pure fuel and pure air. The second option is more CPU

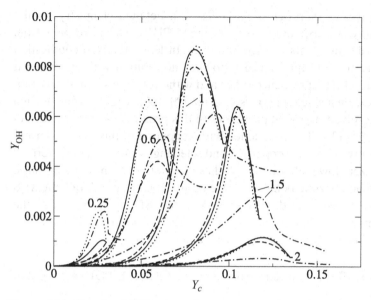

Figure 2.11. Tabulated OH trajectories vs. progress of reaction using a 2D composition space ($M = 2$, mixture fraction and reaction progress), from [101]. Equivalence ratios are given in the graph, and various progresses of reaction scalar-dissipation-rate levels are considered, Dotted, 1 s^{-1}; solid, 5 s^{-1}; dashed, 20 s^{-1}; dot–dashed, 1D diffusion flamelet response for comparison (scalar dissipation rate of 40 s^{-1}).

time consuming, but potentially more precise because the boundary conditions used to solve (2.69) and (2.70) can be constructed from the local flow conditions feeding the reaction zones. In the first option, the self-similar behaviour of flames is of great help to reduce the size of the chemical database [131].

With the rapidly growing computer capabilities, modern simulations of real burners involving many billions of computational cells will soon feature a mesh resolution of about 100 μm, which is not sufficient to resolve the intermediate radical-species signals, but enough to fully capture the φ_j signal based on major species (CO_2, CO, H_2O), specifically in the lean premixed combustion regime. Therefore all ingredients of the tabulation (φ_j, χ) will be fully resolved in these coming and advanced simulations, and no submodel will be needed for simulating the control parameters of the chemical table. In this future context, turbulent combustion modelling will be reduced to the building of accurate chemistry reduction, for instance by means of Eqs. (2.69) and (2.70). These highly refined simulations may become the standard tool for engineering design within the next decade, following the rapid development of massively parallel computing, but meanwhile, there is still room for the modelling of unresolved fluctuations of flame properties to conduct simulations of many turbulent combustion applications. In this framework, both FSD and the G equation are useful tools to couple with a given detailed chemistry look-up table.

2.2.4 FSD as a PDF Ingredient

FSD provides information on flame-front position and flame wrinkling. This information on flame topology can be introduced in coarse-grid simulations to presume

the unresolved properties of progress variables of a detailed chemistry tabulation [41, 132], usually approximated from presumed PDFs, which are built from
first- and second-order moments of their variables. In lean premixed combustion
without any stratification of equivalence ratio (i.e., no dilution with air or burnt
products having a different equivalence ratio) and without heat loss, chemistry tabulation may be based on a single progress variable [99, 100]. The challenge is then
to estimate the unresolved statistical properties of this progress variable as its SGS
variance, $\varphi_v = \overline{\varphi^2} - \overline{\varphi}^2$ [132]. The production and dissipation of this SGS scalar energy are highly sensitive to flame properties, and selecting closures that are derived
from chemically frozen flow scalar-mixing studies may not be the best compromise
for capturing thin reaction zone behaviors. One option consists of relating the SGS
variance to the FSD, to estimate the variance from the FSD equation, so that the
topology of the flame front enters the modelling. The shape of the PDF is then
presumed from the knowledge of $\overline{\varphi}$ and $\overline{\Sigma}$, instead of $\overline{\varphi}$ and φ_v.

DEFINITION OF Σ-PDF. For a thin flame, the premixed flame Σ-PDF may be cast
as [41, 132]

$$\overline{P}(\varphi^*; \mathbf{x}, t) = \alpha(\mathbf{x}, t)\delta(\varphi^*) + \beta(\mathbf{x}, t)\delta(1 - \varphi^*) + \frac{\overline{\Sigma}(\mathbf{x}, t)}{\mathcal{G}(\varphi^*)}H(\varphi^*)H(1 - \varphi^*), \quad (2.73)$$

where H denotes the Heaviside function. This PDF must satisfy

$$\int_0^1 \overline{P}(\varphi^*; \mathbf{x}, t)\mathrm{d}\varphi^* = 1; \quad \int_0^1 \varphi^* \overline{P}(\varphi^*; \mathbf{x}, t)\mathrm{d}\varphi^* = \overline{\varphi}; \quad \int_0^1 \varphi^{*2}\overline{P}(\varphi^*; \mathbf{x}, t)\mathrm{d}\varphi^* = \overline{\varphi^2}. \quad (2.74)$$

With Eq. (2.73), the constraint on the square of the progress variable reads [132] as

$$\overline{\Sigma} = \frac{\varphi_\varphi}{\delta_\varphi}, \quad (2.75)$$

where φ_φ is the difference between the maximum variance level allowed and the
variance φ_v [132],

$$\varphi_\varphi = \overline{\varphi}(1 - \overline{\varphi}) - \varphi_v = \overline{\varphi} - \overline{\varphi^2}, \quad (2.76)$$

and δ_φ is a characteristic flame-length scale defined as

$$\delta_\varphi = \int_\epsilon^{1-\epsilon} \frac{\varphi^*(1 - \varphi^*)}{\mathcal{G}(\varphi^*)}\mathrm{d}\varphi^*. \quad (2.77)$$

Assuming that combustion occurs in the thin-flame-front regime, $\mathcal{G}(\varphi^*)$ may be
computed from the chemistry tabulation based on premixed flamelets. The PDF
second-order moment is then deduced from the FSD with relation (2.75). The α and
β coefficients of (2.73) are conveniently expressed from $\overline{\varphi}$ and $\overline{\Sigma}$ [132] as

$$\alpha = (1 - \overline{\varphi}) - \overline{\Sigma} \int_\epsilon^{1-\epsilon} \frac{1 - \varphi^*}{\mathcal{G}(\varphi^*)}\mathrm{d}\varphi^*, \quad (2.78)$$

$$\beta = \overline{\varphi} - \overline{\Sigma} \int_\epsilon^{1-\epsilon} \frac{\varphi^*}{\mathcal{G}(\varphi^*)}\mathrm{d}\varphi^*, \quad (2.79)$$

respectively.

The FSD may be obtained by the solution of Eq. (2.60), thus ensuring a full coupling between FSD and detailed chemistry tabulation, the species mass fraction being approximated as

$$\overline{Y}_i(\mathbf{x}, t) = \int_0^1 Y_i^{\text{MFM}}(\varphi^*)\overline{P}(\varphi^*; \mathbf{x}, t)d\varphi^*, \tag{2.80}$$

where $Y_i^{\text{MFM}}(\varphi^*)$ denotes the low-dimensional manifold trajectory relating the progress variable to the ith species and obtained with Eqs. (2.69) and (2.70). The progress-variable source term is, however, preferably obtained from relation (2.52) to ensure correct flame propagation [133].

With this simple procedure, the FSD modeling is enriched with the possibility of addressing pollutants emissions, and the presumed PDF approach is improved because it accounts for the existence of a thin-flame and propagating-flame interface.

RELATION BETWEEN SGS VARIANCE AND SCALAR GRADIENT. The specific premixed-type Σ-PDF is useful to explore SGS scalar variance behaviour. Relation (2.75) suggests that

$$\varphi_v = \overline{\varphi}(1 - \overline{\varphi}) - \delta_\varphi \overline{\Sigma}$$
$$= \overline{\varphi}(1 - \overline{\varphi}) - \delta_\varphi \Xi |\nabla\overline{\varphi}|, \tag{2.81}$$

where the generalized FSD $\overline{\Sigma} = \overline{|\nabla\varphi|} = \Xi |\nabla\overline{\varphi}|$ is expressed with the magnitude of the resolved gradient, thanks to the wrinkling factor Ξ.

Experimental results have shown that the following relation holds [134]:

$$|\nabla\overline{\varphi}| = \frac{1}{C_s} \frac{\overline{\varphi}(1 - \overline{\varphi})}{\Delta}, \tag{2.82}$$

where C_s is a model constant and Δ is the characteristic length of the LES filter. From the previous relations, the link between the SGS scalar variance and the resolved gradient reads

$$\varphi_v \propto \left(C_s - \frac{\delta_\varphi \Xi}{\Delta}\right) \Delta |\nabla\overline{\varphi}|. \tag{2.83}$$

The SGS variance then scales as $\Delta |\nabla\overline{\varphi}|$, a relation that is in line with results from experimental measurements [134] and the method of manufactured solution (MMS) [135], which reported that the time average of the SGS variance was indeed scaling as

$$\langle\varphi_v\rangle \propto \langle\Delta|\nabla\overline{\varphi}|\rangle, \tag{2.84}$$

where Δ is the LES filter size and the angle brackets $\langle\cdot\rangle$ denote time averaging. Notice that this scaling differs from the usual one obtained from a chemically frozen flow production–dissipation equilibrium hypothesis of the SGS variance [132, 136], which leads to the quadratic law $\langle\varphi_v\rangle \simeq \langle(\Delta|\nabla\overline{\varphi}|)^2\rangle$, but was not observed after space filtering of both experiments [134] and MMS [135].

2.2.5 Conclusion

Turbulent premixed flames are easily identified as surfaces separating fresh and burnt gases. The FSD and field equation (or *G* equation) were developed to collect information on the behaviour of a given isoscalar surface, or a collection of isosurfaces in the case of the generalized FSD. These quantities provide a generic framework for developing flame modelling that reproduces basic premixed flame properties as the response of their propagation speed and characteristic thickness to various chemical and turbulence parameters.

In future highly resolved simulations, aside from the mandatory accurate numerics and high performance of the computing tools, it is mostly detailed chemistry tabulation that will actually be important because of it ability to include all effects related to the complex interaction between reactants and their local environment before they enter the reaction zone (viz., dilution by hot burnt gases, or mixing with vitiated air). Within this context, multidimensional flamelets appear as a promising approach to incorporate more physics into the chemical look-up table, which will then be more able to reproduce complex flame features observed in the forthcoming lean premixed flame burners developed to reduce pollutant emissions.

In simulations using coarser meshes, the modelling of unresolved fluctuations of flame parameters is still the main concern. There, the modelling of thin reaction zones from FSD or the *G* equation is of great help to complete closures reproducing experimentally observed turbulent flame properties.

2.3 Scalar-Dissipation-Rate Approach
By N. Chakraborty, M. Champion, A. Mura, and N. Swaminathan

The objective of turbulent combustion modelling is to seek a way to express the averaged reaction rate and reactive scalar fluxes in terms of known quantities such as turbulence length and time scales, usually of the turbulence integral scales, as has been noted in Chapter 1. The exact form of the relationship depends on the modelling approach. In this section, we discuss the mean-reaction-rate modelling using the SDR and its relationship to other methods presented in this chapter.

The SDR is an important quantity in turbulent combustion. This quantity denotes the mixing rate of fuel and oxidiser in turbulent non-premixed flames, whereas it denotes the rate of mixing of hot and cold fluids on the flame surface required to sustain combustion in turbulent premixed flames. The instantaneous SDR is defined as $N_c = \hat{\alpha} (\partial c / \partial x_k)(\partial c / \partial x_k)$, where c is a reaction progress variable with diffusivity $\hat{\alpha}$. It is clear that the gradient of the reaction progress variable and thus the physical processes affecting the gradient $\partial c / \partial x_k$ will influence the SDR. This will be elucidated based on strong theoretical foundations later in this section using data from physical and numerical experiments on turbulent premixed flames. The SDR is directly related to the heat release rate in premixed [21, 62] as well as in non-premixed [124, 137] flames. A wealth of information on the SDR of mixture fraction, which is a chemically conserved scalar (see Section 1.3.2), is available in the literature [70, 138, 139]. However, the information on N_c is limited, but it is growing. One of the aims of this section is to bring the available information together to show and discuss the role of the SDR approach and its success in recent studies [140–145].

The progress variable c may be defined using any reactive scalars; a temperature or fuel mass fraction is commonly used, but alternative choices [146] are also available. The instantaneous c, irrespective of how it is defined, is governed by

$$\rho \frac{Dc}{Dt} = \dot{\omega}_c + \mathcal{D}, \tag{2.85}$$

where ρ is the fluid density and D/Dt is the total or substantial derivative defined by Eqs. (1.1) in Chapter 1. The first term on the right-hand side, $\dot{\omega}_c$, is the chemical source term, and the second term represents the diffusive flux of c, $\mathcal{D} \equiv \partial[\rho \hat{\alpha} \partial c/\partial x_k]/\partial x_k$. Here, as in the earlier sections, the repeated indices imply summation over them. In RANS simulations (the current workhorse for engineering calculations), a balance equation [see Eq. (1.29)] for the Favre-averaged progress variable $\tilde{c}$ is solved along with models for the mean reaction rate $\overline{\omega}_c$ and the turbulent flux $\widetilde{u_k'' c''}$. In the LES approach, the SGS reaction rate and scalar flux are to be modelled (see Chapter 1). In the following discussion, the primary focus is on the RANS modelling of the SDR and its relation to mean-reaction-rate closure. The LES modelling of the SDR is discussed briefly in a later part of this section.

The challenging problem of mean-reaction-rate closure can be simplified to a tractable form if one assumes the flame scales to be smaller than typical fluid dynamic (Kolmogorov) scales of relevance [147]. This assumption implies high-Damköhler-number or thin-flamelets combustion (see Section 1.6) and allows the PDF of c to be approximated as a double-delta function, representing the unburnt and burnt mixtures [148]. This approximation leads to many attractive simplifications, as noted in Section 2.1. A couple of relevant points for our discussion here are (1) the progress-variable variance is given by $\widetilde{c''^2} = g\,\tilde{c}\,(1 - \tilde{c})$ and g is equal to one when the combustion occurs in thin flamelets and (2) the mean reaction rate in Eq. (1.36) is rewritten here for convenience as

$$\overline{\omega}_c = \frac{2}{2C_m - 1}\,\overline{\rho}\,\tilde{\epsilon}_c. \tag{2.86}$$

The Favre-averaged SDR $\tilde{\epsilon}_c$ is defined as $\tilde{\epsilon}_c \equiv \overline{\rho \hat{\alpha} (\partial c''/\partial x_k)(\partial c''/\partial x_k)}/\overline{\rho}$, and it is one half of the dissipation rate of the progress-variable variance. Modelling of this dissipation rate with the turbulent time scale alone is insufficient for premixed flames [142, 149, 150] because of strong coupling among turbulence, chemical, and molecular-diffusion processes in premixed flames, and an SDR-based approach can capture this coupling naturally as one will see later in this section.

The mean-reaction-rate model in Eq. (2.86) is for situations in which the turbulent eddies do not enter the reaction zone and they create only large-scale wrinkling [124, 151]. When the small length and time scales of turbulence are comparable to flame scales, then the small-scale turbulent eddies can enter the reaction zone and disturb it, as noted in Section 1.6. For these situations, the burning mode part of the BM PDF will be non-negligible, and it must be included in the analysis. Bray et al. [41] showed how to account for these effects by including an additional variable that quantifies the deviation from the situation of thin flamelets. This additional variable is related to the Favre variance of the progress variable. Bradley [152] suggested modelling the mean reaction rate by

$$\overline{\dot{\omega}_c} = \int \int \dot{\omega}_c \left(\zeta, \psi \right) \overline{P} \left(\zeta, \psi \right) \, \mathrm{d}\zeta \, \mathrm{d}\psi, \tag{2.87}$$

where ζ and ψ are the sample space variables for c and the stretch rate κ induced by the turbulent eddies, respectively. The JPDF $\overline{P} \left(\zeta, \psi \right)$ is obtained by use of presumed forms for the marginal PDFs $\overline{P}_c(\zeta)$ and $\overline{P}_\kappa(\psi)$ and treating them as statistically independent. Details of these two approaches are discussed in Section 2.1. The relevant point for our discussion here is that either of these approaches requires a solution to the Favre variance equation, given by Eq. (1.34), which requires a model for the mean SDR, $\widetilde{\epsilon}_c$. The reaction-rate contribution, the first term on the right-hand side of Eq. (1.34), can be closed exactly by use of

$$\overline{c''\dot{\omega}_c''} = \left(C_m - \widetilde{c} \right) \overline{\dot{\omega}_c}, \tag{2.88}$$

and the last two terms are negligible in large-Reynolds-number premixed flames. The models for the turbulent fluxes are discussed in Sections 1.7 and 2.1.

If one chooses to avoid the flamelets description to model lean premixed combustion and proposes to employ alternative methods such as transported PDFs [95] or CMC [96], then again a model for the SDR is required, but the SDR must be conditional on the reaction progress variable. Despite some attempts in this direction [153], it is to be noted that these two approaches have not evolved to the same level as for the non-premixed flames primarily because of issues surrounding the (conditional) SDR modelling. Recent developments in the JPDFs approach are discussed in Section 2.4.

Few experimental studies [154–157] have been carried out to explain the behaviour of N_c. These studies, except [157], show that the values of N_c in turbulent premixed flames are low compared with unstrained laminar flame values. Analysis [158] of results from DNS shows an increase in N_c compared with the laminar flame values. A reconciliation of these differences is not attempted here, but it is hoped that the concepts and ideas provided here can help such an investigation.

2.3.1 Interlinks among SDR, FSD, and Mean Reaction Rate

Equation (2.86) clearly shows the direct relationship between the mean reaction rate and the averaged SDR when the combustion occurs in thin flamelets. Also, for this situation the average reaction rate may be modelled [147] as the product of mean reaction rate per unit flame surface area and the FSD, Σ. Thus $\overline{\dot{\omega}_c} = \left(\rho_u S_L^0 I_0 \right) \Sigma$, where ρ_u is the unburnt mixture density, S_L^0 is the unstretched laminar burning velocity, and I_0 represents the effect of flame stretch on the laminar burning velocity. This effect is captured by means of a correlation involving Karlovitz and Markstein numbers [159]. It is apparent from the preceding expression that the amplification of reaction rate by turbulence is predominantly represented by Σ.

Various approaches have been suggested [70] to model Σ. The BML approach in Section 2.1 yields an algebraic expression for Σ. In an another approach discussed in detail in Section 2.2, a balance equation for Σ is solved. This balance equation, postulated by Marble and Broadwell [89], was rigorously derived by Pope [94] and Candel and Poinsot [103]. The FSD is related to the SDR by $\widetilde{\epsilon}_c = K_\Sigma S_L \Sigma$, where K_Σ is a constant [160]. Pope [95] deduced that the SDF of an isosurface $c = \zeta$ is

$\Sigma_p(\zeta; \mathbf{x}, t) = \langle \sqrt{N/\hat{\alpha}} | c = \zeta \rangle \, \overline{P}_c(\zeta)$ without invoking the assumption of thin flame. There should not, however, be any local extinction on the surface $c = \zeta$ as this would invalidate the continuity of c on the particular surface. The quantity within the angle brackets denotes conditional average. The FSD Σ can be obtained from the SDF Σ_p either by choosing a particular value of ζ or by averaging $\overline{\Sigma} = \int \Sigma_p \, d\zeta$ [70]. Under the assumption of thin flamelets, one gets $\Sigma \simeq \overline{\Sigma}$ because the FSD evaluation is not sensitive to the choice of the c isosurface. In the preceding discussion, one sees that the SDR is closely related to the FSD and to the reaction rate. The dissipation-rate-based and FSD-based approaches are closely related to each other, and these methods have their own merits.

The relevance of SDR for turbulent combustion modelling being established from the preceding discussion, we now introduce the background works on this topic. In the forthcoming sections, the transport equation for the SDR is derived first. The analysis of this equation is taken up next in the order of increasing level of complexity: non–reactive scalar, reactive with negligible thermal-expansion effects, and finally the case of premixed flames in which the scalar and velocity fields are coupled by thermal-expansion effects. Lumley and his co-workers [161–163] derived a balance equation for the dissipation rate of non-reactive scalar in homogeneous isotropic turbulence. Borghi and his co-workers [160, 164–166] adapted the approach for turbulent flows with chemical reactions, which was the first attempt to capture the effects of chemical reactions on SDR but with a constant-density approximation. Jones and Musonge [167] included the effects of inhomogeneities in the non-reactive flow. Swaminathan and Bray [150] included thermal-expansion effects of combustion. In the following subsections, emphasis is placed on the significance of various processes that would influence the SDR in various situations, and order-of-magnitude analysis (OMA) is presented in detail for each situation. The strengths and weaknesses of OMA are also briefly stressed, thus emphasising the important role of DNS and laser diagnostic data to investigate such crucial small-scale quantities.

2.3.2 Transport Equation for the SDR

The transport equation for the instantaneous SDR N_c can be written as

$$\rho \frac{DN_c}{Dt} - \frac{\partial}{\partial x_k}\left(\rho\hat{\alpha}\frac{\partial N_c}{\partial x_k}\right) = -2\rho\hat{\alpha}^2\left(\frac{\partial}{\partial x_k}\frac{\partial c}{\partial x_\ell}\right)^2 - 2\rho\hat{\alpha}\frac{\partial c}{\partial x_k}\frac{\partial u_k}{\partial x_\ell}\frac{\partial c}{\partial x_\ell}$$
$$- 2\hat{\alpha}\frac{\dot{\omega}_c + \mathcal{D}}{\rho}\frac{\partial c}{\partial x_k}\frac{\partial \rho}{\partial x_k} + 2\hat{\alpha}\frac{\partial \dot{\omega}_c}{\partial x_k}\frac{\partial c}{\partial x_k}, \qquad (2.89)$$

after Eq. (2.85) is differentiated with respect to x_k and then multiplied by $\hat{\alpha}\,\partial c/\partial x_k$. The diffusivity $\hat{\alpha}$ is taken to be weakly dependent on temperature. Following this procedure and using a transport equation for the Favre-averaged reaction progress variable $\tilde{c}$, a balance equation for $\Psi = \hat{\alpha}\,(\partial\tilde{c}/\partial x_k)(\partial\tilde{c}/\partial x_k)$ can be obtained. Subtracting this balance equation from the Favre-averaged form of Eq. (2.89), one obtains a transport equation for $\tilde{\epsilon}_c = \tilde{N}_c - \Psi$. This equation is written as [150, 168]

$$\underbrace{\frac{\partial\overline{\rho\epsilon_c}}{\partial t}}_{(\text{I})} + \underbrace{\frac{\partial\tilde{u}_k\overline{\rho\epsilon_c}}{\partial x_k}}_{(\text{II})} - \underbrace{\frac{\partial}{\partial x_\ell}\left(\rho\hat{\alpha}\frac{\partial\tilde{\epsilon}_c}{\partial x_\ell}\right)}_{(\text{III})} = (\text{IV}) + (\text{IV-b}) + (\text{V})$$

$$+ (\text{VI}) + (\text{VII}) - (\text{VIII}) + (\text{IX}) + (\text{X}), \qquad (2.90)$$

where

$$(\text{IV}) = -\frac{\partial \overline{\rho \, u_k'' \epsilon_c}}{\partial x_k},$$

$$(\text{IV-b}) = -2\overline{\rho \, \hat{\alpha} \, u_\ell'' \left(\frac{\partial c''}{\partial x_k}\right)} \left(\frac{\partial^2 \tilde{c}}{\partial x_\ell \partial x_k}\right),$$

$$(\text{V}) = -2\rho \, \hat{\alpha} \overline{\frac{\partial u_\ell''}{\partial x_k} \frac{\partial c''}{\partial x_k}} \frac{\partial \tilde{c}}{\partial x_\ell},$$

$$(\text{VI}) = -2\rho \, \hat{\alpha} \overline{\frac{\partial c''}{\partial x_\ell} \frac{\partial c''}{\partial x_k}} \frac{\partial \tilde{u}_\ell}{\partial x_k},$$

$$(\text{VII}) = -2\rho \, \hat{\alpha} \overline{\frac{\partial c''}{\partial x_\ell} \frac{\partial c''}{\partial x_k} \frac{\partial u_\ell''}{\partial x_k}},$$

$$(\text{VIII}) = 2\,\rho \, \hat{\alpha}^2 \overline{\frac{\partial^2 c''}{\partial x_\ell \partial x_k} \frac{\partial^2 c''}{\partial x_\ell \partial x_k}},$$

$$(\text{IX}) = 2\left(\overline{\hat{\alpha} \, \frac{\partial c''}{\partial x_k} \frac{\partial \dot{\omega}''}{\partial x_k}}\right),$$

$$(\text{X}) = -2\,\overline{\hat{\alpha} \left(\frac{\dot{\omega}_c + \mathcal{D}}{\rho}\right) \frac{\partial c}{\partial x_k} \frac{\partial \rho}{\partial x_k}}$$

$$+2\frac{\overline{\rho \, \hat{\alpha}}}{\overline{\rho} \, \overline{\rho}} \frac{\partial \tilde{c}}{\partial x_k} \frac{\partial \overline{\rho}}{\partial x_k} \left(\overline{\dot{\omega}_c + \mathcal{D}} - \frac{\partial \overline{\rho \, u_\ell'' c''}}{\partial x_\ell}\right). \tag{2.91}$$

The first term on left-hand side of Eq. (2.90) denotes the unsteady evolution, the second term represents the mean advection, and the third term represents the molecular diffusion. The various terms on the right-hand side are defined in Eq (2.91) and the physical meanings of these terms are as follows. The turbulent transport of the dissipation rate is denoted by (IV) and the influence of curvature of the mean scalar field is denoted by (IV-b). The turbulence–scalar interaction effects are denoted by three terms, viz., (V), (VI), and (VII). The influence of the molecular-dissipation process is denoted by (VIII), which is a positive-definite term. The last two terms appear specifically for turbulent premixed flames because they involve a chemical reaction rate. Specifically, the production of scalar gradients by the chemical reactions is denoted by (IX), and the thermal-expansion (dilatation) effects are denoted by (X). Equation (2.90) is valid for general premixed combustion in turbulent flows and it can be simplified for specific cases such as premixed combustion with unity Lewis number, negligible thermal-expansion effects, etc. Many of the terms in the preceding transport equation need to be modelled, which is quite challenging, and, as already noted, the role of DNS on this modelling front is inevitable. Before studying the combustion situations, one would like to establish a reference situation of non-reactive scalars.

2.3.3 A Situation of Reference – Non-Reactive Scalars

If one takes the scalar c to be non-reactive then $\dot{\omega}_c = 0$ in Eq. (2.85). Using ρ_u, a reference value for the density and the scalar Taylor microscale $\lambda_c = \Lambda_c \, \text{Re}_T^{-1/2}$ with

Λ_c as the scalar integral length scale, one gets a scaling [169] for the dissipation rate as

$$\overline{\rho \epsilon_c} \equiv \overline{\rho \hat{\alpha} \left(\frac{\partial c''}{\partial x_k} \right) \left(\frac{\partial c''}{\partial x_k} \right)} \simeq \mathcal{O} \left(\frac{\rho_u \, u_{\text{rms}}}{\Lambda} \, \widetilde{c''^2}; \text{Re}_T^0 \right), \qquad (2.92)$$

where the Schmidt number $\text{Sc} = \nu / \hat{\alpha}$ is taken to be unity and the scalar integral length scale is equal to the turbulence integral length scale Λ. This result confirms that $\overline{\rho \epsilon_c}$ is of the order of Re_T^0 – that is, Re_T raised to the power zero – and consequently it remains finite as the turbulence Reynolds number becomes very large ($\text{Re}_T \to \infty$).

The transport equation for this situation can be obtained from Eq. (2.90) by approximating the molecular-diffusion coefficients to be constant and the Lewis numbers for all species to be close to unity. The density is taken to be constant but, for convenience, the equation is presented within the Favre-averaging formalism. This will ease subsequent comparisons with situations involving a reactive scalar (1) under isothermal conditions corresponding to a 'cold flame' and (2) in combustion including thermal-expansion effects. It is understood that the fluctuations involved will not be Favre fluctuations, and the transport equation will then be exactly the same as Eq. (2.90) with $(IX) = (X) = 0$, because there are no chemical reactions and dilatation.

ORDER-OF-MAGNITUDE ANALYSIS. The classical rules of OMA proposed by Tennekees and Lumley [169] are used. The spatial derivatives of mean quantities and fluctuating quantities are scaled by the turbulence integral length scale Λ and Taylor microscale λ, respectively. The time derivative of the mean quantity is scaled using the turbulence time scale, $\tau_T \equiv \Lambda / u_{\text{rms}}$. This gives

$$(I) \simeq \mathcal{O} \left(\rho_u \left(\frac{u_{\text{rms}}}{\Lambda} \right)^2 \widetilde{c''^2}; \quad \text{Re}_T^0 \right),$$

$$(II) \simeq \mathcal{O} \left(\rho_u \left(\frac{u_{\text{rms}}}{\Lambda} \right)^2 \widetilde{c''^2}; \quad \text{Re}_T^0 \right),$$

$$(III) \simeq \mathcal{O} \left(\rho_u \left(\frac{u_{\text{rms}}}{\Lambda} \right)^2 \widetilde{c''^2}; \quad \text{Re}_T^{-1} \right),$$

$$(IV) \simeq \mathcal{O} \left(\rho_u \left(\frac{u_{\text{rms}}}{\Lambda} \right)^2 \widetilde{c''^2}; \quad \text{Re}_T^0 \right),$$

$$(IV\text{-b}) \simeq \mathcal{O} \left(\rho_u \left(\frac{u_{\text{rms}}}{\Lambda} \right)^2 \widetilde{c''^2}; \quad \text{Re}_T^{-1/2} \right),$$

$$(V) \simeq \mathcal{O} \left(\rho_u \left(\frac{u_{\text{rms}}}{\Lambda} \right)^2 \widetilde{c''^2}; \quad \text{Re}_T^0 \right),$$

$$(VI) \simeq \mathcal{O} \left(\rho_u \left(\frac{u_{\text{rms}}}{\Lambda} \right)^2 \widetilde{c''^2}; \quad \text{Re}_T^0 \right),$$

$$(VII) \simeq \mathcal{O} \left(\rho_u \left(\frac{u_{\text{rms}}}{\Lambda} \right)^2 \widetilde{c''^2}; \quad \text{Re}_T^{1/2} \right), \qquad (2.93)$$

in standard notations. Because there is no a priori scaling rule for the second-order derivative of scalar fluctuations, it is not possible to derive an order of magnitude from the expression of term (VIII) in Eqs. (2.91). However, because for a passive scalar (VII) is $\mathrm{Re}_T^{1/2}$, one can expect (VIII) to be of the same order of magnitude in such a manner that, at leading order, (VII) and (VIII) balance together; further details of these arguments are provided in [166].

Let us now envisage the simplest linear closures for each of the terms just discussed. The turbulence–scalar interaction term is represented by $(\mathrm{VII}) = A_e \overline{\rho \epsilon_c}/\tau_T$, where A_e is a modelling constant and τ_T is the integral time scale of turbulence; the closure for the dissipation term (VIII) is similar except that the time scale is the one associated with the scalar field $(\mathrm{VIII}) = \beta \overline{\rho \epsilon_c}/\tau_s$, with β being a modelling constant and $\tau_s = \widetilde{c''^2}/\widetilde{\epsilon_c}$. Invoking the balance between (VII) and (VIII), a classical algebraic model, also known as the linear relaxation closure, for the mean SDR is recovered:

$$\overline{\rho \epsilon_c} = C_D \, \overline{\rho c''^2}/\tau_T, \tag{2.94}$$

where the constant C_D is found to be equal to the ratio A_e/β. This model can also be obtained by simply balancing the dissipation and the turbulence production of the scalar variance; see Eq. (1.34). Unfortunately, the turbulence-to-scalar integral time-scale ratio C_D is unlikely to have a single value for all types of flow, even for the passive scalar situation [170–174].

MODELLED SDR TRANSPORT EQUATION. The modelled transport equation proposed by Lumley and his co-workers [161, 162] and studied by Newman et al. [163] and Jones and Musonge [167] is written as

$$\overline{\rho} \frac{D\widetilde{\epsilon}_c}{Dt} = \frac{\partial}{\partial x_\ell}\left(\overline{\rho}\hat{\alpha}_T \frac{\partial \widetilde{\epsilon}_c}{\partial x_\ell}\right) - \overline{\rho \epsilon_c}\left(C_{D_1}\frac{\widetilde{\epsilon}_c}{\widetilde{c''^2}} + C_{D_2}\frac{\widetilde{\varepsilon}}{\widetilde{k}}\right)$$
$$+ C_{D_3}\overline{\rho}\hat{\alpha}_T \frac{\widetilde{\varepsilon}}{\widetilde{k}}\frac{\partial \widetilde{c}}{\partial x_\ell}\frac{\partial \widetilde{c}}{\partial x_\ell} + C_{D_4}\overline{\rho}\nu_T \frac{\widetilde{\epsilon}_c}{\widetilde{k}}\frac{\partial \widetilde{u}_k}{\partial x_\ell}\frac{\partial \widetilde{u}_k}{\partial x_\ell}, \tag{2.95}$$

where gradient models are used for the scalar fluxes and Reynolds stresses.

In this equation it must be recalled that the last three terms are supposed to represent the sum of the two largest terms, namely (VII) and (VIII). Thus, these terms, of the order of $\mathrm{Re}_T^{1/2}$, compensate themselves and the remainder is of the order of Re_T^0. Moreover, terms proportional to mean gradients of concentration and of velocity appearing in the unclosed equation are supposed to be negligible because of the assumption of small-scale isotropy and of small-scale independence (see [166]). Now, it must be also recalled that the modelled equation proposed in [165, 166] differs from Eq. (2.95) as (1) the terms related to mean gradients of c and of velocity, i.e., (V) and (VI), are retained but not only as a contribution coming from (VII) and (VIII) because the assumption of small-scale isotropy is not supported by experiments [170]; and (2) terms (VII) and (VIII) are modelled separately, and the result is one production term proportional to $\widetilde{\epsilon}_c$ in addition to a destruction term proportional to $(\widetilde{\epsilon}_c)^2$.

Accordingly, following the OMA proposed in [165, 166], the modelled transport equation can be written as

$$\overline{\rho} \frac{D\widetilde{\epsilon}_c}{Dt} = \frac{\partial}{\partial x_k}\left(\overline{\rho}\hat{\alpha}_T \frac{\partial \widetilde{\epsilon}_c}{\partial x_k}\right) + C_{P_c}\overline{\rho}\hat{\alpha}_T \frac{\widetilde{\varepsilon}}{\widetilde{k}}\frac{\partial \widetilde{c}}{\partial x_k}\frac{\partial \widetilde{c}}{\partial x_k}$$
$$+ C_{P_U}\overline{\rho}\nu_T \frac{\widetilde{\epsilon}_c}{\widetilde{k}}\frac{\partial \widetilde{u}_\ell}{\partial x_k}\frac{\partial \widetilde{u}_\ell}{\partial x_k} + \overline{\rho \epsilon_c}\left(A_e\frac{\widetilde{\varepsilon}}{\widetilde{k}} - \beta\frac{\widetilde{\epsilon}_c}{\widetilde{c''^2}}\right), \tag{2.96}$$

where gradient models have been used for the scalar fluxes and Reynolds stresses. It is to be noted that the last contribution of Eq. (2.96) differs from previous proposals of Zeman and Lumley [162] and Jones and Musonge [167]: the sign of the third term on the right-hand side of Eq. (2.95) is negative.

2.3.4 SDR in Premixed Flames and Its Modelling

EFFECTS OF CHEMICAL REACTION. Let us now consider the mean dissipation rate of a reactive scalar $c(\mathbf{x}, t)$:

$$
\frac{\partial \overline{\rho \epsilon_c}}{\partial t} = - \text{(II)} + \text{(III)} + \text{(IV)} + \text{(V)} + \text{(VI)} + \text{(VI-b)} + \text{(VII)} - \text{(VIII)}
$$

$$
+ \underbrace{2\left(\overline{\hat{\alpha} \frac{\partial c''}{\partial x_k} \frac{\partial \dot{\omega}''}{\partial x_k}}\right)}_{\text{(IX)}}
$$

$$
\underbrace{- 2\overline{\hat{\alpha}\left(\frac{\dot{\omega}_c + \mathcal{D}}{\rho}\right) \frac{\partial c}{\partial x_k} \frac{\partial \rho}{\partial x_k}} + 2\frac{\overline{\rho} \hat{\alpha}}{\overline{\rho} \, \overline{\rho}} \frac{\partial \widetilde{c}}{\partial x_k} \frac{\partial \overline{\rho}}{\partial x_k} \left(\overline{\dot{\omega}_c + \mathcal{D}} - \frac{\partial \overline{\rho u_\ell'' c''}}{\partial x_\ell}\right)}_{\text{(X)}}. \quad (2.97)
$$

There are two additional contributions, (IX) and (X). Term (IX) corresponds to the influence of passive chemical reaction only whereas the term (X) includes the effects of variable density that are due to heat release.

OMA FOR REACTIVE SCALAR IN THE FAST CHEMISTRY LIMIT. In the special case of turbulent premixed combustion within the flamelet regime, the OMA is modified as follows: the instantaneous quantities involving the species gradient must be scaled by the flamelet thickness, δ_L^0. This is because such quantities are non-zero only inside flamelets. Moreover, for a quantity that is zero for $c = 1$ (fully burned products) and $c = 0$ (fresh reactants), the averaging procedure is nothing else but a product by the reactive scalar PDF $\widetilde{P}(c)$, which is of the order of Da^{-1} for intermediate values $c \in [0 : 1]$; see [21, 62]. The Damköhler number is defined as $\text{Da} \equiv \tau_T/\tau_c = \Lambda S_L/u_{\text{rms}}\delta_L^0$. The previous point is of particular importance and should be kept in mind. If this point is not taken into account, then the scaling of the mean reactive scalar dissipation term is found to be Da,[1] an unrealistic feature because Eq. (2.86) clearly shows that, in the flamelet regime of turbulent premixed combustion, the mean reaction rate is proportional to this quantity and it must remain finite when $\text{Da} \to \infty$, as emphasised in the early analyses of Bray, Champion, and Libby [175].

Using the definition of SDR and recalling that both molecular viscosity and diffusivity scale as $S_L^0 \delta_L^0$, one obtains the order of magnitude of the mean reactive SDR as follows:

$$
\overline{\rho \epsilon_c} = \int_{c=0^+}^{c=1^-} \rho D \frac{\partial c''}{\partial x_k} \frac{\partial c''}{\partial x_k} \widetilde{P}(c) \mathrm{d}c
$$

$$
\simeq \mathcal{O}\left(\frac{\rho_u u_{\text{rms}}}{\Lambda} \widetilde{c''^2}; \text{Da}^0 \text{Re}_T^0\right). \quad (2.98)
$$

This expression is similar to Eq. (2.92) obtained for a passive scalar; as expected, $\overline{\rho \epsilon_c}$ is $\mathcal{O}\left(\mathrm{Re}_T^0 \, \mathrm{Da}^0\right)$ and remains finite whatever, the values of the Damköhler and Reynolds numbers, i.e., as Da or $\mathrm{Re}_T \to \infty$. This behaviour confirms the relevance of the proposed scalings.

The OMA of the different terms (II)–(XI) appearing on the right-hand side of Eq. (2.97) can be performed with the new scalings. Several studies have been already performed in this direction, and interested readers are referred to [150, 165, 176, 177] for further details. In fact, even if attention is focused on the flamelet regime, the results can differ from one reference to another. This is because, there is not only a priori one single possible scaling for the fluctuating velocity gradient. The choice should depend on the level of exothermicity [178]. This question was not raised in the early work of Mantel and Borghi because thermal expansion across a flamelet was not considered by taking the fluid density to be constant. The corresponding issue is rather general and largely exceeds the restricted scope of the SDR transport-equation modelling. It concerns the description of reactive scalar transport (see for instance [179]), and it is also related to the closure of certain terms in the scalar flux transport equation [28, 180]. We now see how the analysis is modified by the chemical reaction and associated thermal expansion.

For the thin-flamelet combustion regime, one may write the progress-variable transport equation as $\rho_u S_L n_k \partial c / \partial x_k = \dot{\omega}_c + \mathcal{D}$, where n_k is kth component of the flame normal pointing towards the product side. The dissipation term (VIII) in Eqs. (2.91) can be rewritten as

$$(\text{VIII}) = \overline{\rho \hat{\alpha}^2 (c''_{,n})_{,n} (c''_{,n})_{,n}} + 4\overline{\rho \hat{\alpha}^2 (c''_{,n})_{,\hat{\imath}} (c''_{,n})_{,\hat{\imath}}} + 4\overline{\rho \hat{\alpha}^2 (c''_{,\hat{\imath}})_{,\hat{\imath}} (c''_{,\hat{\imath}})_{,\hat{\imath}}}, \qquad (2.99)$$

where local isotropy has been assumed in the flamelet tangential directions that are indiscriminately denoted by the subscript $\hat{\imath}$. The symbol $c_{,k}$ denotes the derivative of c in direction k. Combining the instantaneous progress-variable budget with Eq. (2.99), it can be shown that the reactive contribution (IX) simplifies with a part of the dissipation contribution (VIII), as first established in [165]. As a result, the reactive contribution (IX) no longer deserves to be considered, and the remaining part of the dissipation term, denoted as (VIII)*, is written as

$$(\text{VIII})^* = 8\overline{\rho \hat{\alpha}^2 (c''_{,n})_{,\hat{\imath}} (c''_{,n})_{,\hat{\imath}}} + 8\overline{\rho \hat{\alpha}^2 (c''_{,\hat{\imath}})_{,\hat{\imath}} (c''_{,\hat{\imath}})_{,\hat{\imath}}}. \qquad (2.100)$$

In this manner, the consequence of the persistence of flamelet structures that are sustained in the high-Da limit is that the complex correlation that involves the chemical rate does not require further consideration. This peculiarity was evidenced in [165], proposing the following modelled SDR transport equation:

$$\frac{\partial \overline{\rho \tilde{\epsilon}_c}}{\partial t} + \frac{\partial}{\partial x_k} \left[\overline{\rho} (\tilde{u}_k - \tilde{U}_{L_k}) \tilde{\epsilon}_c \right] = \frac{\partial}{\partial x_\ell} \left(\overline{\rho} \tilde{\alpha}_T \frac{\partial \tilde{\epsilon}_c}{\partial x_\ell} \right)$$

$$+ C_{P_c} \, \overline{\rho} \, \hat{\alpha}_T \frac{\tilde{\varepsilon}}{\tilde{k}} \frac{\partial \tilde{c}}{\partial x_k} \frac{\partial \tilde{c}}{\partial x_k} + C_{P_U} \, \overline{\rho} \, \nu_T \frac{\tilde{\epsilon}_c}{\tilde{k}} \frac{\partial \tilde{u}_\ell}{\partial x_k} \frac{\partial \tilde{u}_\ell}{\partial x_k}$$

$$+ \overline{\rho} A_e \, \mathrm{Re}_T^{1/2} \frac{\tilde{\varepsilon}}{\tilde{k}} \tilde{\epsilon}_c$$

$$- \frac{2}{3} \overline{\rho} \beta \, \mathrm{Re}_T^{1/2} \frac{\tilde{\epsilon}_c^2}{\widetilde{c''^2}} \left(\frac{3}{2} - C_{\epsilon_c} \frac{S_L}{k^{1/2}} \right), \qquad (2.101)$$

where the term representing local flame propagation is

$$\widetilde{U}_{L_i} = \frac{S_L}{1 + u_{\text{rms}}/S_L} \frac{\partial \widetilde{c}}{\partial x_k} \Big/ \sqrt{\frac{\partial \widetilde{c}}{\partial x_\ell} \frac{\partial \widetilde{c}}{\partial x_\ell}}.$$

This proposal was revisited later in [166], and it was noted that the factors $\text{Re}_T^{1/2}$ are no longer present in the last two terms on the right-hand side that represent turbulent stretching and dissipation–curvature effects.

DEPARTURE FROM THE FLAMELET REGIME – BRIDGING MODEL. One can also be interested in a closure that is valid for the situation of finite-rate chemistry. In the case of highly turbulent flames (thickened-flame regime), the derivation of a transport equation for the SDR of a reaction progress variable c was proposed first by Borghi and Dutoya [164]. In this specific case of thickened turbulent flames, because diffusion and chemical processes are not as strongly coupled as in the flamelet regime of turbulent combustion [166], the reactive contribution (IX) of Eq. (2.97) can be evaluated separately without considering the dissipation term (VIII). This means that the previous modeled Eq. (2.96) can be used but with the additional reactive term (IX) of Eq. (2.97), which must be closed:

$$\frac{\partial \overline{\rho}\widetilde{\epsilon}_c}{\partial t} + \frac{\partial}{\partial x_k}(\overline{\rho}\widetilde{u}_k\widetilde{\epsilon}_c) = \frac{\partial}{\partial x_\ell}\left(\overline{\rho}\hat{\alpha}_T \frac{\partial \widetilde{\epsilon}_c}{\partial x_\ell}\right) + C_{P_Y}\overline{\rho}\hat{\alpha}_T \frac{\widetilde{\varepsilon}}{\widetilde{k}} \frac{\partial \widetilde{c}}{\partial x_k} \frac{\partial \widetilde{c}}{\partial x_k}$$

$$+ C_{P_U}\overline{\rho}\, \nu_T \frac{\widetilde{\epsilon}_c}{\widetilde{k}} \frac{\partial \widetilde{u}_\ell}{\partial x_k} \frac{\partial \widetilde{u}_\ell}{\partial x_k} + \overline{\rho}\widetilde{\epsilon}_c\left(A_e \frac{\widetilde{\varepsilon}}{\widetilde{k}} - \beta \frac{\widetilde{\epsilon}_c}{\widetilde{c''^2}}\right)$$

$$+ 2D\overline{\frac{\partial c''}{\partial x_k}\frac{\partial \dot{\omega}_c}{\partial x_k}}. \tag{2.102}$$

In principle, a two-point statistical description is required for closing this reactive contribution. As, in general, only single-point information is available, an additional hypothesis is required. Using, for instance, the linear mean-square estimation (LMSE) formalism of Dopazo and O'Brien [181], one obtains the following closure:

$$2\rho\hat{\alpha}\overline{\frac{\partial c''}{\partial x_k}\frac{\partial \dot{\omega}_c}{\partial x_k}} = \frac{\overline{\rho}}{\tau_c}\int_c \frac{\partial \dot{\omega}_c''}{\partial c}c''^2\widetilde{P}(c)\mathrm{d}Y, \tag{2.103}$$

where $\widetilde{P}(c)$ is the one-point, one-time scalar Favre PDF of c.

Equation (2.102) for $\widetilde{\epsilon}_c$ is strictly valid for small values of the Damköhler number Da. For vanishingly small values of Da, the reactive contribution (IX) expressed by Eq. (2.103) is negligible – the gradients of both the chemical rate and progress variable become very small in the physical space – and it can be dropped from the analysis and finally, because at large Re_T, the instantaneous gradients of c are fixed by turbulence rather than by chemical reaction; thus Eq. (2.94) can be used instead of Eq. (2.102) to obtain $\widetilde{\epsilon}_c$.

The previous result can be used to obtain a more general algebraic closure of the SDR applicable to a wide range of conditions in terms of the Damköhler number. Indeed, in the case of premixed flames when combustion occurs in the flamelet regime of turbulent premixed combustion, it is well known that there is a direct proportionality between the mean chemical rate and the mean dissipation rate of reactive scalar fluctuations. This result, which is strictly valid for infinite values of the Damköhler number, was recalled at the beginning of this section through Eq. (2.86).

By deriving a transport equation for $c(1 - c)$ and considering that this is zero in the limit of a very large Damkohler number, one obtains the following identity for the instantaneous dissipation rate:

$$\rho N_c = \rho \hat{\alpha} \frac{\partial c}{\partial x_k} \frac{\partial c}{\partial x_k} = \frac{\dot{\omega}_c}{2}(2c - 1). \tag{2.104}$$

After averaging, the following result is obtained:

$$\overline{\rho \epsilon_c} = -\overline{\rho \hat{\alpha} \frac{\partial \tilde{c}}{\partial x_k} \frac{\partial \tilde{c}}{\partial x_k}} + \frac{(2C_m - 1)}{2} \overline{\dot{\omega}_c}, \tag{2.105}$$

which is nothing but the BML result in Eq. (2.86). In the other limit, that of a very low Damköhler number, turbulence dominates the chemical reaction to generate scalar gradients, and this is expected to contribute in the non-reactive regions [160], commonly known as the 'out-of-flamelet' contribution. This contribution is given by Eq. (2.94). Hence the simplest strategy to recover these two limits is to use a linear bridging function, depending on the value of the fluctuation-level parameter g by

$$\overline{\rho \epsilon_c} = g \left[-\overline{\rho \hat{\alpha} \frac{\partial \tilde{c}}{\partial x_k} \frac{\partial \tilde{c}}{\partial x_k}} + \frac{(2C_m - 1)}{2} \overline{\dot{\omega}_c} \right] + (1 - g)C_D \frac{\overline{\rho c''^2}}{\tau_T}. \tag{2.106}$$

This closure was extended to partially premixed situations in [182] and successfully applied to stratified conditions in [141]. Such conditions are of special interest for lean premixed flames that often propagate in imperfectly premixed mixtures of fuel and oxidiser.

Having discussed the complexities associated with the chemical reaction, we should now focus our attention on the effects induced by thermal expansion.

EFFECTS OF THERMAL EXPANSION. The heat release will duly influence turbulence, which in turn will affect the reactions. Thus there is a close coupling among chemical reactions, turbulence, and molecular diffusion in turbulent premixed flames [70, 183], which needs to be captured for successful modelling. Swaminathan and Bray [150] showed that the turbulent flame propagates significantly faster when the density change across the flame front, represented by (X) in Eqs. (2.91), is considered in the analysis. Bychkov [184] and Akkerman and Bychkov [185] made similar observations about the influence of heat release on the flame propagation speed. DNS [177, 186–188] and experimental [189] studies showed that the dilatation changes the characteristics of a turbulence–scalar interaction, resulting in the dissipation of the scalar gradient, rather than its production, by turbulence. It will be shown later in this section that these effects appear at leading order. Because of this, the scalings used earlier in this section are likely to be modified and it is possible to use the laminar flame speed S_L^0 or the turbulence rms velocity u_{rms} to scale the fluctuating velocity. However, if one uses the rms velocity then the dilatation-related term is found [190] to be a lower-order term, which is contrary to intuition and observations in recent studies. Thus the planar laminar flame speed is used in the following analyses.

For the situation with heat release from chemical reactions, one cannot neglect any of the terms in Eq. (2.90); however the dilatation-related term (X) can be simplified if one considers the combustion to be adiabatic with unity Lewis number in low-Mach-number flows. These are reasonable approximations for typical practical situations. Now, by use of the equation of state, the instantaneous and averaged densities can be simply related to the progress variable and its Favre average by

$$\rho = \frac{\rho_u}{1 + \tau c}, \quad \bar{\rho} = \frac{\rho_u}{1 + \tau \tilde{c}},$$

where τ is the heat release parameter (see Section 2.1). These relations allow one to write [150]

$$(\mathbf{X}) = 2\overline{\rho \epsilon_c \frac{\partial u_\ell}{\partial x_\ell}}. \tag{2.107}$$

This term involves the correlation between the SDR and dilatation, which turns out to be a leading-order term, as one will see in the OMA presented next. It is to be noted that this term was not hitherto considered in the analysis.

ORDER-OF-MAGNITUDE ANALYSIS. The turbulent integral length and time scales and reactant density are used to scale the spatial derivative of mean quantities, time derivative, and the density. The mean velocity is scaled with a reference velocity U_{ref} and the diffusivity is scaled with $S_L^0 \delta_L^0$, as noted earlier. Because the gradient of the progress variable is non-zero only inside the flame front, the quantities involving this gradient are scaled with the laminar flame scales. Defining $\text{Re}_T \equiv u_{\text{rms}} \Lambda/(S_L^0 \delta_L^0)$ and $\text{Da} \equiv \Lambda S_L^0/u_{\text{rms}} \delta_L^0$, and using the previous scalings, one obtains

$$(\text{I}) \simeq \mathcal{O}\left(\rho_u \left(\frac{S_L^0}{\delta_L^0}\right)^2 \widetilde{c''^2}; \quad \text{Da}^{-1}\right),$$

$$(\text{II}) \simeq \mathcal{O}\left(\rho_u \left(\frac{S_L^0}{\delta_L^0}\right)^2 \widetilde{c''^2}; \quad \left(\frac{\text{Da}\, u_{\text{rms}}}{U_{\text{ref}}}\right)^{-1}\right),$$

$$(\text{III}) \simeq \mathcal{O}\left(\rho_u \left(\frac{S_L^0}{\delta_L^0}\right)^2 \widetilde{c''^2}; \quad \text{Re}_T^{-1}\,\text{Da}^{-1}\right),$$

$$(\text{IV}) \simeq \mathcal{O}\left(\rho_u \left(\frac{S_L^0}{\delta_L^0}\right)^2 \widetilde{c''^2}; \quad \text{Da}^{-1}\right),$$

$$(\text{IV-b}) \simeq \mathcal{O}\left(\rho_u \left(\frac{S_L^0}{\delta_L^0}\right)^2 \widetilde{c''^2}; \quad \text{Re}_T^{-1}\,\text{Da}^{-1}\right),$$

$$(\text{V}) \simeq \mathcal{O}\left(\rho_u \left(\frac{S_L^0}{\delta_L^0}\right)^2 \widetilde{c''^2}; \quad \text{Re}_T^{-1/2}\,\text{Da}^{-1/2}\right),$$

$$(\text{VI}) \simeq \mathcal{O}\left(\rho_u \left(\frac{S_L^0}{\delta_L^0}\right)^2 \widetilde{c''^2}; \quad \left(\frac{\text{Da} u'}{U_{\text{ref}}}\right)^{-1}\right),$$

$$(\text{VII}) \simeq \mathcal{O}\left(\rho_u \left(\frac{S_L^0}{\delta_L^0}\right)^2 \widetilde{c''^2}; \quad \text{Re}_T^0\,\text{Da}^0\right),$$

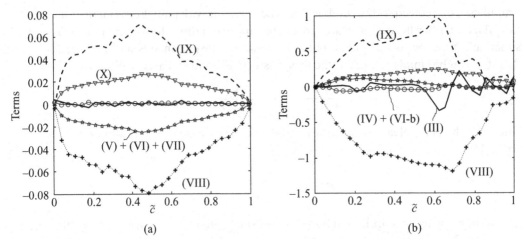

Figure 2.12. Variations of terms in Eq. (2.90) for (a) high-Da (Da = 6.84, Ka = 0.54, Re$_T$ = 56.7, τ = 2.3) and (b) low-Da (Da = 0.32, Ka = 13.2, Re$_T$ = 47, τ = 3.0) flames considered in [191]. These DNS flames are statistically planar, have a Zeldovich number of 6.0, and a unity Lewis number. The values given above are normalised by $\rho_u S_L^{0^4}/\hat{\alpha}^2$.

$$(\text{VIII}) \simeq \mathcal{O}\left(\rho_u \left(\frac{S_L^0}{\delta_L^0}\right)^2 \widetilde{c''^2}; \quad \text{Re}_T^0 \text{ Da}^0\right),$$

$$(\text{IX}) \simeq \mathcal{O}\left(\rho_u \left(\frac{S_L^0}{\delta_L^0}\right)^2 \widetilde{c''^2}; \quad \text{Re}_T^0 \text{ Da}^0\right),$$

$$(\text{X}) \simeq \mathcal{O}\left(\rho_u \left(\frac{S_L^0}{\delta_L^0}\right)^2 \widetilde{c''^2}; \quad \text{Re}_T^0 \text{ Da}^0\right). \tag{2.108}$$

It is evident from the preceding equations that (VII), (VIII), (IX) and (X) are the leading-order terms for high-Damköhler-number flames and the other terms are likely to be non-negligible when combustion occurs at low Da. This can be verified by use of DNS or laser diagnostics data. However, spatial and temporal resolutions required for resolving gradients of the progress-variable fluctuation and its reaction rate preclude the use of laser diagnostics data at this time and the use of DNS data is inevitable. Figure 2.12 shows the behaviour of various terms across a statistically planar flame brush [191]. The values are appropriately normalised by use of the reference density, planar laminar flame speed, and the reference diffusivity. The case of a high-Da flame (Da ≈ 6.84) is shown in Fig. 2.12(a), and the low-Da (≈ 0.32) case is shown in Fig. 2.12(b). A positive value in this figure implies a source, whereas a negative value indicates a sink for the Favre-averaged SDR, $\tilde{\epsilon}_c$. The following points are evident from this figure [191]:

- the most dominant contributions come from the turbulence–scalar interaction, dilatation, dissipation, and chemical reaction processes for both the flames.
- The chemical reactions (IX) and dilatation (X) provide dominant sources whereas the turbulence–scalar interaction and dissipation processes provide dominant sinks. More detailed analyses [176, 186, 191, 192] clearly showed that major contributions, by nearly an order of magnitude, of the turbulence–scalar

interaction effects come from (VII) for high-Da flames whereas it is quite likely to have equal contributions from all of the three terms for low-Da situations.

- The most apparent difference is on the physical behaviour of the turbulence–scalar interaction process. It is clear from Fig. 2.12 that this process destroys the (negative-value) scalar gradient in high-Da situations whereas it produces the (positive-value) Scalar gradient in most parts of the flame brush in low-Da situations. Small negative values can also be noted in Fig. 2.12(b) for $\tilde{c} \geq 0.65$, implying that strong heat release in those regions is destroying the scalar gradients. This is intuitively correct, as one would expect, but requires further close understanding to help model construction.

These relative behaviours among the various terms and their components in the SDR transport equation are also observed for non-unity Lewis number situations [168, 193] and the dominant terms noted in Fig. 2.12 remain dominant. However, their magnitudes increase as the Lewis number decreases [193] because thermodiffusive (TD) instability, noted in Section 3.1, significantly increases the reaction rate when the Lewis number is smaller than unity. The increased reaction rate results in an increase of scalar and velocity gradients, yielding larger magnitude for all the terms shown in Fig. 2.12. When Le > 1 the magnitudes of these terms are smaller than for the Le = 1 case; their relative behaviours, however, remain the same [193]. It is clear from the previous discussion that the dominant terms shown in Fig. 2.12 are likely to remain dominant for a wide range of flow and thermochemical conditions, spanning from the corrugated-flamelets to the thin-reaction-zones combustion regimes when Da > 1. Hence one can envisage developing an algebraic model for the SDR by balancing the dominant terms, which can be used in Eq. (2.86) to obtain the mean reaction rate. However, one should be careful for low-Da cases, and it is quite likely that the SDR transport equation, Eq. (2.90), will then need to be solved. Before an attempt is made to derive an algebraic model for $\tilde{\epsilon}_c$, the closure models proposed in the earlier discussion on the dissipation-rate transport equation are to be revised, where necessary, to capture the close coupling among turbulence, heat release, scalar mixing, diffusion, and molecular dissipation. Hence the modelling of unclosed terms is considered next.

MODELLING OF TURBULENT TRANSPORT TERM (IV). Scalings in Eqs. (2.108) give (IV)/(IV-b) $\sim \mathcal{O}\,(\mathrm{Re}_T)$, which indicates that the modelling of the turbulent flux of the SDR, $\overline{\rho u_k'' \epsilon_c}$, which is often modelled with the gradient hypothesis, is essential. This flux is counter-gradient when $\overline{\rho u_k'' \epsilon_c}(\partial \tilde{\epsilon}_c / \partial x_k) > 0$, and it is gradient when $\overline{\rho u_k'' \epsilon_c}(\partial \tilde{\epsilon}_c / \partial x_k) < 0$. A typical behaviour of this flux in statistically planar flames is shown in Fig. 2.13. By writing the gradient of $\tilde{\epsilon}_c$ as $(\partial \tilde{\epsilon}_c / \partial \tilde{c})(\partial \tilde{c} / \partial x_k)$ and noting the typical behaviours of these two gradients, one observes that turbulent flux is counter-gradient in high-Da and small-Le flames. A strong flame-normal acceleration that is due to greater heat release in the low-Le as well as in the large-Da flames acts to promote counter-gradient transport, and the magnitude of the flame-normal acceleration decreases with either increasing Lewis number or decreasing Damkohler number, promoting gradient transport. This is evident from the DNS results shown in Fig. 2.13.

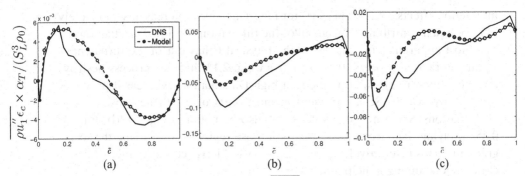

Figure 2.13. Variations of turbulent flux $\overline{\rho u_1'' \epsilon_c}$ in statistically planar turbulent premixed flames of (a) high Da (Da = 6.84, Ka = 0.54, Re_T = 56.7, τ = 2.3, Le = 1), (b) low Da (Da = 0.32, Ka = 13.2, Re_T = 47, τ = 4.5, Le = 1.0); (c) Da = 0.32, Ka = 13.2, Re_T = 47, τ = 4.5, Le = 0.34. These flames are studied in [193] and have a Zeldovich number of 6.0.

Further analyses [193] revealed that the counter-gradient transport for the turbulent flux is observed when the scalar flux $\overline{\rho u'' c''}$ is counter-gradient. Earlier studies by Veynante et al. [1] and Chakraborty and Cant [194] also demonstrated that the turbulent transport of a scalar gradient is closely related to the scalar flux in premixed flames. Following these studies, a model for $\overline{\rho u_1'' \epsilon_c}$ is proposed as [193]

$$\overline{\rho u_1'' \epsilon_c} = \hat{A} \left(\Phi - \tilde{c} \right) \frac{\overline{\rho u_1'' c''}}{\left[\widetilde{c''^2} + \tilde{c}(1 - \tilde{c}) \right]} \widetilde{\epsilon}_c, \tag{2.109}$$

where $\hat{A} = 2$ and $\Phi = 0.5$ are model parameters. A typical performance of this model is compared with the DNS result in Fig. 2.13. It is evident that this model captures the counter-gradient as well as gradient transport without any modification to the model parameters, $\hat{A}$ and Φ. The quantitative agreement between the model prediction and the DNS data is satisfactory. It is to be noted, however, that proper modelling of the scalar flux $\overline{\rho u_1'' c''}$ is required, which will be discussed later (see also Section 2.1).

MODELLING OF SCALAR–TURBULENCE INTERACTION. The sum of three terms, (V) + (VI) + (VII) in Eq. (2.90), is collectively called the scalar–turbulence interaction term because they originate from $\mathcal{T} \equiv -\overline{2 \rho \hat{\alpha} \, \partial c / \partial x_\ell \, (\partial u_\ell / \partial x_k) \, \partial c / \partial x_k}$, which is the averaged form of the second term on the right-hand side of Eq. (2.89). This term signifies the interaction of turbulence and scalar gradients, and it is to be noted that only the symmetric part, $e_{\ell k} \equiv 0.5(\partial u_\ell / \partial x_k + \partial u_k / \partial x_\ell)$, of the strain tensor $\partial u_\ell / \partial x_k$ is involved because the contribution of the rotational part $R_{\ell k}$ to $\mathcal{T}$ is exactly zero. Using eigendecomposition, one writes

$$\mathcal{T} = -\overline{2 \rho N_c (e_\alpha \cos \theta_\alpha + e_\beta \cos \theta_\beta + e_\gamma \cos \theta_\gamma)}, \tag{2.110}$$

where e_α, e_β, and e_γ are the eigenvalues of $e_{\ell k}$ ordered as $e_\alpha > e_\beta > e_\gamma$, with e_γ being the most compressive principal strain rate; the three eigenvalues are orthogonal to one another. It is apparent that the statistical behaviour of $\mathcal{T}$ is governed by the statistics of the alignment angles, denoted previously by θ, between the scalar-gradient vector and directions of the principal strain rates. It is well known that the scalar gradient aligns most probably with the compressive strain rate e_γ in cold turbulence [195–200] and even in flows with passive chemical reactions [203, 204] (without dilatation). This situation, shown schematically in Fig. 2.14(a), implies that the turbulence

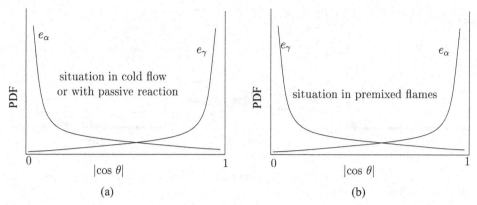

Figure 2.14. Schematic of scalar-gradient alignment with the most extensive and compressive principal strain rates e_α and e_γ, for (a) passive scalar mixing in a turbulent flow and (b) reactive scalar in a premixed flame.

produces scalar gradients by bringing isoscalar surfaces together and this production is balanced by the molecular-diffusion process. When there is strong heat release, the dilatation usually occurring in the local normal direction overwhelms the turbulence effects, resulting in alignment PDFs shown in Fig. 2.14(b) for premixed flames – the scalar gradient aligns most probably with the most extensive strain rate, e_α, resulting in the destruction of the scalar gradient by the turbulence [168, 186–189]. This destruction, aided by the molecular diffusion, balances the production of scalar gradients by chemical reactions. As one can see, this is different from the passive scalar or cold-flow scenario, and it is important to understand and to account for the change in physics for correct modelling of turbulent premixed flames.

The local production or destruction of scalar gradients by the turbulence is determined by the competing effects of heat-release-induced straining and turbulence straining. The former is related to the dilatation rate, which scales as $a_{\text{chem}} \equiv \partial u_\ell/\partial x_\ell \sim \tau f(\text{Le}) S_L^0/\delta_L^0$, and the turbulent strain rate scales as $a_{\text{turb}} \equiv u_{\text{rms}}/\Lambda$. A function of the Lewis number is included to account for the non-unity Le effects [168] on dilatation, and its specific form, introduced later, is not important at this time. However, it is worth noting that this function should take a value of unity for Le $= 1$ to be consistent with the classical scaling, and it increases with decreasing Le [168]. The ratio of these two strain rates is $a_{\text{chem}}/a_{\text{turb}} \sim \tau f(\text{Le}) \text{Da}_L$, where Da_L is local Damköhler number. Because the turbulence involves a range of scales, one can also use the Kolmogorov time to scale the turbulent strain rate. If one uses this time scale, then $a_{\text{chem}}/a_{\text{turb}} \sim \tau f(\text{Le})/\text{Ka}_L$, where Ka_L is the local Karlovitz number. One has $a_{\text{chem}}/a_{\text{turb}} \gg 1$ when the heat release effects are dominant, resulting in the scalar gradient aligning with the extensive strain rate e_α. This happens when (1) τ is large, (2) Le is small, (3) the local Damköhler number is large, and (4) the local Karlovitz number is small. All of these conditions, except (2), are met, even for Le ≥ 1 flames when the combustion is in the corrugated-flamelets as well as wrinkled-flamelets regimes, which are signified by Da $\gg 1$ and Ka < 1 (see Fig. 1.3). Physically, in these regimes the flame scales are smaller than typical large and small scales of turbulence. The alignment characteristics shown in Fig. 2.15(a) for this situation indicates that the turbulence dissipates the scalar gradient. On the other hand, one expects the turbulence to produce the scalar gradients when Da < 1 and Ka > 1 because the

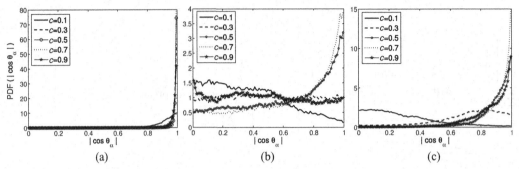

Figure 2.15. PDF of scalar-gradient alignment angle with the most extensive principal strain rate e_α in the flames considered in Fig. 2.13.

turbulence straining dominates the dilatation-induced straining, but the DNS results in Figs. 2.15(b) and 2.15(c) show that the heat release effects dominate the turbulence locally in the regions of intense heat release. For Da > 1 and Ka > 1 conditions, one may argue that the turbulence dissipates the scalar gradient because Da > 1 by just considering the large scales of turbulence. But Ka > 1 implies that the small-scale fluid dynamic processes are faster than the chemical reaction and thus the alignments may be those of a passive scalar situation. It seems that the bulk behaviour of the turbulence–scalar interaction will be determined by the preponderance of one of these competing processes at different scales. Experimental investigation by Hartung et al. [189] showed that the scalar gradient aligns with the most extensive strain rate even in laboratory flames with Da > 1 and Ka > 1. Although results for statistically planar flames with unity and non-unity Le flames are shown here, a de-tailed analysis [205] of flame kernels (curved flame brushes) showed that the scalar gradients are likely to align with the most extensive strain rate when the local heat release effects are strong.

Out of the three terms of turbulence–scalar interaction, (V) and (VII) are influenced by the alignment characteristics previously discussed. One can note from the preceding discussion that the alignment characteristics are influenced by both Da and Ka. Thus the modelling of these terms should explicitly include Da and Ka to capture the heat release effects; such models were developed in [168, 191, 193, 205]. In the following these models are briefly discussed, highlighting their salient features, and the readers are referred to appropriate works for further details. However, the information provided here is sufficient for model calculations using the SDR approach.

The predominant contribution to the turbulence–scalar interaction term, i.e., the sum (V) + (VI) + (VII), comes from (VII) when the thermochemical or heat release effects are significant. The contributions of (V) and (VI) are likely to be important in low-Da combustion. These observations are noted in [168, 186, 191, 192, 205].

Modelling of (V)

Mantel and Borghi [165] proposed a model for (V) as

$$(\text{V}) \sim -\bar{\rho} \frac{\widetilde{u''_\ell c''}}{t_{\text{scal}}} \left(\frac{\partial \tilde{c}}{\partial x_\ell} \right) \simeq -C_{P_c} \bar{\rho} \widetilde{u''_\ell c''} \left(\frac{\tilde{\varepsilon}}{\tilde{k}} \right) \left(\frac{\partial \tilde{c}}{\partial x_\ell} \right), \qquad (2.111)$$

with $\widetilde{k}$ and $\widetilde{\varepsilon}$ denoting the turbulent kinetic energy and its dissipation rate, respectively, and the model constant C_{P_c} is of the order of unity. The time scale t_{scal} is taken to be the turbulence time scale τ_T. This model notes a direct relationship between the scalar flux and (V), which was noted in a number of DNS studies under a variety of flame and flow conditions [191–193, 205]. However, if one follows the arguments [1] behind Eq. (2.22) then the scalar flux can be expressed as

$$\widetilde{u_\ell'' c''} \approx (2\alpha u_{rms} - \tau S_L)\widetilde{c}(1 - \widetilde{c})M_\ell, \qquad (2.112)$$

where α is an appropriate efficiency function and M_ℓ is the component of the unit normal vector of $\widetilde{c}$ field pointing towards the reactant side. After substituting this expression into relation (2.111), one sees that the resultant expression has scalings $\mathcal{O}(\rho_u \widetilde{c''^2} u_{rms}^2/\Lambda^2; \mathrm{Re}_T^0)$ and $\mathcal{O}(\rho_u \widetilde{c''^2} S_L^{0^2}/\delta_L^{0^2}; \mathrm{Re}_T^{-1/2} \mathrm{Da}^{-3/2})$ for the non-reacting and reacting parts, respectively. The scaling of the non-reacting part is consistent with that in Eqs. (2.93) but that of the reacting part is not consistent with Eqs. (2.108). This is because the turbulent time scale is also used to represent t_{scal} in the reactive regions. Clearly this is inadequate because the heat release affects the turbulence–scalar interaction, as previously discussed, and also it is inconsistent with the scalings suggested in [150]. Thus the preceding model is revised [191, 192] as

$$(V) \simeq -[C_1 + C_2\, \mathrm{Da}_L^*]\,\overline{\rho}\,\widetilde{u_\ell'' c''}\left(\frac{\widetilde{\varepsilon}}{\widetilde{k}}\right)\left(\frac{\partial \widetilde{c}}{\partial x_\ell}\right), \qquad (2.113)$$

where $\mathrm{Da}_L^* \equiv \rho_u\, \mathrm{Da}_L/\overline{\rho}$ is the density-weighted local Damkohler number and the density weighting is used to minimise the effect of the kinematic viscosity change that is due to heat release on $\widetilde{\varepsilon}$. The model parameters are $C_1 = 0.5$ and $C_2 = 1.3\mathrm{Ka}_L^2/(1 + \mathrm{Ka}_L)^2$. The second term involving C_2 inside the square brackets accounts for the contributions from the reacting regions, and it explicitly includes the local Damköhler and Karlovitz numbers to account for the heat release effects on the turbulence–scalar interactions. This is because δ_L^0/S_L^0 is used to represent the time scale t_{scal}. This model is noted to work well in [192] and [191] for high- and low-Da flames without having to alter the numerical values of the model parameters C_1 and C_2.

Local flame propagation is strongly influenced by the TD instability in non-unity Le flames and also by the flame stretch effects induced by flame straining and bending. Flame straining by turbulence is already included in relation (2.113). Although this model qualitatively predicts (V) in non-unity Le cases satisfactorily [168], its quantitative predictions may be improved by including Le effects on flame straining by use of $f(\mathrm{Le})$, as noted earlier. This gives a model

$$(V) \simeq \left[\hat{C}_1 + \hat{C}_2 f(\mathrm{Le})\,\mathrm{Da}_L^*\right]\overline{\rho}\,\widetilde{u_\ell'' c''}\left(\frac{\widetilde{\varepsilon}}{\widetilde{k}}\right)\left(\frac{\partial \widetilde{c}}{\partial x_\ell}\right); \qquad (2.114)$$

alternative ways are also available [193]. The flame-curvature-induced stretch effects can be included [205] following the Markstein diffusivity analogy [56] [see Eq. (3.1) also] for curved flames. The turbulent scalar flux expression given in relation (2.112) can be written as

$$\widetilde{u_\ell'' c''} \approx (2\alpha u_{rms} - \tau S_L')\widetilde{c}(1 - \widetilde{c})M_\ell, \qquad (2.115)$$

where a modified flame speed $S'_L = S^0_L - \hat{D}\partial M_\ell/\partial x_\ell$ is used, with $\hat{D}$ as an appropriate diffusivity to include the mean-curvature effects. Now, some algebraic rearrangements yield

$$(V) \simeq -[C_1 + C_2\,\mathrm{Da}^*_L]\,\overline{\rho}\,\widetilde{u''_\ell c''}\left(\frac{\widetilde{\varepsilon}}{\widetilde{k}}\right)\left(\frac{\partial \widetilde{c}}{\partial x_\ell}\right)$$

$$+\hat{C}_2\,\overline{\rho}\,\tau\,\mathrm{Da}^*_L\,\widetilde{c}(1-\widetilde{c})\left(\frac{\widetilde{\varepsilon}}{\widetilde{k}}\right)\left(\frac{\partial \widetilde{c}}{\partial x_\ell}\right)\left(\hat{D}\frac{\partial M_\ell}{\partial x_\ell}\right). \qquad (2.116)$$

This relation becomes relation (2.113) for statistically planar flames because the last term in relation (2.116) becomes zero. This model is shown in [205] to capture the variations of (V) in kernel DNS satisfactorily with $\hat{C}_2 \approx 1.0$.

Mura et al. [176] recently proposed the following models:

$$(V) = -C_M\overline{\rho}\frac{\widetilde{\epsilon}_c}{\widetilde{c''^2}}\,\widetilde{u''_\ell c''}\left(\frac{\partial \widetilde{c}}{\partial x_\ell}\right),$$

$$= -\tau S_L\overline{\rho}\,\widetilde{\epsilon}_c\langle \vec{n}_f \cdot \vec{x}_\ell\rangle\frac{\partial \widetilde{c}}{\partial x_\ell}, \qquad (2.117)$$

where C_M is a model constant of the order of unity, $\vec{n}_f$ is the local flame-normal vector, and the average value of the direction cosine is taken to be 0.8. If one uses the scaling $\widetilde{\epsilon}_c \sim \widetilde{c''^2}\widetilde{\varepsilon}/\widetilde{k}$ then the first of the two models in Eqs. (2.117) becomes similar to that in relation (2.111). These two models perform satisfactorily for high-Da flames considered by Mura et al. [176]; however, their performance for low-Da combustion is yet to be assessed. One shall recall an earlier observation that the contribution of (V) in high-Da flames is negligible, and it is likely to be important in low-Da situations.

Modelling of (VI)

Mantel and Borghi [165] proposed a model for this term as

$$(VI) \simeq -C_{P_U}\,\overline{\rho}\,\widetilde{\epsilon}_c\frac{\widetilde{u''_\ell u''_k}}{\widetilde{k}}\frac{\partial \widetilde{u}_\ell}{\partial x_k}. \qquad (2.118)$$

The scaling estimate for this model is consistent with the order-of-magnitude estimate in Eqs. (2.108), but it relates gradients of scalar fluctuation to the Reynolds stress with an assumption

$$H_{\ell k} \equiv \frac{\overline{\rho D\,\frac{\widetilde{\partial c''}}{\partial x_\ell}\frac{\partial c''}{\partial x_k}}}{\overline{\rho}\widetilde{\epsilon}_c} \simeq \frac{\widetilde{u''_\ell u''_k}}{2\widetilde{k}}, \qquad (2.119)$$

which is unclear. Recently, a proposition [192] that is similar to the concept of the Cant, Pope, and Bray (CPB) model [206] for surface averages of the products of the flame-normal components, $\overline{(n_\ell n_k)}_s = \overline{n_\ell n_k|\nabla c|}/\Sigma$, appearing in the FSD approach is made as

$$H_{\ell k} = \psi_\ell\psi_k + \frac{1}{3}\delta_{\ell k}(1 - \psi_m\psi_m), \quad \text{where} \quad \psi_k = -\frac{\partial \overline{c''}}{\partial x_k}\sqrt{\frac{\rho_u\,\mathcal{D}_u}{\overline{\rho}\,\widetilde{\epsilon}_c}}. \qquad (2.120)$$

The Reynolds average of Favre fluctuations is $\overline{c''} = \overline{c} - \widetilde{c} = \tau\widetilde{c}(1-\widetilde{c})/(1+\tau\widetilde{c})$ when the combustion occurs in thin flamelets, and the PDF of c can be approximated as the BML PDF [148]. A model for (VI) can be written as [191, 192]

$$(VI) \simeq -C_{T3}\,\overline{\rho}\,\widetilde{\epsilon}_c \left[\psi_\ell \psi_k + \frac{1}{3}\delta_{\ell k}(1 - \psi_m \psi_m) \right] \frac{\partial \widetilde{u}_\ell}{\partial x_k}, \qquad (2.121)$$

where the model parameter is $C_{T3} = (1 + 2\,\mathrm{Ka}^{-0.23})$, which ensures that it reaches a positive constant value in the limit of Ka becoming very large. The comparison of the preceding two models with DNS results in [191–193, 205] showed that the model in relation (2.121) performs satisfactorily for a variety of flames, without having to change the model constants previously given. Recently Mura et al. [176] discussed several alternatives to model (VI) and concluded that these alternative models captured the qualitative behaviour of (VI) in their DNS data with $u_{\mathrm{rms}}/S_L \sim 1$, but the quantitative predictions were not satisfactory for all the flames in their DNS database. A similar observation is reported in [205]. It is to be noted that the contribution of (VI) is important in low-Da flames, as suggested by the scaling results in Eqs. (2.108).

Modelling of (VII)

The analyses [168, 186, 187, 192, 193, 205] of DNS data showed that a significant contribution to the turbulence–scalar interaction term comes from (VII) in agreement with the scaling in Eqs. (2.108). Hence variations of the turbulence–scalar interaction term shown in Fig. 2.12 is primarily that of (VII). This term is qualitatively as well as quantitatively different in high- and low-Da flames, as shown in this figure. Nevertheless, this is a leading-order term, and thus its correct modelling is crucial for flame calculation. As we noted earlier, the scalar-gradient alignment with the principal strain rate influences this terms and this alignment depends on the local Da and Ka. Thus a proper modelling of this term should include these non-dimensional numbers, signifying the relative role of thermochemical and fluid dynamic processes.

If one writes (VII) similar to that in Eq. (2.110) by using the eigendecomposition and presuming a passive scalar scenario then one obtains

$$(VII) \sim \overline{\rho}\,\widetilde{\epsilon}_c\,|e_\gamma| \simeq A_e\,\overline{\rho}\,\widetilde{\epsilon}_c \left(\frac{\widetilde{\varepsilon}}{\widetilde{k}} \right). \qquad (2.122)$$

One writes the second part by assuming that the magnitude of the most compressive principal strain rate e_γ is proportional to the large-scale turbulence strain rate. This model was proposed by Mantel and Borghi [165] and predicts a positive value for (VII). This is acceptable because the turbulence is expected to produce a scalar gradient when the scalar is passive. This model will not give negative values for (VII) resulting from the change in the alignment characteristics that is due to heat release effects (see Fig. 2.14), which suggests [186] that (VII) $\sim -\rho_u\,\widetilde{\epsilon}_c\,\tau\,S_L^0/\delta_L^0$ in the reacting regions. This scaling uses laminar flame time to scale the most extensive principal strain rate e_α. This is consistent with the scales used in the order-of-magnitude estimates in Eqs. (2.108) and the observed [186, 187, 189] physical behaviour. Thus one can combine the preceding two scalings to capture the behaviour of (VII) in the reacting and non-reacting regions. Such a model was proposed in [186] and developed further in [192] to include the influence of local Da and Ka. This model

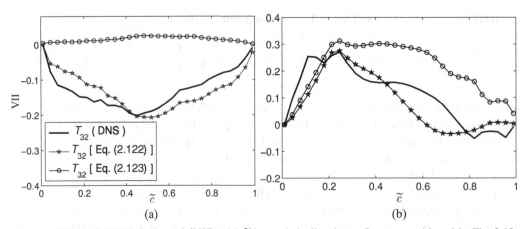

Figure 2.16. Variation of (VII) with $\tilde{c}$ in statistically planar flames considered in Fig. 2.12. The solid line is the DNS result; the line marked with stars is relation (2.122); the line marked with circles is relation (2.123).

for statistically planar flames with unity Le is given as [192]

$$\text{(VII)} \simeq [C_3 - C_4 \, \tau \text{Da}_L^*] \, \overline{\rho} \left(\frac{\tilde{\varepsilon}}{\tilde{k}} \right) \tilde{\epsilon}_c, \tag{2.123}$$

where the model parameters are $C_3 = 1.5$ and $C_4 = 1.1(1 + \text{Ka}_L)^{-0.4}$. A few points to note: (1) The classical model in Eq. (2.94) is recovered when $\text{Da}_L^* \to 0$ or $\text{Ka}_L \to \infty$ or $\tau = 0$ (passive chemical reactions); (2) in the other limit of $\text{Da}_L^* \to \infty$ or $\text{Ka}_L \to 0$ the preceding model yields $\text{(VII)} \approx -\tau \rho_u S_L^0 \tilde{\epsilon}_c / \delta_L^0$, which is independent of the Damköhler number and is consistent with the scaling results in Eqs. (2.108); and (3) comparisons of these model values with the DNS results [191, 192] show that it satisfactorily captures the behaviour of (VII) in high- as well as low-Da situations, as shown in Fig. 2.16. This figure also shows the values given by relation (2.122).

Mura et al. [177] proposed two alternative models:

$$\text{(VII)} \simeq \left[A_1 \frac{\tilde{\varepsilon}}{\tilde{k}} - 2C_A \tau \text{Da} \, \tilde{\epsilon}_c \right] \overline{\rho} \tilde{\epsilon}_c,$$

$$\simeq \left[A_1 \frac{\tilde{\varepsilon}}{\tilde{k}} - 2C_B \ln(\tau + 1)\text{Da} \, \tilde{\epsilon}_c \right] \overline{\rho} \tilde{\epsilon}_c, \tag{2.124}$$

where $A_1 = 1.0$, $C_A = 0.6$, and $C_B = 1.6$ are the values of model constants proposed by Mura et al. [177]. These models are shown to predict the sink contribution of (VII) satisfactorily for high-Da flames but its performance for low-Da combustion is yet to be tested.

The models for (VII) previously given are for statistically planar flames with unity Le. The non-unity Le effects can be included by revising the scaling for heat-release-induced straining, as noted earlier. The revised scaling [168] is (VII) $\sim$ $-\overline{\rho} \tilde{\epsilon}_c \tau f_1(\text{Le}) S_L^0 / \delta_L^0$, which gives [193]

$$\text{(VII)} \simeq \left[\hat{C}_3 - \hat{C}_4 \, \tau f_1(\text{Le}) \text{Da}_L^* \right] \overline{\rho} \tilde{\epsilon}_c \left(\frac{\tilde{\varepsilon}}{\tilde{k}} \right), \tag{2.125}$$

where $\hat{C}_3 = 2$ and $\hat{C}_4 = 1.2(1 + \mathrm{Ka}_L)^{-0.4}$ are model parameters and $f_1(\mathrm{Le}) = (1 - \tilde{c})^{\Phi(\mathrm{Le})}/\mathrm{Le}^p$, with $p = 2.57$ and $\Phi(\mathrm{Le}) = 0.2 + 1.5(1 - \mathrm{Le})$. These model parameters are found [193] to work for Le ranging from 0.3 to 1.2.

The effects of mean curvature on the modelling of (VII) are considered in [205] although the preceding models work satisfactorily when the mean curvature of the flame brush is small. When the mean curvature is large, turbulent eddies of a comparable scale, usually smaller than the large-scale eddies, impart predominant straining. This straining is known [47] to be weak and this diminished effect of turbulent strain is typically considered by means of an efficiency function Γ_k, which depends on u_{rms}^*/S_L^0 and Λ^*/δ_L^0, where u_{rms}^* and Λ^* are the local quantities. Hence the turbulent strain part of (VII) is modelled as $\hat{C}_3 \, \Gamma_k \, \overline{\rho} \, \tilde{\epsilon}_c \, \tilde{\varepsilon}/\tilde{k}$. The influence of mean curvature on the heat-release-induced straining part is modelled as $-C_4 \, \rho_u \, \tilde{\epsilon}_c \, \tau \, S_L'/\delta_L$, where S_L' is a modified flame speed, noted earlier, following the Markstein diffusivity analogy. Now the modelling for (VII) in turbulent kernels is [205]

$$(\mathrm{VII}) = \left[\hat{C}_3 \, \Gamma_k - C_4 \, \tau \left(1 - \frac{\hat{D}}{S_L} \frac{\partial M_\ell}{\partial x_\ell}\right) \mathrm{Da}_L^*\right] \overline{\rho} \, \tilde{\epsilon}_c \left(\frac{\tilde{\varepsilon}}{\tilde{k}}\right), \qquad (2.126)$$

with $\hat{C}_3 = 3$, when Le $= 1$. It is to be noted that this model is for unity Le and it becomes relation (2.123) when the mean curvature is zero.

MODELLING OF MOLECULAR DISSIPATION AND REACTION, (VIII) AND (IX). The dissipation term, $-(\mathrm{VIII})$, is negative semi-definite and it is a dominant sink for evolution of $\tilde{\varepsilon}_c$, as shown in Fig. 2.12, whereas the reactive term (IX) is a dominant source. Thus the combined effect of these two terms, $-(\mathrm{VIII}) + (\mathrm{IX})$, would be a relevant quantity to model. As noted in Eqs. (2.99) and (2.101), Borghi and his co-workers [165, 166] proposed a model for the combined reaction, dissipation, and molecular-diffusion terms by analysing the progress-variable equation for flamelets. This model is written as

$$T_4^* \equiv \frac{\partial}{\partial x_\ell}\left(\overline{\rho \tilde{\alpha}} \frac{\partial \tilde{\epsilon}_c}{\partial x_\ell}\right) - (\mathrm{VIII}) + (\mathrm{IX}) = -\frac{2}{3}\beta \overline{\rho} \frac{\tilde{\epsilon}_c^2}{\tilde{c}(1 - \tilde{c})}\left[\frac{3}{2} - C_{\epsilon_c}\frac{S_L}{\sqrt{\tilde{k}}}\right], \qquad (2.127)$$

where the model constants are $\beta = 4.2$ and $C_{\epsilon_c} = 0.1$. The positive term decreases with increasing Da, which is consistent with the scaling arguments in Eqs. (2.108). The first term, however, is negative semi-definite and it is related to the dissipation process. This model was found [191] to underpredict the DNS results of T_4^* unless the model parameter β is increased by 70%. The reason for this was found [191] to be the positive contribution from the second term in the previous model by analysing the relation between T_4^* and the correlations among the scalar-gradient magnitude, displacement speed, and flame curvatures, using kinematic form of N_c transport equation. This equation is given in [168, 191], and the analyses are presented in [191]. Furthermore, the diffusion term does not require a model as it is closed. Hence a model for $-(\mathrm{VIII}) + (\mathrm{IX})$ is proposed in [191] as

$$-(\mathrm{VIII}) + (\mathrm{IX}) = -\beta_2 \, \overline{\rho} \frac{\tilde{\epsilon}_c^2}{\tilde{c}(1 - \tilde{c})}, \qquad (2.128)$$

following the methodology of Mantel and Borghi [165], with $\beta_2 = 6.7$. This model was shown to work for a variety of flames; planar flames with high and low Da [191], unity and non-unity Lewis numbers [193], and flame kernels [205] with the same values for β_2.

MODELLING OF DENSITY VARIATION EFFECT, (X). The order-of-magnitude estimates in Eqs. (2.108) suggests that this term is of leading order when Da is large. The results shown in Fig. 2.12 confirm this, and a careful study of this figure suggests that this term is of equal magnitude and positive in high-Da as well as low-Da flames. Obviously the positive value implies that this is a source, and the other behaviour is easily understood if one considers $(X) = \overline{2\rho \epsilon_c \, \partial u_k/\partial x_k}$ for the unity Le flames (see the end of the subsection on the effects of thermal expansion). Now the scaling for (X) is $\rho_u \tau (S_L^0/\delta_L^0)^2$, which implies that this term, normalised as in Fig. 2.12, depends only on τ. One can obtain more insight by normalising this term using the turbulence integral time scale τ_T, which gives

$$(X)\tau_T^2 \sim \rho_u \tau \frac{\text{Da}\,\text{Re}_T^{1/2}}{\text{Ka}} \tag{2.129}$$

after $\text{Ka}\,\text{Da} = \text{Re}_T^{1/2}$ is used. One can expect (X) to be small for $\text{Ka} \gg 1$ for a given turbulence Reynolds number, which also implies that $\text{Da} \ll 1$. In the corrugated, wrinkled, and thickened regimes of turbulent combustion, which are broadly denoted by $\text{Da} \gg 1$, this term is important when $\text{Re}_T^{1/2} > \text{Ka}$, which is commonly met in practical flames.

A careful scrutiny of the definition of (X) in Eqs. (2.91) shows that the correlation between the gradients of temperature T and the progress variable influences this term in general. This can be seen clearly if one writes the density gradient $\partial \rho/\partial x_k$ in terms of temperature gradient. The part involving this correlation can be scaled as $\overline{\bar{\alpha}\, \partial c/\partial x_k\, \partial T/\partial x_k} \sim \tilde{\epsilon}_c \delta_L^0 \sigma_T/(\delta_{\text{th}}^0 \sigma_c)$, where δ_L^0 is the slope thickness for c and δ_{th}^0 is the slope thickness for T in unstrained laminar flames. The rms of temperature and progress-variable fluctuations are respectively denoted by σ_T and σ_c. As noted earlier, the dilatation in non-unity Le flames can be written as $\partial u_\ell/\partial x_\ell \sim \tau f(\text{Le}) S_L^0/\delta_L^0 = (\tau/\text{Le}^m) S_L^0/\delta_L^0$. Using the previous two scalings, one can write a model for (X) in general as [193]

$$(X) \simeq 2\, C_X\, \overline{\rho}\, \tilde{\epsilon}_c \left(\frac{\tau S_L^0}{\text{Le}^m \delta_L^0} \right) \left(\frac{\delta_L^0}{\delta_{\text{th}}^0} \frac{\sigma_T}{\sigma_c} \right), \tag{2.130}$$

with C_X as a model parameter.

One has $\delta_L^0 = \delta_{\text{th}}^0$ and $\sigma_T = \sigma_c$ for unity Le and thus the above model becomes

$$(X) \simeq 2\, C_X\, \overline{\rho}\, \tilde{\epsilon}_c \left(\frac{\tau S_L^0}{\delta_L^0} \right), \tag{2.131}$$

with [191] $C_X = B_X/(1 + \text{Ka}_L)^{1/2}$. The value of B_X is of the order of unity, and it is sensitive to combustion kinetics modelling and reactant mixture [142, 207, 209] because of the strong dependence [49, 208] of the SDR and the dilation to the preceding two factors. If a single irreversible chemical reaction is used to model combustion chemistry, then B_X takes a value of about 2 [193]. This model also works for flame kernels, and the correction to S_L by means of Markstein diffusivity analogy is not required because the positive correlation between dilatation and gradient of c seems to compensate for the effects of mean curvature on the dilatation [205].

Swaminathan and Bray [150] noted that (X) must be proportional to the burning mode part of the BML PDF for c [see Eq. (2.4)]. This gives

$$(X) \simeq 2\rho_u \left(\frac{S_L^0}{\delta_L^0}\right) \gamma^* \int_o^1 \left[\frac{\rho \delta_L^{02}}{\rho_u S_L^{02}} N_c \frac{\partial u_\ell}{\partial x_\ell}\right]_L f(c)\, dc$$

$$= 2K_c \overline{\dot{\omega}} \left(\frac{S_L^0}{\delta_L^0}\right) = \frac{4K_c}{(2C_m - 1)} \left(\frac{S_L^0}{\delta_L^0}\right) \overline{\rho} \widetilde{\epsilon}_c, \qquad (2.132)$$

after Eq. (2.86) is used, and the definition of K_c is evident. A point to note is that the parameters K_c and C_m cannot be chosen arbitrarily once the flame structure is prescribed. The value of the model parameter B_X previously given is also fixed because $B_X = 4K_c/\tau(2C_m - 1)$. However, K_c/τ calculated from laminar flames [142] and computed with a complex chemical kinetic mechanism such as GRI–3.0 varies by nearly 60% when the equivalence ratio of the methane–air mixture is changed from 0.6 to 1.4.

Because the contribution of the mean gradient is small in high-Da and Re_T flames, one can write $\overline{\rho}\widetilde{\epsilon}_c \approx \overline{\rho N_c}$. This allows one to obtain the mean dissipation rate from the joint PDF of c and its gradient Δ as [142]

$$\overline{\rho \epsilon_c} = \int_0^1 \int_0^\infty \rho N_c \overline{P}(c, \Delta)\, d\Delta\, dc = \int_0^1 \langle \rho N_c | c \rangle \overline{P}(c)\, dc, \qquad (2.133)$$

where $\langle \cdot | \cdot \rangle$ denotes a conditional average. Here also, one can realise that the contributions come from only the burning mode (internal parts) of $\overline{P}(c)$ because N_c is zero at the ends. Now one can write [142, 209]

$$(X) \simeq 2K_c^* \overline{\rho} \widetilde{\epsilon}_c \left(\frac{S_L^0}{\delta_L^0}\right), \qquad (2.134)$$

where

$$K_c^* \equiv \frac{\delta_L^0}{S_L^0} \frac{\int_0^1 [\rho N_c\, \partial u_\ell/\partial x_\ell]_L\, \overline{P}(c)\, dc}{\int_0^1 [\rho N_c]_L\, \overline{P}(c)\, dc},$$

after taking the conditional averages to be the laminar flamelets values, which an acceptable for high-Da flames. The values of K_c^*/τ vary between 0.8 and 0.9 for methane–air and propane–air mixtures over a broad range of equivalence ratios [142]. Even for hydrogen–air flames, it varies from 0.7 to 0.8 when the equivalence ratio is varied from 1 to 0.4 [210].

2.3.5 Algebraic Models

One can derive algebraic models for the SDR for high-Da flames by balancing the leading-order terms identified by means of OMA and by using their models discussed in the previous subsection. Before an algebraic model is derived, other available models are discussed. The classical model in Eq. (2.94) with $C_D = 1$ includes only a large-scale turbulence time scale. To include the effects of small scales and to account for the contribution of (IX), Kuan et al. [211] suggested

$$\widetilde{\epsilon}_c \simeq \frac{C_\phi}{4} \left[1 + C_\phi^* \frac{\rho_u\, S_L}{\overline{\rho}\, v_\eta} \right] \frac{\widetilde{\varepsilon}}{k} \widetilde{c''^2}, \tag{2.135}$$

based on fractal analysis. The model constants were suggested to be $C_\phi = 4$ and $C_\phi^* = 1.2$ [212]. The Kolmogorov velocity v_η is equal to u_{rms} times $\text{Re}_T^{-1/4}$. This model degenerates to the classical model when c is a passive scalar (with no propagation velocity). One can also see that values predicted by this model will be larger than the classical model value.

By using Eq. (2.86) and its equivalence to $\rho_u\, S_L\, \overline{\Sigma}$ in the FSD approach, Vervisch et al. [213] proposed a model for the mean dissipation rate as

$$\overline{\rho}\widetilde{\epsilon}_c \simeq \frac{(2C_m - 1)}{2} \rho_u\, S_L\, \hat{\mathcal{G}}\, \Xi \left| \frac{\partial \overline{c}}{\partial x_k} \right|, \tag{2.136}$$

where the FSD is related to the flame-wrinkling factor Ξ by $\overline{\Sigma} = \Xi\, |\partial \overline{c}/\partial x_k|$ and the factor $\hat{\mathcal{G}} \equiv \overline{c''^2}/\overline{c}(1 - \overline{c})$ is introduced to ensure that the dissipation rate goes to zero when the progress-variable variance goes to zero. The flame-wrinkling factor needs to be modelled (see Section 2.2).

The balance among the leading-order terms of Eq. (2.90) gives

$$(\text{VII}) - (\text{VIII}) + (\text{IX}) + (\text{X}) \approx 0. \tag{2.137}$$

If one were to ignore the density change across the flame front, then $(\text{X}) = 0$. Using the models in relations (2.122) and (2.127), one obtains

$$\widetilde{\epsilon}_c \simeq \left(1 + \frac{2C_{\epsilon_c} S_L^0}{3\sqrt{k}} \right) \left(C_B \frac{\widetilde{\varepsilon}}{k} \right) \widetilde{c''^2}; \qquad C_{\epsilon_c} = 0.1; \; C_B = \frac{A_e}{\beta} = 0.21, \tag{2.138}$$

when $\sqrt{k}/S_L^0 > 0.067$. If the density change is included through (X) and its model in Eq. (2.132) is done as in [150], then another model for the mean SDR results as

$$\widetilde{\epsilon}_c \simeq \left(1 + \frac{2C_{\epsilon_c} S_L^0}{3\sqrt{k}} \right) \left(C_{B1} \frac{S_L^0}{\delta_L^0} + C_B \frac{\widetilde{\varepsilon}}{k} \right) \widetilde{c''^2}; \qquad C_{B1} = \frac{4K_c}{(2C_m - 1)\beta} = 0.24. \tag{2.139}$$

The chemical time scale appears naturally and explicitly through the term signifying the density change. If one uses the laminar flame scales to normalise Eq. (2.139) then it follows that $\widetilde{\epsilon}_c$ scales as S_L^0/δ_L^0 for large Da [207].

The heat release is noted to change the scalar-gradient alignment, with the principal strain rate resulting in the destruction of scalar gradient by turbulence. This physics is not included in the model given by relation (2.122) for (VII), and thus the preceding algebraic models do not account for this change in scalar mixing physics. One can use relation (2.123) to account for the correct physics. However, the density weighting for Da does not guarantee the physical realisability of $\widetilde{\epsilon}_c$, and thus Kolla et al. [142] proposed a model $(\text{VII}) \simeq [C_{32} - C_4\, \tau\, \text{Da}]\, \overline{\rho}\, \widetilde{\epsilon}_c\, \widetilde{\varepsilon}/\widetilde{k}$, with $C_{32} = 1.5\sqrt{\text{Ka}_L}/(1 + \sqrt{\text{Ka}_L})$ to account for the change in the time-scale ratio across the flame brush. By using this model for (VII), Eq. (2.134) for (X), and Eq. (2.128)

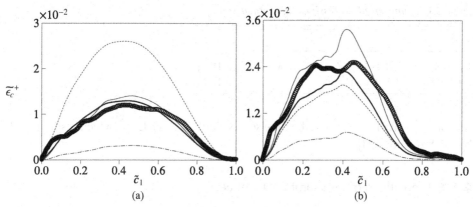

Figure 2.17. Comparison of modelled SDR with DNS results; the values are normalised with S_L and δ_L of a stoichiometric hydrogen–air laminar flame with reactants preheated to 700 K, the mixture used in the DNS. The flame conditions are (a) $\hat{D}a \approx 92$, $Ka = 0.2$, $Re_T \approx 107$ and (b) $\hat{D}a \approx 6$, $Ka = 3$, $Re_T \approx 106$. Symbols, DNS values; thin solid line, Eq. (2.94)/4; dash–dotted line, relation (2.138); dashed line, relation (2.139); thick solid line, Eq. (2.140).

in the leading-order balance, one gets [142]

$$\widetilde{\epsilon}_c \simeq \frac{1}{\beta_2}\left([2K_c^* - \tau C_4]\frac{S_L^0}{\delta_L^0} + C_{32}\frac{\widetilde{\varepsilon}}{\widetilde{k}} \right)\widetilde{c''^2}. \qquad (2.140)$$

The realisability of $\widetilde{\epsilon}_c$ requires that the mean dissipation rate be bounded and greater than or equal to zero. All of the preceding algebraic models clearly satisfy the first condition, and it is also obvious that they, except the last model, satisfy the physical realisability condition $\widetilde{\epsilon}_c \geq 0$. The sufficient condition for the physical realisability is $2K_c^*/\tau - C_4 \geq 0$, and it is straightforward to verify that this condition is always satisfied for the values of K_c^* and C_4 discussed previously. Thus the model in Eq. (2.140) is unconditionally realisable [142]. The values of the mean SDR, normalised by planar laminar flame scales and computed with the preceding models are compared with DNS results in Fig. 2.17. The DNS data of Nada et al. [214] used for this comparison are different from those (shown earlier) used for model development. Nada et al. [214] simulated stoichiometric hydrogen–air flames with preheated reactants by using a complex chemical kinetic mechanism. These flame conditions are given in Fig. 2.17, and the Damkohler number is defined by the Zeldovich flame thickness ($\delta \equiv \hat{\alpha}/S_L^0$) as $\hat{D}a \equiv \Lambda\, S_L^0/(\delta\, u_{\mathrm{rms}})$.

Although the algebraic models previously given are for unity Le cases, a similar approach can be followed for non-unity Le cases and flame kernels by use of the respective models presented earlier. This exercise is left to the reader. However, a simple relationship between the dissipation rate $\widetilde{\epsilon}_T$ of normalised temperature fluctuations and that of the progress-variable (normalised fuel mass fraction) fluctuations $\widetilde{\epsilon}_c$ can be found as follows [193]: The ratio of these two dissipation rates is simply $\widetilde{\epsilon}_T/\widetilde{\epsilon}_c = t_c/t_T$, where the respective time scales are denoted by t. If one defines these time scales by using the respective dissipation cut-off scales and the diffusivities as $t_c = \eta_c^2/\hat{D}$ and $t_T = \eta_T^2/\hat{\alpha}$ then the cut-off scales are the respective Obukhov–Corrsin scales when $Pr < 1$ and $Sc < 1$. The Obukhov–Corrsin scales are related to the Kolmogorov scale by $\eta = \eta_T\, Pr^{3/4} = \eta_c\, Sc^{3/4}$. Hence a simple

Table 2.1. *Flame-speed expressions for comparison*

S_T/S_L^0	Source
$1 - 0.195\Lambda/\delta + \left[(0.195\Lambda/\delta)^2 + 0.78 u_{\mathrm{rms}} \Lambda/s_L^0 \delta \right]^{0.5}$	[108]
$1 + 0.62 (u_{\mathrm{rms}}/s_L^0)^{0.75} (\Lambda/\delta)^{0.25}$	[216]

rearrangement of the ratio of the time scales yields $\widetilde{\epsilon}_c = \widetilde{\epsilon}_T \, \mathrm{Le}^{1/2}$, which is supported by DNS results [194].

2.3.6 Predictions of Measurable Quantities

Although the comparison shown in Fig. 2.17 validates the algebraic model in relation (2.140) and supports the arguments behind this model, it is necessary to verify this model by use of experimental measurements. Reliable and quantitative measurements of the mean SDR, along with the quantities required in relation (2.140), are not yet available because of the challenges involved in accurately measuring the progress-variable gradient and the local turbulence quantities simultaneously. However, one can make an indirect validation of this model by deriving an expression for the propagation speed of the turbulent flame-brush leading edge in the local normal direction. This is achieved by use of the Kolmogorov, Petrovskii, and Piskunov (KPP) analysis [215], used in many earlier studies [142, 147], which gives

$$S_T^2 = 4 \frac{\nu_T}{\rho_u \, \mathrm{Sc}_c} \left(\frac{\partial \overline{\widetilde{\omega}}}{\partial \widetilde{c}} \right)_{\widetilde{c} \to 0} . \tag{2.141}$$

This equation gives the propagation speed as

$$S_T^2 = \frac{8 \nu_T}{(2C_m - 1) \, \mathrm{Sc}_c} \left(\frac{\partial \widetilde{\epsilon}_c}{\partial \widetilde{c}} \right)_{\widetilde{c} \to 0} , \tag{2.142}$$

after Eq. (2.86) is used. If one uses the model in relation (2.140) for the mean SDR, then

$$\left(\frac{S_T}{S_L^0} \right)^2 = \frac{18 \, C_\mu}{(2C_m - 1)\beta_2} \left\{ [2K_c^* - \tau C_4] \left(\frac{u_{\mathrm{rms}} \Lambda}{S_L^0 \delta_L^0} \right) + \frac{2C_{32}}{3} \left(\frac{u_{\mathrm{rms}}}{S_L^0} \right)^2 \right\}, \tag{2.143}$$

where the classical closure for turbulent viscosity $\nu_T = C_\mu \widetilde{k}^2 / \widetilde{\varepsilon}$, where $\widetilde{k} = 3 u_{\mathrm{rms}}^2 / 2$ and $\widetilde{\varepsilon} = u_{\mathrm{rms}}^3 / \Lambda$ are used and $\mathrm{Sc}_c \approx 1$, $C_\mu = 0.09$ [142]. This flame-speed expression is compared with commonly used flame-speed expressions, listed in Table 2.1, and experimental data for a range of flow and flame conditions in Fig. 2.18. The data of Savariananadam and Lawn [217] correspond to lean methane–air flames $(0.75 \leq \phi \leq 0.95)$ stabilised in a wide-angled diffuser for very low turbulence intensities $(u_{\mathrm{rms}}/S_L^0 < 1)$ because the interest was to study the effect of laminar flame instabilities on turbulent flame propagation. The experimental data in Fig. 2.18 are the compilation of Abdel-Gayed et al. [218] for the flame stretch parameter $K = 0.157$, $\mathrm{Ka} = 0.053$, and it covers moderate values of u_{rms}/S_L. The experiments of Il'yashenko and Talantov [219] considered u_{rms}/S_L^0 up to 50. This comparison and those shown in [142, 221] support the flame-speed expression previously given above and its physical basis discussed earlier. Comprehensive analysis and validation of this

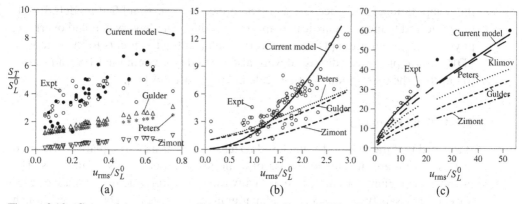

Figure 2.18. Comparison of measured and calculated, by use of Eq. (2.143), turbulent flame speed for a range of flame and flow conditions. The experimental data shown are from (a) Savarianandam and Lawn [217], (b) Abdel-Gayed et al. [218], and (c) Il'yashenko and Talantov [219]. The Klimov line [220] is $S_T/S_L^0 = 3.5(u_{rms}/S_L^0)^{0.7}$. Other flame-speed expressions compared are from Peters [108], Zimont [222], and Gulder [216].

flame-speed expression are given in [142, 221]. The influences of the algebraic closure for the mean dissipation rate in relation (2.140) on the predictions of turbulent premixed flame structures are discussed in [143–145].

2.3.7 LES Modelling for the SDR Approach

In comparison with RANS, the essential advantage of LES is that it allows one to compute explicitly large turbulent scales. Nevertheless the heat release itself occurs at small scales that remain unresolved in LES computations as well as in RANS computations, and important modelling efforts remain to be done to close the filtered reaction rate. Together with numerical developments, the experience that has been gained over the years in the field of turbulent combustion modelling within the RANS context will undoubtedly contribute to improve the fidelity of such numerical simulations of turbulent reactive flows. LESs of turbulent reactive flows provide very challenging issues for the modellers because the chemical reactions take place at the smallest unresolved scales and affect the largest scales of the flow by means of the heat release. SGS transport and chemical reaction remain to be modelled, including the possible occurrence of counter-gradient diffusion, (CGD), flame-generated turbulence (FGT), and all unresolved thermal-expansion phenomena in general, as discussed earlier. The corresponding expansion effects will influence both the unresolved and resolved scalar and velocity transports, which emphasize (if necessary) the importance of providing a satisfactory estimate of the filtered chemical reaction term. Even if some previous studies [53, 85] have confirmed that the relative influence of expansion through CGD and FGT phenomena tends to decrease with the grid resolution, it should be stressed that both the unresolved and the resolved contributions to scalar fluxes and stresses are largely driven by the heat release that takes place at unresolved scales. With this perspective all the recent insights that have been gained about the SDR dynamics will undoubtedly be helpful to improve the the modelling of the filtered reaction rate. Nevertheless, within the LES approach, the computational costs become a critical issue, and oversimplified models are often preferred to more sophisticated approaches that include very detailed overlying physics; but recent attempts that are very similar in their contents to the algebraic SDR closure previously discussed were also carried out within the LES framework to improve the description of small-scale mixing processes [63].

2.3.8 Final Remarks

It is noted that the models for unclosed terms in the $\tilde{\varepsilon}_c$ transport equation discussed in this section are developed by use of a limited number of DNS databases. The proposed modelling parameters, specifically used to capture the heat release effects, are chosen to satisfy their expected behaviour in the limit of large Da or Ka. Although these models are validated by measured turbulent flame speed and the results from DNS data different from those used for model development, more comprehensive validation is needed. The sensitivity of the proposed model parameters to a turbulent Reynolds number needs to be studied through experimental investigations.

The modelling techniques previously discussed are strictly valid for fully premixed flames under adiabatic conditions. Some of these constraints need to be relaxed to make the modelling technique more versatile. Recently these modellings were successfully extended to turbulent partially premixed combustion [141, 222–225]. The RANS modelling discussed in this section needs to be extended for LES, as noted earlier.

Finally, attention in this section was focused on the closure of the averaged SDR. Conditionally averaged dissipation rates are required in approaches such as the CMC [96] and JPDF approach [91]. Although a simple approach is proposed [145], a rigorous closure for the conditional dissipation rate still remains challenging because of strong coupling among micromixing phenomena, chemical reaction, and turbulence [153, 226]. The corresponding issue is discussed in the next section which is devoted to the transported JPDF approach.

Acknowledgements

The research of M. Champion and A. Mura is funded by the CNRS. Part of the research work presented here benefited from the financial support of INTAS Program 2000-353, 'Development and comparative analysis of different approaches to micromixing processes in turbulent reactive flows', from the ANR Program NT05-2-42482, 'Micro-Mélange', and from the support of industrial partners (Snecma DMS, EDF). The authors also benefited from interesting discussions with R. Borghi, K.N.C. Bray, M. Gonzalez, and V. Robin. The support of EPSRC is acknowledged by Chakraborty and Swaminathan. The help of Dr Kolla in generating Figs. 2.17 and 2.18 is acknowledged.

2.4 Transported Probability Density Function Methods for Premixed Turbulent Flames
By R. P. Lindstedt

The transported PDF approach has the significant advantage of avoiding difficulties associated with the inclusion of chemical kinetic effects into computational methods for turbulent reacting flows [91, 227]. The terms relating to chemical reaction appear in closed form as convection in scalar space, and, accordingly, kinetically controlled or influenced phenomena can in principle be readily included. A particularly attractive consequence is that independently developed chemical mechanisms can be applied without further simplification. The formulation of more complete closures is non-trivial as inconsistencies in submodels may lead to accuracy or realisability problems (e.g. [17, 228, 229]), be it in the context of LES- or moment-based methods. The transported PDF approach has the potential to overcome difficulties related to the coupling of chemical reaction and spatial transport (turbulent convection) and may be used as subfilter-scale models for LES by means of filtered density function (FDF) methods. The coupling between turbulent transport and chemical reaction becomes significant in view of the delicate balance of gradient and non-gradient scalar transport in turbulent premixed flames [7, 17, 44, 228].

Most previous work on transported PDF closures for flames featured non-premixed combustion in which, typically, scalar transport is more readily addressed. For example, work on piloted diffusion flames (e.g., the Sandia flame series [230]) enabled stable flames and those undergoing extinction or reignition to be studied through a combination with augmented reduced [231–234] or detailed [235] chemical reaction mechanisms. Studies featuring more complex flow fields were also performed. For example, Muradoglu et al. [236] modelled a bluff-body stabilized flame using a joint scalar approach by applying time averaging after a statistically stationary solution was obtained. Kuan and Lindstedt [237] performed computations of a bluff-body flame [238, 239] by using a hybrid Monte Carlo–Finite-volume algorithm with the velocity field closed at the second-moment level. The thermochemistry was computed by means of a systematically reduced mechanism featuring 300 reactions, 20 solved, and 28 steady-state species. Studies of flames featuring auto-ignition [235, 240], significant local extinction [232–234], or both, suggest that ignition tends to occur in a partially premixed mode and that the transported PDF approach can reproduce experimental data with encouraging accuracy. Bidaux et al. [241] used a scalar PDF approach and a particle-interaction mixing model with the mean reaction rate determined from the PDF solution to predict a partially premixed Bunsen flame.

The attraction of extending moment-based methods to include finite-rate chemistry effects through the application of transported PDF methods to premixed turbulent flames is of increasing practical interest. For example, flows found in burners, gas turbine combustion chambers, and reciprocating engines cover a wide range of Damköhler (Da) and Karlovitz (Ka) numbers and are increasingly characterised by strong interactions between turbulence and finite-rate chemistry because of changes in burning modes [242–244]. One implication is the need to extend the knowledge gained from the application of high-Damköhler-number-based approaches to encompass an extended range of conditions. The latter may include (re)ignition phenomena, dilution effects, local or global flame extinction, non-adiabatic conditions, and the presence of a multitude of chemical time-scales (e.g., the fast heat release vs. the relatively slow formation of pollutants).

Comparatively few studies of premixed turbulent combustion were performed with transported-PDF-based methods. Pope and Cheng [245] considered the evolution of spherical flame kernels by using a simplified Langevin model [91] to account for the effects of viscous dissipation and the fluctuating pressure gradient. The mean pressure gradient acted upon the fluid irrespective of the instantaneous density, and thus preferential acceleration effects could not be observed. Anand and Pope [66, 226] extended the work and applied a joint velocity–reaction progress-variable formulation to calculate steady premixed flames in the flamelet regime of combustion. The viscous dissipation of velocity and the fluctuating pressure gradient were incorporated by means of a particle-interaction model featuring stochastic reorientation [91]. The chemical source term incorporated the coupled effects of chemistry and molecular diffusion. The work showed that the transported PDF approach can reproduce key features of the BML flamelet approach developed by Bray et al. [7], as reviewed in earlier sections.

Hůlek and Lindstedt [246] computed steady and transient 1D flames for a range of turbulent Reynolds (Re_T) numbers by using an elliptic formulation closed at the joint velocity–scalar level. The work extended past efforts through the application of the comprehensive Haworth and Pope [247, 248] and Speziale, Sarkar, and Gatski [249] closures for velocity statistics combined with the binomial Langevin model [250] for molecular mixing. It was shown that the approach can reproduce conditional velocities of burned and unburned gases as well as the change in anisotropy across the turbulent flame brush. It was also shown that the predicted slip velocity ($u_b - u_u$) at the leading edge ($\tilde{c} \to 0$) compared reasonably well with the theoretical analysis of Bray [147]. The impact of alternative formulations for the contributions made by small-scale diffusion processes was analysed, and it was found that direct extraction of source terms from laminar flamelet computations provided a suitable means of predicting the heat release at moderate turbulence intensities. It was also shown that the conventional model for the SDR leads to the expected linear scaling of the turbulent burning velocity with turbulence fluctuations.

Lindstedt and Váos [251] subsequently investigated the effects of mixing models on the relationship between computed turbulent burning velocities and flame structures. The closures considered included the Linear–Mean–Square–Estimate (LMSE) model [252], the nonlinear integral binomial sampling [253], and modified Curl's [254] models as well as a binomial Langevin-based formulation [250, 255]. It was shown that the modified Curl's and binomial Langevin models can be expected to produce agreement with experimental data whereas the other models suffered from significant shortcomings.

More recently, Lindstedt and Váos [211] extended their past work through an investigation of the effects of closure approximations for the SDR on premixed turbulent flame structures and on the scaling relationship between turbulence fluctuations and the turbulent burning velocity. It was shown that a multiscale extension to the conventional closure was capable of reproducing turbulent burning velocities across a wide range of turbulence intensities. The study also considered the piloted methane–air flames investigated experimentally by Chen et al. [256], with the most turbulent flame featuring a Re $\simeq 52\,500$ and a Da ~ 1. The transported PDF approach was closed at the joint scalar level with the velocity field computed at the second-moment level [249]. The thermochemistry was closed using an augmented systematically reduced C–H–O mechanism featuring 141 reactions, 15 solved, and 14 steady-state species. The extended multiscale mixing time-scale closure [211] was also applied by Lindstedt et al. [257] in the context of partially premixed flames at high Reynolds numbers. It was found that the standard time scale expression had a tendency to produce excessive extinction.

Few studies featuring more complex flow fields were pursued. However, Cannon et al. [258] computed a bluff-body stabilised lean premixed methane–air flame by using an elliptic calculation procedure combined with the k-ε model for the velocity field and with the molecular mixing term closed by use of the LMSE model. Sabel'nikov and Soulard [259] applied a stochastic Eulerian field method to compute a stoichiometric methane–air flame stabilised in a backward-facing step configuration using the k–ε closure coupled with the LMSE model for molecular mixing and a one-step chemical reaction. In summary, the majority of past studies focussed on the application of simplified chemistry combined with different levels of closure for the velocity and scalar fields. In the subsequent subsections, some of the key closure aspects are reviewed and their potential impact assessed.

2.4.1 Alternative PDF Transport Equations

The brief outline just presented suggests that work remains on formulating approaches for the treatment of terms appearing as part of the application of the transported PDF approach to premixed turbulent flames. It is evident that additional complexities associated with turbulent transport arise for premixed flames and that in many cases very well-established closures and simple approximations for the modelling of chemical reaction and molecular mixing have been applied. The subsequent discussion considers the impact of closure assumptions on premixed and partially premixed flames at low and high Da in the context of joint scalar and joint velocity–scalar methods. The impact of turbulent transport, molecular mixing models, unclosed pressure-related terms, and different levels of accuracy in the applied thermochemistry are considered.

The JPDF of composition and velocity in low-Mach-number flows is considered for premixed and partially premixed flows. In the latter case, the composition is described by two (c, Z) scalars, where Z denotes a conserved and c a chemically reacting scalar. The pressure is assumed to be thermochemically and thermodynamically constant. Under such conditions the evolution equation for the mass density function (MDF) $\mathcal{F}_{zcu}(\xi, \zeta, \mathbf{U}, \mathbf{x}; t) = \rho(\xi, \zeta) f_{zcu}(\xi, \zeta, \mathbf{U}; \mathbf{x}, t)$, where f_{zcu} is the JPDF of composition and velocity, can be derived from basic conservation laws [91]:

$$\frac{\partial \mathcal{F}_{zcu}}{\partial t} + \frac{\partial u_k \mathcal{F}_{zcu}}{\partial x_k} = \frac{1}{\rho(\xi, \zeta)} \frac{\partial \overline{p}}{\partial x_k} \frac{\partial \mathcal{F}_{zcu}}{\partial U_k} - \frac{\partial S_c(\xi, \zeta) \mathcal{F}_{zcu}}{\partial \zeta}$$

$$+ \frac{\partial}{\partial U_k} \left[\frac{1}{\rho(\xi, \zeta)} \left\langle -\frac{\partial \tau_{k\ell}}{\partial x_\ell} + \frac{\partial p'}{\partial x_k} \middle| \xi, \zeta, \mathbf{U} \right\rangle \mathcal{F}_{zcu} \right]$$

$$+ \frac{\partial}{\partial \xi} \left[\frac{1}{\rho(\xi, \zeta)} \left\langle \frac{\partial J^z}{\partial x_k} \middle| \xi, \zeta, \mathbf{U} \right\rangle \mathcal{F}_{zcu} \right]$$

$$+ \frac{\partial}{\partial \zeta} \left[\frac{1}{\rho(\xi, \zeta)} \left\langle \frac{\partial J^c}{\partial x_k} \middle| \xi, \zeta, \mathbf{U} \right\rangle \mathcal{F}_{zcu} \right]. \tag{2.144}$$

In the MDF equation, p is the pressure, $\tau_{k\ell}$ is the viscous stress tensor, S_c is the chemical reaction source term of the scalar c, and J_k^z and J_k^c are the molecular-diffusion flux vectors of the scalars Z and c, respectively. The notation $\langle \mathbf{A} | Q \rangle$ expresses the expected value of **A** conditioned on the event Q, and the prime $'$ indicates the difference from the expected value (i.e., $\phi' = \phi - \langle\phi\rangle$). Mean and fluctuating quantities based on Favre averaging are denoted as $\phi = \tilde{\phi} + \phi'' = \overline{\rho\phi}/\overline{\rho} + \phi''$. The quantities ξ, ζ, and U are sample-space counterparts of the actual values Z, c, and u, respectively. The MDF conservation equation contains, in closed form, the effects of mean and turbulent convection, the mean pressure gradient acceleration, and chemical reaction. Unclosed terms express the molecular mixing of scalars, viscous dissipation of momentum, and the fluctuating pressure gradient. In addition, the mean pressure field has to be solved in a manner that guarantees consistency of the MDF with mass conservation. The corresponding MDF equation for a fully premixed case is [the notation is the same as applied in Eq. (2.144)]

$$\frac{\partial \mathcal{F}_{cu}}{\partial t} + \frac{\partial u_l \mathcal{F}_{cu}}{\partial x_l} = \frac{1}{\rho(\zeta)} \frac{\partial \overline{p}}{\partial x_l} \frac{\partial \mathcal{F}_{cu}}{\partial U_l} - \frac{\partial S(\zeta)\mathcal{F}_{cu}}{\partial \zeta}$$

$$+ \frac{\partial}{\partial U_k}\left[\frac{1}{\rho(\zeta)}\left\langle -\frac{\partial \tau_{k\ell}}{\partial x_\ell} \middle| c = \zeta, \mathbf{u} = \mathbf{U}\right\rangle \mathcal{F}_{cu}\right]$$

$$+ \frac{\partial}{\partial U_k}\left[\frac{1}{\rho(\zeta)}\left\langle \frac{\partial p'}{\partial x_k} \middle| c = \zeta, \mathbf{u} = \mathbf{U}\right\rangle \mathcal{F}_{cu}\right]$$

$$+ \frac{\partial}{\partial \zeta}\left[\frac{1}{\rho(\zeta)}\left\langle \frac{\partial J_k}{\partial x_k} \middle| c = \zeta, \mathbf{u} = \mathbf{U}\right\rangle \mathcal{F}_{cu}\right]. \tag{2.145}$$

The application of the preceding equations to different cases is reported in order to explore the importance of different closure elements. The further simplification of the approach is also considered through the application of the corresponding joint composition PDF equation in combination with alternative closures for the velocity field. The joint composition PDF results from integration of Eq. (2.145) over the full velocity space and the decomposition of convection in physical space into an unconditional mean part and a conditional intensity part:

$$\frac{\partial \overline{\rho}\tilde{P}(\Psi)}{\partial t} + \frac{\partial \overline{\rho}\tilde{u}_l\tilde{P}(\Psi)}{\partial x_l} + \sum_{\alpha=1}^{N} \frac{\partial}{\partial \Psi_\alpha}\left[\overline{\rho}S_\alpha(\Psi)\tilde{P}(\Psi)\right]$$

$$= -\frac{\partial}{\partial x_l}\left[\overline{\rho} < u_l''|\varphi = \Psi > \tilde{P}(\Psi)\right] + \sum_{\alpha=1}^{N} \frac{\partial}{\partial \Psi_\alpha}\left[\left\langle \frac{1}{\rho}\frac{\partial J_{l,\alpha}}{\partial x_l} \middle| \varphi = \Psi\right\rangle \langle\rho\rangle\tilde{P}(\Psi)\right]. \tag{2.146}$$

At this closure level, a model is required for the turbulent transport of the PDF in physical space and, typically, a gradient-diffusion-type closure is used,

$$\langle u_l''|\varphi = \Psi\rangle\tilde{P}(\Psi) = \frac{v_t}{\sigma_t}\frac{\partial \tilde{P}(\Psi)}{\partial x_l}, \tag{2.147}$$

where v_t is the 'standard' kinematic eddy-viscosity coefficient and σ_t is the turbulent Prandtl or Schmidt number. It is also possible to formulate a second-moment-based closure for the transport of the PDF in physical space. Dopazo [260] suggested a closure in which the conditional intensity terms are approximated with a deterministic LMSE-type model:

$$\langle u_l''|\varphi = \Psi\rangle\tilde{P}(\Psi) = \frac{\widetilde{u_l''\varphi_m''}(\Psi_n - \tilde{\varphi}_n)}{\widetilde{\varphi_m''\varphi_n''}}\tilde{P}(\Psi). \tag{2.148}$$

This model was also proposed in the context of moment-based closures by Borghi and Dutoya [164] and Libby [261], with possible extensions outlined by Pope [227]. For a single scalar, the conditional expectation may be written in the familiar form as

$$\langle u_l'' | c = \zeta \rangle \tilde{P}(\zeta) = \frac{\widetilde{u_l'' c''}(\zeta - \tilde{c})}{\widetilde{c''^2}} \tilde{P}(\zeta). \tag{2.149}$$

Swaminathan and Bilger [262] evaluated the preceding linear model against DNS data obtained for lean hydrogen flames with $u'/u_L \simeq 30$. Two alternative formulations based on gradient and PDF models were also considered. The work showed that the preceding model captures DNS data comparatively well, whereas the alternative formulations produced less accurate results – possibly because of insufficient samples. The application of the model in the context of transported PDF closures was very limited and is explored in the next subsection. The impact of the treatment of molecular mixing is also discussed further in the context of specific examples.

2.4.2 Closures for the Velocity Field

The velocity field may be obtained through LES or at the equivalent second-moment or eddy-viscosity levels. In view of the anisotropy typical of premixed turbulent flames, the latter is considered mainly for comparison purposes, and the treatment of pressure-related terms has to be consistent across scales. Analysis of the Reynolds stress equations show [263] that, in variable-density, low-Mach-number turbulent flows, the effects of the anisotropic part of the fluctuating pressure gradient and turbulence dissipation are *redistributive*. In other words, the terms are traceless in tensorial form and therefore do not appear in the turbulent kinetic energy ($\tilde{k} = \widetilde{u_k'' u_k''}/2$) conservation equation. The latter condition is automatically satisfied by the generalised Langevin model (GLM) of Haworth and Pope [264], and the velocity statistics (viscous dissipation of momentum and fluctuating pressure gradient) presented here are based on their approach and other suitably modified closures, such as the Lagrangian version [265] of the Speziale–Sarkar–Gatski (LSSG) model [249]. However, the intention is not to evaluate specific second-moment methods, but to study the impact of closures at the equivalent first- and second-moment levels alongside other model aspects. The k-ε and GLM models are thus used as representative examples of each class.

In the presence of volumetric changes, the so-called 'pressure–dilatation' term $\langle p' \partial u_k'' / \partial x_k \rangle$ appears. There is some evidence that pressure–dilatation may not be important when the turbulence intensity u_{rms} (and thus the turbulent burning velocity u_T) is much larger than the laminar burning velocity u_ℓ [266]. Furthermore, issues with respect to consistency and realizability remain [17], and the matter is subsequently treated separately. These considerations lead to a Lagrangian model of the form

$$du_k^{(p)} = \left[-\frac{1}{\rho} \frac{\partial \langle p \rangle}{\partial x_k} + G_{k\ell}(u_\ell^{(p)} - \tilde{u}_\ell) \right] dt + (C_0 \tilde{\varepsilon})^{1/2} \, dW_k, \tag{2.150}$$

where the superscript $^{(p)}$ denotes the properties of a Lagrangian particle (solution methods are discussed in Subsection 2.4.5), W_k is an isotropic Wiener process [267], $\tilde{\varepsilon}$ is the dissipation rate of $\tilde{k}$, and C_0 is assumed to be a constant $C_0 = 2.1$ [265].

The latter value was discussed by Pope [268], and it has been shown, by comparison with DNS data, that some uncertainties prevail. However, in this section the 'standard' value is retained, and the tensor of modelling coefficients $G_{k\ell}$ is given in Appendix 2.A.

A modification of the 'standard' Lagrangian velocity statistics model is also investigated in order to assess the importance of solving the joint composition–velocity statistics by introducing a modification so that the acceleration by the mean pressure gradient depends on the mean rather than on the local instantaneous density. Equation (2.150) then changes to

$$du_k^{(p)} = \left[-\frac{1}{\langle \rho \rangle} \frac{\partial \langle p \rangle}{\partial x_k} + G_{k\ell}(u_\ell^{(p)} - \widetilde{u}_\ell) \right] dt + (C_0 \widetilde{\varepsilon})^{1/2} \, dW_k. \tag{2.151}$$

This formulation is equivalent to a second-moment closure in which (1) the pressure–work term $- \langle u_k'' \rangle \, \partial \langle p \rangle / \partial x_\ell$ in the Reynolds stress conservation equations – and (2) the terms $- \langle c'' \rangle \, \partial \langle p \rangle / \partial x_k$ and $- \langle z'' \rangle \, \partial \langle p \rangle / \partial x_k$ – in the turbulent scalar flux conservation equations are neglected. The terms are mainly responsible for the observed 'counter-gradient' (or non-gradient) turbulent transport in flames [44], and comparisons of the result obtained with models (2.150) and (2.151) should indicate the importance of this effect in the context of premixed and partially premixed flames, as subsequently discussed.

2.4.3 Closures for the Scalar Dissipation Rate

The closure for the mixing frequency is directly related to the closure for the SDR ($\widetilde{\varepsilon}_c$). Fox [269] argues that in premixed turbulent flames a strong dependence between the joint SDR $\langle \epsilon_\varphi | \Psi \rangle$ and the chemical source term can be expected and that the majority of mixing models omit this feature. The customary scaling relationship that has a direct impact on the treatment of the molecular mixing term is given by

$$\tau_c^{-1} = \frac{\widetilde{\varepsilon}_c}{\widetilde{c''^2}} = \frac{C_\phi}{2} \frac{\widetilde{\varepsilon}}{\widetilde{k}} = \frac{C_\phi}{2} \tau_t^{-1}. \tag{2.152}$$

This relationship provides a direct link to moment-based closures for premixed turbulent flames. Bray [147] showed that any SDR model corresponds to a specific mean-reaction-rate model at the high-Damköhler-number limit. Hence the assumption of a constant value for the velocity–scalar time-scale ratio results in an EBU-type expression for the mean reaction rate, as suggested by Spalding [270]:

$$\langle \rho \rangle S_c \propto \langle \rho \rangle \widetilde{\varepsilon}_c = \langle \rho \rangle \tau_c^{-1} \widetilde{c''^2} = \langle \rho \rangle \frac{C_\phi}{2} \tau_t^{-1} \widetilde{c''^2} = \langle \rho \rangle C_R \tau_t^{-1} \widetilde{c''^2}. \tag{2.153}$$

The assumption of a bimodal PDF ($\widetilde{c''^2} = \widetilde{c}(1 - \widetilde{c})$) leads to the standard BML form [147] with values of $C_R \simeq 4$ [17]. Transport-based closures for the SDR are typically based on non-reacting flows and have been derived by Jones and Musonge [167] and Borghi and Mantel [165], among others. A detailed review of such approaches can be found in Section 2.3 of this book. The scaling is typically exclusively based on large-scale (integral) properties with added terms featuring a dependency on laminar flame properties. Kuan et al. [210] used a fractal-based approach [271] to derive a multiscale expression for the time-scale ratio to alleviate the former limitation:

$$\tau_c^{-1} = \frac{\widetilde{\varepsilon}_c}{\widetilde{c''^2}} \propto \frac{\rho_u}{\langle \rho \rangle} \frac{u_L}{v_K} \frac{\widetilde{\varepsilon}}{\widetilde{k}}. \qquad (2.154)$$

Yeung et al. [272] investigated the effects of $\widetilde{\varepsilon}_c$ on material surfaces for cases in which $u_L \ll v_K$ and suggested that the Kolmogorov velocity provides an appropriate scaling. Bray et al. [273] considered closures for the chemical reaction rate and concluded that the majority of algebraic and transport-equation-based models could not be calibrated to perform satisfactorily irrespective of the choice of modelling constants. The simple form of Eq. (2.154) was found to be an exception. The model implies a monotonic increase of the u_L/v_K ratio with the Da number in agreement with the experimental study of O'Young and Bilger [157], which also indicated that the velocity–scalar ratio of time scales increases by almost an order of magnitude as Da $\rightarrow \infty$. A disadvantage is that τ_c^{-1} becomes zero for a passive scalar, and Lindstedt and Váos [211] proposed a simple modification as

$$\tau_c^{-1} = \frac{\widetilde{\varepsilon}_c}{\widetilde{c''^2}} = \frac{C_\phi}{2} \left[1.0 + C_\phi^* \frac{\rho_u}{\langle \rho \rangle} \frac{u_L}{v_K} \right] \frac{\widetilde{\varepsilon}}{\widetilde{k}}. \qquad (2.155)$$

The form of Eq. (2.155) ensures a consistent scaling behaviour of turbulent burning velocities for flames in the flamelet regime of combustion. The expression also complies with the standard approach for the limiting case of a passive scalar. The constant C_ϕ^* was assigned a value of 1.2 [210]. An extensive review of alternative algebraic expressions was presented in Section 2.3, though these have not as yet been evaluated in the context of serving as closure elements for transported PDF models.

2.4.4 Reaction and Diffusion Terms

The terms responsible for reaction and diffusion may be expressed in the customary form as

$$\frac{\partial}{\partial \psi_\alpha} \left[\left\langle -\rho(\psi)S_\alpha(\psi) + \frac{\partial J_k^\alpha}{\partial x_k} \middle| \psi \right\rangle f_\phi \right] = \frac{\partial}{\partial \psi_\alpha} \left[h_\alpha(\psi) f_\phi \right]. \qquad (2.156)$$

The quantity $h_\alpha(\psi)$ appears in closed form for laminar flamelets. In general, two different aspects need to be considered in the modelling of the conditional expectation of the molecular scalar flux term in Eq. (2.156), as discussed by Haworth [275]. The first is transport in physical space, and the second is mixing in composition space. At high Reynolds numbers and for Sc and Pr numbers of the order of unity, the term describing transport in physical space is typically small. Furthermore, the implicit coupling present in the laminar flamelet assumption will weaken at lower Damköhler numbers and can be expected to break down in the well-stirred reactor regime. However, for premixed combustion at high Damköhler numbers the term can be important and the effect may be included by means of the addition of a random-walk term in the equation for the particle position in physical space [275]. The approach provides a correction for the mean scalar composition at the expense of a spurious production term in the scalar variance equation [275, 276]. An alternative approach, which does not lead to such difficulties [275], is to add an explicit deterministic molecular-transport term in the manner implied by Eq. (2.156). The impact of such approximations at the leading edge of the flame was explored by

Lindstedt et al. [277], and McDermott and Pope [278] provided an extensive analysis of diffusion approximations in the context of FDF methods.

2.4.5 Solution Methods

Solution techniques for transported PDF methods were recently reviewed by Haworth [275], and the following discussion is confined to the approach applied to obtain the illustrative examples shown. The PDF evolution equations were solved with a Lagrangian-particle-based Monte Carlo approach combined with fractional step method [91] featuring a first-order (Euler) time integration. The integration of particle position and velocity follows the procedure of Haworth and Pope [279]. The masses of modelling particles were modified at each time step to maintain a uniform particle density. For axisymmetric flows the mass of each particle (p) is therefore approximately proportional to the value $r^{(p)}\rho^{(p)}$ [r is the distance from the symmetry axis and $\rho^{(p)}$ is the local instantaneous density].

The mean momentum and mass conservation equations were solved with a Pressure Implicit with Split Operator (PISO)-based [17, 280] finite-volume solver, and the governing equations are given in Appendices 2.B and 2.C. The time discretisation was first-order implicit and second-order central differencing was used for the diffusive terms. The convective terms were discretised by van Leer's total variation diminishing (TVD) scheme with a limited diffusion antiflux modified for implicit time discretisation [281]. The stochastic and finite-volume parts of the algorithm were coupled as follows. The scalar statistics obtained from the Monte Carlo PDF solution were used to evaluate the mean density (i.e., to close the mean equation of state) in the pressure–momentum solver. Depending on the method used (first- or second-moment closure), the PDF solution was used to provide the unclosed correlations in the momentum as well as $\widetilde{k}$ and $\widetilde{\epsilon}$ equations ($\widetilde{u''_k u''_\ell}$ and $\langle u''_k \rangle$). The finite-volume solution for the mean velocity field was then used in the Monte Carlo part to provide the mean pressure gradient and turbulence time scales.

Following each time step, the particle velocities were adjusted so that the mean velocity and $\widetilde{k}$ in the finite-volume and Monte Carlo solutions were consistent [282]. To decrease the statistical error resulting from the use of a finite number of stochastic particles, time averaging was used in evaluating mean profiles such as $\widetilde{c}$, the anisotropy tensor $\widetilde{c'' u''_k}$, or $\langle p \rangle$. The number of particles used in the simulations, ≥ 60 per cell combined with the additional time-averaging procedure, resulting in excess of an order-of-magnitude increase in effective particle numbers, is arguably sufficient to reduce statistical errors to an acceptable level [231, 283]. For cases featuring joint velocity–scalar closures, the mean profiles obtained from the PDF solution were approximated with tensor-product cubic smoothing splines [284] with the smoothing coefficient obtained with a cross-validation technique [285]. Reynolds stress profiles are not suitable for fitting with smooth interpolants because values vary by orders of magnitude within the flow field and because sharp peaks may be observed in shear layers and similar regions. Therefore, instead of fitting the Reynolds stresses directly, the smoothing splines were fitted to the Reynolds stress anisotropy tensor. The anisotropy is of the order of 10^{-1}, and the profiles (typically), do not feature sharp gradients.

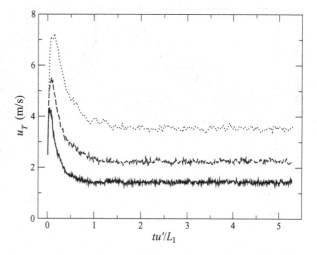

Figure 2.19. The computed evolution of u_T [211] for stoichiometric CH_4–air mixtures ($\Lambda = 20$ mm) with $u_{rms} = 0.5$ m/s (bottom), $u_{rms} = 1.0$ m/s (middle), and $u_{rms} = 2.0$ m/s (top). Transported PDF calculations feature the modified Curl's model [254] and Eq. (2.155) with $C_\phi = 4$ and $C_\phi^* = 1.2$.

2.4.6 Freely Propagating Premixed Turbulent Flames

The impact of the mixing frequency closures given by Eqs. (2.152) and (2.155) on predictions of the turbulent burning velocity was evaluated by Lindstedt and Váos [211] by using freely propagating planar flames. It was shown by Hakberg and Gosman [286], through the application of a KPP-type [214] analysis, that the governing equations should yield a linear scaling of the turbulent burning velocity with the turbulence intensity. The same result is obtained numerically with moment-based methods, provided that the intrinsic length and time scales corresponding to the governing equations are well resolved [287], as confirmed by Sabel'nikov et al. [288]. For transient propagating flames, the flame thickness scales on the integral length scale of turbulence. The need to resolve the intrinsic length and time scales inherent in the solution of laminar reacting flows and in DNSs is well established, and it is unsurprising that similar considerations apply in the context of model-based simulations of turbulent flames. For transported PDF approaches, the functional form of Arrhenius-type reaction-rate expressions at the leading edge of the flame ($dS_c/dc \to 0$, $c \to 0$) obviates the need to apply a Heaviside function to prevent the growth of numerical round-off errors [287]. On the other hand, the solution procedure features an explicit and first-order-accurate (overall) fractional step method, and it was shown by Lindstedt and Váos [211] that computed values of the turbulent burning velocities are consistent, provided the time step does not exceeds a certain value (here $\simeq 10^{-5}$ s). Examples of the temporal evolution of turbulent burning velocities are shown for methane flames at different Reynolds numbers in Fig. 2.19. The chemical source term was extracted from a laminar flame computation, and the modified Curl's model [254] was used to represent mixing in scalar space. The transport of the PDF in physical space was closed by use of the gradient approximation given in Eq. (2.147).

Hulek [283] showed that the interaction between the mixing model and the reaction rate is mainly confined to a region close to the "cold" boundary and that the width of the region is inversely proportional to the Damköhler number. In this region, the mixing model induces transport of the discrete PDF in scalar space so that the reaction rate becomes effective. Accordingly, the predicted burning velocities will

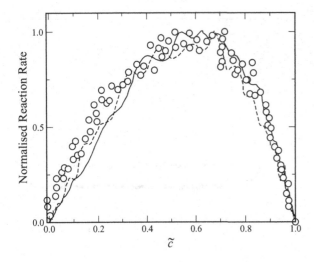

Figure 2.20. Normalised mean-reaction-rate profiles in scalar space [211] along with the probability of reaction (normalised crossing frequency) measurements (o) of Cheng and Shepherd [35]. Lines: (—) modified Curl's model [254] and Eq. (2.155) with $C_\phi = 4$ and $C_\phi^* = 1.2$; (– – –) modified Curl's model [254] and Eq. (2.155) with $C_\phi = 4$ and $C_\phi^* = 0$. In both cases $u_{rms} = 0.5$ m/s and $\Lambda = 20$ mm.

depend, first, on the structure of the model for turbulent transport in scalar space (i.e., the structure of the prescribed *mixing* transition density function) and, second, on the magnitude of the relevant (mixing) time scale. Examples of computed mean-reaction-rate profiles obtained with both the standard Eq. (2.152) and extended Eq. (2.155) algebraic closures for the SDR are shown in Fig. 2.20. The profiles were normalised by the maximum mean reaction rate to facilitate comparisons and essentially show the probability of reaction. A comparison with the experimental measurements by Cheng and Shepherd [35], obtained for a stoichiometric ethylene–air mixture ($u_L \simeq 0.67$ m/s) with $u_{rms} \sim 0.5$ m/s, reveals that the computed shape is in reasonable agreement with experimental data for the crossing frequency. The statistical noise visible in Fig. 2.20 does not influence the computed turbulent burning velocities strongly, as is evident from examples of the temporal evolution of u_T shown for $u_{rms} = 0.5$, 1.0, and 2.0 m/s in Fig. 2.19.

The customary (near-) linear relationship [66, 246] of u_T/u_{rms} is not satisfactory at the high-Damköhler-number limit [17]. Comparison with the experimental data of Abdel-Gayed et al. [274], presented in Fig. 2.21, indicates that the extended closure for the SDR given in Eq. (2.155) results in significantly improved agreement for $1 < u_{rms}/u_\ell < 7$. The closure results in a burning velocity scaling behaviour that

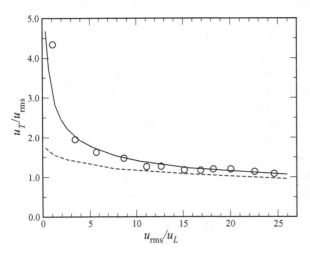

Figure 2.21. Predicted turbulent burning velocities for stoichiometric CH_4–air mixtures ($\Lambda = 20$ mm) [211]. Transported PDF calculations feature (—) the modified Curl's model [254] and Eq. (2.155) with $C_\phi = 4$ and $C_\phi^* = 1.2$; (– – –) modified Curl's model [254] and Eq. (2.155) with $C_\phi = 4$ and $C_\phi^* = 0$. Experimental burning velocity data (o) from Abdel-Gayed et al. [274].

Table 2.2. *Mixing model and turbulent transport closure*
effects on computed u_T (m/s) using the 'standard'
constant-time-scale ratio shown in Eq. (2.152)

Mixing model	PDF transport via Eq. (2.147)	PDF transport via Eq. (2.149)
Binomial Langevin [250]	1.24	1.62
Binomial sampling [253]	0.75	1.39
Modified Curl's [254]	1.32	1.65
LMSE [252]	0.54	0.84

Note: Experimental data suggest that $u_T \approx 2$ m/s (see Fig. 2.21)
for the conditions considered ($u_{rms} = 1$ m/s, $\Lambda = 20$ mm, $C_\phi = 2$,
and $u_L = 0.385$ m/s).

asymptotically approaches that of the standard time-scale expression at high u_{rms}/u_ℓ
ratios as $u_\ell/v_k \to 0$. Comparisons as Da $\to 0$ are clearly not meaningful as finite-
rate chemistry effects become increasingly important, and an accurate evaluation of
turbulent burning velocities would require a more comprehensive representation of
the chemistry. Furthermore, it can be expected that the coupling between chemical
and diffusion processes will become weaker. The impact of different SDR models
on the corresponding predictions of turbulent burning velocities in the context of
moment-based methods was further assessed in Section 2.3.

2.4.7 The Impact of Molecular-Mixing Terms

A common feature of transported PDF methods within a single-point single-time
context is that modelling approximations are required to account for *mixing* processes, as previously discussed in the context of Eq. (2.156). Mixing essentially
amounts to transport of the PDF in composition space. Lindstedt and Váos [251]
considered the classical LMSE model (see [252]) and nonlinear integral models [253, 254]. More sophisticated approaches covered include the binomial Langevin
model [250], originally proposed to simulate Brownian motion, which can readily be
recovered from the Fokker–Planck equation for the special case in which the drift
coefficient is a linear function of the state variable and the diffusion coefficient is
constant.

Results have been obtained through the use of gradient-diffusion (Eq. 2.147)
and second-moment-based (Eq. 2.149) closures for the turbulent transport of the
PDF in physical space. The former is based on the classical eddy-diffusivity–two-
equation-type turbulence model and the latter features the GLM coupled with a
LMSE-type formulation for the conditional turbulence intensity terms in the PDF
equation, as shown in Eq. (2.149). The applied thermochemistry was identical to that
previously presented with the instantaneous source term extracted from a calculation
of an unstrained CH_4–air flame. Results suggest that integral properties, such as the
turbulent burning velocity, are affected by the choice of mixing model *and* the closure
for the transport of the PDF in physical space as shown in Table 2.2.

The absolute values of the predicted turbulent burning velocities are obviously
affected by the chosen conditions. However, the *relative* influence of the mixing and
turbulent transport closures can be expected to be reasonably general. Under *all*
conditions a gradient-diffusion closure results in reduced burning velocities. This

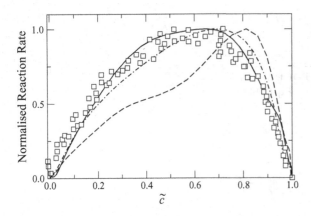

Figure 2.22. Predictions of normalised mean-reaction-rate profiles in scalar space by use of a second-moment closure [251]. (□) Measurements of Cheng and Shepherd [35]; (−) binomial Langevin model; (−·−) modified Curl's model; (−−) binomial sampling model.

observation, combined with the unambiguous manner in which PDF methods allow the incorporation of scalar reaction rates, leads to further concerns regarding the effectiveness of gradient-diffusion closures for premixed turbulent flames. An example of the predictive capability of the more accurate models is shown in Fig. 2.22. The same approach was also applied to the computation of a 2D slightly lean CH_4–air ($\phi = 0.9$) flame stabilised in an opposed jet geometry, as investigated experimentally by Mastorakos [289]. The configuration is identical to that studied by Lindstedt and Váos [17], who used a moment method. The calculation procedure featured a finite-volume part for the integration of the momentum and the second-moment equations and a Monte Carlo part for the integration of the scalar PDF. An average number of 400 weighted Lagrangian particles per cell was used.

The application of the standard expression for the scalar time scale [see Eq. (2.152)] results in a failure to stabilise the flame. The behaviour belies a problem consistent with the preceding discussion. Accordingly, variations in the turbulent transport and mixing model closures were found to be ineffective and yield either the same result (modified Curl's model) or more rapid extinction (all other mixing models and gradient-diffusion closures). Jones and Prasetyo [291] also performed a transported PDF study of the flame considered here – using a gradient-diffusion closure and a global four-step scheme – and they found that, although the LMSE model resulted in flame extinction, the modified Curl's model yielded accurate predictions with respect to the flame location. However, the time-scale ratio had to be significantly increased to promote flame stabilization. Inspection of Figs. 2.23 and 2.24 reveals that application of the extended time-scale expression shown in Eq. (2.155) results in reasonable predictions. Higher values for C_ϕ, C_ϕ^*, or both, would cause the flame to stabilise farther upstream.

2.4.8 Closure of Pressure Terms

The balance equation for the Favre-averaged kinetic energy of turbulence in a high-Re_T flow can, for a 1D steady-state flame, be written as

$$
\langle\rho\rangle\tilde{u}_1\frac{d\tilde{k}}{dx_1} = -\frac{d\langle\rho\rangle\widetilde{u_1''k}}{dx_1} - \frac{d\overline{u_1''p'}}{dx_1} - \langle\rho\rangle\widetilde{u_1''^2}\frac{d\tilde{u}_1}{dx_1}
$$

$$
- \overline{u_1''}\frac{d\overline{p}}{dx_1} + \overline{p'\frac{\partial u_k''}{\partial x_k}} - \langle\rho\rangle\tilde{\varepsilon}. \tag{2.157}
$$

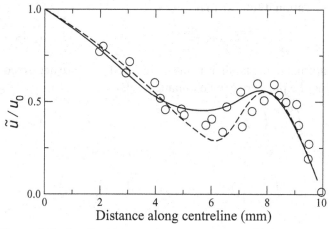

Figure 2.23. Predicted normalised axial velocity ($\tilde{u}/u_0$) profiles along the centreline [291]. Symbols: (○) experimental data by Mastorakos [289]. Lines: (−) presumed PDF approach using the fractal reaction-rate model implied by Eq. 2.155; (−−) transported PDF approach using the binomial Langevin model and the scalar time-scale expression of Eq. (2.155). In both cases a full second-moment closure is adopted for the turbulent transport of scalar and momentum. Velocity normalised by burner exit velocity on centreline (u_0).

The fifth term on the right-hand side of Eq. (2.157) describes pressure–dilatation, which is zero in constant-density turbulence and is usually neglected in models of turbulent combustion. However, Zhang and Rutland [292] performed a DNS of planar premixed turbulent flames propagating upstream in a decaying turbulence field at relatively low (0.67 and 1.5) expansion ratios $\tau\,(=\rho_u/\rho_b - 1)$ and found that pressure–dilatation is dominant under such conditions. The value of the term cannot be extracted from a PDF calculation at the current closure level, though a modelled term may readily be formulated to estimate the magnitude.

In the flamelet regime of combustion, the preceding term is non-zero only within a flamelet. If it is assumed that the latter has a structure close to that of a laminar flame, then the expected value of the term conditioned upon the fluid being within the flamelet can be estimated by integrating an expression based on the momentum

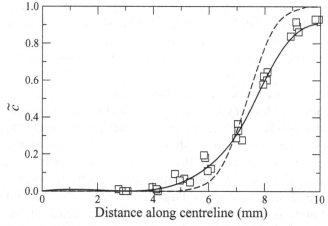

Figure 2.24. Predicted progress variable ($\tilde{c}$) profile along the centreline [291]. Lines, symbols, and closures are as in Figure 2.23.

equation. From mass conservation it follows that

$$\frac{\partial u_k}{\partial x_k} = \frac{\partial u_\zeta}{\partial \zeta} = u_L \tau \frac{\partial c}{\partial \zeta}, \tag{2.158}$$

where ζ is the coordinate normal to the laminar flame. If viscous forces are neglected in the momentum equation, the pressure can be obtained as [283]

$$p|_{c=c^*} \approx p|_{c\to 0} + \int_{\zeta|_{c\to 0}}^{\zeta|_{c=c^*}} \left(-\rho_u \tau u_L^2 \frac{\partial c}{\partial \zeta}\right) d\zeta = p|_{c\to 0} - \rho_u \tau u_L^2 c^*, \tag{2.159}$$

and it follows that

$$\overline{\left(p'\frac{\partial u_k}{\partial x_k}\right)}_{\text{flamelet}} = \left\langle \frac{1}{\delta_F} \int_{\zeta|_{c\to 0}}^{\zeta|_{c\to 1}} \left\{ [p'|_{c\to 0} - \rho_u \tau u_L^2 c(\zeta)] u_L \tau \frac{\partial c}{\partial \zeta} \right\} d\zeta \right\rangle$$

$$= \frac{1}{\delta_F} u_L \tau \left(\langle p'|c \to 0 \rangle - \tfrac{1}{2} \rho_u u_L^2 \tau \right), \tag{2.160}$$

where $\delta_F = \zeta|_{c\to 1} - \zeta|_{c\to 0}$ is the thickness of the flamelet. If the volume fraction of flamelets within the turbulent flame brush is estimated by use of the expected surface-to-volume ratio of the flame sheet as $\delta_F \Sigma$, the following expression for the pressure–dilatation term is obtained:

$$\overline{p'\frac{\partial u_k''}{\partial x_k}} = \rho_u u_L \Sigma (u_L \tau)^2 \left(\frac{\langle p'|c \to 0 \rangle}{\rho_u u_L^2 \tau} - \frac{1}{2} \right). \tag{2.161}$$

To form a model, an approximation has to be found for the conditional expectation $\langle p'|c \to 0 \rangle$. Neglecting viscous dissipation and introducing the simple assumption $\langle p'|c \to 0 \rangle = \langle p'|c = 0 \rangle$, which ignores any pressure rise ahead of the flamelet, the following expression is obtained [246]:

$$\overline{p'\frac{\partial u_k''}{\partial x_k}} = \rho_u u_L \Sigma (u_L \tau)^2 \left(\langle c \rangle - \tfrac{1}{2} \right). \tag{2.162}$$

Zhang and Rutland [292] proposed an alternative expression:

$$\left\langle p'\frac{\partial u_k''}{\partial x_k} \right\rangle = \rho_u u_L \Sigma (u_L \tau)^2 \frac{\tilde{c}}{2}. \tag{2.163}$$

A PDF solution may be used to analyse the budget of Eq. (2.157) and to compare the magnitude of the known terms with the preceding models for pressure–dilatation. For the purpose of order-of-magnitude estimates, the surface-to-volume ratio can be obtained as $\Sigma = \langle \rho \rangle \tilde{S}/(\rho_u u_L)$ [147], with the mean reaction rate $\tilde{S}$ obtained from the PDF solution.

The enhancement of turbulence in the turbulent flame structure is exemplified in Figs. 2.25 and 2.26. Generally, the higher the value u_T/u_{rms}, the stronger the turbulence generation in the flame. This applies to the cases of decreasing u_{rms} (Fig. 2.25) and increasing integral length scale (Fig. 2.26). As shown later, the dominant turbulent production term is the pressure–work term [the fourth term on the right-hand side of Eq. (2.157)]. The mean turbulent scalar flux and its scaling with the turbulence intensity and the integral length scale is shown in Figs. 2.27 and 2.28. The fact that counter-gradient transport prevails in most of the flame brush for the current planar flames, except at the flame tip [7, 66], suggests that gradient-based models for the

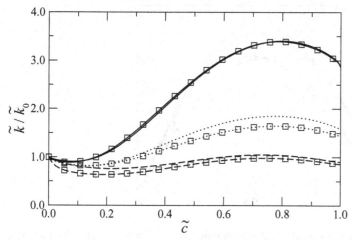

Figure 2.25. Profiles of the kinetic energy of turbulence, normalised with its upstream value, as functions of the reaction progress variable [246]. The integral length scale of turbulence is $\Lambda = 40$ mm, and the profiles are plotted for u_{rms} values of 1, 2.5, and 10 (top) m/s. Comparisons for CH_4–air flames obtained with (no symbols) the Haworth and Pope [248, 264] and (lines with symbols) the Lagrangian [265] variant of the Speziale et al. [249] closure for velocity statistics.

correlation $\widetilde{u_k''c''}$ are inadequate. Results shown in Figs. 2.27 and 2.28 suggest that the width of the region of gradient transport at the cold front (which is responsible for the propagation of the flame front and thus determines the burning velocity) increases for larger u_{rms} and smaller Λ. A budget of the terms in the $\tilde{k}$ balance equation (2.157) for a steady flame is presented in Fig. 2.29. Production is mainly by means of the pressure–work term. The main sink term is the third term on the right-hand side of Eq. (2.157) – traditionally called the 'turbulence production term' in constant-density flows.

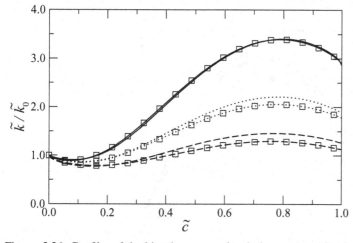

Figure 2.26. Profiles of the kinetic energy of turbulence, normalised with its upstream value, as functions of the reaction progress variable [246]. The upstream turbulence intensity is $u_{rms} = 1$ m s^{-1}, and the profiles are plotted for $\Lambda = 10, 20$ and 40 (top) mm. Lines and symbols are as in Fig. 2.25.

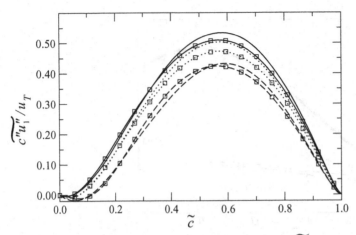

Figure 2.27. Profiles of the velocity–scalar correlation $\widetilde{u_1'' c''}$, normalised with the turbulent burning velocity u_T, as functions of the reaction progress variable [246]. The dependency of the values on u_{rms} is shown for a fixed $\Lambda = 40$ mm. Lines and symbols are as in Fig. 2.25.

The convection, destruction, and generation terms in Fig. 2.29 show a behaviour similar to that obtained by Bradley et al. [228], who used a presumed β-PDF approach closed at the second-moment level. However, other features show qualitative differences [246]. This is not surprising as a turbulent scalar flux ($\widetilde{u_1'' c''}$) consistent with both scalar and velocity fields is obtained naturally in a PDF solution, whereas in moment methods very little information is known about important terms in the scalar flux equation (e.g., correlations between fluctuating velocity components and chemical source terms) such as the $\widetilde{u_i'' S}$ term. The models for the pressure–dilatation term [Eqs. (2.162) and (2.163)] are also shown, and with the current high expansion ratio the term does not dominate the budget. However, models of the type of Eqs. (2.162) and (2.163) are fairly easy to implement in a $\tilde{k}$ equation. Further discussion on the subject in the context of moment-based methods can be found in

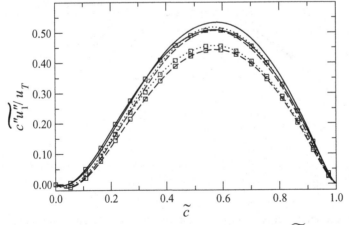

Figure 2.28. Profiles of the velocity–scalar correlation $\widetilde{u_1'' c''}$, normalised with the turbulent burning velocity, as functions of the reaction progress variable [246]. The dependency of the values on Λ is shown for a fixed $u_{\text{rms}} = 1 \text{ m s}^{-1}$. Lines and symbols are as in Fig. 2.26.

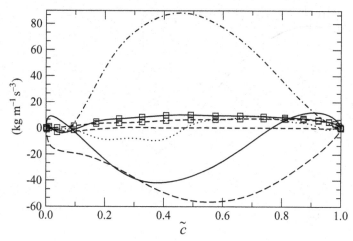

Figure 2.29. Budget of the terms in $\widetilde{k}$ equation (2.157) for $u_{rms} = 1\ \mathrm{m\,s^{-1}}$ and $\Lambda = 20$ mm and a stoichiometric methane–air flame [246]. The extracted values are (—) convection, $-\langle\rho\rangle\,\widetilde{u}_1\,\mathrm{d}\widetilde{k}/\mathrm{d}x_1$; ($\cdots$) diffusion, $1 - 0.5\,\mathrm{d}\,\langle\rho\rangle\,\widetilde{u_1''^3}/\mathrm{d}x_1$; (- - -) diffusion, $2 - \mathrm{d}\,\langle\rho\rangle\,\widetilde{u_1''u_2''^2}/\mathrm{d}x_1$; (– –) destruction, $-\langle\rho\rangle\,\widetilde{u_1''^2}\,\mathrm{d}\widetilde{u}_1/\mathrm{d}x_1$; ($\cdot$ –) pressure work, $-\langle u_1''\rangle\,\mathrm{d}\,\langle p\rangle\,/\mathrm{d}x_1$. Pressure–dilatation models: Eq. (2.162) (square and solid line) and Eq. (2.163) (square and dotted line).

the studies by Lindstedt and Váos [17], Domingo and Bray [18], and Lipatnikov and Chomiak [293].

The effect of adding an approximation of the combustion-induced pressure gradient to the pressure-acceleration term driving stochastic particles in a PDF simulation was investigated by Hůlek and Lindstedt [266], and Eq. (2.150) becomes

$$\mathrm{d}u_k^{(p)} = \left[-\frac{1}{\rho}\frac{\partial p}{\partial x_k}\bigg|^{(p)} + G_{k\ell}(u_\ell^{(p)} - \widetilde{u}_\ell) \right]\mathrm{d}t + (C_0\widetilde{\varepsilon})^{1/2}\,\mathrm{d}W_k, \qquad (2.164)$$

where

$$\frac{\partial p}{\partial x_k}\bigg|^{(p)} = \frac{\partial\,\langle p\rangle}{\partial x_k} + \left(\rho\,u_L\tau\frac{\mathrm{D}c^{(p)}}{\mathrm{D}t}\,n_k^{(p)} - u_L\tau\left\langle \rho\,\frac{\mathrm{D}c}{\mathrm{D}t}\,n_k \right\rangle \right). \qquad (2.165)$$

In Eq. (2.165) $n_k = \mathbf{n}$ is the normal vector of the flame sheet. The expression for the pressure gradient [(2.165)] was formulated so as to preserve the mean pressure gradient value and thus the overall momentum. The last term on the right-hand side of (2.165) may be interpreted as a 'zero-order approximation' of the dispersion of the flamelet-induced pressure gradients in the non-reacting fluid.

The Lagrangian time derivative $\mathrm{D}c^{(p)}/\mathrm{D}t$ in Eq. (2.165) is obtained naturally as part of the scalar statistics solution. The statistics of $\mathbf{n}$ was obtained with the suggestion by Cant et al. [206],

$$\langle n_k\rangle = -\frac{\mathrm{d}\,\langle c\rangle}{\mathrm{d}x_k}\frac{1}{\Sigma}, \qquad (2.166)$$

where Σ is the surface-to-volume ratio of the flame sheet. The PDF of the $\mathbf{n}$ component normal to the turbulent flame brush was approximated with a linear combination of a δ peak and a uniform PDF. The distribution of the components parallel to the turbulent flame was assumed to be isotropic, and their PDF then follows from the

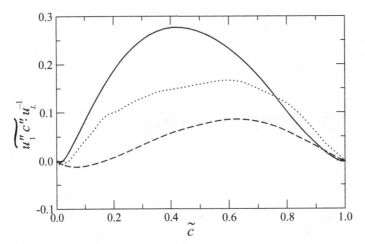

Figure 2.30. Comparison of PDF predictions of the scalar–velocity correlation [266] obtained with (—) and without (- - -) the flamelet-acceleration model of Eq. (2.165) and the DNS data (· · ·) of Rutland and Cant [43].

condition $|\mathbf{n}| = 1$. Equation (2.166) involves division with surface-to-volume density Σ {again estimated as $\Sigma \approx \langle \rho \rangle \widetilde{S}/(\rho_u u_L)$ [147]}. The mean source term $\widetilde{S}$ approaches zero at both the hot and cold ends of the flame, and the value estimated from a PDF simulation contains statistical errors. The singularity at the limits $\widetilde{c} \to 0$ and $\widetilde{c} \to 1$ was overcome by bounding the obtained values for $\langle n_k \rangle$ between 0 and 1.

Two cases of planar premixed turbulent flame propagation are shown in Figs. 2.30 and 2.31. The first is a flame in a decaying turbulence field as solved with a DNS by Rutland and Cant [43]. The flame has an expansion ratio of $\tau = 2.3$, an inlet turbulence intensity of $u_{rms} = \sqrt{2}u_L$, and a turbulence Reynolds number of $\mathrm{Re}_T = 56.7$. Because of the unstable nature of the flame, transient effects exert

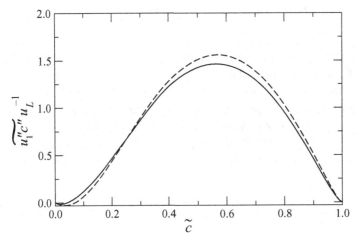

Figure 2.31. Comparison of PDF predictions of the scalar–velocity correlation in a planar premixed steady-state turbulent flame [266]. Stoichiometric methane–air mixture with an upstream turbulence intensity $u_{rms} = 2.5 \ \mathrm{m \, s^{-1}}$ and an integral length scale $\Lambda = 40$ mm. Lines as in Fig. 2.30.

some influence. Nevertheless, the DNS data are expected to provide some guidance. The results obtained show that the joint velocity–scalar statistics (in terms of moments like $\widetilde{u_k'' c''} = \langle \rho u_k'' c'' \rangle / \langle \rho \rangle, \langle \rho u_k'' S \rangle$) certainly are affected by the model. For example, Fig. 2.30 shows comparisons of the quantity $\widetilde{u_1'' c''}/u_L$ against $\tilde{c}$. The second case is a stoichiometric methane–air flame ($\tau \approx 7$ and laminar burning velocity $u_\ell = 0.386 \ \mathrm{m\,s^{-1}}$). The case shown here features $u_{\mathrm{rms}} = 2.5 \ \mathrm{m\,s^{-1}}$ and $\Lambda = 40$ mm. The upstream Reynolds number of turbulence is $\mathrm{Re}_T \approx 6496$ (based on the integral length scale Λ). Results shown in Fig. 2.31 suggest that, at higher expansion ratios and with u_{rms} significantly higher than u_ℓ, the influence of the model is very small. This indicates that for more strongly turbulent flames 'flamelet acceleration' is not necessarily an important mechanism. This finding is not surprising in view of the scaling characteristics of the pressure gradient terms in Eq. (2.165). The mean pressure gradient can for a steady-state planar flame be obtained from the mean momentum equation [66]:

$$\frac{\mathrm{d}\overline{p}}{\mathrm{d}x_1} = -\frac{\mathrm{d}\langle \rho \rangle \widetilde{u_1''^2}}{\mathrm{d}x_1} - \rho_u u_T^2 \tau \frac{\mathrm{d}\tilde{c}}{\mathrm{d}x_1}, \tag{2.167}$$

where the second term on the right-hand side is dominant. Therefore the mean pressure gradient in Eq. (2.165) scales in a steady-state turbulent flame with u_T^2, where u_T is roughly proportional to u_{rms}. At the same time, the current model term in Eq. (2.165) scales effectively with u_L^2:

$$\frac{\mathrm{d}p}{\mathrm{d}\zeta} \approx -\rho \, u_L \tau \frac{\mathrm{D}c}{\mathrm{D}t} = -\rho_u \, u_L^2 \tau \frac{\mathrm{d}c}{\mathrm{d}\zeta}. \tag{2.168}$$

Overall, the relative importance of flamelet acceleration can vary by orders of magnitude, depending on the ratio u_L^2/u_T^2 (or u_L^2/u_{rms}^2). The effect may thus be important for only a subrange of turbulent conditions. However, some caution is required with respect to the current analysis, as Vervisch et al. [294] showed by means of DNS that the flamelet approximation breaks down close to the scalar bounds ($c \to 0$ and $c \to 1$).

2.4.9 Premixed Flames at High Reynolds Numbers

The strongly piloted high-Reynolds-number premixed turbulent flames investigated experimentally by Chen et al. [256] present challenges related to the configuration, as outlined by Pitsch and de Lageneste [76] and Lindstedt and Váos [211]. Heat losses to the burner surface were estimated to reach up to 20%, and the close proximity of the flame to the burner plate was expected to influence the scalar time-scale ratio. The role of entrainment was also highlighted. However, the data sets include a case featuring $\mathrm{Da} \approx 1$ and permit the further evaluation of the influence of modelling parameters. The influence of molecular transport in physical space can be expected to be reduced at lower Damköhler numbers [see Eq. (2.156)] as the coupling between molecular transport and chemical reaction inherent in the laminar flamelet assumption will weaken.

The applied calculation procedure featured enthaply as a solved scalar that permits arbitrary enthalpies to be set for the different reactant streams. Computations featuring a systematic variation of the time-scale ratio ($2 \leq C_\phi \leq 8$) were performed

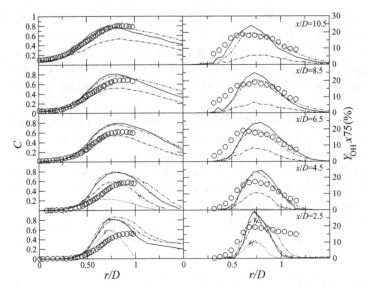

Figure 2.32. Mean temperature and OH mass fraction for a stoichiometric CH_4/–air flame at $Re = 52\,500$ [211]. Experiments: (○) Chen et al. [256], transported PDF calculations with modified Curl's model [254] and Eq. (2.155) with $C_\phi^* = 0$ and $T_P = 2005$ K. $C_\phi = 2$ ($\cdots$), 4 ($-\cdot-$), 6 ($-\cdot\cdot-$), and 8 ($-$).

to illustrate sensitivities. The influence of the time-scale ratio was shown to be modest for stably burning mixtures and significantly close to the extinction of non-premixed piloted diffusion flames [295]. The sensitivity is exemplified here in the context of a premixed turbulent flame close to global extinction. The corresponding joint composition PDF in low-Mach-number flows is considered with the thermochemistry described by up to 12 independent scalars and enthalpy. The applied augmented systematically reduced chemistry is identical to that of previous studies [234] (with the nitrogen-containing species omitted) and features 15 solved species (H, O, OH, HO_2, H_2O, H_2, O_2, CH_4, CH_3, CO, CO_2, C_2H_2, C_2H_4, C_2H_6, and N_2). Steady-state approximations were applied to 14 species (C, CH, $^1CH_2$, $^3CH_2$, CHO, CH_2O, CH_2OH, CH_3O, C_2, C_2H, C_2H_3, C_2H_5, C_2HO, and CH_2CO). The chemical source term was computed by means of a direct integration technique featuring a Newton method with the Jacobian evaluated analytically. The mean mass and momentum conservation equations were solved in a similarity transformed coordinate system in a manner similar to that of Lindstedt et al. [233, 295]. The Reynolds stresses were obtained through the second-moment closure of Speziale et al. [249]. The applied closure level does not account for non-gradient transport. However, it has the benefit of relative simplicity and constitutes the natural starting point for the further investigation of premixed turbulent flames featuring comprehensive closures for the chemical source term.

Results were obtained with a pilot temperature of $T_p = 2005$ K, the standard closure model given in Eq. (2.152), and with the time-scale ratio varied in the range $2 \le C_\phi \le 8$. The effects of the time-scale ratio on the temperature-based reaction progress variable [$\bar{c} = (\overline{T} - T_u)/(T_b - T_u)$ and $T_b = 2248$] [256] and OH concentration are shown in Fig. 2.32. A value of $C_\phi = 2$ leads to extinction shortly after $x/D = 2.5$ and $C_\phi = 4$ shows strong indications of blowoff farther downstream. The

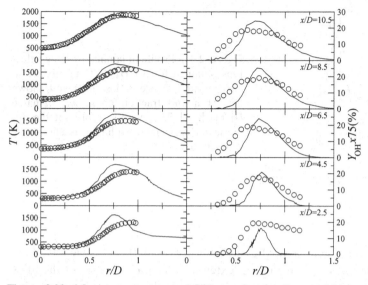

Figure 2.33. Mean temperature and OH mass fraction for a stoichiometric CH_4/–air mixture at Re = 52 500 [211]. Transported PDF calculations with modified Curl's model [254] and Eq. (2.155) with $C_\phi^* = 1.2$ and $C_\phi = 4$ (−) and $T_P = 1785$ K. Experimental data (∘) from Chen et al. [256].

two computations with higher time-scale ratios burn stably. It is also evident from the reaction progress variable that the pilot flame temperature is too high. However, computations with $T_p = 1785$ K lead to extinction over an extended range of values of C_ϕ. The computations thus indicate more extensive flame-stabilisation problems than observed experimentally. The influence of the extended SDR closure [Eq. (2.155)] can, however, be expected to be significant in the early part of the flame. Based on the experimental data, u_L/v_k can be estimated to be $O(1)$ close to the nozzle. It may also be noted that, despite the apparent overprediction of temperature, the computed peak OH concentrations are well reproduced. This is surprising as there is a direct sensitivity of OH levels to temperature, as shown in Fig. 2.32. The comparatively high value of C_ϕ required in the current flame, compared with diffusion flames [295], may at first be somewhat surprising. However, given that typically $2.6 \leq C_R \leq 4.0$ [17], the reported values of C_ϕ are not inconsistent.

The experimental study of O'Young and Bilger [156] indicates that the velocity–scalar time–scale ratio varies significantly as a function of the Damköhler number and increases by almost an order of magnitude as Da → ∞. As just discussed, the influence is likely to be particularly strong in the proximity of the pilot flame and thus close to the burner surface. Accordingly, the functional form of Eq. (2.155) was explored with a pilot temperature of $T_p = 1785$ K and with $C_\phi = 4$ and $C_\phi^* = 1.2$. The choice of modelling constants thus corresponds to that used in the simulations shown in Fig. 2.21. As previously mentioned, the computational flame could not be stabilized with $C_\phi^* = 0$ under these conditions. The level of agreement is arguably significantly improved, though some level of overprediction remains, as shown in Fig. 2.33. The discrepancies between the measured temperature and OH profiles, also previously discussed, are readily apparent. However, the current closure at the joint scalar level has a tendency to produce thinner flames [17] and the presence on

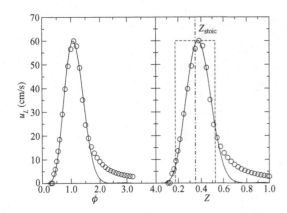

Figure 2.34. Profile of laminar burning velocities obtained for a $CH_4:H_2$ (1:1) mixture at varying equivalence ratios (left) and shown against mixture fraction for the $\phi = 3.2$ flame (right) [257]. The symbols represent the numerical results, and the solid line is a fit using a β-function.

non-gradient transport effects cannot be ruled out – particularly in the vicinity of the burner. However, it appears clear that the extended algebraic closure for the SDR carries potentially significant benefits in the context of the computation of mixing frequencies and that the inclusion of more comprehensive chemistry permits the computation of the thermochemical structure of high-Reynolds-number premixed flames.

2.4.10 Partially Premixed Flames

The closure applied to compute the premixed flames in the previous section was also used by Lindstedt et al. [257] to investigate piloted partially premixed CH_4–H_2–air turbulent jet flames at high Reynolds numbers (Re $\approx$ 60 000 and 67 000) for two equivalence ratios, $\phi = 2.1$ and $\phi = 3.2$. The flames were studied experimentally at Sandia National Laboratories and were well characterised close to the nozzle through multiscalar measurements for $1 \leq x/d \leq 4$. The flames offer the opportunity of computational investigations of their thermochemical structure close to the burner. Moreover, the data sets are finely spaced through the local extinction region. The current aim is to provide further examples of the impact of closure approximations for the SDR. The effects of variations in the time-scale ratio ($2.3 \leq C_\phi \leq 4$) in Eq. (2.152) were thus investigated along with the impact of the extended algebraic relationship shown in Eq. (2.155). The laminar burning velocity required in the latter closure was obtained from the corresponding premixed laminar flames with a fuel stream of 50% H_2 and 50% CH_4, giving a value of $\simeq$0.6 m/s. The value $C_\phi^* = 1.2$ was retained from previous work [210, 211].

Lindstedt et al. [257] evaluated two approaches for the extension of Eq. (2.155) to partially premixed combustion. In the first, the stoichiometric value of the burning velocity was used to determine the time-scale ratio in the flammability region ($Z_{st} \pm \Delta Z$, where $\Delta Z = 0.5\, Z_{st}$) with the term set to zero elsewhere by means of Heaviside functions at the edges of the flammability range. The approach was found to overestimate the influence of the term; $u_L(\phi)$ was also computed, and the resulting shape was represented by a β function [257]. The resulting expressions are shown in Fig. 2.34, and the fitted function was judged to provide a sufficient match in the critical regions of scalar mixing ($\phi \leq 1.6$).

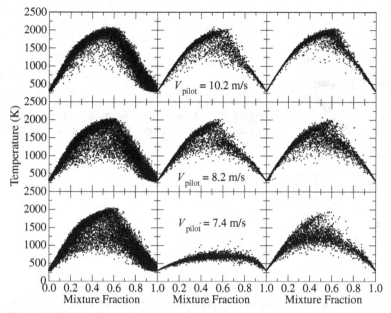

Figure 2.35. (Left) measurements and computations of temperature in mixture fraction space in a $\phi = 2.1$ flame at $x/d = 4$ for three pilot settings using (middle) a constant time–scale ratio of $C_\phi = 3.0$ and (right) the extended algebraic relationship with $C_\phi = 2.3$ and $C_\phi^* = 1.2$ [257].

The value of C_ϕ (with $C_\phi^* = 0$) was varied in the range $2.3 \leq C_\phi \leq 4.0$ as several flames extinguished early with the 'standard' value of 2.3 used in earlier studies, e.g., [233, 234]. Raising the value to 3.0 promoted burning and led to better predictions for most cases, as shown in Figs. 2.35 and 2.36. However, the extinction kernel was still overextended unless a strong pilot was used, and the sensitivity to successive

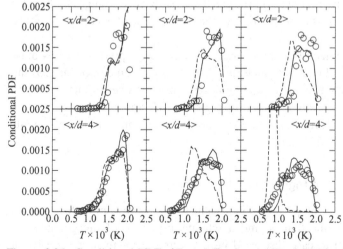

Figure 2.36. Conditional PDFs ($Z_{st} \pm \Delta Z$, where $\Delta Z = 0.5\, Z_{st}$) of temperature for the $\phi = 2.1$ flame at $x/d = 4$ for (left) $V_{pilot} = 10.2$, (middle), $V_{pilot} = 8.2$, and (right) $V_{pilot} = 7.4$ [257]. The symbols represent experimental data, and the lines are predictions of the extended algebraic relationship with ($-$) $C_\phi = 2.3$ and $C_\phi^* = 1.2$ and ($---$) the modified Curl's model with $C_\phi = 3.0$.

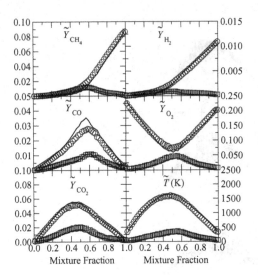

Figure 2.37. Conditional means and rms of temperature and species mass fractions for the $\phi = 2.1$ flame with $V_{pilot} = 7.4$ m/s at $x/d = 4$ [257].

reductions in the pilot velocity was not predicted accurately. A value $C_\phi = 4.0$ caused burning for all cases, though the validity of applying such a high value, compared with the standard $\simeq 2.0$, is questionable and may lead to an inappropriate downstream evolution of the variance.

An advantage of the extended algebraic expression is that the time–scale ratio increases conditionally in the flame zone and that other parts of the flow remain unaffected. The results obtained are also shown in Figs. 2.35 and 2.36. The findings are encouraging in the sense that the progressive extinction patterns are captured. However, the tendency of the extended model to underpredict extinction at the highest pilot velocity is shown by the slightly elevated temperatures on the rich side. There is also a tendency to promote too much extinction at the lowest pilot velocity. Nonetheless, scatter plots of temperature, as shown in Fig. 2.35, clearly show the extent to which the closure is able to reproduce the sensitivity to changes in the pilot velocity. The improvement is also apparent in the conditional PDFs of temperature shown in Fig. 2.36.

The ability of the transported PDF method to correctly predict species mass fractions is highlighted in Fig. 2.37, which shows the most extreme pilot case for the $\phi = 2.1$ flame. The trends are encouraging, though a modest overprediction of CO appears as a result of the elevated temperatures on the rich side of the flame. Clearly some inadequacies remain, though the results do indicate significant improvements compared with the standard closure.

2.4.11 Scalar Transport at High Reynolds Numbers

Ferrão [296] and Duarte et al. [297] studied scalar transport in lean highly sheared premixed turbulent propane flames, and a schematic picture of the geometry is presented in Fig. 2.38. The case discussed here features a mean inlet velocity of $U_0 = 42.4$ m s^{-1} resulting in a Reynolds number of 154 000. The Reynolds number of turbulence, estimated [296] with $u_{rms} \approx 0.14 U_0$ and an integral length scale $\Lambda \approx d/2$, was Re$_T \approx 10\,800$. The ambient temperature was 303 K, and the adiabatic flame temperature for $\phi = \phi_0 = 0.55$ was reported as 1617 K, which agrees well with the

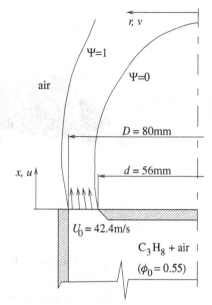

Figure 2.38. The experimental geometry of Ferrão [296] (not to scale) [283]. The flow pattern is indicated by the dividing streamlines and values of the normalised mean stream function Ψ. The streamline defined by $\Psi = 0$ marks the recirculation bubble.

value of 1596 K obtained in laminar flame calculations using detailed chemistry. The transported PDF approach closed at the joint velocity–scalar level was applied [283] with different velocity closures to explore the impact on scalar transport: (1) The k-ϵ model, (2) the GLM of Haworth and Pope [248], and (3) the GLM with the preferential acceleration term removed [see Eq. (2.151)].

The principal aim was to explore the impact of turbulent transport, and hence a simplified chemical source term, based on detailed chemical kinetic laminar flame solutions, was used for the scalar c. However, the flame under consideration is not purely premixed and dilution by ambient entrained air must be considered. Assuming equal diffusivities of species, a conserved scalar z can be defined that takes zero value in air and unity in the reactant mixture ($\phi = \phi_0$). Only very lean mixtures ($\phi \leq \phi_0 = 0.55$) are considered. The reactive scalar c is defined as a 'global' reaction progress variable. Its definition is based on the mass fraction of O_2 converted to products at a particular value of Z and made non-dimensional by the maximum mass fraction of O_2 available for conversion into products (i.e., at $Z = 1$) as outlined by Hůlek and Lindstedt [255]:

$$c = \frac{Y_{O_2,u}|_{Z=\xi} - Y_{O_2}}{Y_{O_2,u}|_{Z=1} - Y_{O_2,b}|_{Z=1}}. \tag{2.169}$$

An analysis of detailed chemical kinetic solutions of relevant lean propane–air flames shows that a reaction progress variable defined in the manner of Eq. (2.169) has the practical advantage in that the gas temperature is, to a very good approximation, linear in c and almost independent of the value of z. If $\tau = \tau|_{Z=1} = (\rho_u/\rho_b)|_{Z=1} - 1$ is the maximum expansion ratio (for $Z = 1$ and thus $\phi = \phi_0$), then the gas density can be written in the customary manner [$\rho = \rho_u/(1 + \tau c)$]. The maximum temperature of the burnt gases is approximately linear in Z and $c_{\max}(\xi) = \xi$. As a consequence, the scalar c has the properties of a standard reaction progress variable (e.g., [7]) except that its upper bound depends on the local stoichiometry. The resulting region

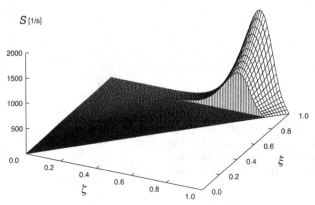

Figure 2.39. A 3D plot of the source term for the reaction progress variable c as a function of ξ and ζ [see Eq. (2.170)] [283].

of allowed compositions forms a triangle in ξ–ζ space, defined by $0 \le \zeta \le \xi$ and $\zeta \le \xi \le 1$.

The practical flammability limit was set to $\phi_q = 0.50$, which corresponds to $z_q \approx 0.91$. Reaction rates based on the depletion rate of O_2 [i.e., consistent with Eq. (2.169)] were extracted from laminar flame calculations for stoichiometries $\phi = 0.50$, 0.52, and 0.55. The shape of the source term does not depend strongly on stoichiometry, and the value for the *local* reaction progress variable $\hat{\zeta} = \zeta/\xi$ scales with the laminar burning velocity as u_L^2 as shown by Spalding [298]. The laminar burning velocity can be approximated as a linear function of ξ with the parameterisation of the chemical source term:

$$S_c(\xi, \zeta) = \begin{cases} ZS_{\hat{c}}(\xi, \hat{\zeta}) & \text{if } \xi \ge \xi_q \\ 0 & \text{if } \xi < \xi_q \end{cases}, \tag{2.170}$$

where $S_{\hat{c}}(\xi, \hat{\zeta})$ is the chemical source term for the local-stoichiometry-based reaction progress variable (bounded between 0 and 1), obtained as

$$S_{\hat{c}}(\xi, \hat{\zeta}) = S_{\hat{c}}(\hat{\zeta})|_{\xi=1} \left(\frac{u_L|_\xi}{u_L|_{\xi=1}} \right)^2. \tag{2.171}$$

The value of $u_L|_\xi$ was fitted as a linear function of Z, and the function $S_{\hat{c}}(\hat{\zeta})|_{Z=1}$ was approximated with a least-squares fit.

$$S_{\hat{c}}(\hat{c})|_{Z=1} = (1 - \hat{c}) \left[\exp \left(\frac{a_0 + a_2\,\hat{c} + a_4\,\hat{c}^2 + a_6\,\hat{c}^3}{1 + a_1\,\hat{c} + a_3\,\hat{c}^2 + a_5\,\hat{c}^3} \right) - \exp(a_0) \right]. \tag{2.172}$$

The shape of $S_c(\xi, \zeta)$ with $a_0 = -2.219641$, $a_1 = -0.566512$, $a_2 = 26.17081$, $a_3 = -1.294479$, $a_4 = -39.28250$, $a_5 = 0.87046$, and $a_6 = 15.44937$ is illustrated in Fig. 2.39. The source term falls rapidly as ξ approaches the limit and at $\xi = \xi_q$ the peak rate is only 40% of the maximum at $\xi = 1$.

The molecular transport in Eq. (2.144) was closed with the binomial Langevin model [250] extended to account for joint velocity–scalar statistics and multiple scalars, as described by Hůlek and Lindstedt [255]. The functional forms for the diffusion (B_c) and drift (G_c) coefficients are given in Appendix 2.D. The decay time scale of scalar fluctuations $[\tau_c = \langle c'^2 \rangle /(2\langle \epsilon_c \rangle)]$ was estimated by Eq. (2.152)

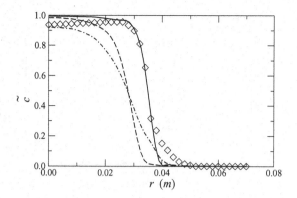

Figure 2.40. Radial profiles of the mean reaction progress variable $\tilde{c}$ at $x = 0.081$ m. Experimental data (symbols) [296] compared with predictions obtained with (– –) the k-ϵ model; (—) the GLM; (– · –) the GLM without preferential acceleration terms (2.151) [283].

with $\tau_z = \tau_c = \tau_u/C_{zc}$, where $\tau_u = \tilde{k}/\tilde{\epsilon}$ and the coefficient had the standard value of $C_{zc} = 2$ [91].

The treatment of boundary conditions was outlined by Hulek [283], and experimental profiles obtained at $x = 10$ mm were used to derive an estimate at $x = 0$ mm. The distance of 10 mm was relatively small compared with the diameter of the jet, and the flow was considered to be locally parabolic. Following translation along the experimentally determined mean streamlines, the profiles measured at $x = 10$ mm were used for in-flow boundary conditions between $d = 56$ mm and $D = 80$ mm at $x = 0$ mm (see Fig. 2.38). Solid boundaries at $x = 0$ mm were treated with the wall-functions approximation in the finite-volume part of the algorithm, and the no–slip specular-reflection boundary condition [299] was used for the Lagrangian particles.

Information relating to the the flame shape can be obtained from radial profiles of $\tilde{c}$. Comparisons of predictions with the experiment at $x = 0.081$ m are shown in Fig. 2.40. Two features are worthy of comment. First, the position of the flame is predicted accurately by the second-moment-based model provided that preferential acceleration is included. By contrast, the k-ε model gives a flame brush that is located too close to the symmetry axis, and the same problem prevails for the second-moment closure with the preferential acceleration term removed. Secondly, the shape of the mean reaction progress variable is not well reproduced by the eddy viscosity closure.

The impact of preferential acceleration on the turbulent scalar flux is clearly noticeable in Fig. 2.41, which shows the radial component of the scalar flux $\widetilde{c''v''}$ at the same locations. Similar observations can be made from the $\tilde{c}(r)$ profiles at $x = 0.114$ m, shown in Fig. 2.42, and the corresponding scalar flux shown in Fig. 2.43.

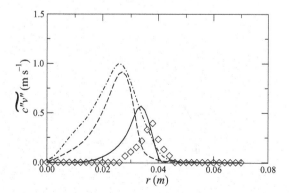

Figure 2.41. Radial profiles of the radial component of the turbulent scalar flux $\widetilde{c''v''}$ at $x = 0.081$ m [283]. Lines and symbols are as in Fig. 2.40.

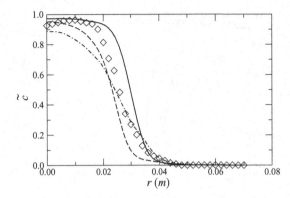

Figure 2.42. Radial profiles of the mean reaction progress variable $\tilde{c}$ at $x = 0.114$ m [283]. Lines and symbols are as in Fig. 2.40.

Measurements of the scalar–velocity correlation are very difficult to perform, and one should exercise caution when making quantitative comparisons. Nevertheless, it is obvious that neglecting preferential acceleration results in the magnitude of the flux being overpredicted by a large margin. Interestingly, the k-ε model also predicts very high values of the turbulent scalar flux (the PDF solution in this case includes the effect of preferential acceleration). An explanation can be found in the turbulent kinetic energy profiles at $x = 0.081$ m in Fig. 2.44 – both the k-ε model and the second-moment model without preferential acceleration overpredict the $\tilde{k}$ values close to the symmetry axis. A higher level of turbulence means a stronger generation of gradient-wise scalar transport (by means of the term $\langle \rho \rangle \widetilde{u_k'' u_\ell''} \, \partial \tilde{c}/\partial x_\ell$) and thus more gradient diffusion. Farther downstream, all predictions are characterised by a too-low value of $\tilde{k}$ at the centreline, as shown in Fig. 2.45. Nevertheless, predicted values in the shear layer where the flame is located are adequate, and comparisons with experimental data suggest that preferential acceleration effects and non–gradient diffusion exert a significant influence at high Reynolds numbers and need to be taken into account to ensure accurate modelling of velocity statistics under the current conditions.

2.4.12 Conclusions

In this chapter the ability of the transported PDF approach to reproduce the characteristics of stable premixed turbulent flames and those close to extinction was assessed at different closure levels. Predicted turbulent burning velocities depend

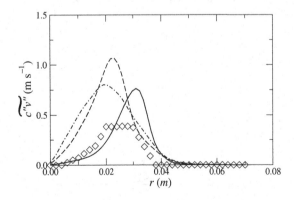

Figure 2.43. Radial profiles of the radial component of the turbulent scalar flux $\widetilde{c''v''}$ at $x = 0.114$ m [283]. Lines and symbols are as in Fig. 2.40.

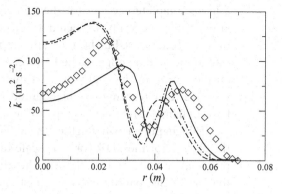

Figure 2.44. Radial profiles of the kinetic energy of turbulence $\tilde{k}$ at $x = 0.081$ m [283]. Lines and symbols are as in Fig. 2.40.

on the structure of the model for turbulent transport in scalar space (i.e., the structure of the prescribed *mixing* transition density function) and on the magnitude of the relevant (mixing) time scale. Examples given suggest that the effects of closure approximations for the SDR do have a significant influence on computed flame properties. An extended multiscale algebraic model, derived to account for small-scale properties, yields significant improvements irrespective of the application of a closure at the joint scalar or joint velocity–scalar level. The success of the modified Curl's model in predicting turbulent burning velocities may perhaps be viewed as surprising, and it can be argued that mixing models that enforce locality in composition space [300, 301] may be better suited to handling combustion in the high-Damköhler-number regime [275]. As noted in Chapter 1, the JPDF approach as implemented in the context of LES or FDF methods needs to be explored further in the context of premixed turbulent flames. However, Yilmaz et al. [302] applied a closure at the joint scalar level to compute a premixed turbulent flame stabilised on a Bunsen burner with encouraging results. Furthermore, the approach does in principle permit a consistent treatment of pressure terms, provided a higher level of closure is applied.

Computations of axisymmetric high-Re_T turbulent flames with the joint scalar PDF approach coupled with augmented reduced chemistry showed that the detailed flame structure can be reproduced with encouraging accuracy – including effects related to ignition and extinction. However, the reported cases also suggest that

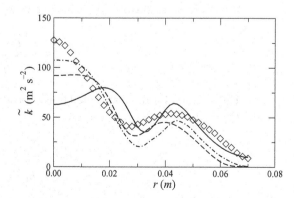

Figure 2.45. Radial profiles of the kinetic energy of turbulence $\tilde{k}$ at $x = 0.134$ m [283]. Lines and symbols are as in Fig. 2.40.

a more complete account for the influence of turbulent transport, including non-gradient effects, can be expected to further improve model predictions over a wide range of conditions. Furthermore, the impact of molecular transport on the reaction rate at the leading edge of the flame can be expected to exert influence depending on the nature of the fuel and the Reynolds number of the flow. The influence of pressure-related terms was highlighted and the impact of models derived on the basis of flamelet-related assumptions assessed.

This chapter also highlighted the influence of boundary conditions and identified the closure for the SDR (or time-scale ratio of scalar to mechanical turbulence) as principal sources of uncertainty. The computations also suggest, by implication, that scalar dissipation and velocity statistics and scalar fluxes along with detailed boundary conditions are essential for evaluating and improving the accuracy of the current models. Hence future progress is inextricably linked to the procurement of appropriate experimental data sets that should preferably cover a wide range of Reynolds and Damköhler numbers. Such data sets should ideally include information relating to SDRs, species concentrations, and conditional PDFS for key parameters such as temperature and pollutants (e.g., CO and NO) exhibiting a range of chemical time scales.

Acknowledgements

The author wishes to acknowledge the contributions of Tomas Hůlek, Evangelos Váos, Henri Ozarovsky, Magnus Persson, and Philipp Geipel. The encouragement and financial support of Gabriel Roy and the Office of Naval Research is gratefully acknowledged.

Appendix 2.A

The general form of the tensor G_{ij} follows directly from the original constant-density form [264]:

$$G_{ij} = (\alpha_1 \delta_{ij} + \alpha_2 b_{ij} + \alpha_3 b_{ij}^2)\frac{1}{\tau_u} + H_{ijkl}\frac{\partial \widetilde{u}_k}{\partial x_l}, \tag{2.173}$$

where τ_u is the time scale of decay of velocity fluctuations ($\tau_u = \widetilde{k}/\widetilde{\varepsilon}$) and the Reynolds stress anisotropy tensor is defined as

$$b_{ij} = \frac{\widetilde{u_i' u_j'}}{\widetilde{u_k' u_k'}} - \frac{\delta_{ij}}{3}. \tag{2.174}$$

The notation for the contracted tensor product $b_{ij}^2 = b_{ik}b_{kj}$ and the tensor coefficient H_{ijkl} is defined as

$$H_{ijkl} = \beta_1 \delta_{ij}\delta_{kl} + \beta_2 \delta_{ik}\delta_{jl} + \beta_3 \delta_{il}\delta_{jk}$$
$$+ \gamma_1 \delta_{ij}b_{kl} + \gamma_2 \delta_{ik}b_{jl} + \gamma_3 \delta_{il}b_{jk}$$
$$+ \gamma_4 b_{ij}\delta_{kl} + \gamma_5 b_{ik}\delta_{jl} + \gamma_6 b_{il}\delta_{jk}. \tag{2.175}$$

In total, there are 12 modelling coefficients: α_1, α_2, α_3, β_1, β_2, β_3, γ_1, γ_2, γ_3, γ_4, γ_5, and γ_6. The strain invariants are denoted as

$$I_0 = \frac{\partial \tilde{u}_k}{\partial x_k}, \tag{2.176}$$

$$I_1 = b_{kl}\frac{\partial \tilde{u}_k}{\partial x_l}, \tag{2.177}$$

$$I_2 = b_{ki}b_{il}\frac{\partial \tilde{u}_k}{\partial x_l}, \tag{2.178}$$

and $\gamma^* = \gamma_2 + \gamma_3 + \gamma_5 + \gamma_6$. The final forms of the coefficients are

$$\alpha_1 = -\left(\frac{1}{2} + \frac{3}{4}C_0\right) - (\alpha_2 + \frac{\alpha_3}{3})b_{ll}^2 - \alpha_3 b_{ll}^3 - \tau_u \gamma^* I_2. \tag{2.179}$$

The remaining coefficients are $\alpha_2 = 3.78$, $\alpha_3 = 0.00$, $\beta_1 = -0.20$, $\beta_2 = 0.80$, $\beta_3 = -0.20$, $\gamma_1 = -1.24$, $\gamma_2 = 1.04$, $\gamma_3 = -0.34$, $\gamma_4 = 0.00$, $\gamma_5 = 1.99$, and $\gamma_6 = -0.76$.

Appendix 2.B

The mean mass and momentum conservation equations were solved to obtain the mean pressure field with turbulence scale information derived from the rate of dissipation and the kinetic energy of turbulence:

$$\frac{\partial \langle \rho \rangle}{\partial t} + \frac{\partial \langle \rho \rangle \tilde{u}_i}{\partial x_i} = 0 \tag{2.180}$$

$$\frac{\partial \langle \rho \rangle \tilde{u}_i}{\partial t} + \frac{\partial \langle \rho \rangle \tilde{u}_i \tilde{u}_j}{\partial x_j} = -\frac{\partial \langle p \rangle}{\partial x_i} - \frac{\partial \langle \rho \rangle \widetilde{u_i'' u_j''}}{\partial x_j}. \tag{2.181}$$

The Reynolds stresses were obtained either through the k-ε turbulence model or directly from the velocity PDF. For the former case, the Reynolds stresses were obtained from the eddy-viscosity hypothesis:

$$-\langle \rho \rangle \widetilde{u_i'' u_j''} = \mu_T \left(\frac{\partial \tilde{u}_i}{\partial x_j} + \frac{\partial \tilde{u}_j}{\partial x_i} - \frac{2}{3}\frac{\partial \tilde{u}_k}{\partial x_k}\delta_{ij}\right) - \frac{2}{3}\langle \rho \rangle \tilde{k}\delta_{ij}. \tag{2.182}$$

The effective turbulent viscosity μ_T was obtained in the customary manner, $\mu_T = C_\mu \langle \rho \rangle \tilde{k}^2/\tilde{\varepsilon}$, with the standard value of $C_\mu = 0.09$.

The mean turbulence kinetic energy dissipation rate was used along with the kinetic energy of turbulence to estimate turbulence scales [16]. The $\tilde{k}$ and $\tilde{\varepsilon}$ equations were thus solved in the following forms:

$$\frac{\partial \langle \rho \rangle \tilde{k}}{\partial t} + \frac{\partial \langle \rho \rangle \tilde{u}_i \tilde{k}}{\partial x_i} = \frac{\partial}{\partial x_i}\left(\frac{\mu_T}{\sigma_k}\frac{\partial \tilde{k}}{\partial x_i}\right) - \langle \rho \rangle \widetilde{u_i'' u_j''}\frac{\partial \tilde{u}_i}{\partial x_j} - \langle \rho \rangle \tilde{\varepsilon} - \langle u_i'' \rangle \frac{\partial \langle p \rangle}{\partial x_i}, \tag{2.183}$$

$$\frac{\partial \langle \rho \rangle \tilde{\varepsilon}}{\partial t} + \frac{\partial \langle \rho \rangle \tilde{u}_i \tilde{\varepsilon}}{\partial x_i} = \frac{\partial}{\partial x_i}\left(\frac{\mu_T}{\sigma_\epsilon}\frac{\partial \tilde{\varepsilon}}{\partial x_i}\right)$$
$$+\frac{\tilde{\varepsilon}}{\tilde{k}}\left(-C_{\epsilon_1}\langle \rho \rangle \widetilde{u_i'' u_j''}\frac{\partial \tilde{u}_i}{\partial x_j} - C_{\epsilon_2}\langle \rho \rangle \tilde{\varepsilon} - C_{\epsilon_3}\langle u_i'' \rangle \frac{\partial \langle p \rangle}{\partial x_i}\right). \tag{2.184}$$

The standard values of the constants were used ($\sigma_k = 1$, $\sigma_\epsilon = 1.3$, $C_{\epsilon_1} = 1.44$, $C_{\epsilon_2} = 1.92$, $C_{\epsilon_3} = 1$). To ensure consistency, the Reynolds stresses in the production source term $\langle \rho \rangle \widetilde{u_i'' u_j''} \, \partial \tilde{u}_i / \partial x_j$ were closed in the same way as in the case of the mean momentum equations. The pressure–work term $(-\langle u_i'' \rangle \, \partial \langle p \rangle / \partial x_i)$ was included only if preferential acceleration effects were considered. Therefore, if the Lagrangian velocity statistics were modelled with Eq. (2.151), the pressure–work term was omitted from both the $\tilde{k}$ and the $\tilde{\epsilon}$ equations to ensure consistency of the $\tilde{k}$ equation with the Lagrangian velocity model. The value of $\langle u_i'' \rangle$ was obtained as [16]

$$\langle u_i'' \rangle = \frac{\tau \, \widetilde{c'' u_i''}}{1 + \tau \tilde{c}}. \tag{2.185}$$

The scalar flux $\widetilde{c'' u_i''}$ was obtained from the joint velocity–scalar PDF solution. The turbulent convection of both $\tilde{k}$ and $\tilde{\epsilon}$ was approximated with a simple eddy-diffusivity model.

Appendix 2.C

Governing equations are used to compute the velocity field in cases in which a joint scalar transported PDF approach is applied (in order): Reynolds stress closure and turbulent kinetic energy dissipation closure:

$$\frac{\partial \overline{\rho} \widetilde{u_i'' u_j''}}{\partial t} + \frac{\partial \overline{\rho} \widetilde{u_i'' u_j'' u_l''}}{\partial x_l} = \frac{\partial}{\partial x_k} \left[C_s \overline{\rho} \frac{\tilde{k}}{\tilde{\epsilon}} \widetilde{u_k'' u_l''} \frac{\partial \widetilde{u_i'' u_j''}}{\partial x_l} \right] + P_{ij} + \phi_{ij} - \overline{\rho} \delta_{ij} \frac{2}{3} \tilde{\epsilon},$$

$$P_{ij} = -\overline{\rho} \left[\widetilde{u_i'' u_l''} \frac{\partial \tilde{u}_j}{\partial x_l} + \widetilde{u_j'' u_l''} \frac{\partial \tilde{u}_i}{\partial x_l} \right],$$

$$\phi_{ij} = -\left(C_1 \tilde{\epsilon} + C_1^* P_{kk} \right) b_{ij} + C_2 \tilde{\epsilon} \left(b_{ik} b_{kj} - \frac{1}{3} b_{mn} b_{mn} \delta_{ij} \right) + \left(C_3 - C_3^* II^{1/2} \right) \tilde{k} S_{ij}$$

$$+ C_4 \tilde{k} \left(b_{ik} S_{jk} + b_{jk} S_{ik} - \frac{2}{3} b_{mn} S_{mn} \delta_{ij} \right) + C_5 \tilde{k} \left(b_{ik} W_{jk} + b_{jk} W_{ik} \right),$$

$$b_{ij} = \frac{\widetilde{u_i'' u_j''}}{\widetilde{u_k'' u_k''}} - \frac{1}{3} \delta_{ij}, \qquad\qquad II = b_{ij} b_{ji},$$

$$S_{ij} = \frac{1}{2} \left(\frac{\partial \tilde{u}_i}{\partial x_j} + \frac{\partial \tilde{u}_j}{\partial x_i} \right), \qquad\qquad W_{ij} = \frac{1}{2} \left(\frac{\partial \tilde{u}_i}{\partial x_j} - \frac{\partial \tilde{u}_j}{\partial x_i} \right),$$

$$C_1 = 3.4, \qquad C_1^* = 4.2, \qquad C_2 = 4.2, \qquad C_3 = 4/5,$$

$$C_3^* = 1.3, \qquad C_4 = 1.25, \qquad C_5 = 1.25, \qquad C_S = 0.15,$$

$$\frac{\partial \overline{\rho} \tilde{\epsilon}}{\partial t} + \frac{\partial \overline{\rho} \tilde{u}_l \tilde{\epsilon}}{\partial x_l} = \frac{\partial}{\partial x_k} \left[C_{S\epsilon} \overline{\rho} \frac{\tilde{k}}{\tilde{\epsilon}} \widetilde{u_k'' u_l''} \frac{\partial \tilde{\epsilon}}{\partial x_l} \right] - C_{\epsilon 1} \frac{\tilde{\epsilon}}{\tilde{k}} P_{kk} - C_{\epsilon 2} \frac{\tilde{\epsilon}}{\tilde{k}} \tilde{\epsilon},$$

$$C_{\epsilon 1} = 1.44, \qquad\qquad C_{\epsilon 2} = 1.83, \qquad\qquad C_{S\epsilon} = 0.18.$$

Appendix 2.D

The form of the Langevin model for scalar statistics is subsequently given. The same expressions [i.e., the equivalents of Eq. (2.193)] are also applied in the scalar direction z:

$$dc^{(p)} = \frac{G_c}{2\tau_c}(c^{(p)} - \langle c \rangle)\, dt + \left(B_c \frac{\langle c'^2 \rangle}{\tau_c}\right)^{\frac{1}{2}} dW_{bin}, \qquad (2.186)$$

where the diffusion (B_c) and drift (G_c) coefficients are evaluated as

$$B_c = K_c \left[1 - \left(\frac{c'^{(p)}}{c'_*}\right)^2\right], \qquad (2.187)$$

$$G_c = -\left\{K_c \left[1 - \left\langle \left(\frac{c'^{(p)}}{c'_*}\right)^2\right\rangle\right] + 1\right\}, \qquad (2.188)$$

$$K_c = K_0 \left(1 - \frac{\theta_c}{|\theta_c| + 1}\right), \qquad (2.189)$$

$$\theta_c = C_K \frac{\left[(c^{(p)} - \langle c \rangle)(u_i^{(p)} - \langle u_i \rangle) - \langle c' u_i' \rangle\right]\langle c' u_i' \rangle}{\frac{2}{3}\langle k \rangle \langle c'^2 \rangle}. \qquad (2.190)$$

The values of the constants K_0 and C_K are 2.1 and 0.76, repectively.

The evaluation of the term $c'^{(p)}/c'_*$ is consistent with the intention of the original single-scalar model [250]. In the present implementation, this leads to

$$c'^{(p)} = c^{(p)} - \langle c \rangle^{(p)}, \qquad (2.191)$$

$$c'_* = \begin{cases} c'^{(p)}_{max} & \text{if } c'^{(p)} \text{ is positive} \\ c'^{(p)}_{min} & \text{if } c'^{(p)} \text{ is negative} \end{cases}, \qquad (2.192)$$

where the quantity $\langle c \rangle^{(p)}$ is evaluated as

$$\langle c \rangle^{(p)} = c_{min}|_{Z=z^{(p)}} + (\langle c \rangle - c_{min}|_{Z=\langle z \rangle}) \frac{c_{max}|_{Z=z^{(p)}} - c_{min}|_{Z=z^{(p)}}}{c_{max}|_{Z=\langle z \rangle} - c_{min}|_{Z=\langle z \rangle}}. \qquad (2.193)$$

REFERENCES

[1] D. Veynante, A. Trouvé, K. N. C. Bray, and T. Mantel, Gradient and counter-gradient transport in turbulent premixed flames, *J. Fluid Mech.* **332**, 263–293 (1997).

[2] A. Mura and M. Champion, Relevance of the Bray number in the small-scale modelling of turbulent premixed flames, *Combust. Flame* **156**, 729–733 (2009).

[3] P. A. Libby, *An Introduction to Turbulence* (Taylor & Francis, Washington, D.C., 1996).

[4] S. B. Pope, *Turbulent Flows* (Cambridge University Press, Cambridge, 2000).

[5] K. N. C. Bray and J. B. Moss, A unified statistical model of the premixed turbulent flame, *Acta Astron.* **4**, 291–320 (1977).

[6] P. A. Libby and K. N. C. Bray, Countergradient diffusion in premixed turbulent flames, *AIAA J.* **15**, 1186–1193 (1981).

[7] K. N. C. Bray, P. A. Libby, G. Masuya, and J. B. Moss, Turbulence production in premixed turbulent flames, *Combust. Sci. Technol.* **25**, 127–140 (1981).

[8] P. A. Libby, Theory of normal premixed turbulent flames revisited, *Prog. Energy Combust. Sci.* **11**, 86–96 (1985).

[9] E. Schneider, A. Maltsev, A. Sadiki, and J. Janicka, Study on the potential of BML-approach and *G*-equation concept-based models for predicting swirling partially premixed combustion systems: URANS computations, *Combust. Flame* **152**, 548–572 (2008).

[10] N. Chakraborty and R. S. Cant, Direct numerical simulation analysis of the flame surface density transport equation in the context of large eddy simulation, *Proc. Combust. Inst.* **32**, 1445–1453 (2009).

[11] K. N. C. Bray and P. A. Libby, Recent developments in the BML model of premixed turbulent combustion, in P. A. Libby and F. A. Williams (eds.), *Turbulent Reacting Flows* (Academic, New York, 1994), pp. 115–151.

[12] N. Chakraborty and R. S. Cant, Effects of Lewis number on turbulent scalar transport and its modelling in turbulent premixed flames, *Combust. Flame* **156**, 1427–1444 (2009).

[13] M. Champion and P. A. Libby, The influence of a thermally active wall on premixed turbulent combustion, *Combust. Sci. Technol.* **175**, 2015–2060 (2003).

[14] P. A. Libby, Theoretical analysis of the effect of gravity on premixed turbulent flames, *Combust. Sci. Technol.* **68**, 15–33 (1989).

[15] D. Veynante and T. J. Poinsot, Effects of pressure gradients on turbulent premixed flames, in *Annual Research Briefs* (Center for Turbulence Research, Stanford University, Stanford, CA, 1995), pp. 273–300.

[16] W. P. Jones, Turbulence modelling and numerical solution methods for variable density and combusting flows, in P. A. Libby and F. A. Williams (eds.), *Turbulent Reacting Flows* (Academic, New York, 1994), pp. 309–374.

[17] R. P. Lindstedt and E. M. Váos, Modeling of premixed turbulent flames with second moment methods, *Combust. Flame* **116**, 461–485 (1999).

[18] P. Domingo and K. N. C. Bray, Laminar flamelet expressions for pressure fluctuation terms in second moment models of premixed turbulent combustion, *Combust. Flame* **121**, 555–574 (2000).

[19] K. N. C. Bray, The interaction between turbulence and combustion, *Proc. Combust. Inst.* **17**, 223–233 (1979).

[20] R. Hauguel, Flamme en V turbulente: Simulation numérique directe et modélisation de la combustion turbulente prémélangée, Ph.D. thesis (Institut National des Sciences Appliqués de Rouen, Rouen, France, 2003).

[21] P. A. Libby and K. N. C. Bray, Implications of the laminar flamelet model in premixed turbulent combustion, *Combust. Flame* **39**, 33–41 (1980).

[22] Y.-H. Im, K. Y. Huh, S. Nishiki, and T. Hasegawa, Zone conditional assessment of flame-generated turbulence with DNS database of turbulent premixed flame, *Combust. Flame* **137**, 478–488 (2004).

[23] E. Lee and K. Y. Huh, Zone conditional modeling of premixed turbulent flames at high Damköhler number, *Combust. Flame* **138**, 211–224 (2004).

[24] A. N. Lipatnikov, Conditionally averaged balance equations for modeling premixed turbulent combustion in flamelet regime, *Combust. Flame* **152**, 529–547 (2008).

[25] K. N. C. Bray, M. Champion, N. Dave, and P. A. Libby, On the thermochemistry of premixed turbulent combustion in variable enthalpy systems, *Combust. Sci. Technol.* **46**, 31–43 (1986).

[26] K. N. C. Bray and J. B. Moss, A closure model for the Turbulent premixed flame with sequential chemistry, *Combust. Flame* **30**, 125–131 (1977).

[27] K. N. C. Bray, M. Champion, and P. A. Libby, Systematically reduced rate mechanisms and presumed PDF models for premixed turbulent combustion, *Combust. Flame* **157**, 455–464 (2010).

[28] V. Robin, A. Mura, M. Champion, and T. Hasegawa, A new analysis of the modeling of the pressure fluctuations effects in premixed turbulent flames and its validation based on DNS data, *Combust. Sci. Technol.* **180**, 996–1009 (2008).

[29] K. N. C. Bray, M. Champion, and P. A. Libby, Premixed flames in stagnating turbulence. Part II: The mean velocities and pressure and the Damköhler number, *Combust. Flame* **112**, 635–654 (1998).

[30] K. N. C. Bray, M. Champion, and P. A. Libby, Premixed flames in stagnating turbulence. Part IV: A new theory for the Reynolds stresses and Reynolds fluxes applied to impinging flows, *Combust. Flame* **120**, 1–18 (2000).

[31] K. N. C. Bray, M. Champion, and P. A. Libby, Premixed flames in stagnating turbulence. Part V: Evaluation of models for the chemical source term, *Combust. Flame* **127**, 2023–2040 (2001).

[32] K. N. C. Bray, M. Champion, and P. A. Libby, Premixed flames in stagnating turbulence. Part VI: Predicting the mean density and the permitted rates of strain for impinging reactant streams, *Combust. Flame* **156**, 310–321 (2009).

[33] L. W. Kostiuk, I. G. Shepherd, and K. N. C. Bray, Experimental study of premixed turbulent combustion in opposed streams. Part III – Spatial structure of flames, *Combust. Flame* **118**, 129–139 (1999).

[34] S. C. Li, P. A. Libby, and F. A. Williams, Experimental investigation of a premixed flame in an impinging turbulent stream, *Proc. Combust. Inst.* **25**, 1207–1214 (1994).

[35] R. K. Cheng and I. G. Shepherd, The influence of burner geometry on premixed turbulent flame propagation, *Combust. Flame* **85**, 7–26 (1991).

[36] E. Stevens, K. N. C. Bray, and B. Lecordier, Velocity and scalar statistics for premixed turbulent stagnation flames using PIV, *Proc. Combust. Inst.* **27**, 949–955 (1998).

[37] B. O. Ayoola, R. Balachandran, J. H. Frank, E. Mastorakos, and C. F. Kaminski, Spatially resolved heat release rate measurements in turbulent premixed flames, *Combust. Flame* **144**, 1–16 (2006).

[38] P. A. M. Kalt, Y.-C. Chen, and R. W. Bilger, Experimental investigation of turbulent scalar flux in premixed stagnation-type flames, *Combust. Flame* **129**, 401–415 (2002).

[39] G. Coppola, B. Coriton, and A. Gomez, Highly turbulent counterflow flames: A laboratory scale benchmark for practical systems, *Combust. Flame* **156**, 1834–1843 (2009).

[40] Y.-C. Chen, M. Kim, J. Han, S. Yun, and Y. Yoon, Measurement of the heat release rate integral in turbulent premixed premixed flames with particle image velocimetry, *Combust. Flame* **154**, 434–447 (2008).

[41] K. N. C. Bray, M. Champion, P. A. Libby, and N. Swaminathan, Finite rate chemistry and presumed PDF models for premixed turbulent combustion, *Combust. Flame* **146**, 665–673 (2006).

[42] M. V. Heitor, A. M. K. P. Taylor, and J. H. Whitelaw, The interaction of turbulence and pressure gradients in a baffle-stabilized premixed flame, *J. Fluid Mech.* **181**, 287–413 (1987).

[43] C. J. Rutland and R. S. Cant, Studying turbulence using numerical simulation databases V, in *Proceeding of the 1994 Summer Program* (Center for Turbulence Research, Stanford University, Stanford, CA, 1994), pp. 75–94.

[44] J. B. Moss, Simultaneous measuremnts of concentration and velocity in an open premixed turbulent flame, *Combust. Sci. Technol* **22**, 115–129 (1980).

[45] D. Bradley, P. H. Gaskell, and X. J. Gu, Application of a Reynolds stress, stretched flamelet, mathematical model to computations of turbulent burning velocities and comparison with experiments, *Combust. Flame* **96**, 221–248 (1994).

[46] V. L. Zimont and F. Biagioli, Gradient, counter-gradient transport and their transition in turbulent premixed flames, *Combust. Theory Model.* **6**, 79–101 (2002).

[47] C. Meneveau and T. Poinsot, Stretching and quenching of flamelets in premixed turbulent combustion, *Combust. Flame* **86**, 311–332 (1991).

[48] N. Swaminathan, R. W. Bilger, and G. R. Ruetsch, Interdependence of the instantaneous flame front structure and the overall scalar flux in turbulent premixed flames, *Combust. Sci. Technol.* **128**, 73–97 (1997).

[49] N. Swaminathan, R. W. Bilger, and B. Cuenot, Relationship between turbulent scalar flux and conditional dilatation in premixed flames with complex chemistry, *Combust. Flame* **126**, 1764–1779 (2001).

[50] Y.-C. Chen, P. A. M. Kalt, R. W. Bilger, and N. Swaminathan, Effects of mean flow divergence on turbulent scalar flux and local flame structure in premixed turbulent combustion, *Proc. Combust. Inst.* **29**, 1863–1871 (2002).

[51] G. Masuya and P. A. Libby, Nongradient theory for oblique turbulent flames with premixed reactants, *AIAA J.* **19**, 1590–1599 (1981).

[52] G. Troiani, M. Marrocco, S. Giammartini, and C. M. Casciola, Counter-gradient transport in the combustion of a premixed CH4/air annular jet by combined PIV/OH-LIF, *Combust. Flame* **156**, 608–620 (2009).

[53] S. Tullis and R. S. Cant, Scalar transport modelling in large eddy simulation of turbulent premixed flames, Proc. *Combust. Inst.* **29**, 2097–2104 (2002).

[54] G. Damköhler, The effect of turbulence on the flame velocity in gas mixtures. NACA Tech. Rep. TM 1112 (National Advisory Committee for Aeronautics, 1947) [also Z. Elektrochem. **46**, 601–652 (1940).

[55] N. Peters, *Turbulent Combustion* (Cambridge University Press, New York, 2000).

[56] T. Poinsot and D. Veynante, *Theoretical and Numerical Combustion* (Edwards, Philadelphia, 2001).

[57] J. F. Driscoll, Turbulent premixed combustion: Flamelet structure and its effect on turbulent burning velocities, *Prog. Energy Combust. Sci.* **34**, 91–134 (2008).

[58] T. Poinsot, D. Veynante, and S. Candel, Quenching processes and premixed turbulent combustion diagrams, *J. Fluid Mech.* **228**, 561–605 (1991).

[59] C. Mueller, J. F. Driscoll, D. Reuss, M. Drake, and M. Rosalik, Vorticity generation and attenuation as vortices convect through a premixed flame, *Combust. Flame* **112**, 342–358 (1998).

[60] H. Kolla and N. Swaminathan, Strained flamelets for turbulent premixed flames, I: Formulation and planar flame results, *Combust. Flame* **157**, 943–954 (2010).

[61] S. Candel, D. Veynante, F. Lacas, E. Maistret, N. Darabiha, and T. Poinsot, Coherent flamelet model: Applications and recent extensions, in B. Larrout-turou (ed.), *Recent Advances in Combustion Modeling*, Vol. 6 of Advances in Mathematics for Applied Sciences Series (World Scientific, Singapore, 1990), pp. 19–64.

[62] K. N. C. Bray, Turbulent flows with premixed reactants, in P. A. Libby and F. A. Williams (eds.), *Turbulent Reacting Flows* (Springer-Verlag, Berlin, 1980), pp. 115–183.

[63] P. Domingo, L. Vervisch, and D. Veynante, Large eddy simulation of a lifted methane–air jet flame in a vitiated coflow, *Combust. Flame* **152**, 415–432 (2008).

[64] S. B. Pope, The evolution of suirfaces in turbulence, *Int. J. Eng. Sci.* **26**, 445–469 (1988).

[65] L. Vervisch, E. Bideaux, K. N. C. Bray, and W. Kollmann, Surface density function in premixed turbulent combustion modeling, similarities between probability density function and flame surface approaches, *Phys. Fluids* **7**, 2496–2503 (1995).

[66] M. S. Anand and S. B. Pope, Calculations of premixed turbulent flames by PDF methods, *Combust. Flame* **67**, 127–142 (1987).

[67] D. Bradley, P. H. Gaskell, X. J. Gu, and A. Sedaghat, Premixed flamelet modelling: Factors influencing the turbulent heat release rate source term and the turbulent burning velocity, *Combust. Flame* **143**, 227–245 (2005).

[68] C. K. Law, *Combustion Physics* (Cambridge University Press, Cambridge, 2006).

[69] F. A. Williams, *Combustion Theory* (Benjamin/Cummings, Redwood City, CA, 1985).

[70] D. Veynante and L. Vervisch, Turbulent combustion modeling, *Prog. Energy Combust. Sci.* **28**, 193–266 (2002).

[71] J. Duclos, D. Veynante, and T. Poinsot, A comparison of flamelet models for premixed turbulent combustion, *Combust. Flame* **95**, 101–118 (1993).

[72] R. Prasad, R. Paul, Y. Sivathanu, and J. Gore, An evaluation of combined flame surface density and mixture fraction models for nonisenthalpic premixed turbulent flames, *Combust. Flame* **117**, 514–528 (1999).

[73] F. Fichot, F. Lacas, D. Veynante, and S. Candel, One-dimensional propagation of a premixed turbulent flame with a balance equation for the flame surface density, *Combust. Sci. Technol.* **90**, 35–60 (1993).

[74] F. A. Tap, R. Hilbert, D. Thévenin, and D. Veynante, A generalized flame surface density modelling approach for the auto-ignition of a turbulent non-premixed system, *Combust. Theory Model.* **8**, 165–193 (2004).

[75] W.-W. Kim, S. Menon, and H. C. Mongia, Large-eddy simulation of a gas turbine combustor flow, *Combust. Sci. Technol.* **143**, 25–62 (1999).

[76] H. Pitsch and L. Duchamp de Lageneste, Large-eddy simulation of premixed turbulent combustion using a level-set approach, *Proc. Combust. Inst.* **29**, 2001–2008 (2002).

[77] E. R. Hawkes and R. S. Cant, Implications of a flame surface density approach to large eddy simulation of premixed turbulent combustion, *Combust. Flame* **126**, 1617–1629 (2001).

[78] F. Charlette, C. Meneveau, and D. Veynante, A power-law flame wrinkling model for LES of premixed turbulent combustion part I: Non-dynamic formulation, *Combust. Flame* **131**, 159–180 (2002).

[79] F. Charlette, C. Meneveau, and D. Veynante, A power-law flame wrinkling model for LES of premixed turbulent combustion part II: Dynamic formulation, *Combust. Flame* **131**, 181–197 (2002).

[80] S. Richard, O. Colin, O. Vermorel, A. Benkenida, C. Angelberger, and D. Veynante, Towards large eddy simulation of combustion in spark ignition engines, *Proc. Combust. Inst.* **31**, 3059–3066 (2007).

[81] V. Moureau, B. Fiorina, and H. Pitsch, A level set formulation for premixed combustion LES considering the turbulent flame structure, *Combust. Flame* **156**, 801–812 (2009).

[82] A. Trouvé and T. Poinsot, The evolution equation for the flame surface density, *J. Fluid Mech.* **278**, 1–31 (1994).

[83] W. Kollmann and J. H. Chen, Dynamics of the flame surface area in turbulent non-premixed combustion, *Proc. Combust. Inst.* **25**, 1091–1098 (1994).

[84] E. Van-Kalmthout, D. Veynante, and S. Candel, Direct numerical simulation analysis of flame surface density equation in non-premixed turbulent combustion, *Proc. Combust. Inst.* **26**, 35–42 (1996).

[85] M. Boger, D. Veynante, H. Boughanem, and A. Trouvé, Direct numerical simulation analysis of flame surface density concept for large eddy simulation of turbulent premixed combustion, *Proc. Combust. Inst.* **27**, 917–925 (1998).

[86] E. Van-Kalmthout and D. Veynante, Direct numerical simulation analysis of flame surface density models for nonpremixed turbulent combustion, *Phys. Fluids* **10**, 2347–2368 (1998).

[87] I. G. Shepherd, Flame surface density and burning rate in premixed turbulent flames, *Proc. Combust. Inst.* **26**, 373–379 (1996).

[88] D. Veynante, J. Piana, J. M. Duclos, and C. Martel, Experimental analysis of flame surface density models for premixed turbulent combustion, *Proc. Combust. Inst.* **26**, 413–420 (1996).

[89] F. Marble and J. Broadwell, The coherent flame model of non-premixed turbulent combustion. Project Squid TRW-9-PU (Project Squid Headquarters, Chaffee Hall, Purdue University, West Lafayette, IN, 1977).

[90] T. Lundgren, Distribution function in the statistical theory of turbulence, *Phys. Fluids* **10**, 969–975 (1967).

[91] S. B. Pope, Pdf method for turbulent reacting flows, *Prog. Energy Combust. Sci.* **11**, 119–195 (1985).

[92] R. Borghi, Turbulent combustion modelling, *Prog. Energy Combust. Sci.* **14**, 245–292 (1988).

[93] F. Gao and E. E. O'Brien, A large-eddy simulation scheme for turbulent reacting flows, *Phys. Fluids* **5**, 1282–1284 (1993).

[94] S. B. Pope, The evolution of surfaces in turbulence, *Int. J. Eng. Sci.* **26**, 445–469 (1988).

[95] S. Pope, Computations of turbulent combustion: Progress and challenges, *Proc. Combust. Inst.* **23**, 591–612 (1990).

[96] A. Y. Klimenko and R. W. Bilger, Conditional moment closure for turbulent combustion, *Prog. Energy Combust. Sci.* **25**, 595–687 (1999).

[97] A. Kronenburg, Double conditioning of reactive scalar transport equations in turbulent nonpremixed flames, *Phys. Fluids* **16**, 2640–2648 (2004).

[98] D. Bradley, L. K. Kwa, A. K. C. Lau, M. Missaghi, and S. B. Chin, Laminar flamelet modeling of recirculating premixed methane and propane-air combustion, *Combust. Flame* **71**, 109–122 (1988).

[99] O. Gicquel, N. Darabiha, and D. Thevenin, Laminar premixed hydrogen/air counterflow flame simulations using flame prolongation of ILDM with differential diffusion, *Proc. Combust. Inst.* **28**, 1901–1908 (2000).

[100] J. A. van Oijen, F. A. Lammers, and L. P. H. de Goey, Modeling of complex premixed burner systems by using flamelet-generated manifolds, *Combust. Flame* **127**, 2124–2134 (2001).

[101] P. Nguyen, L. Vervisch, V. Subramanian, and P. Domingo, Multidimensional flamelet-generated manifolds for partially premixed combustion, *Combust. Flame* **157**, 43–61 (2010).

[102] K. N. C. Bray, The challenge of turbulent combustion, *Proc. Combust. Inst.* **26**, 1–26 (1996).

[103] S. M. Candel and T. J. Poinsot, Flame stretch and the balance equation for the flame area, *Combust. Sci. Technol.* **70**, 1–15 (1990).

[104] P. Clavin, Premixed combustion and gasdynamics, *Annu. Rev. Fluid Mech.* **26**, 321–52 (1994).

[105] V. L. Zimont, Analytical formulation of the problem of turbulent premixed combustion and modeling of the chemical source for the flamelet mechanism, in W. J. Carey (Ed.), *New Development in Combustion Research* (Nova Science, New York, 2006), pp. 95–134.

[106] O. Colin, F. Ducros, D. Veynante, and T. Poinsot, A thickened flame model for large eddy simulations of turbulent premixed combustion, *Phys. Fluids* **12**, 1843–1863 (2000).

[107] C. Nottin, R. Knikker, M. Boger, and D. Veynante, Large eddy simulations of an acoustically excited turbulent premixed flame, *Proc. Combust. Inst.* **28**, 67–73 (2000).

[108] N. Peters, The turbulent burning velocity for large-scale and small-scale turbulence, *J. Fluid Mech.* **384**, 107–132 (1999).

[109] H. Pitsch, Large eddy simulation of turbulent combustion, *Annu. Rev. Fluid Mech.* **38**, 453–482 (2006).

[110] S. P. R. Muppala, N. K. Aluri, F. Dinkelacker, and A. Leipertz, Development of an algebraic reaction rate closure for the numerical calculation of turbulent premixed methane, ethylene and propane/air flames for pressures up to 1.0 MPa, *Combust. Flame* **140**, 257–266 (2005).

[111] N. Peters, A spectral closure for premixed turbulent combustion in the flamelet regime, *J. Fluid Mech.* **242**, 611–629 (1992).

[112] C. Stone and S. Menon, Swirl control of combustion instabilities in a gas turbine combustor, *Proc. Combust. Inst.* **29**, 155–160 (2002).

[113] D. Q. Nguyen, R. P. Fedkiw, and M. Kang, A boundary condition capturing method for incompressible flame discontinuities, *J. Comput. Phys.* **172**, 71–98 (2001).

[114] V. Moureau, P. Minot, C. Bérat, and H. Pitsch, A ghost-fluid method for large-eddy simulations of premixed combustion in complex geometries, *J. Comput. Phys.* **221**, 600–614 (2007).

[115] V. Smiljanovski, V. Moser, and R. Klein, A capturing – tracking hybrid scheme for deflagration discontinuities, *Combust. Theory Model.* **1**, 183–215 (1997).

[116] U. Maas and S. Pope, Simplifying chemical kinetics: Intrinsic low-dimensional manifolds in composition space, *Combust. Flame* **88**, 239–264 (1992).

[117] S. H. Lam and D. A. Goussis, The CSP method for simplifying kinetics, *Int. J. Chem. Kinet.* **26**, 461–486 (1994).

[118] C. Hasse and N. Peters, A two mixture fraction flamelet model applied to split injections in a DI diesel engine, *Proc. Combust. Inst.* **30**, 2755–2762 (2005).

[119] B. Fiorina, O. Gicquel, L. Vervisch, N. Darabiha, and S. Carpentier, Approximating the chemical structure of partially premixed and diffusion counterflow flames using FPI flamelet tabulation, *Combust. Flame* **140**, 147–160 (2005).

[120] Z. Ren and S. B. Pope, The use of slow manifolds in reactive flows, *Combust. Flame* **147**, 243–261 (2006).

[121] V. Bykov and U. Maas, The extension of the ILDM concept to reaction–diffusion manifolds, *Combust. Theory Model.* **11**, 839–862 (2007).

[122] S. Delhaye, L. Somers, J. van Oijen, and L. de Goey, Incorporating unsteady flow-effects beyond the extinction limit in flamelet-generated manifolds, *Proc. Combust. Inst.* **32**, 1051–1058 (2009).

[123] C. Felsch, M. Gauding, C. Hasse, S. Vogel, and N. Peters, An extended flamelet model for multiple injections in DI diesel engines, *Proc. Combust. Inst.* **32**, 2775–2783 (2009).

[124] N. Peters, Laminar flamelet concepts in turbulent combustion, *Proc. Combust. Inst.* **21**, 1231–1250 (1986).

[125] J. Galpin, A. Naudin, L. Vervisch, C. Angelberger, O. Colin, and P. Domingo, Large-eddy simulation of a fuel lean premixed turbulent swirl burner, *Combust. Flame* **155**, 247–266 (2008).

[126] J. Galpin, C. Angelberger, A. Naudin, and L. Vervisch, Large-eddy simulation of H_2–air auto-ignition using tabulated detailed chemistry, *J. Turbulence* **9**(13) (2008). Doi: 10.1080/14685240801953048

[127] A. W. Vreman, B. A. Albrecht, J. A. van Oijen, and R. J. M. Bastiaans, Premixed and nonpremixed generated manifolds in large-eddy simulation of Sandia flame D and F, *Combust. Flame* **153**, 394–416 (2008).

[128] H. Pitsch and H. Steiner, Large-eddy simulation of a turbulent piloted methane/air diffusion flame (Sandia flame D), *Phys. Fluids* **12**, 2541–2554 (2000).

[129] V. Subramanian, P. Domingo, and L. Vervisch, Large eddy simulation of forced ignition of an annular bluff-body burner, *Combust. Flame* **157**, 579–601 (2010).

[130] S. F. Ahmed, R. Balachandran, T. Marchione, and E. Mastorakos, Spark ignition of turbulent nonpremixed bluff-body flames, *Combust. Flame* **151**, 366–385 (2007).

[131] D. Veynante., B. Fiorina, P. Domingo, and L. Vervisch, Using self-similar properties of turbulent premixed flames to downsize chemical tables in high-performance numerical simulations, *Combust. Theory Model.* **12**, 1055–1088 (2008).

[132] P. Domingo, L. Vervisch, S. Payet, and R. Hauguel, DNS of a premixed turbulent V-flame and LES of a ducted-flame using a FSD–PDF subgrid scale closure with FPI tabulated chemistry, *Combust. Flame* **143**, 566–586 (2005).

[133] G. Lecocq, S. Richard, O. Colin, and L. Vervisch, An hybrid presumed pdf and flame surface density approach for large-eddy simulation of premixed turbulent combustion, part 1: Formalism and simulations of a quasi-steady burner, *Combust. Flame*, Doi: 10.1016/j.combustflame.2010.09.23.

[134] D. Veynante and R. Knikker, Comparison between LES results and experimental data in reacting flows, *J. Turbulence* **7** (35)(online only, DOI: 10.1088/1468-5248/5/1/037) (2006).

[135] L. Vervisch, P. Domingo, G. Lodato, and D. Veynante, Scalar energy fluctuations in large-eddy simulation of turbulent flames: Statistical budget and mesh quality criterion, *Combust. Flame* **157**, 778–789 (2010).

[136] C. Pera, J. Réveillon, L. Vervisch, and P. Domingo, Modeling subgrid scale mixture fraction variance in LES of evaporating spray, *Combust. Flame* **146**, 635–648 (2006).

[137] R. W. Bilger, The structure of diffusion flames, *Combust. Sci. Technol.* **13**, 155–170 (1976).

[138] P. A. Libby and F. A. Williams, Fundamental aspects and a review, in *Turbulent Reacting Flows* (Academic, New York, 1994), pp. 1–62.

[139] S. H. Starner, R. W. Bilger, M. B. Long, J. H. Frank, and D. F. Marran, Scalar dissipation measurements in turbulent jet diffusion flames of air diluted methane and hydrogen, *Combust. Sci. Technol.* **129**, 141–164 (1997).

[140] V. Robin, A. Mura, M. Champion, and P. Plion, A multi-Dirac presumed pdf model for turbulent reactive flows with variable equivalence ratio, *Combust. Sci. Technol.* **118**, 1843–1870 (2006).

[141] V. Robin, A. Mura, M. Champion, O. Degardin, B. Renou, and M. Boukhalfa, Experimental and numerical analysis of stratified turbulent V-shaped flames, *Combust. Flame* **153**, 288–315 (2008).

[142] H. Kolla, J. W. Rogerson, N. Chakraborty, and N. Swaminathan, Scalar dissipation rate modelling and its validation, *Combust. Sci. Technol.* **181**, 518–535 (2009).

[143] H. Kolla and N. Swaminathan, Strained flamelets for turbulent premixed flames, II: Laboratory flame results, *Combust. Flame* **157**, 1274–1289 (2010).

[144] H. Kolla and N. Swaminathan, Strained flamelets for turbulent premixed flames, I: Formulation and planar flame resutls, *Combust. Flame* **157**, 943–954 (2010).

[145] H. Kolla, Scalar dissipation rate based flamelet modelling of turbulent premixed flames, Ph.D. thesis (Engineering Department, Cambridge University, Cambridge, 2010).

[146] R. W. Bilger, Conditional moment closure modelling and advanced laser measurements, in T. Takeno (ed.), *Turbulence and Molecular Processes in Combustion* (Elsevier Science, New York, 1993), pp. 267–285.

[147] K. N. C. Bray, Studies of the turbulent burning velocity, *Proc. R. Soc. London A* **431**, 315–335 (1990).

[148] K. N. C. Bray, P. A. Libby, and J. B. Moss, Unified modeling approach for premixed turbulent combustion – part I: General formulation, *Combust. Flame* **61**, 87–102 (1985).

[149] T. Mantel and R. W. Bilger, Some conditional statistics in a turbulent premixed flame derived from direct numerical simulations, *Combust. Sci. Technol.* **110–111**, 393–417 (1995).

[150] N. Swaminathan and K. N. C. Bray, Effect of dilatation on scalar dissipation in turbulent premixed flames, *Combust. Flame* **143**, 549–565 (2005).

[151] R. Borghi, On the structure and morphology of turbulent premixed flames, in C. Casci (ed.), *Recent Advances in the Aerospace Sciences* (Plenum, New York, 1985), pp. 117–138.

[152] D. Bradley, How fast can we burn?, *Proc. Combust. Inst.* **24**, 247–262 (1992).

[153] A. Mura, F. Galzin, and R. Borghi, A unified pdf-flamelet model for turbulent premixed combustion, *Combust. Sci. Technol.* **175**, 1573–1609 (2003).

[154] Y. C. Chen and N. Mansour, Investigation of flame broadening in turbulent premixed flames in the thin-reaction-zones regime, *Proc. Combust. Inst.* **27**, 811–818 (1998).

[155] Y. C. Chen and R. W. Bilger, Experimental investigation of three-dimensional flame-front structure in premixed turbulent combustion-I: Hydrocarbon/air Bunsen flames, *Combust. Flame* **131**, 400–435 (2002).

[156] F. O'Young and R. W. Bilger, Scalar gradient and related quantities in turbulent premixed flames, *Combust. Flame* **109**, 682–700 (1997).

[157] A. Soika, F. Dinkelacker, and A. Leipertz, Measurement of the resolved flame structure of turbulent premixed flames with constant Reynolds number and varied stoichiometry, *Proc. Combust. Inst.* **27**, 785–792 (1998).

[158] N. Swaminathan and R. W. Bilger, Scalar dissipation, diffusion and dilatation in turbulent H_2–air flames with complex chemistry, *Combust. Theory Model.* **5**, 429–446 (2001).

[159] K. N. C. Bray and N. Peters, Laminar flamelets in turbulent flames, in P. A. Libby and F. A. Williams (eds.), *Turbulent Reacting Flows* (Academic, New York, 1994), pp. 63–113.

[160] R. Borghi, Turbulent premixed combustion: Further discussions on the scales of fluctuations, *Combust. Flame* **80**, 304–312 (1990).

[161] J. L. Lumley and B. Khajeh-Nouri, Computational modelling of turbulent transport, in F. N. Frenkiel and R. M. Munn (eds.), Vol. 18A of *Advances in Geophysics* Series (Academic, New York, 1974), pp. 169–192.

[162] O. Zeman and J. Lumley, Modelling buoyancy driven mixed layers, *J. Atmos. Sci.* **33**, 1974–1988 (1976).

[163] G. Newman, B. Launder, and J. Lumley, Modelling the behaviour of homogeneous scalar turbulence, *J. Fluid. Mech.* **111**, 217–232 (1981).

[164] R. Borghi and D. Dutoya, On the scales of the fluctuations in turbulent combustion, *Proc. Combust. Inst.* **17**, 235–244 (1979).

[165] T. Mantel and R. Borghi, A new model of premixed wrinkled flame propagation based on a scalar dissipation equation, *Combust. Flame* **96**, 443–457 (1994).

[166] A. Mura and R. Borghi, Towards an extended scalar dissipation equation for turbulent premixed combustion, *Combust. Flame* **133**, 193–196 (2003).

[167] W. P. Jones and P. Musonge, Closure of the Reynolds stress and scalar flux equations, *Phys. Fluids* **31**, 3589–3604 (1988).

[168] N. Chakraborty, M. Klein, and N. Swaminathan, Effects of Lewis number on the reactive scalar gradient alignment with local strain rate in turbulent premixed flames, Proc. *Combust. Inst.* **32**, 1409–1417 (2009).

[169] H. Tennekees and J. L. Lumley, *A First Course in Turbulence* (MIT University Press, Cambridge, MA, and London, 1972).

[170] R. A. Antonia and L. Browne, The destruction of temperature fluctuations in a turbulent plane jet, *J. Fluid. Mech.* **134**, 67–83 (1983).

[171] C. Béguier, I. Dekeyser, and B. Launder, Ratio of scalar and velocity time scales in shear flow turbulence, *Phys. Fluids* **21**, 307–310 (1978).

[172] V. Eswaran and S. B. Pope, Direct numerical simulations of the turbulent mixing of a passive scalar, *Phys. Fluids* **31**, 506–520 (1988).

[173] J. R. Chasnov, Simulation of the inertial-conductive subrange, *Phys. Fluids A* **3**, 1164–1168 (1991).

[174] J. R. Chasnov, The viscous–convective subrange in nonstationary turbulence, *Phys. Fluids* **10**, 1191–1205 (1998).

[175] M. Champion, Modelling the effects of combustion on a premixed turbulent flow: a review, in R. Borghi and S. N. B. Murthy (eds.), *Turbulent Reactive Flows*, Vol. 40 of Lecture Notes in Engineering Series (Springer-Verlag, New York, 1989), pp. 732–753.

[176] A. Mura, V. Robin, M. Champion, and T. Hasegawa, Small scales features of velocity and scalar fields in turbulent premixed flames, *Flow Turbulence Combust.* **82**, 339–358 (2009).

[177] A. Mura, K. Tsuboi, and T. Hasegawa, Modelling of the correlation between velocity and reactive scalar gradients in turbulent premixed flames based on dns data, *Combust. Theory Model.* **12**, 671–698 (2008).

[178] A. Mura and M. Champion, Relevance of the Bray number in the small scale modeling of turbulent premixed flames, *Combust. Flame* **156**, 729–733 (2009).

[179] V. Robin, A. Mura, M. Champion, and T. Hasegawa, Accepted for publication in *Combust. Sci. Technol.* **182**, 449–464 (2010).

[180] V. Robin, M. Champion, and A. Mura, A second order model for turbulent reactive flows with variable equivalence ratio, *Combust. Sci. Technol.* **180**, 1707–1732 (2008).

[181] C. Dopazo and E. O'Brien, An approach to the autoignition of a turbulent mixture, *Acta Astron.* **1**, 1238–1266 (1974).

[182] A. Mura, V. Robin, and M. Champion, Modeling of scalar dissipation in partially premixed turbulent flames, *Combust. Flame* **149**, 217–224 (2007).

[183] M. J. Dunn, A. R. Masri, R. W. Bilger, R. S. Barlow, and G. H. Wang, The compositional structure of highly turbulent piloted premixed flames issuing into a hot coflow, *Proc. Combust. Inst.* **32**, 1779–1786 (2009).

[184] V. Bychkov, Velocity of turbulent flamelets with realistic fuel expansion, *Phys. Rev. Lett.* **84**, 6122–6125 (2000).

[185] V. Akkerman and V. Bychkov, Velocity of weakly turbulent flames of finite thickness, *Combust. Theory Model.* **9**, 323–351 (2005).

[186] N. Swaminathan and R. W. Grout, Interaction of turbulence and scalar fields in premixed flames, *Phys. Fluids* **18**, 045102–1-9 (2006).

[187] N. Chakraborty and N. Swaminathan, Influence of the Damköhler number on turbulence scalar interaction in premixed flames. I: Physical insight, *Phys. Fluids* **19**, 045103–1-10 (2007).

[188] S. H. Kim and H. Pitsch, Scalar gradient and small-scale structure in turbulent premixed combustion, *Phys. Fluids* **19**, 115104–1-14 (2007).

[189] G. Hartung, J. Hult, C. F. Kaminski, J. W. Rogerson, and N. Swaminathan, Effect of heat release on turbulence and scalar–turbulence interaction in premixed combustion, *Phys. Fluids* **20**, 035110–1-16 (2008).

[190] T. Mantel, A transport equation for the scalar dissipation in reacting flows with variable density: first results, in *Annual Research Briefs* (Center for Turbulence Research, Stanford University, Stanford, CA, 1993).

[191] N. Chakraborty, J. W. Rogerson, and N. Swaminathan, A priori assessment of closures for scalar dissipation rate transport in turbulent premixed flames using direct numerical simulation, *Phys. Fluids* **20**, 045106–1-15 (2008).

[192] N. Chakraborty and N. Swaminathan, Influence of the Damköhler number on turbulence scalar interaction in premixed flames. II: Model development, *Phys. Fluids* **19**, 045104–1-11 (2007).

[193] N. Chakraborty and N. Swaminathan, Effects of Lewis number on scalar dissipation transport and its modelling in turbulent premixed combustion, *Combust. Sci. Technol.* **182**, 1201–1240 (2010).

[194] N. Chakraborty and R. S. Cant, Physical insight and modelling for Lewis number effects on turbulent heat and mass transport in turbulent premixed flames, *Numer. Heat Transfer A* **55**, 762–779 (2009).

[195] G. K. Batchelor, The effect of homogeneous turbulence on material lines and surfaces, *Proc. R. Soc. London A* **231**, 349–366 (1952).

[196] G. Batchelor, Small scale variations of convected quantities like temperature in turbulent fluid, part 1. General discussion and the case of small conductivity, *J. Fluid Mech.* **5**, 113–133 (1959).

[197] C. H. Gibson, Fine structure of scalar fields mixed by turbulence. I. Zero-gradient points and minimal gradient surfaces, *Phys. Fluids* **11**, 2305–2315 (1968).

[198] W. T. Ashurst, A. R. Kerstein, R. M. Kerr, and C. H. Gibson, Alignment of vorticity and scalar gradient with strain rate in simulated navier-stokes turbulence, *Phys. Fluids* **30**, 2343–2363 (1987).

[199] G. R. Ruetsch and M. R. Maxey, Small-scale features of vorticity and passive scalar fields in homgeneous isotropic turbulence, *Phys. Fluids A* **3**, 1587–1597 (1991).

[200] K. K. Nomura and S. E. Elghobashi, Mixing characteristics of an inhomogeneous scalar in isotropic and homogeneous sheared turbulence, *Phys. Fluids A* **4**, 606–625 (1992).

[201] M. Tabor and I. Klapper, Stretching and alignment in chaotic and turbulent flows, *Chaos Solitons Fractals* **4**, 1031–1055 (1994).

[202] A. Garcia and M. Gonzalez, Analysis of passive scalar gradient alignment in a simplified three-dimensional case, *Phys. Fluids* **18**, 058101 (2006).

[203] W. T. Ashurst, N. Peters, and M. D. Smooke, Numerical simulation of turbulent flame structure with non-unity Lewis number, *Combust. Sci. Technol.* **53**, 339–375 (1987).

[204] C. J. Rutland and A. Trouvé, Direct simulations of premixed turbulent flames with nonunity Lewis numbers, *Combust. Flame* **94**, 41–57 (1993).

[205] N. Chakraborty, J. W. Rogerson, and N. Swaminathan, The scalar gradient alignment statistics of flame kernels and its modelling implications for turbulent premixed combustion, *Flow Turbulence Combust* **85**, 25–55 (2010).

[206] R. S. Cant, S. B. Pope, and K. N. C. Bray, Modelling of flamelet surface-to-volume ratio in turbulent premixed combustion, *Proc. Combust. Inst.* **23**, 809–815 (1990).

[207] K. N. C. Bray and N. Swaminathan, Scalar dissipation and flame surface density in premixed turbulent combustion, *C. R. Mec.* **334**, 466–473 (2006).

[208] R. W. Bilger, Some aspects of scalar dissipation, *Flow Turbulence Combust.* **72**, 93–114 (2004).

[209] J. W. Rogerson and N. Swaminathan, Correlation between dilatation and scalar dissipation in turbulent premixed flames, in *Proceedings of the European Combustion Meeting* Mediterranean Agronomic Institute of Chania, Greece (2007).

[210] T. S. Kuan, R. P. Lindstedt and E. M. Vaos, Higher moment based modeling of turbulence enhanced explosion kernels in confined fuel-air mixtures, in G. D. Roy (ed.), *Advances in Confined Detonations and Pulse Detonation Engines* (Torus Press, Moscow, 2003), pp. 17–40.

[211] R. P. Lindstedt and E. M. Vaos, Transpoted pdf modeling of high-Reynolds-number premixed turbulent flames, *Combust. Flame* **145**, 495−511 (2006).

[212] L. Vervisch, R. Hauguel, P. Domingo, and M. Rullaud, Three facets of turbulent combustion modelling: DNS of premixed V-flame, LES of lifted nonpremixed flame and rans of jet-flame, *J. Turbulence* **5**, (online only, DOI: 10.1088/1468–5248/5/1/004) (2004).

[213] Y. Nada, M. Tanahashi, and T. Miyauchi, Effect of turbulence characteristics on local flame structure of H_2–air premixed flames, *J. Turbulence* **5**, (online only, DOI: 10.1088/1468–5248/5/1/016) (2004).

[214] A. N. Kolmogorov, I. G. Petrovskii, and N. S. Piskunov. *Bull. Moscow State Univ.*, **1**(7), 1–72 (1937).

[215] V. R. Savarianandam and C. J. Lawn, Burning velocity of premixed turbulent flames in the weakly wrinkled regime, *Combust. Flame* **146**, 1–18 (2006).

[216] R. G. Abdel-Gayed, D. Bradley and M. Lawes, Turbulent burning velocities: A general correlation in terms of straining rates, *Proc. R. Soc. London A* **414**, 389–413 (1987).

[217] S. M. Il'yashenko and A. V. Talantov, Theory and analysis of straight-through-flow combustion chambers, edited machine translation (Wright-Patterson Air Force Base, OH, 1964).

[218] H. Kolla, J. W. Rogerson and N. Swaminathan, Validation of a turbulent flame speed model across combustion regimes, *Combust. Sci. Technol.* **182**, 284–308 (2010).

[219] V. L. Zimont, Gas premixed combustion at high turbulence. turbulent flame closure combustion model, *Exp. Therm. Fluid Sci.* **21**, 179–186 (2000).

[220] O. L. Gülder, Turbulent premixed flame propagation models for different combustion regimes, *Proc. Combust. Inst.* **23**, 743–750 (1990).

[221] A. M. Klimov, Premixed turbulent flames – interplay of hydrodynamic and chemical phenomena, in J. R. Bowen, N. Manson, A. K. Oppenheim, and R. I. Soloukhin (eds.), *Flames, Lasers and Reactive Systems* (American Institute of Aeronautics and Astronautics, New York, 1983), pp. 133–146.

[222] S. P. Malkeson and N. Chakraborty, Scalar dissipation rate transport modelling of partially premixed flames using DNS, in *Proceedings of 4th European Combustion Meeting*, paper P810083, 2009.

[223] V. Robin, A. Mura, M. Champion, P. Plion, Modélisation de la combustion turbulente des mélanges hétérogènes en richesse. Des flammes de prémélange aux flammes de diffusion, *C. R. Mec.* **337**, 362–372 (2009).

[224] V. Robin, N. Guilbert, A. Mura, and M. Champion, Modélisation de la combustion turbulente des mélanges hétérogènes en richesse. Application au calcul d'une flamme stabilisée par l'élargissement brusque d'un canal bidimensionnel, *C. R. Mec.* **338**, 40–47 (2010).

[225] O. Darbyshire, N. Swaminathan, and S. Hochgreb, Effects of small scale mixing models on prediction of turbulent premixed and stratified combustion. Accepted in *Combust. Sci. Technol.*, **182**, 1141–1170 (2010).

[226] S. Pope and M. S. Anand, Flamelet and distributed combustion in premixed turbulent flames, *Proc. Combust. Inst.* **20**, 403–410 (1984).

[227] S. B. Pope, The statistical theory of turbulent flames, *Philos. Trans. R. Soc. London A* **291**, 529–568 (1979).

[228] D. Bradley, P. H. Gaskell, and X. J. Gu, Application of a Reynolds stress, stretched flamelet, mathematical model to computations of turbulent burning velocities and comparisons with experiments, *Combust. Flame* **96**, 221–248 (1994).

[229] K. N. C. Bray, M. Champion, and P. A. Libby, Premixed flames in stagnating turbulence: Part I. The general formulation for counterflowing streams and gradient models for turbulent transport, *Combust. Flame* **84**, 391–410 (1991).

[230] R.S. Barlow, www.ca.sandia.gov/TNF.

[231] J. Xu and S. B. Pope, Assessment of numerical accuracy of PDF/Monte Carlo methods for turbulent reacting flows, *J. Comput. Phys.* **152**, 192–230 (1999).

[232] Q. Tang, J. Xu, and S. B. Pope, PDF calculations of local extinction and NO production in piloted jet turbulent methane/air flames, *Proc. Combust. Inst.* **28**, 133–139 (2000).

[233] S. A. Louloudi, R. P. Lindstedt, and E. M. Vaos, Joint scalar PDF modelling of pollutant formation in piloted turbulent jet diffusion flames with comprehensive chemistry, *Proc. Combust. Inst.* **28**, 149–156 (2000).

[234] R. P. Lindstedt, S. A. Louloudi, J. R. Driscoll, and V. Sick, Finite rate chemistry effects in turbulent reacting flows, *Flow Turbulence Combust.* **72**, 407–426 (2004).

[235] K. Gkagkas and R. P. Lindstedt, Transported PDF modelling with detailed chemistry of pre- and auto-ignition in CH_4/air mixtures, *Proc. Combust. Inst.* **31**, 1559–1566 (2007).

[236] M. Muradoglu, K. Liu, and S. B. Pope, PDF modeling of a bluff-body stabilized turbulent flame, *Combust. Flame* **132**, 115–137 (2003).

[237] T. S. Kuan and R. P. Lindstedt, Transported probability density function modelling of a bluff body stabilized turbulent flame, *Proc. Combust. Inst.* **30**, 767–774 (2005).

[238] R. S. Barlow, B. B. Dally, A. R. Masri, and G. J. Fiechtner, Instantaneous and mean compositional structure of bluff-body stabilised nonpremixed flames, *Combust. Flame* **114**, 119–148 (1998).

[239] A. R. Masri, J. B. Kelman, and B. B. Dally, The instantaneous spatial structure of the recirculation zone in bluff-body stabilised flames, *Proc. Combust. Inst.* **27**, 1031–1037 (1998).

[240] R. R. Cao, S. B. Pope, and A. R. Masri, Turbulent lifted flames in a vitiated coflow investigated using joint PDF calculations, *Combust. Flame* **142**, 438–453 (2005).

[241] E. Bidaux, M. Gorokhovski, and R. Borghi, Simulation of Bunsen burner methane–air flames using combined PDF–equation–KIVA-2 numerical model, in *Statistical Properties of Turbulent Gaseous Flames* (Delft University of Technology, Delft, The Netherlands, 1995), Euromech Colloquium 340.

[242] R. P. Lindstedt, Modelling of the chemical complexities of flames, *Proc. Combust. Inst.* **27**, 269–285 (1998).

[243] R. W. Bilger, S. B. Pope, K. N. C. Bray, and J. F. Driscoll, Paradigms in turbulent combustion, *Proc. Combust. Inst.* **30**, 21–42 (2005).

[244] N. Peters, Multiscale combustion and turbulence, *Proc. Combust. Inst.* **32**, 1–25 (2009).

[245] S. B. Pope and W. K. Cheng, Statistical calculations of spherical turbulent flames, *Proc. Combust. Inst.* **21**, 1473–1481 (1986).

[246] T. Hulek and R. P. Lindstedt, Computations of steady-state and transient premixed turbulent flames using PDF methods, *Combust. Flame* **104**, 481–504 (1996).

[247] S. B. Pope and D. C. Haworth, The mixing layer between turbulent fields of different scales, in *Turbulent Shear Flows* (Springer-Verlag, Berlin, 1987), Vol. 5, pp. 44–53.

[248] D. C. Haworth and S. B. Pope, A generalized Langevin model for turbulent flows, *Phys. Fluids A* **29**, 387–405 (1986).

[249] C. G. Speziale, S. Sarkar, and T. B. Gatski, Modelling the pressure–strain correlation of turbulence: An invariant dynamical system approach, *J. Fluid Mech.* **227**, 245–272 (1991).

[250] L. Valiño and C. Dopazo, A binomial Langevin model for turbulent mixing, *Phys. Fluids A* **3**, 3034–3037 (1991).

[251] R. P. Lindstedt and E. M. Vaos, Modeling of mixing processes in non-isothermal and combusting flows, in C. Dopazo (ed.), *Advances in Turbulence* (Int. Center Numerical Methods Engineering, Barcelona, Spain, 2000), Vol. 8, pp. 493–496.

[252] C. Dopazo, Probability density function approach for a turbulent axisymmetric heated jet, centerline evolution, *Phys. Fluids A* **18**, 397–404 (1975).

[253] L. Valiño, C. Dopazo, A binomial sampling model for scalar turbulent mixing, *Phys. Fluids A* **2**, 1204–1212 (1990).

[254] J. Janicka, W. Kolbe, and W. Kollmann, Closure of the transport equation for the probability density function of scalar fields, *J. Non-Equilib. Thermodyn.* **4**, 47–66 (1979).

[255] T. Hulek and R. P. Lindstedt, Joint scalar-velocity PDF modelling of finite rate chemistry in a scalar mixing layer, *Combust. Sci. Technol.* **136**, 303–331 (1998).

[256] Y.-C. Chen, N. Peters, G. A. Schneemann, N. Wruck, U. Renz, and M. S. Mansour, The detailed flame structure of highly stretched turbulent premixed methane–air flames, *Combust. Flame* **107**, 223–226 (1996).

[257] R. P. Lindstedt, H. C. Ozarovsky, R. S. Barlow, and A. N. Karpetis, Progression of localised extinction in high Reynolds number turbulent jet diffusion flames, *Proc. Combust. Inst.* **31**, 1551–1558 (2007).

[258] S. M. Cannon, B. S. Brewster, L. D. Smoot, Pdf modeling of lean premixed combustion using in situ tabulated chemistry, *Combust. Flame* **119**, 233–252 (1999).

[259] O. Soulard and V. Sabel'nikov, Eulerian Monte Carlo method for the joint velocity and mass fraction probability density function in turbulent reactive gas flows, *Flow Turbulence Combust.* **77**, 333–357 (2006).

[260] C. Dopazo, Non-isothermal turbulent reacting flows: Stochastic approaches, Ph.D. dissertation (New York State University, New York, 1973).

[261] P. A. Libby, A non-gradient theory for premixed turbulent flames, in *Mechanics Today* (Pergamon, New York, 1980), Vol. 5.

[262] N. Swaminathan and R. W. Bilger, Analyses of conditional moment closure for turbulent premixed flames, *Combust. Theory Model.* **5**, 241–260 (2001).

[263] J.-Y. Chen and W. Kollmann, Comparison of prediction and measurement in non-premixed turbulent flames, in P. A. Libby and F. A. Williams (eds.), *Turbulent Reacting Flows* (Academic, London, 1994), pp. 211–308.

[264] D. C. Haworth and S. B. Pope, A PDF modeling study of self–similar turbulent free shear flows, *Phys. Fluids A* **30**, 1026–1044 (1987).

[265] S. B. Pope, On the relationship between stochastic Lagrangian models of turbulence and second-moment closures, *Phys. Fluids* **6**, 973–985 (1994).

[266] T. Hulek and R. P. Lindstedt, Modelling of unclosed non-linear terms in a PDF closure for turbulent flames, *Math. Comput. Model.* **24**, 137–147 (1996).

[267] G. R. Grimmett and D. R. Stirzaker, *Probability and Random Processes*, 2nd ed. (Clarendon, Oxford, 1992).

[268] S. B. Pope, A stochastic Lagrangian model for acceleration in turbulent flows, *Phys. Fluids* **14**, 2360–2375 (2002).

[269] R. O. Fox, *Computational Models for Turbulent Reacting Flows* (Cambridge University Press, Cambridge, 2003), p. 207.

[270] D. B. Spalding, Mixing and chemical reaction in steady confined turbulent flames, *Proc. Combust. Inst.* **13**, 649–657 (1970).

[271] R. P. Lindstedt, V. Sakthitharan, Modelling of compressible reacting flows, in *Proceedings of the Eighth Symposium on Turbulent Shear Flows* (Technical University of Munich, Munich 1991), pp. 1–6.

[272] P. K. Yeung, S. S. Girimaji, and S. B. Pope, Straining and scalar dissipation on material surfaces in turbulence: Implications for flamelets, *Combust. Flame* **79**, 340–365 (1990).

[273] K. N. C. Bray, M. Champion, and P. A. Libby, Pre-mixed flames in stagnating turbulence – Part V – Evaluation of models for the chemical source term, *Combust. Flame* **127**, 223–240 (2001).

[274] R. G. Abdel-Gayed, K. J. Al-Khishali, and D. Bradley, Turbulent burning velocities and flame straining in explosions, *Proc. R. Soc. London A* **391**, 393–414 (1984).

[275] D. C. Haworth, Progress in probability density function methods for turbulent reacting flows, *Prog. Energy Combust. Sci.* **36**, 168–259 (2010).

[276] S. Viswanathan and S. B. Pope, Turbulent dispersion behind line sources in grid turbulence, *Phys. Fluids* **20**, 101514 (2008).

[277] R. P. Lindstedt, V. D. Milosavljevic, and M. Persson, Turbulent burning velocity predictions using transported PDF methods, *Proc. Combust. Inst.* DOI: 10.1016/j.proci.2010.05.092.

[278] R. McDermott and S. B. Pope, A particle formulation for treating differential diffusion in filtered density function methods, *J. Comput. Phys.* **226**, 947–993 (2007).

[279] D. C. Haworth and S. B. Pope, Monte Carlo solutions of a joint PDF equation for turbulent flows in general orthogonal coordinates, *J. Comput. Phys.* **72**, 311–346 (1987).

[280] R. I. Issa, Solution of the implicitly discretised fluid flow equations by operator-splitting, *J. Comput. Phys.* **62**, 40–65 (1986).

[281] T. Hulek and R. P. Lindstedt, Adapting van Leer's limited diffusion antiflux method for the PISO algorithm for solving reacting turbulent flows, Technical Report TF/91/26 (Mechanical Engineering Department, Imperial College of Science, Technology and Medicine, London, 1991).

[282] D. C. Haworth and S. H. El Tahry, Probability density function approach for multidimensional turbulent flow calculations with application to in-cylinder flows in reciprocating engines, *AIAA J.* **29**, 208–218 (1991).

[283] T. Hulek, Modelling of turbulent combustion using transported PDF methods, Ph.D. thesis (University of London, London, 1996).

[284] C. de Boor, *A Practical Guide to Splines* (Springer-Verlag, New York, 1978).

[285] S. B. Pope, R. Gadh, Fitting noisy data using cross-validated cubic smoothing splines, *Commun. Stat.-Simul. Comput.* **17**, 349–376 (1988).

[286] E. M. Vaos, Second moment methods for turbulent flows with reacting scalars, Ph.D. thesis (University of London, London, 1998).

[287] E. Mastorakos, Turbulent combustion in opposed jet flows, Ph.D. thesis (University of London, London, 1993).

[288] B. Hakberg and A. D. Gosman, Analytical determination of turbulent flame speed from combustion models, *Proc. Combust. Inst.* **20**, 225–232 (1984).

[289] C. A. Catlin and R. P. Lindstedt, Premixed turbulent burning velocities derived from mixing controlled reaction models with cold front quenching, *Combust. Flame* **85**, 427–439 (1991).

[290] V. A. Sabel'nikov, C. Corvellec, and P. Bruel, Analysis of the influence of cold front quenching on the turbulent burning velocity associated with an eddy-break-up model, *Combust. Flame* **113**, 492–497 (1998).

[291] W. P. Jones and Y. Prasetyo, Probability density function modelling of premixed turbulent opposed jet flames, *Proc. Combust. Inst.* **26**, 275–282 (1996).

[292] S. Zhang and C. J. Rutland, Premixed flame effects on turbulence and pressure-related terms, *Combust. Flame* **102**, 447–461 (1995).

[293] A. N. Lipatnikov and J. Chomiak, Effects of premixed flames on turbulence and turbulent scalar transport, *Prog. Energy Combust. Sci.* **36**, 1–102 (2010).

[294] L. Vervisch, W. Kollmann, and K. N. C. Bray, PDF modeling for premixed turbulent combustion based on the properties of iso-concentration surfaces, in *Proceedings of the Summer Program* (Center for Turbulence Research, Stanford University, Stanford, CA, 1994).

[295] R. P. Lindstedt and S. A. Louloudi and Joint scalar transported probability density function modeling of turbulent methanol jet diffusion flames, *Proc. Combust. Inst.* **29**, 2147–2154 (2002).

[296] P. M. C. Ferrão, Análise Experimental de Chamas Turbulentas com Recirculação, PhD thesis (Universidade Técnica de Lisboa, Lisbon, Portugal, 1993).

[297] P. Ferrão, D. Duarte, and M.V. Heitor, Turbulence statistics and scalar transport in highly sheared premixed flames, *Flow, Turbulence Combust.* **60**, 361–376 (1999).

[298] D. B. Spalding, I. Predicting the laminar flame speed in gases with temperature-explicit reaction rates, *Combust. Flame* **1**, 287–295 (1957).

[299] M. S. Anand, S. B. Pope, and H. C. Mongia, A PDF method for turbulent recirculating flows, in B. Borghi and S. N. B. Murthy (eds.), *Turbulent Reactive Flows* (Springer-Verlag, Berlin, 1989), pp. 672–693.

[300] S. Subramaniam and S. B. Pope, A mixing model for turbulent reactive flows based on Euclidean minimum spanning trees, *Combust. Flame* **115**, 487–514 (1998).

[301] A. P. Wandel and R. P. Lindstedt, Hybrid binomial Langevin-MMC modeling of a reacting mixing layer, *Phys. Fluids* **21**, 015103 (2009).

[302] S. L. Yilmaz, M. B. Nik, P. Givi, and P. A. Strakey, Scalar filtered density function for large eddy simulation of a Bunsen burner, *J. Propul. Power* **26** 84–93 (2010).

3 Combustion Instabilities

3.1 Instabilities in Flames
By D. Bradley

The past half century has seen impressive advances in reacting flow computational fluid dynamics (CFD), initially based on Reynolds-averaged Navier–Stokes (RANS) turbulent flow modelling and detailed chemical kinetic modelling of laminar flames. These two strands were combined in the modelling of turbulent combustion. In general, the systems analysed were hydrodynamically and chemically stable. On the other hand, it is well known from non-reacting flows that there is a rich variety of hydrodynamic instabilities. Probably the best known is laminar flow in a tube stabilised by viscous stresses. As the flow rate increases, the flow is eventually destabilised by the formation of vortices and the onset of turbulence at a critical Reynolds number. Viscosity does not feature in Kelvin–Helmholtz instabilities, which arise at the interface between two fluids of different densities flowing at different velocities. When the ratio of inertia to gravitational forces attains a critical value, the flow becomes unstable, with the generation of waves at the interface. Another example is the unstable interaction of gravitational and surface tension forces at the water–air interface of a droplet that generates capillary waves of small wavelength. The various combustion instabilities are broadly discussed in Section 3.1.1.

Transition to instability is characterised by quite minor perturbations of small amplitude, often at the same level as minor physical or numerical noise. At a critical value of a dimensionless group, formulated from the conflicting trends, the destabilising perturbations are rapidly amplified at particular frequencies, with the transfer of kinetic or thermal energy into the instability. In reacting flows the rapid release of chemical energy into an acoustic wave can create a new destructive entity, a detonation wave. In contrast, changes in the composition of the mixture to a burner can make a planar flame unstable, only for it to restabilise through the generation of a benign stable cellular structure over its entire surface, a structure that, along with flames oscillations, was demonstrated more than a hundred years ago by Smithells and Ingle [1] and more recently reviewed by Hertzberg [2]. The computing power necessary to obtain numerical solutions of these developing combustion instabilities exceeds that currently employed for large-eddy simulation (LES) and direct numerical simulation (DNS). Consequently codes are generally unable to offer detailed solutions, and practical guidance is often sought through relatively simple hydrodynamic models that yield analytic solutions for dispersion relationships [3, 4].

This section of Chapter 3 deals with some of the important combustion instabilities. Their onset is first discussed, then how the changing burning rate can generate pressures waves and the feedback mechanism by which such waves accelerate the flame still further. Localised flame quenching and reignition in turbulent flames can be important, probably initiating thermoacoustic oscillations. Unintended auto-ignition creates instabilities, and the severity of the accompanying acoustic waves is discussed. If they are sufficiently strong, they can lead to the onset of detonation, which can arise also from an accelerating turbulent flame. Finally, there is discussion of benign auto-ignitive burning through the control of reactivity gradients.

3.1.1 Flame Instabilities

ORIGIN OF INSTABILITIES. It is first necessary to describe some characteristics of stable laminar flames, with a laminar, stretch-free, burning velocity u_ℓ. If A is the localised area of a flame front, its stretch rate α^* is $(1/A)(dA/dt)$. The burning velocity that expresses the mass rate at which burned gas is formed in a stretched flame is u_{nr}. The effect on u_{nr} of the contributions to the overall flame stretch rate from the aerodynamic strain rate and the flame curvature is expressed by [5]

$$\frac{u_\ell - u_{nr}}{u_\ell} = K_s \, \text{Ma}_{sr} + K_c \, \text{Ma}_{cr}. \tag{3.1}$$

Here K_s and K_c are laminar Karlovitz stretch factors for aerodynamic strain rate and flame curvature. They are the products of the respective stretch rates α_s^*, α_c^* and the chemical time given by the flame thickness divided by the laminar burning velocity δ_ℓ/u_ℓ. The associated Markstein numbers are Ma_{sr} and Ma_{cr}. The first term on the right is usually dominant, and variations of Ma_{sr}, uncertainty of about ± 2, with equivalence ratio ϕ, are shown by the broken curves in Fig. 3.1. Mixtures are with air, under atmospheric conditions for H_2 [6], C_3H_8, [7] and CH_4 [8]. The linearised Eq. (3.1) is a first-order approximation. Detailed laminar flame models reveal nonlinearities in the relationship, particularly at higher pressures [8].

A hydrodynamic theory of flame instabilities was developed first by Darrieus [9] and Landau and Lifshitz [10] for a planar flame front. Figure 3.2 shows diagrammatically the effect of a wave-like perturbation of such a planar front. The perturbation generates vortices at the developing line interfaces between burned and unburned gas, with the rotations indicated near the streamlines, shown by the broken curves [4]. This and the expanding hot gases contract and expand the streamtubes, as indicated by the streamlines. These perturbations create relative pressure changes, indicated by $+$ and $-$ in the unburned gas, the gradients of which add to the original perturbation of the planar front. The perturbation is fed by the expansion of hot gas, and the theory suggests that planar flames are unconditionally unstable. The model described by Fig. 3.2 suggests that, if the flame sheet were to be positively stretched, it would, at least partially, neutralise the developing instability.

Historically, improved understanding of a laminar flame structure, including molecular thermal–diffusive (TD) effects, emerged as factors that could either

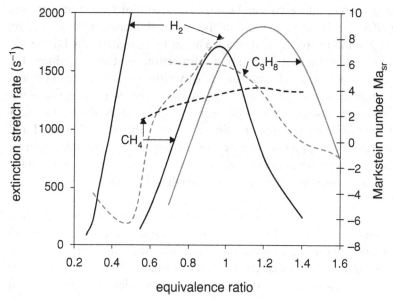

Figure 3.1. Markstein numbers (broken lines) and extinction stretch rates (solid lines) under atmospheric conditions.

counter, or enhance, the Darrieus–Landau (DL) instability. These comprise a conductive energy flux from burned to unburned gas, indicated by the broken arrowed lines at the leading edge of the flame in Fig. 3.2, and a diffusion flux of the deficient reactant from the unburned gas into the flame reaction zone, indicated by the solid lines. The deficient reactant is fuel in the case of a lean mixture and oxygen in the case of a rich one. The Lewis number, $Le = \hat{\lambda}/\rho c_p \mathcal{D}$, is a key parameter, where $\mathcal{D}$ is the diffusion coefficient of the deficient reactant. With $Le < 1$, the flame curvature focuses diffusion of enthalpy into the flame, shown by the arrowed solid lines. This exceeds the conductive energy loss out of the flame, shown by the arrowed broken lines. As a consequence, the local burning velocity increases. An opposite effect occurs in the valley, with a reduction in local burning velocity. As a consequence, the flame instability is increased further. This TD instability reinforces the DL instability.

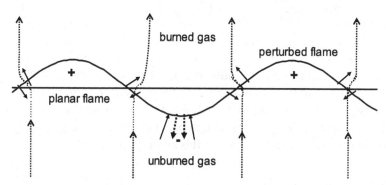

Figure 3.2. Development of DL and TD instabilities, after perturbation from a planar flame. $+$ and $-$ between perturbed streamlines indicate relative changes of unburned gas pressure. Arrowed solid lines at the wave crest indicate deficient reactant diffusion; arrowed dotted lines indicate thermal conduction.

Similar reasoning shows that Lewis numbers greater than unity exert an opposite and stabilising effect, as does the increase in viscosity with increasing temperature. The two types of instability are coupled and are hereafter designated by DLTD.

The additional TD term in the instability dispersion relation introduces the influences of Le, the unburned to burned gas density ratio σ, flame stretch rate, and the overall activation energy for the reaction, all acting over a range of wavelengths. These further complexities make even guideline solutions difficult to obtain. A successful one was achieved through the linear stability theory of Bechtold and Matalon [11] for spherical explosion flames. This demonstrates how flames can be unstable to both types of instability between inner and outer wavelength limits. It led to a dispersion relationship that expressed the growth of the perturbation in terms of wave number.

The general approach in [11], now outlined, analyses the perturbation of a spherical flame, including the effect of the changing global flame stretch rate. Perturbed variables are expanded in spherical harmonic series, and flame instability is investigated over a spectrum of wave numbers. For a flame radius r, with an initial value r_0 significantly greater than the flame thickness δ_ℓ, the dimensionless amplitude a of the perturbation relative to r develops according to

$$a = a_0 \, R^{\omega(1 + \Omega/\mathrm{Pe} \ln R)}. \tag{3.2}$$

Here a_0 is the initial value of a and R is r/r_0, Pe is the Peclet number defined as $\mathrm{Pe} = r/\delta_\ell$, ω is a growth-rate parameter, which depends on only σ and wave number n. The expression for Ω is complex [11]:

$$\Omega = \frac{Q_1}{\omega} + \frac{\hat{\beta}(\mathrm{Le} - 1)}{\omega(\sigma - 1)} \, Q_2, \tag{3.3}$$

where Q_1, Q_2 are functions of σ, ω, and n, given in [12], and $\hat{\beta}$, the Zeldovich number, is defined as $\hat{\beta} = T_a (T_b - T_u)/T_b^2$.

The growth rate $\overline{A}(n)$ of the amplitude of the perturbation for a wave number n is expressed by the differential of the natural logarithm of amplitude of the perturbation with respect to that of Pe. It can be derived from Eq. (3.2) and is

$$\overline{A}(n) = \frac{\mathrm{d} \ln(a/a_0)}{\mathrm{d} \ln(\mathrm{Pe})} = \omega \left(1 - \frac{\Omega}{\mathrm{Pe}} \right). \tag{3.4}$$

The first term on the right gives the contribution of the DL instability to the growth rate, and the second, $\omega \Omega/\mathrm{Pe}$, gives the contribution of the TD instability. The sign of $\overline{A}(n)$ depends on that of Ω/Pe and indicates whether the flame is stable. If it is negative, there is no growth and the flame is stable. If it is positive, the amplitude of the perturbation grows and the flame is unstable. The critical Peclet number, $\mathrm{Pe}_{\mathrm{cl}}$, is reached when, at some value of $n(= n_{\mathrm{cl}})$, $\overline{A}(n)$ first becomes equal to zero. This is the condition that defines both $\mathrm{Pe}_{\mathrm{cl}}$ and n_{cl}. From this and Eq. (3.4),

$$\mathrm{Pe}_{\mathrm{cl}} = \Omega. \tag{3.5}$$

It follows from Eqs. (3.4) and (3.5) that the relative contributions of the DL and TD instabilities to $\overline{A}(n)$ are in the ratio $\mathrm{Pe}/\mathrm{Pe}_{\mathrm{cl}}$. In the absence of instabilities, the influence of flame stretch rate on the stretched flame speed, $\mathrm{d}r/\mathrm{d}t$, is well approximated

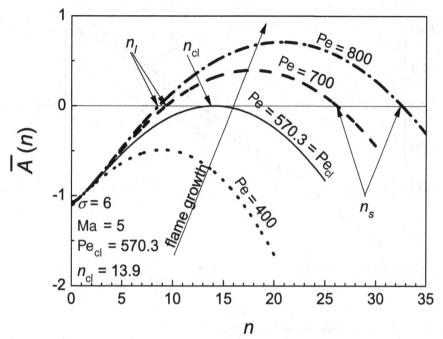

Figure 3.3. Growth rate of perturbation as a function of wave number, $\sigma = 6$, Ma $= 5$.

by a variant of Eq. (3.1), for the flame speed of a spherical flame:

$$S_s - S = L_b \alpha^*, \quad \text{with} \quad \alpha^* = \frac{2}{r} \frac{dr}{dt}. \tag{3.6}$$

Here S is the stretched flame speed, S_s is the flame speed at zero stretch rate, and L_b is the burned-gas Markstein length, which when divided by δ_ℓ gives the flame-speed Markstein number Ma_b. When S is linearly extrapolated to $\alpha^* = 0$ ($r = \infty$), it yields S_s.

Reference [11] does not explicitly employ a Markstein number, based on the burning velocity as in Eq. (3.1), but in Eq. (3.3) the term $\hat{\beta}(\text{Le} - 1)$ can be linked to a single Markstein number Ma, as proposed by Clavin [13]. Computed relationships between Markstein numbers based on flame speed and burning velocity are presented in [14]. From [13],

$$\text{Ma} = \frac{\sigma}{\sigma - 1} \ln \sigma + \frac{\hat{\beta}(\text{Le} - 1)}{2(\sigma - 1)} \int_0^{\sigma-1} \frac{\ln(1 + x)}{x} \, dx. \tag{3.7}$$

An improved expression, with allowance for the temperature dependence of the thermal conductivity $\hat{\lambda}$, is given in [4].

Variations of $\overline{A}(n)$ with n for four different values of Pe are shown in Fig. 3.3 for $\sigma = 6$ and Ma $= 5$. The dotted line, Pe $= 400$, shows $\overline{A}(n)$ to be always negative for the full range of n and the flame is stable. As Pe increases, eventually the amplitude of the perturbation begins to increase and it first becomes positive at the critical wave number n_{cl}. The Peclet number at which this occurs is Pe_{cl}, which is 570.3 for the present conditions, with $n_{cl} = 13.9$. When Pe increases above Pe_{cl}, there is a range of wave numbers, between the lower limit n_l and the higher limit n_s, within which

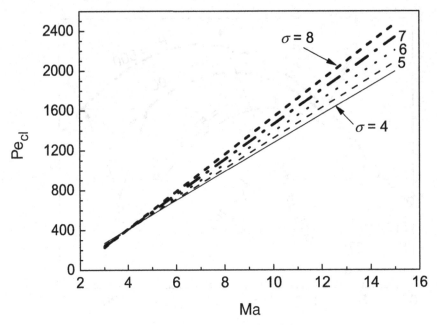

Figure 3.4. Theoretical dependence of Pe_{cl} on Ma and σ.

the flame is unstable. The range of unstable wave number broadens as Pe increases, with the greatest rate of growth of amplitude about midway between these limits.

Computed theoretical values of Pe_{cl} for different values of Ma are shown in Fig. 3.4. The higher the values of Ma and σ, the higher are those of Pe_{cl} and, at a given Pe, the lower are those of n_s [15]. All these are factors making for a more stable flame. The full Bechtold–Matalon theory is valid only for $\hat{\Lambda} > 30$, where $\hat{\Lambda}$ is the wavelength normalised by δ_ℓ. This which generally corresponds to Ma > 3 [15]. The effect of σ is greater at higher values of Ma. When Le is unity, the last term in Eq. (3.7) becomes zero and only σ has an influence on Ma. As a result, Ma attains its lowest values. Furthermore, over the full range of ϕ for a given fuel, the changes in σ are relatively small, and the first term on the right of Eq. (3.7) changes no more than a few percent. With little change in Ma, there is negligible change in Pe_{cl} with ϕ, as indicated by the theoretical plot in Fig. 3.4. This was confirmed experimentally for near-equidiffusive, spherical, acetylene explosion flames with Le close to unity [16].

Theoretically predicted values of Pe_{cl} are expressed as a function of $Ma\,\sigma^{0.25}$ by the bold line in Fig. 3.5. Experimental values, identified by the initial development of a full cellular structure, are given by the symbols. These were obtained from measurements in a spherical explosion bomb with mixtures of methane [8], hydrogen [6], and i-octane [17] with air at different pressures. There is much scatter in these experimental values, not only because of the difficulty in obtaining accurate values of Ma, but also in defining precisely the onset of instability, particularly for very negative Ma_{sr}. Such flames become unstable almost immediately. Because of this, the observed instability of lean H_2 flames at high pressures questions the concept of a stable laminar burning velocity under such conditions [6]. Values of δ_ℓ are given by v/u_ℓ. For both experiment and theory, the effect of σ is embodied in the $Ma\,\sigma^{0.25}$ term. The non-bold, solid line is the best fit to the experimental results. Both lines have

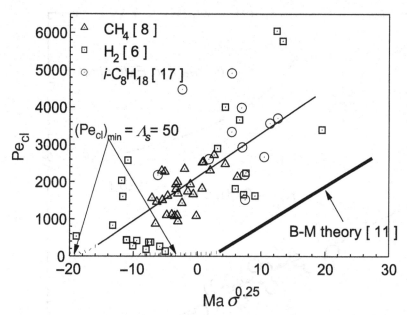

Figure 3.5. Measured values of Pe_{cl} (symbols) compared with theoretical predictions (bold line).

almost the same gradient, and the order-of-magnitude difference between them is attributed to the time lag in the development of a fully cellular flame from an initially 'cracked' flame [12]. The lower theoretical value of Pe_{cl} is close to the point at the first 'cracking' on the flame surface. The higher experimental value is identified by the full development of a cellular structure and the increase in the flame speed that results from the wrinkling of the flame surface.

Because of the importance of flame stretch rate in stabilising a flame, the critical value of a dimensionless group based on this might be of more general utility than one based on flame radius. Hence the laminar Karlovitz stretch factor at criticality, K_{cl}, equal to $\alpha_{cl}^*(\delta_\ell/u_\ell)$, was evaluated, where α_{cl}^* is the total stretch rate at Pe_{cl}. With Eq. (3.6) applied to this condition and the relationship $S_s = \sigma u_\ell$ at constant pressure and zero stretch rate, it is readily shown that

$$K_{cl} = \frac{2\sigma}{Pe_{cl}}\left(1 + 2\frac{Ma_b}{Pe_{cl}}\right)^{-1}.\qquad(3.8)$$

Values of K_{cl}, found from the experimental values of Pe_{cl} in Fig. 3.5, are indicated by crosses and plotted against Ma_{sr} in Fig. 3.6. The larger circle symbols show additional ethanol–air data from [18]. A best-fit curve to all the data is shown in Fig. 3.6 by the lower solid line curve in the plot of laminar Karlovitz stretch factor, $K_\ell = \alpha^*u_\ell/\delta_\ell$, against Ma_{sr} and given by

$$K_{cl} = 0.0075 \exp\left(-0.123\,Ma_{sr}\right).\qquad(3.9)$$

The regime of unstable flames lies below this curve. Above it, the value of K_ℓ is sufficient to stabilise the flame. Also shown are the values of the turbulent Karlovitz stretch factor, $K_{0.8}$, at which the probability of a turbulent flame being able

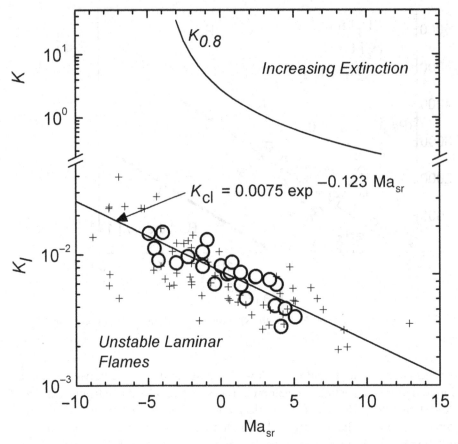

Figure 3.6. Limit of instability for laminar flames, K_{cl} on K_ℓ scale, and onset of flame extinction in turbulent flames, $K_{0.8}$ on K scale.

to propagate is reduced to 0.8. This is due to the onset of flame extinction arising from the higher stretch rate. This is discussed in Subsection 3.1.2.

PENINSULA OF INSTABILITY AND INCREASING FLAME SPEED. It was shown in [12] that the theoretical results from [11] could be presented as a peninsula of instability, bounded by n_l and n_s and also dependent on Pe and Ma_{sr}, as indicated in Fig. 3.7. As Ma_{sr} decreases, f n_s increases, Pe_{cl} decreases, and the area of the peninsula increases. The relationship between wave number n and the wavelength normalised by δ_ℓ, $\hat{\Lambda}$, is

$$n = \frac{2\pi \, \text{Pe}}{\hat{\Lambda}}. \tag{3.10}$$

Allowance for the experimentally observed lag between the theoretical value of Pe_{cl} and the onset of a fully cellular flame was made in [12] by multiplying the theoretical value of the upper-limit wave number n_s by a lag factor of less than unity, f. This was evaluated at the *measured* Pe_{cl} by equating $fn_{s,cl}$ to the known theoretical value of the lower-limit wave number $n_{l,cl}$, where the subscript cl indicates a value at the measured Pe_{cl}.

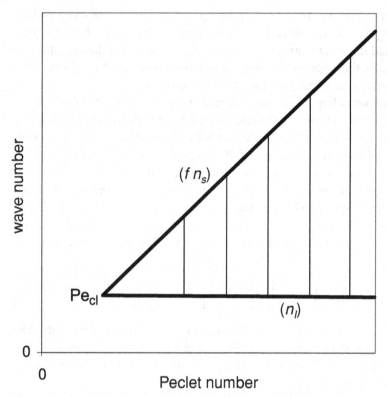

Figure 3.7. Instability peninsula, with limiting wave numbers at constant Ma_{sr}. As Ma_{sr} decreases, so does Pe_{cl}, whereas $(df\, n_s/d\, Pe)$ increases.

For $Pe > Pe_{cl}$ the theory shows n_s to increase close to linearly with Pe. This linearity is maintained for the modified wave number, $f n_s$, as shown in Fig. 3.7 and given by [12]

$$f n_s = f\, n_{s,cl} + (Pe - Pe_{cl})\, \frac{\mathrm{d} f\, n_s}{\mathrm{d}\, Pe}. \tag{3.11}$$

With Eq. (3.10) applied to experimental conditions, $\mathrm{d} f\, n_s/\mathrm{d}\, Pe = 2\pi/\hat{\Lambda}_s$, and this increases as Ma_{sr} decreases. A limit to the gradient is set by the finite thickness of the flame, and experiments suggest a lower limit to the value of $\hat{\Lambda}_s$ of about 50 at very negative values of Ma_{sr} [19]. This results in a maximum value of $\mathrm{d} f\, n_s/\mathrm{d}\, Pe = 2\pi/50$. At this limit, the experiments show that a high negative flame stretch causes localised flame extinctions at the cusps. They also suggest a threshold for the onset of localised extinctions that begin to reduce the expected burning rate of the wrinkled flame at about $\hat{\Lambda}_s = 167$.

As a flame grows in size, $\hat{\Lambda}_s$, the shortest wavelength, remains constant, whereas, from Eq. (3.10), the longest wavelength, $\hat{\Lambda}_l$, might be expected to be proportional to Pe, with consequently very little change in n_l. As the flame propagates, it is wrinkled by an ever-increasing range of wavelengths, which is due entirely to the increase in $\hat{\Lambda}_l$. The evolution of the associated cellular structure that exists through the peninsula was traced in detail in [17]. Newly formed cells are initially stabilised by the localised stretch rate, but, as they grow in size, their stretch rate decreases because of the decreasing curvature. Eventually, at a critical localised Peclet number or Karlovitz stretch factor, an instability arises at the growing cell and the original cell fissions into smaller cells. The newly formed smaller cells are initially stabilised by their high curvature and the greater stretch rate. They grow in size, and the evolutionary cycle is repeated. By these means, the original instability is re-stabilised through the formation of a cellular structure over the entire flame surface.

The increased wrinkling of the flame surface with increasing Pe results in an increasing flame speed that can be estimated from fractal considerations [12]. These suggest that the ratio of the fractal surface area with a resolution of the inner cut-off to that with a resolution of the outer cut-off, at Pe, is $(fn_s/n_l)_{\mathrm{Pe}}^{\hat{D}-2}$, where $\hat{D}$ is the fractal dimension. If Pe is large enough for flame stretch rates to be relatively small, flame speeds are proportional to the surface area. Consequently this ratio is equal to the ratio F of flame speeds, with and without instabilities. The former is indicated by S_n and the latter by S_s $(= u_\ell\sigma)$. The unstable flame speed S_n is equal to dr/dt, where r is the mean flame radius of the wrinkled surface. Hence, at a given Pe, with $\hat{D} = 7/3$,

$$F = \left(\frac{S_n}{S_s}\right)_{\mathrm{Pe}} = \frac{1}{\sigma\,u_\ell}\frac{dr}{dt} = \frac{1}{\sigma}\frac{d\,\mathrm{Pe}}{d\bar{t}} = \left(\frac{fn_s}{n_l}\right)_{\mathrm{Pe}}^{1/3}, \tag{3.12}$$

in which $S_s = u_\ell\sigma$, $\delta_\ell = \nu/u_\ell$, and $\bar{t} = tu_\ell^2/\nu$, where ν is the kinematic viscosity. The value of F increases with the range of unstable wavelengths, which increases as Ma_{sr} becomes more negative and Pe increases. The last is further increased with increasing pressure because of the decrease in δ_ℓ.

With these conditions and $d(fn_s/n_l)/d\,\mathrm{Pe}$ assumed to be constant, Eqs. (3.11) and (3.12), after appropriate integration, yield, in the unstable regime [12, 15],

$$\mathrm{Pe} = \mathrm{Pe}_0 + B\bar{t}^{3/2}, \tag{3.13}$$

where Pe_0 is the value of Pe at $\bar{t} = 0$ and it can be relatively close to Pe_{cl}. The numerical constant B depends on σ and Ma_{sr}, particularly through the gradient $df\,n_s/d\,\mathrm{Pe}$. From [15],

$$B = \left[\left(\frac{2}{3}\right)^{3/2}\sigma^{3/2}\left(\frac{f}{n_l}\frac{dn_s}{d\mathrm{Pe}}\right)^{0.5}\right]. \tag{3.14}$$

It is readily shown from Eq. (3.13) that

$$\frac{d\mathrm{Pe}}{d\bar{t}} = \frac{3}{2}B^{2/3}\left(\mathrm{Pe} - \mathrm{Pe}_0\right)^{1/3}, \tag{3.15}$$

and also from Eqs. (3.12) and (3.13) that

$$F = \frac{3}{2\sigma}B\sqrt{\bar{t}} = \frac{3}{2\sigma}B^{2/3}\left(\mathrm{Pe} - \mathrm{Pe}_0\right)^{1/3}. \tag{3.16}$$

In the earlier nonlinear theory of [20] the flame radius also varied as $t^{3/2}$, and Eq. (3.13) was applied to large-scale explosions [21, 22]. There are problems in the DNS of the self-acceleration of unstable flames because of the generation of numerical noise [23]. The simulations in [23] showed that, after an initial acceleration, 2D expanding flames slow down to a constant speed, whereas spherical flames continue to accelerate. For the latter, over the computed period of time of the acceleration, the optimal exponent for $\bar{t}$ was 1.32.

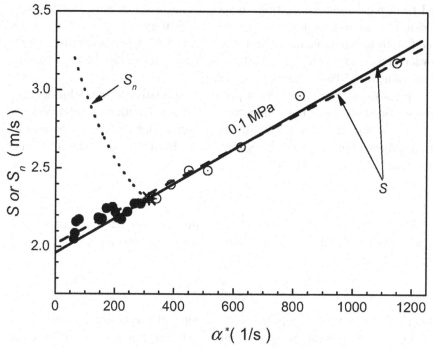

Figure 3.8. Variations of flame speeds S and S_n with flame stretch rate, $\alpha^* = (2/r)S$, for 0.1 MPa, $\phi = 0.4$, $T = 365$ K, from [6].

The changes in flame speed with α^* during an experimental spherical explosion of a lean H_2–air mixture at 0.1 MPa, $T = 365$ K, $\phi = 0.4$, are shown in Fig. 3.8 [6]. The Markstein number is negative, initially the flame is stable, and the open symbols show the stretched flame speed S decreasing with α^* prior to the development of instabilities when α^* decreases below about 350 s^{-1}. The unstable flame speed S_n, shown by the dotted curve, increases sharply as the flame propagates more rapidly as a result of its increasing cellularity. The enhancement of flame speed that is due to cellularity was found with Eq. (3.12), from which the theoretical stable flame speed, shown by the black circular symbols in Fig. 3.8, could be found. Extrapolation of the stable, stretched, flame speed to zero stretch rate gave S, which when divided by σ yielded u_{ell}.

ACOUSTIC WAVES GENERATED BY BURNING-RATE ACCELERATION. The higher the burning velocity, the higher the net rate of volume generation at the reaction front, dV/dt. A rapid rate of change in this rate can induce appreciable acoustic waves. At a sufficient distance d from a part of a flame or an entire small flame surface, the noise approximates that from a monopole sound source. Simple acoustic theory gives the associated instantaneous sound pressure $\Delta p(t)$ above the ambient at time t. From Hurle et al. [24], this is

$$\Delta p(t) = \frac{\rho}{4\pi d} \left| \frac{d}{dt} \left(\frac{dV}{dt} \right) \right|_{t-t_d},\tag{3.17}$$

where t_d is the time for the sound wave to propagate the distance d from the source to its measurement point through a non-reacting gas of density ρ. Values of $\Delta p(t)$, predicted from measurements of the rate of change of C_2 emission intensity in an explosion of a 4-cm-diameter soap bubble of stoichiometric ethylene–air, were confirmed [24] by microphone measurements 0.33 m away. A pressure pulse of 3 Pa was recorded at the end of combustion, after which $\Delta p(t)$ fell to zero before reaching a minimum value of -5 Pa in a rarefaction wave.

If A is the area of the reaction front propagating relative to the unburned gas at a burning velocity u_c, then the volumetric rate of consumption of unburned gas is $A u_c$. The corresponding volumetric rate of production of burned gas is greater in the ratio of unburned-gas to burned-gas densities σ. Hence the net rate of volume generation that is due to reaction is

$$\frac{dV}{dt} = A u_c (\sigma - 1). \tag{3.18}$$

Arising from this expression, mass conservation and the definition of u_c lead to a propagation speed *relative to the fixed radial stationary coordinates* of $u_\ell \sigma$ and, following Eq. (3.12),

$$\frac{dr}{dt} = u_c \sigma = u_\ell \frac{d\,\mathrm{Pe}}{d\bar{t}}. \tag{3.19}$$

In Eq. (3.17) for a truly monopole source, in which the total sound pressure is a simple sum of the component pressures from all flame elements, the overall size of the flame, or part of it, must be restricted. In [24], the size is restricted to a quarter wavelength of the acoustic wave. In addition, the compression and expansion of the acoustic wave are assumed isentropic. This becomes less justifiable as $\Delta p(t)$ becomes large. For a spherical flame of radius r, Eqs. (3.18) and (3.19) give

$$\frac{dV}{dt} = 4\pi r^2 u_c (\sigma - 1) = 4\pi r^2 \frac{\sigma - 1}{\sigma} \frac{dr}{dt}. \tag{3.20}$$

For large flames there can be differences in the time intervals t_a originating at different points in the flame and their arrival time at the fixed measurement point. Allowance can be made for this by expressing Eq. (3.17) in integral form. Here simplified solutions are presented. These assume the same value of dr/dt throughout the flame at a particular instant and a fixed value of d, measured from the centre of the spherical flame. Starting with Eq. (3.20) and involving Eqs. (3.13) and to (3.19), it can be shown that at the average dimensionless distance, $\bar{d} = d/\delta_\ell$, the averaged overpressure in an unconfined explosion is

$$\frac{\Delta p}{\rho u_\ell^2} = \frac{3}{2\bar{d}} \mathrm{Pe}\, B^{4/3} \left(\frac{\sigma - 1}{\sigma} \right) \left[3(\mathrm{Pe} - \mathrm{Pe}_0)^{2/3} + \frac{\mathrm{Pe}}{2(\mathrm{Pe} - \mathrm{Pe}_0)^{1/3}} \right]. \tag{3.21}$$

The first term within the square brackets arises from the rate of change of flame area, the second from the rate of change of flame speed. The mean dimensionless time for an acoustic wave to propagate the average distance d from the centre of the spherical flame at the acoustic velocity $\hat{c}$ is

$$\bar{t}_d = \frac{d}{\hat{c}} \frac{u_\ell^2}{\nu} = \bar{d} \frac{u_\ell}{\hat{c}}. \tag{3.22}$$

The duration of flame propagation in the unstable regime to Pe is found from Eq. (3.13). This time, added to $\bar{t}_d$, gives the time from the onset of the instability to the elevation Δp at the distance $\bar{d}$. The expressions for Δp and F were evaluated

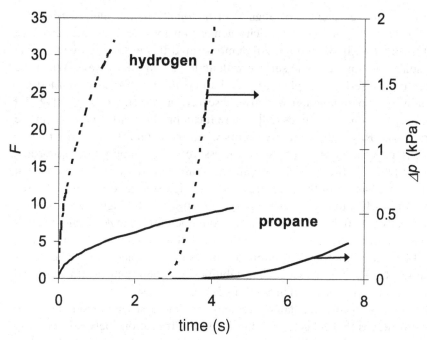

Figure 3.9. Computed F and Δp 1 km from the centre of the fireball for large spherical atmospheric explosions of hydrogen (broken lines), $\phi = 0.5$ and propane–air (solid lines), $\phi = 1.06$, up to maximum radii of 100 m.

for the propane–air flame, $\phi = 1.06$, that was studied in the large-scale experimental, atmospheric explosions of [22]. Relevant properties are $Ma_{sr} = 5.5$, $u_\ell = 0.41$ m/s, $B = 0.242$, $\sigma = 8.303$, and $\hat{c} = 333.7$ m/s. The computed values of F and Δp are plotted against the time from the onset of the instability by the solid lines in Fig. 3.9. The time lag in the arrival of the pressure pulse is apparent. Values of Δp are for a distance of 1 km from the centre of the fireball. The computed flame radii extend to a maximum of 100 m, when the flame terminates. With $\sigma = 8.303$, for this radius of a completely burned mixture, the original radius over which the initial unburned mixture extended, prior to ignition, would be 49.4 m. The maximum computed pressure of about 0.3 kPa is not severely damaging, although it would be higher closer to the flame and is rising rapidly as the flame propagates. It is close to a value at which glass windows would fail. In the experiments of [22], the maximum flame radius was 3.75 m after 0.51 s.

In contrast, the broken lines give computed values of F and Δp in an atmospheric explosion of a more unstable mixture of hydrogen–air, $\phi = 0.5$, $Ma_{sr} = -8$, $u_\ell = 0.65$ m/s [6], $B = 0.59$, $\sigma = 5.01$, and $\hat{c} = 379.8$ m/s. Again, the maximum flame radius was 100 m and d was 1 km. This explosion is more rapid, with F reaching a maximum value of 32 and Δp one of about 2 kPa. Clearly, larger explosions would be even more damaging and there is no apparent limit to the acceleration of large unstable fireballs, other than that provided by an increasing radiative energy loss from the burned gases [22]. This is an area worthy of further study.

RAYLEIGH–TAYLOR INSTABILITIES. The pressure waves generated by DLTD instabilities can create additional instabilities and overpressures. Taylor instabilities arise when pressure gradients are aligned orthogonally to the high-density gradients at the flame surface, with vorticity generation by means of the baroclinic term, $\nabla(1/\rho) \times \nabla p$ [25]. Markstein [26] demonstrated this instability by wrinkling a smooth spherical laminar flame surface with an impacting shock wave. When the planar pressure pulse hit the sphere, it accelerated the lower-density burned gas preferentially. This phenomenon was also observed in vented explosions [14, 27] and flame propagation along tubes [28], where it can produce large increases in the burning rate. The Rayleigh instability occurs when the combustion energy release feeds directly into the positive phase of a pressure wave, amplifying it according to Rayleigh's [29, 30] criterion: The amplitude of an acoustic wave will increase if heat is added in phase with the pressure. The Rayleigh source term $\int p'q'dV$ [see Eqs. (3.39) and (3.40)] in the acoustic energy equation can be highly destabilising, and the unsteady rate of heat release arising from fluctuations in ϕ during the combustion of lean premixtures also can affect the oscillation frequency [31]. In practice, Rayleigh–Taylor thermoacoustic instabilities are usually coupled and are hereafter designated by RT. The thermoacoustic instabilities and their control strategies and simulation methods are discussed in Sections 3.2 and 3.3, respectively.

Confined and semiconfined flames create overall increases in pressure and generate pressure waves that reflect at containing walls. Transitions between different instabilities in confined explosion flames were studied in an explosion bomb of 385-mm diameter with three pairs of orthogonal windows of 150-mm diameter [32]. Simultaneous ignition at two diametrically opposite sparks at the wall of the bomb created two near-identical imploding flames that could be viewed at the central window. The advantage of this configuration was that the flames propagated at higher pressures and Peclet numbers than would have been possible with central ignition. The technique was used to measure laminar burning velocities at high pressures in [33]. Some experimental results from [32] for explosions of three different mixtures, initially at 0.5 MPa and 358 K but with different degrees of instability, are presented in Table 3.1. These values were measured at the maximum values of u_c. The three mixtures ranged from a relatively stable stoichiometric i-octane–air mixture to a very unstable rich mixture, $\phi = 1.6$, of the same gases.

It was possible to explain the maximum enhanced burning velocity $(u_c)_{max}$ of the first and most stable stoichiometric mixture entirely in terms of DLTD instabilities after Pe_{cl} was attained. The value of Δp was too small to be measured. Values of F were found from Eqs. (3.14) and (3.16) and of u_c from Eq. (3.19). The second mixture was of H_2–air with $\phi = 0.4$, and this created a small primary thermoacoustic instability, with $\Delta p = 0.8$ kPa at $(u_c)_{max}$. After Pe_{cl} was attained, burning velocities initially could be attributed entirely to DLTD instabilities until just after the onset of flame oscillations, when the associated cellularity began to decline. The oscillations developed in tandem at both of the imploding flames and seemed to cause the reduction in both flame cellularity and u_c.

These changes are explained by reference to Fig. 3.2. The Taylor effect suggests that, with this configuration, a pressure higher on the unburned than on the burned side, the leading edge of burned gas would be preferentially accelerated towards the burned gas, reducing the cellularity. Conversely, with a higher pressure on the burned side, the same leading edge of burned gas would be preferentially accelerated towards the unburned gas, now increasing the cellularity. Thus the Taylor generation of vorticity can either stabilise or destabilise the flame. Laser-sheet observations and

Table 3.1. *Summary of experimental findings [32] at maximum u_c, with all values for this condition. PA is primary acoustic and SA is secondary acoustic oscillations. estimated value, RT^s is Rayleigh-Taylor stabilisation*

Mixture	p (MPa)	T (K)	u_ℓ at $(u_c)_{max}$ (m/s)	Ma_{sr}	$(u_c)_{max}$ (m/s)	F at $(u_c)_{max}$	Δp at $(u_c)_{max}$ (kPa)	Type of instability
i-C$_8$H$_{18}$–air $\phi = 1.0$	1.8	502	0.295	2	0.66	2.2	0	DLTD
H$_2$–air $\phi = 0.4$	0.97	440	0.34	−11	2.19	6.4	0.8	DLTD, PA, RT^s
i-C$_8$H$_{18}$–air $\phi = 1.6$	2.06	508	0.10	−20*	3.75	37.5	300	DLTD, RT, PA SA

the changing values of u_c suggest the stabilising, or laminarising, tendency to be the stronger, with initially a pronounced reduction in cellularity. Subsequently this was followed by the development of a 'finger-like' surface structure. Particle image velocimetry also showed that, with unburned-gas flow into the flame surface, the surface was smoothed and with the flow away from the surface it tended to form the finger-like structure [32]. Kaskan [34] observed similar stabilising and destabilising area changes in near-flat flames propagating along a tube and subjected to pressure oscillations of small amplitude. Later experiments showed cellular flames to be laminarised by low-intensity primary acoustic oscillations with a maximum Δp of about 0.5 kPa during flame propagation along an open tube [35].

There is no such effect with the more severe, secondary thermoacoustic oscillations generated in the third, highly unstable, rich *iso*-octane–air mixture in Table 3.1. This is clearly manifest as a severe secondary thermoacoustic instability, encouraged by a very negative Ma_{sr} and triggered by the original DLTD instabilities. It yields high values of $\Delta p = 300$ kPa and of $F = 37.5$ at $(u_c)_{max}$. The values of u_c are more than five times greater than those predicted for DLTD instabilities alone. This and the high value of Δp confirm the generation of strong RT instabilities, which are often referred to as thermoacoustic instabilities. These are addressed in detail in the following two sections of this chapter.

Figure 3.10 shows the developing amplitude of the pressure oscillations as the pressure rises, as well as increases in u_c and the changing rate of change of the heat release rate. In the initial stages of the DLTD instabilities, the theory of the earlier subsection on the peninsula of instability was able to predict the values of u_ℓ, indicated by the diamond symbols. This instability seeded the RT instabilities through the rate of increase of u_ℓ and associated increase in Δp, which further increased values of u_c. The further growth of Δp provided a powerful feedback mechanism for further increases. Diagrams similar to Fig. 3.10 are provided in [32] for the other two mixtures of Table 3.1.

In flame propagation from the end of a closed tube, pressure waves are generated by the impulsive start to flame propagation at the closed end. These were analysed numerically in [36]. A leading shock was found to be reflected at the open end and the reflected wave then reflected at both ends, generating an increasingly complex sequence of reflections. With high burning velocities, a rapid acceleration to a high flame speed generates higher values of Δp in secondary thermoacoustic oscillations.

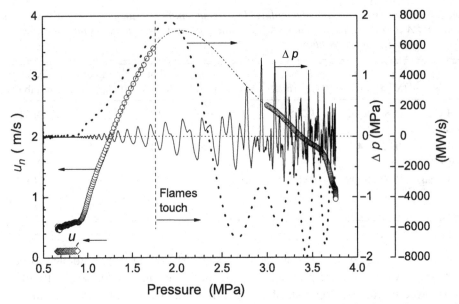

Figure 3.10. Changes in u_c (u_n in figure) and pressure oscillations during inward propagation of two near-simultaneous i-octane–air, $\phi = 1.6$, flames. The dotted curve shows the rate of change of heat release rate [32].

Experiments showed that high flame speeds eliminate reversals in the direction of flame propagation that arise from the interaction of the flame with the oscillating column of gas at lower flame speeds [37]. These reversals led to the laminarisation of the flame. At the higher flame speeds, secondary thermoacoustic instablities with higher-pressure amplitudes Δp of up to 20 kPa ultimately destroyed the cellular structure. This was succeeded by one more akin to that of a highly turbulent flame, with a burning velocity many times greater than the laminar burning velocity [35]. Theoretically, Bychkov [38] and Aldredge [39] identified a range of acoustic amplitudes in which flames are stabilised by primary acoustic velocity oscillations. This regime is intermediate between those of primary and secondary thermoacoustic instabilites.

Unconfined flames also can be subjected to RT intabilities, arising from the generation of oscillations in the burner manifold ahead of the flame. Strong noise can radiate because of the acoustic–combustion coupling between the flame and the manifold's resonant modes. For the configurations considered in [40], pressure perturbations in the manifold were almost in phase with perturbations of the heat release rate. The power generated by the Rayleigh source term minus the small damping losses in the system determined the acoustic radiation to the far field, which is explored more in Section 3.2.

3.1.2 Turbulent Burning, Extinctions, Relights, and Acoustic Waves

When mild velocity fluctuations are imposed on an unstable laminar flame, the burning velocity enhanced by instabilities persists [41, 42]. Paul and Bray [43] developed a flamelet model for moderately turbulent combustion, based on the numerical simulations in [41]. It accounts for the persistence of DLTD instability, together with stretch rate and Markstein number effects. The predicted burning velocities are in good agreement with experimental measurements. As the spectrum of turbulence develops with an increasing rms turbulent velocity u_{rms}, the length scales increasingly subsume those of the original unstable wavelengths. In the experiments in [32] this process seemed to be completed at about $u_{\text{rms}}/u_\ell = 3$ and the original pressure fluctuations that are due to instabilities largely disappeared. At higher turbulence intensities, values of Markstein numbers continue to influence the burning rate and flame extinction [44].

A practical definition of the turbulent burning velocity u_t is that when multiplied by the unburned-gas density and the appropriate flame area it gives the mass rate at which burned gas is formed. Defining this area is problematic because of the very different flow geometries in the various burners and explosion bombs that are used to measure u_t. In addition, as correlations involve u_{rms}, this should be the same over the entire flame surface. Vital information on the turbulent flame bush can be revealed by laser-sheet measurements of the distribution of reactants, products, their interfaces, and the mean reaction progress variable $\bar{c}$. Values of $\bar{c}$ have been used to define appropriate flame surface areas for u_t [45–47]. Further progress on this topic is discussed in Section 2.3.

An earlier correlation of experimental turbulent burning velocities expressed u_t/u_{rms} as a function of the turbulent Karlovitz stretch factor K and Le [48]. For isotropic turbulence, K is based on the rms strain rate on a randomly orientated surface, u_{rms}/λ, where λ is the Taylor scale of turbulence. When normalised by the chemical time, δ_ℓ/u_ℓ, this gives

$$K = \frac{u_{\text{rms}}\, \delta_\ell}{\lambda\, u_\ell}. \tag{3.23}$$

With δ_ℓ approximated by ν/u_ℓ and λ related empirically to the integral length scale Λ by $\lambda^2/\Lambda = A'\nu/u_{\text{rms}}$, where A' is an empirical constant, here assigned a value of 16, Eq. (3.23) gives

$$K = 0.25\, \text{Re}_T^{-1/2} \left(\frac{u_{\text{rms}}}{u_\ell} \right)^2, \tag{3.24}$$

where Re_T is the turbulent Reynolds number based on Λ.

When the flame surface area of spherical explosion flames is defined by a surface at which the mean reaction progress variable is $\bar{c}$, the associated burning velocity measured $u_{t\bar{c}}$ in the isotropic turbulence of a fan-stirred bomb was correlated by [47]:

$$\frac{u_{t\bar{c}}}{u_{\text{rms}}} = A_{\bar{c}}\, \alpha\, K^\beta, \tag{3.25}$$

in which $A_{\bar{c}}$ is a constant that is dependent solely on the surface selected to define the turbulent burning velocity based on the mass rate of burning. For the surface at which the mass unburned behind it is equal to the mass burned ahead of it, $A_{\bar{c}}$ is unity and $\bar{c}$ is usually close to 0.59. For a leading edge with $\bar{c} = 0.05$, the value of $A_{\bar{c}}$ is 0.75. Values of α and β are expressed by first-order equations in Ma_{sr}:

$$\alpha = \begin{cases} 0.022\,(30 - \text{Ma}_{\text{sr}}) & \text{for } +\text{ve Ma}_{\text{sr}} \\ 0.0311\,(30 - \text{Ma}_{\text{sr}}) & \text{for } -\text{ve Ma}_{\text{sr}} \end{cases},$$

$$\beta = \begin{cases} 0.0105\,(\text{Ma}_{\text{sr}} - 30) & \text{for } +\text{ve Ma}_{\text{sr}} \\ -0.0075\,(\text{Ma}_{\text{sr}} + 30) & \text{for } -\text{ve Ma}_{\text{sr}} \end{cases}. \tag{3.26}$$

A limit to the ever-increasing turbulent burning velocity with u_{rms} is provided by localised flame extinctions at sufficiently high values of K [49].

Flame-extinction stretch rates in laminar flow at different ϕ were obtained with experimental opposed jet flames. Values are plotted in Fig. 3.1 by the solid lines for C_3H_8 and CH_4 [50] and H_2 [51], with air, under atmospheric T and p. The high values for lean H_2 mixtures are particularly noteworthy. When turbulent flames make very brief, localised excursions into stretch rates that are higher than those that would extinguish a laminar flame, they do not necessarily extinguish [52]. Because of this apparent time lag and other theoretical complexities [44], data on turbulent flame quenching are best found experimentally. The probability of an initial flame kernel continuing to propagate, p_f, was measured in fan-stirred bombs at different rms turbulent velocities, u_{rms}. These experiments were with mixtures of CH_4, C_3H_8, and i-octane with air and very lean H_2–air mixtures, all in the pressure range 0.1–1.5 MPa and at room temperature. Values of K for probabilities of 0.8 and 0.2 were expressed in terms of Ma_{sr} by [49]

$$K_{0.8} \left(Ma_{sr} + 4\right)^{1.8} = 34.4 \quad \text{for} \quad p_{0.8}, \tag{3.27a}$$

$$K_{0.2} \left(Ma_{sr} + 4\right)^{1.4} = 37.1 \quad \text{for} \quad p_{0.2}, \tag{3.27b}$$

for $-3.0 \le Ma_{sr} \le 11.0$. Values of $K_{0.8}$ are plotted against Ma_{sr} in Fig. 3.6.

Localised turbulence-induced extinctions followed by relights emerge as a problem when lean hydrocarbon–air mixtures are leaned off in low-NO_x burners. The associated rapid changes in dV/dt can induce pressure oscillations with frequencies as high as 100 Hz [53]. A combined computational and experimental study of premixed turbulent combustion of CH_4–air in a swirl burner showed a region of high stretch rate between inner and outer recirculation zones [54]. As the mixture was leaned off and K increased, reaction in this region was locally extinguished at about $\phi = 0.59$. However, reaction continued elsewhere, stabilised by the hot gas in an inner recirculation zone in a different flame configuration. Video films revealed oscillations at a frequency of about 20 Hz between the two quasi-steady solutions with the generation of acoustic waves. With a further reduction in ϕ, the redistribution of strain rates generated a final flame configuration, after which further reduction in ϕ below 0.56 extinguished the flame completely.

In another important area, localised flame extinctions by high stretch rates limit the continuing acceleration of runaway turbulent flames in ducts with turbulence-inducing obstacles. Under these conditions, the turbulent burning velocity attains a maximum value [55]. If this and the generated shock waves [36] are high and strong enough to auto-ignite the unburned mixture between the shock wave and the flame, a transition from deflagration to detonation can ensue [55].

3.1.3 Auto-Ignitive Burning

AUTO-IGNITIVE PROPAGATION VELOCITY. Where the intended combustion mode is either a premixed laminar or turbulent flame, a stochastic transition to auto-ignitive burning can be perceived as an instability. A familiar example is knocking combustion in a gasoline engine. Efficient engine operation, with a high burn rate on the fringe of severe knock, can be nullified by occasional cyclic excursions into such a knock as a result of the sensitivity of auto-ignition to comparatively small changes. In laminar and turbulent flames, molecular-transport processes are integral parts of the mechanism of flame propagation. This is not the case with auto-ignition.

In a homogeneous mixture, auto-ignition manifests itself as a spontaneous reaction throughout the mixture, without any propagating front and occurs after the auto-ignition delay time τ_i has elapsed. Where there are spatial gradients of τ_i in a radial direction r, because of those of either T or ϕ, or both, an auto-ignition propagation velocity can arise, given by

$$u_a = \frac{dr}{d\tau_i}. \tag{3.28}$$

When u_a is close to the acoustic velocity $\hat{c}$, the two can become coupled in a strong shock wave, with the creation of a new entity, a detonation wave, with a distinctive high-pressure peak and a very high detonation velocity. An important parameter is the ratio $\xi = \hat{c}/u_a$. Because of inevitable inhomogeneities in mixtures, auto-ignitive propagation usually originates at a hot spot at which the initial radius of the hot spot must exceed the critical radius given by thermal-explosion theory [56]. This condition is usually satisfied in engines. The rate of change of the rate of volume generation again generates pressure waves, as given by Eq. (3.17).

From Eqs. (3.17) and (3.20), with r equal to the radius of an auto-ignitively propagating spherical hot spot, replacing u_c with u_a, and with $\xi = \hat{c}/u_a$,

$$\Delta p(t) = \frac{\rho}{4\pi d}\left| \frac{d}{dt}\left(\frac{dV}{dt}\right)\right|_{t-t_d} = \frac{\rho}{d}(\sigma-1)\frac{dr^2 u_a}{dt} = \frac{\rho r c^2}{d}(\sigma-1)\left[\frac{2r}{\xi^2} + \frac{r}{\hat{c}}\frac{d\xi^{-1}}{dt}\right]_{t-t_a}. \tag{3.29}$$

With $\hat{c} = \sqrt{\gamma p/\rho}$, Eq. (3.29) becomes

$$\frac{\Delta p(t)}{p} = \frac{r\gamma}{d}(\sigma-1)\left[\frac{2r}{\xi^2} + \frac{r}{\hat{c}}\frac{d\xi^{-1}}{dt}\right]_{t-t_d}. \tag{3.30}$$

The last term within the square brackets, unlike that with unstable flames, is likely to be relatively small, and the fractional pressure increase that is due to the auto-ignition, characterised as knock, is inversely proportional to ξ^2. This can be as high as 0.4 MPa at 1.4 MPa for a single hot spot [57]. To derive this magnitude requires values of u_a. These can be found most conveniently if the mixture composition is assumed uniform and the gradient of reactivity arises entirely as a temperature gradient. In that case, Eq. (3.28) can be re-expressed as

$$u_a = \left(\frac{\partial T}{\partial r}\right)^{-1}\left(\frac{\partial \tau_i}{\partial T}\right)^{-1}. \tag{3.31}$$

Values of $\tau_i(T, p)$ differ between different fuels and change in different ways with T and p. This is illustrated by the plots of $\tau_i(T, p)$, at the indicated pressures, against $1000/T$ in Fig. 3.11, taken from [58]. Four of the curves are for primary reference fuels (PRFs), which are used in attempts to simulate the knocking behaviour of gasolines. The octane number, ON, is the percentage by volume of i-octane mixed with n-heptane in a PRF. Only at the high temperatures in Fig. 3.11 do they exhibit a similar straight-line Arrhenius relationship to the other three fuels. The variation with pressure is an inverse one, expressed by p^{-n}, where n is a pressure exponent.

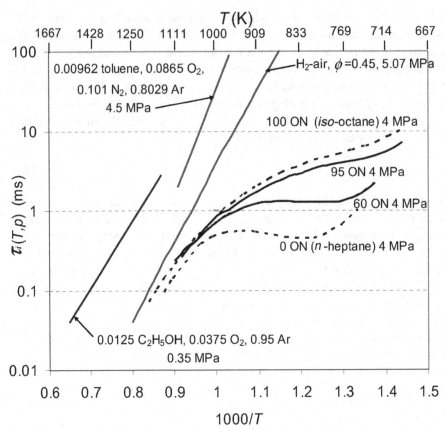

Figure 3.11. Ignition delay times $\tau_i(T, p)$ of various fuels with air at $\phi = 1.0$, unless otherwise stated. See [58] for source references.

It has a value of about 1.7 for PRFs and one closer to unity for many non-PRFs, for $T < 1000$ K [59]. At higher temperatures, values of $\tau_i(T, p)$ become closer for different PRFs.

Values of $\partial \tau_i / \partial T$ were derived for the PRF with ON = 60 for a stoichiometric mixture at 4 MPa from the τ_i versus $1/T$ data in [60]. To obtain u_a, it is necessary to assume a value of $\partial T / \partial r$. Temperature gradients arise at cooled surface boundary layers and, in the body of the charge, due to turbulence. An assumed value of $\partial T / \partial r = -1$ K/mm gave the values of u_a shown by the cross symbols in Fig. 3.12 and, from the corresponding values of $\hat{c}$, it also gave those of ξ. The 60 ON curve in Fig. 3.11 shows a region where $\partial \tau_i / \partial T$ is close to zero between about 800 and 860 K, which implies an infinite value of u_a and $\xi = 0$. It is apparent from Fig. 3.11 that, for the PRFs, $\partial \tau_i / \partial T$ increases numerically with ON. Hence u_a will decrease and ξ will increase. Clearly, from Eq. (3.30), low values of ξ imply large values of $\Delta p(t)/p$, and severe knock would be anticipated for the conditions of Fig. 3.12 in a gasoline engine. It would become less severe at about 950 K, but severe high-temperature knock would begin to develop above 1100 K.

The computations of u_a are for the given initial boundary conditions of T and p. In practice, these correspond best to auto-ignitions in shock tubes and rapid-compression machines. For slower compressions in engines when auto-ignition is

60 ON

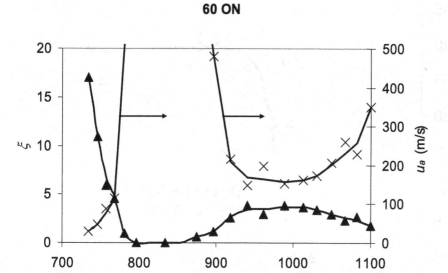

Figure 3.12. Variations of ξ and u_a with temperature at 4 MPa, stoichiometric 0.60 i-octane, 0.40 n-heptane, and air, derived from τ_i values in [55], $dT/dx = -1$ K/mm. From [58].

not completed at fixed values of T and p, the mixture will have partially reacted during compression. Consequently the mixture will auto-ignite rather more rapidly than is suggested by the tabulated values of τ_i for T and p. With a reduced effective τ_i in Eq. (3.31), u_a would be increased.

MODES OF AUTO-IGNITION. A value of ξ close to unity is particularly significant because it suggests a resonance between the acoustic speed $\hat{c}$ of the pressure wave and the propagation velocity u_a of the auto-ignitive front. The auto-ignition delay time is followed by a much shorter excitation time τ_e, during which most of the heat release rate occurs [61]. A high rate of chemical heat release at small τ_e, continually feeding into the pressure front, will reinforce the front and increase the probability of a detonation. The residence time of a pressure pulse in the hot spot is $r_0/\hat{c} = t_{r0}$, and a dimensionless measure of this energy input into the pulse is

$$\varepsilon = \frac{t_{r0}}{\tau_e}. \tag{3.32}$$

Here ε is the number of excitation times feeding into the pressure pulse during its residence in the hot spot.

In a detonation wave the coupling of the chemical reaction with the shock wave results in a very damaging high-pressure spike, and this single entity propagates at the detonation velocity. In addition to this mode of auto-ignitive propagation, there are a variety of other modes, and these depend on the values of ξ and ε. They were identified in DNSs of auto-ignitions at spherical hot spots. The simulations involved detailed chemical kinetics over wide ranges of initial values of $\partial T/\partial r$ and r_0 [62, 63]. Equimole CO–H_2 mixtures with air at different ϕ were chosen for this study because of their relatively simple detailed chemical kinetics.

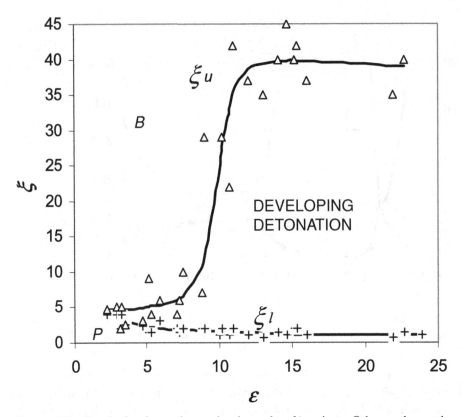

Figure 3.13. Developing detonation peninsula on plot of ξ against ϵ. Other modes are thermal explosion below ξ_l, supersonic and subsonic auto-ignitive fronts propagate in regions P and B, respectively. From [62].

It was first necessary to compute values of τ_i and τ_e for different ϕ, p, and T. The ensuing simulations ultimately identified the upper and lower limits of detonation, ξ_u and ξ_l. These are indicated in Fig. 3.13, from [62], by the open triangle and cross symbols. These define the boundaries of a developing detonation peninsula within which detonations develop within a hot spot. Peak detonation pressures, when normalised by the ambient pressure, increase with ε. Under these conditions Eq. (3.30) cannot be used to predict the pressure pulse. At very low ε, as its value increases, so does the range of ξ over which a detonation can develop at a hot spot. At the highest values of ε this range does not appear to change. The smallest values of ε are associated with the least-reactive lean mixtures and the largest values with the most reactive, with $\phi = 1.0$ and $T = 1200$ K.

With even more reactive mixtures, with values of ε of about 70, the detonation can continue to propagate outside the hot spot in the absence of any gradient of reactivity [64]. In addition, as the reactivity increases, so does the pressure ratio at the detonation front and the hot-spot number density. Initially separate waves from hot spots interact and are reinforced, inducing further auto-ignitions that can generate severe engine knock [65]. Intersecting shock waves from closely packed neighbouring hot spots can create enlarged (though not necessarily spherical) hot spots that, because of their increased size and higher temperature and pressure, have higher values of ε. Further study is needed to identify the initiating conditions, presumably with high values of ε and low values of ξ, under which a detonation can propagate in the absence of a reactivity gradient [66].

Other modes of auto-ignition also are indicated in Fig. 3.13. These involve auto-ignitive propagation in a supersonic mode below ξ_l, with $\xi \leq 1.0$, in region P. From Eq. (3.31), as $(\partial \tau_i / \partial T) \rightarrow 0, u_a \rightarrow \infty$ and $\xi \rightarrow 0$. The reaction front moves so rapidly, it runs ahead of the acoustic wave and Eq. (3.30) is again invalid. There is no sharp pressure peak, as there is with a developing detonation, and the maximum pressure tends towards that of constant-volume combustion. In the limit of $\xi = 0$ a thermal explosion occurs instantaneously throughout the mixture, without any propagating front. In region B, with $\xi > 1.0$ and also outside the peninsula, with $\xi \geq \xi_u$, auto-ignitive propagation is subsonic and more benign.

An important practical aspect of this mode is that it can support combustion in mixtures that are so lean that the turbulence would extinguish normal flame propagation. For non-stratified mixtures, the practical lean-burn limits for turbulent flame propagation in spark-ignition engines are about $\phi = 0.71$ for gasoline–air [67] and 0.62 for natural gas [68]. The addition of 60 vol% H_2 to CH_4 lowers this to $\phi = 0.34$ [69], whereas for H_2–air the limit is about $\phi = 0.2$, with effectively zero-NO_x emission [70]. In sharp contrast, in the misnamed *homogeneous* charge-compression ignition (HCCI) engine, auto-ignitive burning of gasoline–air mixtures is possible down to $\phi = 0.25$ and even lower, notwithstanding the increase in τ_i as ϕ is reduced.

Controlled combustion is, in fact, dependent on a *non-homogeneous* charge, in which gradients of reactivity drive auto-ignitive fronts. Multiple isolated auto-ignitions are followed by combustion at distributed sites [71]. Chemiluminescent imaging, with ϕ in the region of 0.24, showed sequential auto-ignition of progressively cooler regions, with luminescence spreading through the charge with a propagation speed of about 100 m s^{-1} [72]. This is greater than that for normal turbulent flame propagation in spark-ignition engines. DNSs have shown that mixed types of combustion are possible, in which auto-ignitive propagation coexists with laminar or turbulent flames [73].

Acknowledgement

To the late Arthur Smithells [1], in whose name a research studentship was awarded to the author, enabling him subsequently to enjoy a lifetime in combustion research.

3.2 Control Strategies for Combustion Instabilities
By A. P. Dowling and A. S. Morgans

Lean premixed combustion systems are especially prone to thermoacoustic instabilities, as noted in the discussion on RT instabilities in Subsection 3.1.1 and the associated pressure oscillations and possibly enhanced heat transfer can lead to a deterioration in the system performance and may even become sufficiently intense as to cause structural damage. This means that control strategies are an important element of the design of lean premixed combustors.

Combustion oscillations in confined configurations usually arise because of coupling between the unsteady heat release and the acoustic waves, and control strategies need to interrupt this coupling. Passive control methods [74–76] may either seek to reduce the susceptibility of the combustion process to acoustic excitation through ad hoc hardware design changes, such as modifying the fuel injection system or combustor geometry [77–79], or to remove energy from the sound waves by using acoustic dampers such as Helmholtz resonators [80–82], quarter-wave tubes [83], perforated plates, or acoustic liners [84, 85]. The problem with passive approaches is that they tend to be effective over only a limited range of operating conditions. Also they may be ineffective at the low frequencies at which some of the most damaging instabilities occur and the changes of design involved are usually costly and time consuming. The range of operation of a passive device can be extended by introducing an element of variable geometry, a variable volume in a Helmholtz resonator for example, and altering the geometry to retune the device as the instability frequency changes with changes in operating condition.

Active feedback control provides another means of interrupting the coupling between acoustic waves and unsteady heat release. An actuator modifies some system parameter in response to a measured signal. The aim is to design the controller (the relationship between the measured signal and the signal used to drive the actuator) such that the unsteady heat release and acoustic waves interact differently, leading to decaying, rather than growing, oscillations. In the past, approaches to controller design were somewhat empirical, but more systematic approaches such as robust control and adaptive control are now promoted.

In Subsection 3.2.1 the physics of these combustion oscillations is discussed and a generalised energy equation is derived to identify what is needed for control. Passive control techniques are described in Subsection 3.2.2, including factors that affect their practical implementation in combustion systems. Techniques for tuning passive devices are introduced in Subsection 3.2.3, and active control is discussed in Subsection 3.2.4.

3.2.1 Energy and Combustion Oscillations

Self-excited combustion oscillations occur because of interactions between unsteady combustion and acoustic waves. Rayleigh's criterion [29, 30], noted in Subsection 3.1.1, states that an acoustic wave gains energy when heat is added in phase with pressure, but loses energy when heat is added out of phase with pressure. This clearly explains the energy exchange between acoustic waves and heat release, which forms the basis for the majority of combustion oscillations. Chu [86] has since generalised this to incorporate the effect of boundary conditions, and this provides a useful insight into combustion oscillations and methods for controlling them.

Following Chu [86], let us consider a perfect gas burning within a combustor of volume V, bounded by the surface S, as illustrated in Fig. 3.14. For simplicity in this illustrative example, the gas is considered linearly disturbed from rest with no mean heat release (extensions accounting for mean flow and mean heat release can be made [87]). Viscous forces are neglected. The pressure, density, heat release rate per unit volume, particle velocity, speed of sound, and ratio of specific heat capacities are denoted by p, ρ, q, $\mathbf{u}$, $\hat{c}$, and γ, respectively, and a mean value is indicated by an overbar and a fluctuating value by a prime.

We consider the influence of unsteady heat $q(x, t)$ on pressure perturbations. The density ρ varies through changes in both the pressure p and specific entropy s.

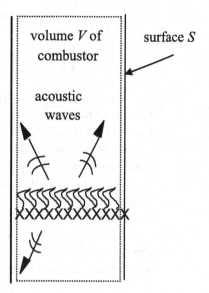

Figure 3.14. Combustion within a resonator.

The chain rule of differentiation shows that

$$\frac{D\rho}{Dt} = \frac{1}{\hat{c}^2} \frac{Dp}{Dt} + \left.\frac{\partial \rho}{\partial s}\right|_p \frac{Ds}{Dt}. \qquad (3.33)$$

When viscous and heat conduction effects are neglected, $\rho T\, Ds/Dt = q(\mathbf{x}, t)$, where T is the absolute temperature. Moreover, for a perfect gas, $\partial \rho/\partial s|_p = -\rho/c_p = -\rho T\,(\gamma - 1)/\hat{c}^2$, where c_p is the specific heat at constant pressure. Substitution into Eq. (3.33) leads to

$$\frac{D\rho}{Dt} = \frac{1}{\hat{c}^2}\left[\frac{Dp}{Dt} - (\gamma - 1)\,q\right]. \qquad (3.34)$$

Equation (3.34) may be applied to a combusting gas, provided that the reactants and products behave as perfect gases, and there is no molecular weight change during the chemical reaction [88]. After linearisation it becomes

$$\frac{D\rho}{Dt} = \frac{1}{\bar{c}^2}\left[\frac{\partial p'}{\partial t} - (\gamma - 1)\,q\right]. \qquad (3.35)$$

The first term on the right-hand side describes the change in density that is due to compressibility, and the second term accounts for the combustion effects. Substitution from Eq. (3.35) into the equation of mass conservation leads to

$$\frac{1}{\bar{\rho}\bar{c}^2}\frac{\partial p'}{\partial t} + \frac{\partial u_k}{\partial x_k} = \frac{(\gamma - 1)}{\bar{\rho}\bar{c}^2}q, \qquad (3.36)$$

where $q = q'$ and $u = u'$ in the present approximation. Multiplying Eq. (3.36) by p' and adding it to the scalar product of the linearised momentum and the velocity vector $\mathbf{u}$ leads to

$$\frac{\partial}{\partial t}\left(\frac{1}{2}\bar{\rho}u^2 + \frac{1}{2}\frac{p'^2}{\bar{\rho}\bar{c}}\right) + \frac{\partial (p'u_k)}{\partial x_k} = \frac{(\gamma - 1)}{\bar{\rho}\bar{c}^2}p'q. \qquad (3.37)$$

Finally, after integration over the volume V, this yields

$$\frac{\partial}{\partial t}\int_V \left(\frac{1}{2}\bar{\rho}u^2 + \frac{1}{2}\frac{p'^2}{\bar{\rho}\bar{c}^2}\right) \mathrm{d}V = \int_V \frac{(\gamma-1)}{\bar{\rho}\bar{c}^2}p'q\mathrm{d}V - \int_S p'\mathbf{u}\cdot\mathrm{d}\mathbf{S}. \qquad (3.38)$$

The term on the left-hand side of Eq. (3.38) represents the rate of change of the sum of the kinetic and potential energies within the volume V. The first term on the right-hand side describes the exchange of energy between the combustion and acoustic waves; as noted by Rayleigh, when the pressure and the rate of heat release have a component that is in phase (i.e., when the phase difference lies between $-90°$ and $+90°$), the acoustic energy tends to increase. The final surface term accounts for energy loss across the bounding surface S, which occurs because the fluid within S does work on its surroundings. Viscous dissipation has been neglected.

Equation (3.38) states that disturbances grow if their net energy gain from the combustion is greater than their energy losses across the boundary. Therefore an acoustic mode grows in amplitude if

$$\int_V \frac{(\gamma-1)}{\bar{\rho}\bar{c}^2}\overline{p'q}\ \mathrm{d}V \ > \ \int_S \overline{p'\mathbf{u}}\cdot\mathrm{d}\mathbf{S}, \qquad (3.39)$$

where the overbar denotes an average over one period of the acoustic oscillation. This is a generalised form of Rayleigh's criterion. When it is satisfied, the combustor has a thermoacoustic instability. Linear waves increase in amplitude until limited by nonlinear effects. If the nonlinearity appears primarily in the heat release rate with the acoustic waves remaining linear, it is clear that, for a system in which acoustic waves are initially growing, heat release saturation or phase-change effects may cause the terms to become equal at a certain pressure amplitude. This is the amplitude at which limit cycle oscillations occur [89–91].

In combustors there have traditionally been a large number of inlet ports, both primary and secondary, together with many cooling holes and rings. Across each of these, there is a pressure drop, and the perturbations in pressure and velocity tend to be in phase, resulting in damping according to the last term in Eq. (3.38). However, to achieve lean burn, most of the airflow is routed through the premixing ducts, leaving little air available for cooling. This means that there is little natural acoustic damping in the combustor. Moreover, in lean premixed combustion systems, the flame responds significantly (i.e., q' is nonzero) across a broad range of frequencies [77]. We might therefore expect such combustors to be particularly susceptible to combustion instability.

As well as offering insight into the energy transfers that give rise to and limit the size of combustion oscillations, relation (3.39) also suggests ways in which combustion instabilities can be eliminated. Either the energy source term, $\int_V \overline{p'q}\mathrm{d}V$, must be decreased or the surface loss term, $\int_S \overline{p'\mathbf{u}}\cdot\mathrm{d}\mathbf{S}$, must be increased. Passive and active control methods can target either term.

3.2.2 Passive Control

MODIFYING THE FLAME RESPONSE. One means of passive control is to seek to modify the fuel input or the flame shape to reduce the energy source term on the left-hand

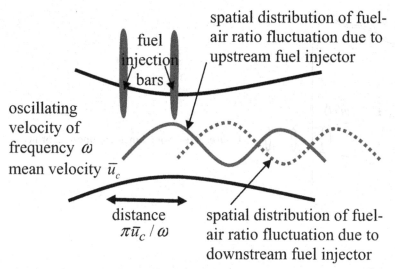

Figure 3.15. Fuel–air ratio fluctuations from fuel bars $\pi\bar{u}_c/\omega$ apart cancel in the combustor.

side of relation (3.39) [4, 5]. Design changes might, for example, choose to alter the position of the flame or of the fuelling system to change the phase relationship between oscillations in combustor pressure and the rate of heat release so that

$$\int_V \overline{p'q} \ dV < 0, \tag{3.40}$$

ensuring that the interaction between the acoustics and the combustion tends to dampen the pressure oscillations. But there are more subtle ways of changing this driving term. For example, for a plane-wave instability, one might choose to introduce some asymmetry into $q(\mathbf{x},t)$: Then there is some cancellation in the integral of $\overline{p'q}$ over the combustor cross section. Such asymmetry could be introduced by ensuring an angular variation in fuel distribution or in flame position.

Convection time delays from fuel input to combustion are important in determining the flame transfer function (FTF), and this can be exploited to reduce the flame response to flow unsteadiness. Staging the fuel injection locations axially within the premixing ducts, so that the fuel–air mixture from different fuelling bars reaches the combustor after different time delays, can reduce the susceptibility to combustion instability [78]. The situation is illustrated in Fig. 3.15. If the two fuel injection bars are close compared with the *acoustic* wavelength, they both experience the same air velocity and pressure perturbation and so instantaneously produce the same fluctuation in fuel–air ratio. If their axial separation corresponds to half the convection wavelength (i.e., is equal to $\pi\bar{u}_c/\omega$, where $\bar{u}_c$ is the mean convection velocity and ω is the radian frequency of the flow perturbations), these fuel–air ratio fluctuations arrive at the combustion zone half a period apart, thereby reducing perturbations in fuel–air ratio and hence in the related fluctuations in the rate of heat release. Nearly complete cancellation is possible for a single frequency of oscillation. By the axial distribution of many injection points, reductions in the flame transfer function and hence in the susceptibility to oscillation can be achieved across a wide frequency range [92].

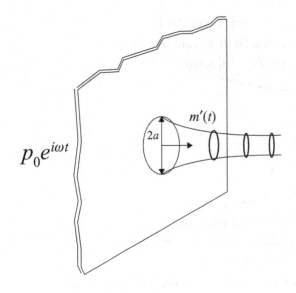

Figure 3.16. Sound incident upon an orifice.

$p_0 e^{i\omega t}$

SOUND ABSORPTION. Passive acoustic dampers aim to increase the energy loss on reflection at the combustor boundaries, typically by converting energy in the acoustic wave into vorticity. If there is a hole in the bounding surface of the combustor such that when p' is high u is outward, according to relation (3.39), this tends to dampen the oscillation.

PERFORATED PLATES. Perforated plates exploit this mechanism, which can be enhanced if the hole is at the entrance to an acoustic resonator causing small pressure perturbations to induce large velocity fluctuations. This is usually done by attaching a resonator, for example a Helmholtz resonator [7, 8] or quarter-wave tube [83], to the walls of the combustor or by inserting perforated plates [85]. When used in combustors, the neck of a Helmholtz resonator or the holes of a perforate need to be cooled by blowing cooler air through them. The interaction between acoustic waves and the orifice leads to unsteady vortex shedding, which is convected away by the mean flow. If the Strouhal number of the flow in the orifice is chosen appropriately, this can be used to enhance the absorption of acoustic energy [93–95].

Howe [93] considered a sound wave that causes an incident pressure perturbation $p_0 e^{i\omega t}$ at a circular aperture in a rigid plate [as usual, the actual pressure perturbation is the real part of the complex amplitude $p'(t) = \Re\left(p_0 e^{i\omega t}\right)$]. This leads to a fluctuating mass flux $m'(t)$ through the aperture (see Fig. 3.16). Howe determines K such that

$$\frac{\mathrm{d}m'}{\mathrm{d}t} = K p_0 e^{i\omega t}. \tag{3.41}$$

In the absence of a mean flow, $K = 2a$, where a is the radius of the aperture. With a mean flow through the aperture, $K = 2a(\eta + i\delta)$, where η and δ are functions of the Strouhal number $\omega a / \overline{u}$. Howe determined the functional forms of $\eta(\omega a / \overline{u})$ and $\delta(\omega a / \overline{u})$ in terms of Bessel functions, and these are plotted in Fig. 3.17.

The sound transmitted and reflected by a perforated plate can be determined by combining the results for individual holes [93]. Consider the plate illustrated

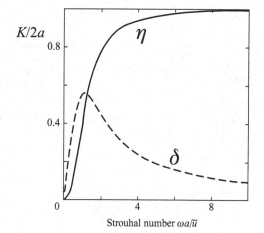

Figure 3.17. The dependence of the aperture conductivity $K = 2a(\eta + i\delta)$ on the Strouhal number $\omega a/\bar{u}$ [93].

in Fig. 3.18, with holes of radius a separated by distance d. There are hence d^{-2} apertures/area. The plate is illuminated with sound of frequency ω, resulting in reflected and transmitted waves. Provided d is small in comparison with $\bar{c}/\omega$, we can work with smoothed flow variables averaged over many holes. The smoothed velocity normal to the plate, $u'(\mathbf{x}, t)$, is related to the mass flow rate through an orifice by $\bar{\rho} u' = m'/d^2$.

Continuity of mass across the plate shows that the area-averaged velocities $u'(\mathbf{x}, t)$ are the same on the two sides of the plate, i.e.,

$$[u']_{x=0^+}^{x=0^-} = 0. \tag{3.42}$$

From Eq. (3.41),

$$\frac{\partial m'}{\partial t} = K[p']_{x=0^+}^{x=0^-} = 2a(\eta + i\delta)\Delta p', \tag{3.43}$$

Figure 3.18. Geometry of a perforated plate.

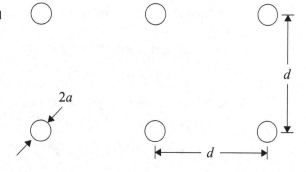

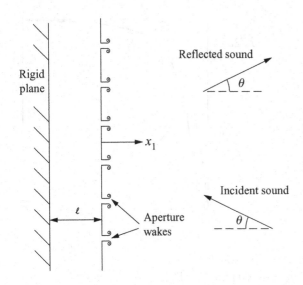

Figure 3.19. Sound incident at angle θ on a perforated plate at $x = 0$ with a rigid backing plate a distance ℓ behind it.

where $\Delta p' = [p']_{x=0^+}^{x=0^-}$, the difference in the smoothed pressure fluctuation $p'(\mathbf{x}, t)$ across the plate, and $K = 2a(\eta + i\delta)$. Rewriting Eq. (3.43) in terms of u' gives

$$\overline{\rho}\frac{\partial u'}{\partial t} = \frac{2a(\eta + i\delta)}{d^2}\Delta p'. \tag{3.44}$$

We see that it is δ, the imaginary part of K, that gives an non-zero value to $\overline{u'\Delta p'}$ that describes the rate of absorption of acoustic energy/unit plate area. In the absence of a mean flow, the Strouhal number $\omega a/\overline{u}$ tends to infinity, δ tends to zero (see Fig. 3.17), and there is no sound absorption by the linear elements of the pressure perturbation. Then it is only the elements of p' of the order of u'^n, where $n \geq 2$, that give the sound absorption. Absorption by this mechanism is discussed further in Subsection 3.2.3. It is effective only for large-amplitude sound waves. Ideally we would like good absorption at low amplitudes so that the pressure perturbations in a combustion oscillation could be controlled to be at a low level. This is possible when there is a bias or mean flow through the apertures. Then the fraction of incident sound energy absorbed is independent of amplitude but relies on a nonzero value of δ. We see from Fig. 3.17 that the maximum δ occurs for $\omega a/\overline{u} \approx 1$.

When perforated plates are used in combustors they are usually as wall liners. Hughes and Dowling [84] used Howe's results to determine the sound absorbed in an incident plane wave by a perforated liner has a rigid backing plate a distance ℓ behind it, as illustrated in Fig. 3.19, after neglecting the axial evolution of pressure wave.

If I and R denote the pressure amplitude of the incident and reflected waves in $x = 0$,

$$p'(\mathbf{x}, t) = e^{i\omega(t-y\sin\theta/\overline{c})}(I\,e^{i\omega x\cos\theta/\overline{c}} + R\,e^{-i\omega x\cos\theta/\overline{c}}),$$

$$u'(\mathbf{x}, t) = e^{i\omega(t-y\sin\theta/\overline{c})}(-I\,e^{i\omega x\cos\theta/\overline{c}} + R\,e^{-i\omega x\cos\theta/\overline{c}})\cos\theta/(\overline{\rho}\,\overline{c}). \tag{3.45}$$

While in $-\ell \leq x \leq 0$, the rigid-wall condition $u' = 0$ on $x = -\ell$ gives

$$p'(\mathbf{x}, t) = A\,e^{i\omega(t-y\sin\theta/\overline{c})}[e^{i\omega(x+\ell)\cos\theta/\overline{c}} + e^{-i\omega(x+\ell)\cos\theta/\overline{c}}]$$

$$u'(\mathbf{x}, t) = A\,e^{i\omega(t-y\sin\theta/\overline{c})}[-e^{i\omega(x+\ell)\cos\theta/\overline{c}} + e^{-i\omega(x+\ell)\cos\theta/\overline{c}}]\cos\theta/(\overline{\rho}\,\overline{c}), \tag{3.46}$$

where A is a complex constant.

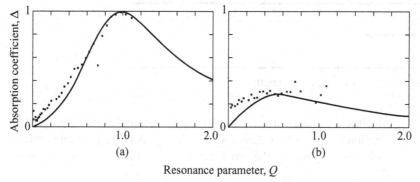

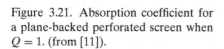

Figure 3.20. The variation of the absorption coefficient Δ with Q for (a) $M = 0.014$, $a/\ell = 0.15$, and $a\ell/d^2 = 0.052$, and (b) $M = 0.052$, $a/\ell = 0.032$, and $a\ell/d^2 = 1.06$ (from [11]).

The reflected-wave amplitude R can be found in terms of the incident-wave amplitude I after substitution from Eqs. (3.45) and (3.46) into (3.42) and (3.44). The fraction of the incident energy absorbed Δ then follows from

$$\Delta = 1 - |R|^2/|I|^2. \tag{3.47}$$

The backing screen introduces an additional non-dimensional parameter ℓ/a into the problem. Hughes and Dowling [84] found it convenient to use a 'resonance' parameter, $Q = \omega^2/\omega_0^2$, where $\omega_0 = \bar{c}[2a/(\cos\theta\,\ell d^2)]^{1/2}$. The absorption coefficient is then a function of three non-dimensional parameters, Q, $\omega\cos\theta\,\ell/\bar{c}$, and $\omega a/\bar{u}$. The variation of absorption coefficient with Q for fixed Mach number $M = \bar{u}/\bar{c}$ and two different plate geometries is shown in Fig. 3.20. The predictions are compared with experimental results for a normally incident sound wave. There is excellent agreement between predictions and experiment.

Hughes and Dowling [84] went on to develop design rules: Typically the peak sound absorption is near $Q = 1$ (the exceptions to this are when the gap is non-compact or the mean velocity is very high). Then Fig. 3.21 can be used to choose an appropriate Strouhal number. It is evident from Fig. 3.21 that very high levels of absorption can be achieved if plate parameters are chosen appropriately.

Figure 3.21. Absorption coefficient for a plane-backed perforated screen when $Q = 1$. (from [11]).

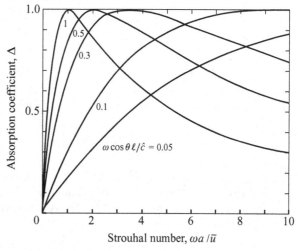

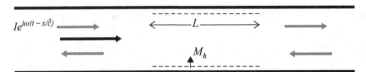

Figure 3.22. Lined section of duct of length L.

However, this theory gives no absorption when the incident wave is tangential to the plane of the liner ($\theta = \pm 90°$). Then, although Eq. (3.41) is still appropriate to describe an individual aperture, the axial evolution of the waves needs to be considered.

Eldredge and Dowling [85] considered the geometry shown in Fig. 3.22. They used the perforated-plate impedance Eq. (3.44) to determine the axial evolution of $\partial p' / \partial x$ over the lined section. They then went on to determine the absorption of an incident plane wave $I e^{i\omega(t-x/\bar{c})}$ and to develop design criteria to optimise the absorption.

Figure 3.23(a) shows that over 80% of the incoming sound energy can be absorbed by a perforated lining plate, provided the plate parameters are chosen appropriately. An inappropriate choice for parameters leads to little absorption [see Fig. 3.23(b)]. For optimal absorption the perforated plate should be located in the duct near a pressure antinode, and M_h, the Mach number of the flow through the holes, should be chosen to be near $\sigma CL/(A_p 2\sqrt{2})$, where σ is the open-area ratio, C is the circumference of the lined duct, L is its length, and A_p denotes its cross-sectional area.

HELMHOLTZ RESONATORS. A Helmholtz resonator is another passive damping device. Figure 3.24 shows an idealised system consisting of a Hemholtz resonator connected to the side of a duct. A Helmholtz resonator consists of a standard acoustic device with a short neck opening out into a large volume [96, 97]. The absorption obtained from a Helmholtz resonator can also be enhanced by cooling flows through its neck [82, 98]. Again the location of the Helmholtz resonator(s) within the combustor has a major influence on its effectiveness. The absorption of sound by a Helmholtz resonator can be analysed in a straightforward way.

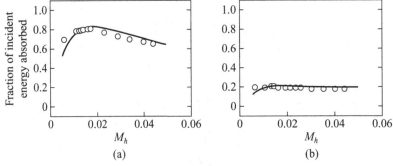

Figure 3.23. Fraction of incident sound energy absorbed by a perforated plate of length L: (a) $\omega L / \bar{c} = 0.89$, (b) $\omega L / \bar{c} = 1.22$, M_h is the Mach number of the mean flow through the holes (from [85]).

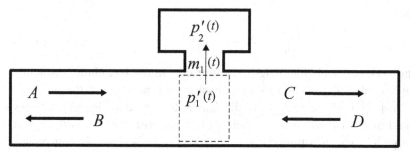

Figure 3.24. A Helmholtz resonator connected to the side of a duct.

We consider a Helmholtz resonator connected to a duct of cross-sectional area S_1. The Helmholtz resonator has a neck of cross-sectional area S_2 and length ℓ and a bulb of volume V. There are incoming sound waves in the duct from the left and right of the Helmholtz resonator of strengths A and D, and reflected and transmitted waves B and C, as shown in Fig. 3.24.

We denote the pressure perturbation at the open end of the neck of the Helmholtz resonator by $p_1'(t)$ and the mass flow into the Helmholtz resonator by $m_1(t)$. Although there may be a mean flow through the neck of the Helmholtz resonator, we assume that any mean flow in the main duct is negligible. $p_2'(t)$ and $\rho_2'(t)$ denote the pressure and density perturbations within the Helmholtz resonator and are assumed to be independent of position because the bulb of the resonator is compact (i.e., its size is small in comparison with $\bar{c}/\omega$) at the frequencies of interest.

The pressure perturbations match at the neck of the Helmholtz resonator:

$$p_1'(t) = (A + B)e^{i\omega t} = (C + D)e^{i\omega t}. \tag{3.48}$$

Conservation of mass for the dotted control volume gives

$$S_1(A - B)e^{i\omega t} = S_1(C - D)e^{i\omega t} + \bar{c}m_1'(t). \tag{3.49}$$

As mass flows into the cavity, the density there must rise

$$m_1' = V\frac{d\rho_2'}{dt}.$$

Associated with this density change is a corresponding change in $p_2'(t)$, the pressure in the cavity. For isentropic perturbations, $p_2'(t) = \bar{c}^2 \rho_2'(t)$, so that

$$\frac{dp_2'}{dt} = \frac{\bar{c}^2}{V}m_1', \tag{3.50}$$

i.e., for harmonic disturbances of frequency ω,

$$p_2' = \frac{\bar{c}^2 m_1'}{V\, i\omega}. \tag{3.51}$$

When there is a mean velocity through the neck of the Helmholtz resonator, the fluctuations in mass flow are related to the pressure difference across the neck by the conductivity K_ℓ, which is independent of amplitude but depends on the length of the neck and on the Strouhal number of mean flow through the neck. We write this in the form

$$\frac{\mathrm{d}m'}{\mathrm{d}t} = K_\ell(p'_1 - p'_2). \tag{3.52}$$

When the neck has negligible length, the entrance to the Helmholtz resonator just being an circular aperture in a thin combustor wall, we can use the previous result: $K_\ell \approx 2a(\eta + i\delta)$, where η and δ are the functions plotted in Fig. 3.17 in terms of the Strouhal number $\omega a/\bar{u}$, a is the radius of the aperture, and $\bar{u}$ is the mean velocity through it. Then substitution for $p'_2(t)$ from Eq. (3.51) into (3.52) gives for disturbances of frequency ω:

$$m'_1 = \frac{2a\omega(-i\eta + \delta)}{\omega^2 - 2a(\eta + i\delta)\bar{c}^2/V} p'_1. \tag{3.53}$$

Large mass flow fluctuations are possible near the resonant frequency $\omega_0 = \bar{c}\sqrt{2a(\eta + i\delta)/V}$.

We saw in energy balance equation (3.38) that the rate of acoustic energy loss at a bounding surface S is equal to $\int_S p'\mathbf{u} \cdot \mathbf{dS}$. Hence the rate of absorption of acoustic energy by a Helmholtz resonator is equal to

$$\frac{\overline{m'_1 p'_1}}{\bar{\rho}} = \frac{\overline{p'^2_1}}{\bar{\rho}} \, \Re\left[\frac{2a\omega(-i\eta + \delta)}{\omega^2 - 2a(\eta + i\delta)\bar{c}^2/V}\right]. \tag{3.54}$$

This is large near the resonant frequency of the Helmholtz resonator provided $p'_1(t)$ has a non-negligible amplitude.

Substitution from Eq. (3.53) into (3.48) and (3.49) gives B and C in terms of A and D. The fraction of incident energy absorbed Δ then follows from

$$\Delta = 1 - \frac{|B|^2 + |C|^2}{|A|^2 + |D|^2}. \tag{3.55}$$

Extension to a longer neck is straightforward [96, 97]: Then the pressure drop in Eq. (3.52) has an additional term $(\ell/S_2)\mathrm{d}m'_1/\mathrm{d}t$. This describes the pressure force needed to overcome the inertia of the fluid in the neck.

Figure 3.25 shows a comparison between theory and experiment. In general, the theory works well. The peak absorption near resonance is clear. Enhancement of this peak absorption can be obtained by 'tuning' the mean flow appropriately and, of course, the Helmholtz resonator should be placed near a pressure antinode, that is, at a position in the duct near a maximum of $p'_{1\text{rms}}$.

If there is no mean flow through the neck of the Helmholtz resonator, the pressure drop across the neck is related to the instantaneous velocity through the neck $u'_1(t)[= m'_1(t)/(\bar{\rho}S_2)]$ by the nonlinear relationship

$$p'_1 - p'_2 = \ell_e\frac{\mathrm{d}u'_1}{\mathrm{d}t} + \frac{c_d}{2}u'_1\left|u'_1\right|, \tag{3.56}$$

where ℓ_e is the effective length of the fluid accelerated in the neck and c_d is the discharge coefficient [99]. A numerical integration of Eqs. (3.50) and (3.56) can be used to determine $u'_1(t)$ for given neck pressure perturbation $p'_1(t)$ and hence to determine the waves in the duct from Eqs. (3.48) and (3.49). The absorption coefficient then follows from Eq. (3.55). The absorption coefficient depends now on the amplitude of $p'_1(t)$, so-called 'nonlinear absorption'. Sample results are shown

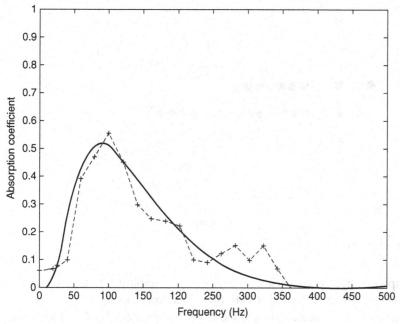

Figure 3.25. Comparison between theory and experiment for a mean flow of Mach number 0.02 through a perforated plate (from [98]).

in Fig. 3.26. In this case, a pressure node at the neck of the Helmholtz resonator for a frequency near 280 Hz results in a minimum absorption coefficient there. Of course, even when there is a mean flow through an aperture, the absorption coefficient depends on the amplitude of the pressure perturbation at the neck of the resonator at high-enough sound amplitude. This is illustrated in Fig. 3.27, which shows absorption from a single aperture. The mean flow velocity through the aperture is characterised by the mean pressure drop $\Delta \overline{p}$ across it. At low sound levels, the absorption coefficient is 'linear', being independent of sound amplitude when there is a mean flow. At high-enough sound amplitudes, the absorption coefficient becomes

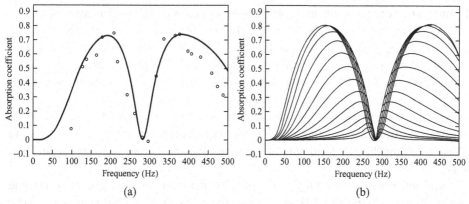

(a) (b)

Figure 3.26. (a) Comparison between the measured and predicted absorption coefficients as a function of frequency for a sound pressure level (SPL) of 165 dB at the neck of a Helmholtz resonator (from [82]). (b) Predicted absorption coefficient as a function of frequency for various SPLs at the neck of a Helmholtz resonator ranging from 120 to 185 dB in increments of 5 dB. The lowest curve is for 120 dB (from [82]).

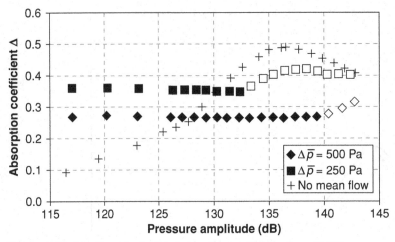

Figure 3.27. Nonlinearity in the absorption coefficient (from [100]).

nonlinear (implying that the coefficient depends on the amplitude), the sound level for the onset of nonlinearity increasing with the mean flow velocity.

The sound absorption achieved by a Helmholtz resonator attached to a combustor will depend on its location and on the acoustic mode shape. An example of this is clear in Fig. 3.26, where the local minimum near 280 Hz is due to the Helmholtz resonator then being near to a pressure node for this axial acoustic mode. The annular combustors of aeroengine gas turbines have an axial length that is short in comparison with their circumference. This means that the lowest resonance frequency of the combustor is often associated with the $n = 1$ circumferential mode. When a combustion instability occurs, it is then close to the resonant frequency of this mode, and to control the instability one needs to understand the effect of one or more Helmholtz resonators on this mode of oscillation. This can be conveniently investigated through a network analysis, with the resonator being treated as a side branch and appropriate conservation conditions applied [101]. The unsteady mass flow from the Helmholtz resonator is again assumed to be related to the pressure at its neck through Eq. (3.53) for the case with a mean flow. The effectiveness of a Helmholtz resonator as an acoustic absorber depends on the overall system.

We noted that, for a plane wave, the Helmholtz resonator is ineffective if placed near a pressure node. For a circumferential travelling wave in a cylindrically symmetric geometry, a single Helmholtz resonator does not absorb energy; instead it just reflects an equal and opposite wave, leading to a standing mode with a pressure node at the neck of the Helmholtz resonator (see Fig. 3.28). There is no absorption of energy. Indeed, a single Helmholtz resonator may have a detrimental effect because, at some locations around the annulus, the incident and reflected waves will be in phase, doubling the amplitude of the pressure perturbation from that in the incident wave.

Multiple resonators are needed to absorb energy in circumferentially travelling waves. Stow and Dowling [102] investigated the absorption that is due to two or more Helmholtz resonators located around the circumference of an annular combustor and give an expression for their optimal locations: If N Helmholtz resonators are to be used to absorb sound with circumferential wave number n, the best configuration

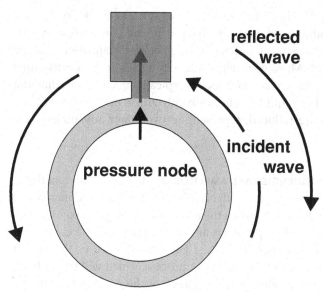

Figure 3.28. A single Helmholtz resonator reflects a travelling circumferential wave, giving a standing wave with a pressure node at the neck of the resonator [102].

is to distribute the Helmholtz resonators evenly over half the wavelength of the mode. Because of the symmetry, moving the position of any Helmholtz resonator by multiples of half a wavelength has no effect. This leads to optimum placement of the jth Helmholtz resonator at

$$\theta_j = \frac{\pi}{n}\left(\frac{j-1}{N} + M\right) \quad \text{for } j = 1, \dots, N, \qquad (3.57)$$

where M is a positive integer.

The passive techniques described in this subsection can work well at modest and high frequencies, provided that the Helmholtz resonator or liner is tuned to give optimal absorption at that frequency. However, the frequency of a combustion instability varies with the operating condition, and so interest has turned to active ways of keeping a passive absorption operating at peak absorption.

3.2.3 Tuned Passive Control

Passive control commonly takes the form of adding acoustic dampers to the combustor. The performance of these tends to be highly frequency dependent; effective damping occurs over only a narrow bandwidth, the frequency of which depends on the geometry of the acoustic damper. Tuned passive control involves using a control system and actuator to vary the geometry of acoustic dampers in time so that they can respond to changes in combustor instability frequency.

For example, Helmholtz resonators (see Subsection 3.2.2) are most effective as acoustic dampers when their resonant frequency is close to the instability frequency present in the combustion chamber. Using the 'thick-neck' form for the pressure drop across the neck, we can approximate their resonant frequency ω_0 by $\omega_0^2 = \dfrac{\bar{c}^2 S_2}{V l_{\text{eff}}}$, where $\hat{c}$ is the speed of sound, S_2 is the neck area, V is the cavity volume, and l_{eff}

is the effective neck length. If any of the geometry parameters, S_2, V, or l_{eff}, are altered, a means of responding to frequency changes in the combustion system is provided. The ability to tune the damping overcomes the main disadvantage of traditional passive control. Moreover, actuation only needs to be on the time scale of the changes in operating condition, which is typically much slower than the time scale of the instability. The required actuator bandwidth is therefore small and the durability requirements are reduced, representing two major advantages over active control.

ONLINE IDENTIFICATION OF FREQUENCIES AND AMPLITUDES. To implement tuned passive control in practice, real-time tracking of the combustor instability frequency and amplitude level are required. Note that, when the acoustic damping of a passive controller is altered, the acoustic eigenfrequencies and mode shapes of the combustor change, causing the instability frequency to shift. The consequence of this is that tuning an acoustic damper needs to be an iterative process, which will typically involve two tuning stages. The first implements a 'guess' of the optimum geometry, based on an initial measurement of the instability frequency. The second involves an online search for the damper geometry that minimises the amplitude level and indirectly takes account of all the instability frequency changes that occur along the way [103, 104].

To rapidly track the instability frequency and amplitude in real time, frequency-domain [fast Fourier transform (FFT)-based] Techniques can be used. However, these provide fairly slow tracking capabilities, particularly at low frequencies. Time-domain integration techniques provide a faster alternative and are based on the premise that integrating the product of a signal and a sinusoid extracts the signal component at the frequency of the sinusoid.

Given a measured pressure signal $p(t)$ with time derivative $\dot{p}(t)$, the aim is to find the instability frequency ω and the (sine and cosine) amplitude components, a and b, such that $a \sin \omega t + b \cos \omega t$ is a good approximation of $p(t)$. The rules for updating the online estimates of ω, a and b, are shown below. Derivations can be found in [104, 105];[1] n denotes the nth iteration and α is a relaxation coefficient:

$$\omega_{n+1} = (1 - \alpha)\omega_n + \alpha \left[\omega_n^2 + \frac{\omega_n^2}{\pi} \frac{p(t) - p(t - 2\pi/\omega_n)}{a(t) \cos \omega_n t + b(t) \sin \omega_n t} \right]^{\frac{1}{2}}, \tag{3.58}$$

$$b(t) = \frac{\omega_n \cos \omega_n t [\dot{p}(t) - \dot{p}(t - 2\pi/\omega_n)] + \omega_n^2 \sin \omega_n t [p(t) - p(t - 2\pi/\omega_n)]}{\pi(\omega_n^2 - \omega_{n+1}^2)}, \tag{3.59}$$

$$a(t) = \frac{\omega_n \sin \omega_n t [\dot{p}(t) - \dot{p}(t - 2\pi/\omega_n)] - \omega_n^2 \cos \omega_n t [p(t) - p(t - 2\pi/\omega_n)]}{\pi(\omega_n^2 - \omega_{n+1}^2)}. \tag{3.60}$$

It is possible to extend these time-domain algorithms to pressure signals containing multiple frequency components. Such a pressure signal might occur in a combustor exhibiting multiple unstable modes [104, 105]. Algorithms based on time-domain tracking are used in several of the examples given in the following subsection.

---

[1] Note that equation 4b in [104] should have the sine and cosine interchanged.

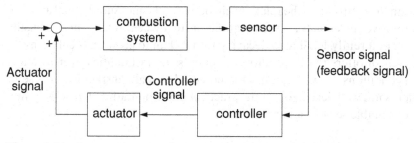

Figure 3.29. Generic system for feedback control of combustion oscillations.

EXAMPLES OF TUNED PASSIVE CONTROL. Returning to the example of the Helmholtz resonator, which is the passive damper on which most tuning investigations have focussed, it is most practical to vary either the cavity volume [103, 106, 107] or the neck area [104]. The cavity volume is normally varied by use of a back-piece plunger, whereas the neck area can be varied by use of an 'iris'-type valve, which operates like a camera lens. Both forms of actuation have been demonstrated to effectively stabilise combustion instabilities over varying operating conditions.

A rather different example of tuned passive control again involves Helmholtz resonators, but this time considers those whose volume is continually oscillated with the frequency present in the combustion system (i.e., the instability frequency)[2] [108, 109]. Whether this volume oscillation enhances or reduces the acoustic damping depends on the phase of the oscillation. When an online tuning of the oscillation phase is performed such that the amplitude level is minimised, peak damping performance can be achieved [109]. This approach has the advantage in that the level of volume oscillation is sufficiently small so that it is possible to avoid the sealing issues that otherwise arise.

Helmholtz resonators are not the only types of passive damper that can be passively tuned. Perforated liners are another type. In Subsection 3.2.2, it was seen that the damping of a perforated liner depends on its bias flow rate (i.e., the flow rate of cooling air through the perforated holes) and its location within the combustor [85]. A two-parameter tuning, in which the bias flow rate and the combustor length were both successively tuned to maintain maximum damping across varying frequencies, was successfully performed in cold-flow experiments [110].

3.2.4 Active Control

Active control provides another means of interrupting the coupling between acoustic waves and unsteady heat release [111–113]. It involves dynamically varying some input to the combustion system and may be either open loop, for which the input does not depend on measurements from the system, or closed loop (feedback), for which the input does depend on system measurements. Closed-loop control offers better opportunities for suppressing instabilities and so will be the form of active control assumed from now on.

For closed-loop active control, an actuator modifies some system parameter in response to a measured signal, as shown in the generic control system in Fig. 3.29. The

---

[2] This can also be thought of as a type of tuned open-loop active control.

aim is to design the controller – the relationship between the measured signal and the signal used to drive the actuator – such that the unsteady heat release and acoustic waves interact differently, leading to decaying rather than to growing oscillations.

Active control offers the possibility for suppressing instabilities over a range of operating conditions. However, it has the potential to make instabilities worse if the controller is not well designed, and its practical implementation depends on the availability of suitable sensors and actuators.

Practical Implementation of Active Control

SENSORS. Sensors are needed to provide measurement of a dynamic quantity related to the combustion oscillations [114]. Although this is a relatively straightforward task for an unstable system, it is more difficult for low-amplitude controlled oscillations.

Microphones and pressure transducers are the most commonly used sensors [115–118]. They have the advantage in that, because acoustic waves propagate throughout the entire combustion system, they do not need to be placed close to the high temperatures of the heat release zone. They have large bandwidth and are reasonably robust. Their location within the combustor is important; attention must be paid to the likely mode shapes so as to avoid placement near a pressure node and ideally to maximise the signal-to-noise ratio. In annular gas turbines, multiple sensors are required for capturing the azimuthal modes [119].

The most common alternative is to measure the light emitted by certain progress chemicals in the flame, most commonly C_2, OH, or CH radicals [120, 121]. Optical filters in front of photomultipliers filter out all but the spectral bands related to these chemicals, with the intensity levels then related to the rate of heat release in the field of view. Such sensors are independent of mode shape, but may be susceptible to changes in flame location as the stability of the system changes. They also require optical access to the system; although optical fibres provide a means of achieving this [121], they have a limited field of view and tend to give rather noisy light-intensity readings.

ACTUATORS. Satisfactory actuation of combustion systems presents a real challenge. The first actuators to be applied to combustion control were loudspeakers [115, 116, 120, 122]. Although these have a good high-frequency response, they are not sufficiently robust for use in industrial systems and their power requirements become prohibitive at larger scales.

An efficient means of actuation exploits the chemical energy released in combustion by modulating the fuel supply. For ideal actuation, a fuel valve should have a linear response (to allow linear control theory to be used), large bandwidth, sufficiently large control authority to affect the limit cycle oscillation, a small-amplitude response that does not exhibit hysteresis, a fast response time, and good robustness and durability. Furthermore, care must be taken to ensure that the acoustics of the pipework connecting the actuator to the fuel injection position results in a large response at the likely control frequencies [117, 123]. Obtaining a satisfactory fuel valve is one of the main challenges facing reliable implementation of feedback control on practical combustion systems. Solenoid valves, which have the advantage of a linear response, have been used in large and full-scale demonstrations of control [117, 121, 124]. However, they suffer from limited bandwidth and control authority, exhibit hysteresis at low amplitudes, and are probably not sufficiently durable for practical applications. Magnetorestrictive valves combine fast response time with increased bandwidth and show significant potential, as long as the valve authority can be maintained at higher frequencies [105].

FEEDBACK CONTROL CHALLENGES. Practical combustors involve turbulence and combustion; these processes are too complex to use either mathematical or numerical modelling as a basis for the design of controllers that will be implemented experimentally. This means that controller design must be based on system measurements. The consequences of stability may be so severe that obtaining these is not straightforward in certain operating regimes. There are likely to be several modes of instability spanning a wide frequency range, and in such cases it is important that the controller must be able to control more than one mode. The system will almost certainly involve substantial time delays caused by factors such as fuel convection and acoustic-wave propagation; these may be larger than an oscillation period and provide a real challenge to controller design. Different control parameters may be needed to cope with different plant-operating conditions, and the presence of turbulent flow will lead to high background-noise levels, which the controller should be robust to.

HISTORICAL DEVELOPMENT. The first experimental demonstration of active control of a combustion instability was for a Rijke tube with a laminar flame burning on gauze [120, 125]. An optical fibre and photomultiplier were used to detect the light emitted by CH free radicals in a Rijke tube flame. This signal was then phase shifted, amplified, and used to drive a loudspeaker near one end of the tube. The optimum gain and phase shift were found by trial and error and gave a sound-pressure-level (SPL) reduction of 35 dB. Control was then demonstrated on a similar combustor, but with the sensor signal being a microphone-measured pressure at a location within the tube [115]; a range of controller phase shifts achieved stability.

These first demonstrations both used a loudspeaker to change the boundary conditions at one end of the tube. In terms of the generalised Rayleigh criterion discussed in Subsection 3.2.1, this primarily increases the energy lost across the tube boundaries per cycle, but also alters the acoustic-wave energetics within the tube, leading to a change in the average energy exchange between the acoustic waves and combustion. For an appropriate range of phase shifts, inequality (3.39) is reversed and the system is stabilised. The fact that a range of stabilising phase shifts exists is consistent with the fact that the inequality only needs to be reversed for instability to be avoided; unlike in antisound, here we are not relying on an exact cancellation of signals [125].

Because the power requirements of mechanical actuators such as loudspeakers become prohibitive at larger scales, attention turned to varying the Rayleigh energy source term, $\int \overline{p'q'}\, dV$, directly using unsteady fuel injection. In the first experimental demonstration of achieving control through modulation of the fuel supply, a ducted premixed flame was stabilised by a phase-shift controller, designed by use of Nyquist methods [126]. An unsteady addition of just 3% excess fuel reduced the spectral peak corresponding to the main instability mode by 12 dB.

The early demonstrations all used simple gain and phase-shift–time-delay controllers. The controller parameters were obtained for a single operating condition, usually empirically. Such approaches are unlikely to be adequate for practical combustion systems, in which there may be multiple instability modes. They offer no guarantees of stability, and a bad choice of controller parameters may increase the amplitude of the unstable oscillations [117]. Furthermore, controllers involving just a fixed time delay or phase shift introduce a new oscillation mode, which becomes unstable as gain is increased and gives rise to a new peak in the pressure spectrum [126–128]. A dynamic controller that can introduce suitable gains and phases at different frequencies is needed, together with a strategy for its design. Some of the common controller design strategies are described in detail in the following subsections.

CONTROL STRATEGIES – MODEL-BASED CONTROL. By increasing the sophistication of the controller structure beyond that of a straightforward gain–time delay, significant improvements in the behaviour of the closed-loop system can be obtained [111]. To design such a controller, some knowledge of the open-loop system is required. Because mathematical or numerical models can only approximate the open-loop behaviour of a combustion system, it is currently preferable to design model-based controllers based on experimental measurements. Model-based controller design was experimentally implemented on several larger-scale combustion rigs [129–132]. Multisensor multi-actuator control strategies were also recently developed for combustion chambers that are annular in shape and may exhibit unstable longitudinal *and* circumferential modes simultaneously [119]. However, we demonstrate the essential concepts of model-based control by applying it to the simplest laboratory-scale example of a combustion instability with a single unstable mode – the Rijke tube.

A Rijke tube is a straight tube, open at both ends, with a heat source inside and a mean flow passing thorough it. By using a vertical tube, a small amount of convection flow arises from the mean heat release and there is no need to drive any additional mean flow through the tube. When the heat source is in the upstream (bottom) half of the tube and heat addition is in phase with the pressure fluctuation and in accordance with Rayleigh's criterion, acoustic waves successively gain energy and a combustion instability occurs. Rijke tubes were used in the first experimental demonstrations of active control of combustion instabilities [115, 116, 120] and, because of their simplicity, were used to experimentally demonstrate many advances in control methodology for combustion instabilities [129, 133, 134].

A vertical quartz Rijke tube, 0.75 m long and with a diameter of 44 mm is considered, as shown in Fig. 3.30. A propane-fuelled Bunsen burner provides a laminar flame, which is stabilised on a grid 0.225 m above the bottom of the tube. In the absence of control, there is an instability at 1533 rad/s (244 Hz). A microphone is fitted to a tube tapping 0.34 m from the bottom of the tube and actuation is provided by a 50-W low-frequency loudspeaker situated close to the lower end of the tube. This has a flat response over the frequency range of interest (500–12 600 rad/s or 80–2000 Hz); above and below this its dynamics is more complicated. For this reason, a bandpass filter with a passband from 500 to 12 600 rad/s is applied to the microphone signal.

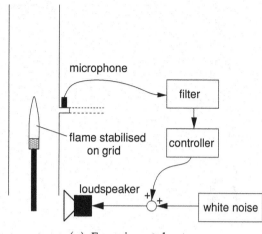

Figure 3.30. The Rijke tube exhibiting
a combustion instability.

(a) Experimental setup

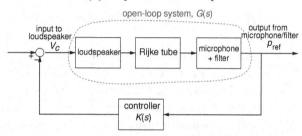

(b) Feedback control system, showing the open-loop
transfer function $G(s)$ and the controller transfer
function $K(s)$

MEASURING THE OPEN-LOOP TRANSFER FUNCTION (OLTF). The transfer function approach to controller design, also known as classical control theory, is used in controller design. For an overview of this refer to [135, 136]

The OLTF needed for control purposes is from the actuator signal to the sensor signal, in this case from the loudspeaker voltage V_c to the filtered microphone reading p_{ref}. This is shown as $G(s)$ in Fig. 3.30(b), where s denotes the Laplace transform variable. As with all naturally unstable systems, characterising $G(s)$ is complicated by the fact that, during instability, nonlinear limit cycle behaviour dominates and masks the linear relationship that would be observed at low oscillation amplitudes. To prevent the growth of small perturbations, it is measurement of the linear behaviour that is required for controller design.

To overcome this problem, a loudspeaker voltage input V_c is used, which comprises two components:

1. a control signal from a trial-and-error controller to eliminate the nonlinear limit cycle, and
2. a wide bandwidth signal, in this case white-noise signal from a white-noise generator, for identification of the OLTF.

One may ask this question: If a trial-and-error controller has been found, why do we need to design another controller? The answer lies in the superior performance

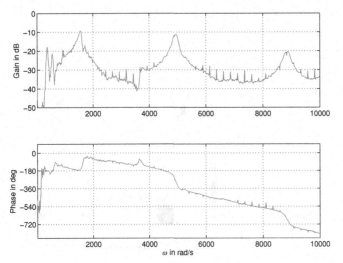

Figure 3.31. The measured Bode diagram of $G(s) = p_{\mathrm{ref}}(s)/V_c(s)$ for the Rijke tube.

of a model-based controller: Feedback control is most successful when an accurate model of the system to be controlled is available.

Measuring the OLTF involves applying the preceding voltage input to the loud-speaker, recording the response of the filtered microphone signal, and using the ratio of Fourier transforms to deduce the open-loop frequency response $G(i\omega)$. Note that it is important to use the total loudspeaker voltage in deducing this, not just the white-noise component. Although the controller causes the loudspeaker input spectrum to have a peak at the unstable frequency, applying sufficiently loud white noise ensures that this peak does not dominate.

The open-loop frequency response $G(i\omega)$ is shown as the Bode diagram of $G(s)$ in Fig. 3.31. The magnitude peaks represent system resonances and correspond to the various 'organ-pipe' modes of the acoustic waves in an open-ended tube. For example, the fundamental frequency of an open–open tube has a wavelength of twice the tube length, in this case 1.5 m. The fundamental frequency is then $c/(2L)$, where $\hat{c}$ is the speed of sound in the tube. Taking this to be 360 m/s gives a fundamental frequency of 240 Hz or approximately 1500 rad/s, which is the frequency of the first modal peak.

By considering each modal peak to be caused by a second-order transfer function,

$$\frac{A}{(s^2 + 2\zeta_n\omega_n s + \omega_n^2)},$$

stability information can be deduced from the phase change across the modal peaks. A phase increase of 180° indicates an unstable conjugate pair of poles ($\zeta_n < 0$) whereas a phase decrease of 180° indicates a stable conjugate pair ($\zeta_n > 0$). The measured transfer function shows that the first (fundamental) mode is unstable, whereas all other modes are stable.

CONTROLLER DESIGN. The system is single-input–single-output, and so Nyquist techniques [135, 136] provide a suitable means of designing robust controllers. The experimentally characterised $G(i\omega)$ can then be used directly in controller design: Subject

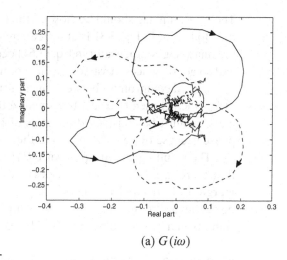

(a) $G(i\omega)$

Figure 3.32. Nyquist diagrams: positive ω, - - - negative ω.

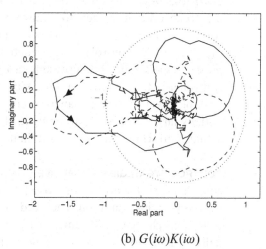

(b) $G(i\omega)K(i\omega)$

to the system having low noise, there is no need to model or approximate what has been measured. The Nyquist diagram for $G(s)$ is shown in Fig. 3.32(a). According to the Nyquist criterion, a negative feedback system such as that in Fig. 3.30 is stable if the number of anticlockwise encirclements of the -1 point shown by the controlled OLTF, $G(s)K(s)$, is equal to the number of unstable poles in the uncontrolled OLTF, $G(s)$ [135, 136]. The phase increase across the first mode confirms that this is the only unstable mode and that it has a pair of unstable poles associated with it. The Nyquist plot for $G(s)K(s)$ therefore needs to encircle the -1 point in an anticlockwise direction twice in order for the closed-loop system to be stable. The Nyquist plot for $G(s)$ has no encirclements of the -1 point. To achieve two encirclements, it is clear that the controller should introduce both gain and additional phase lag near the frequency of the unstable mode.

A good choice of controller structure to achieve this is that of a phase-lag compensator:

$$K(s) = \frac{k(s + \beta a)}{(s + a)}, \quad \text{where } \beta \geq 1. \tag{3.61}$$

The maximum lag should be close to the location of the unstable mode, which from the Bode plot in Fig. 3.31 is at a frequency of $\omega = 1600$ rad/s. The maximum lag of the phase-lag compensator in Eq. (3.61) occurs at an approximate frequency of $a\sqrt{\beta}$, and so good choices of values for the controller are $a = 980$, $\beta = 2.67$. These values give a good compromise between maximising the phase and gain margins. The value of k is then chosen to be 3.1 to maximise the gain margin (note that the gain–phase margins must be considered in terms of both reducing and increasing the gain or phase because two encirclements of the -1 point are required).

The resulting Nyquist plot for $G(s)K(s)$ is shown in Fig. 3.32(b). It can be seen that there are indeed two anticlockwise encirclements of the -1 point. The gain margin is 4.5 dB and the phase margin is 21°; thus the closed-loop system should be stable and reasonably robust to plant uncertainties and changes (if needed, the stability margins could be increased by use of a higher-order controller).

CONTROLLER IMPLEMENTATION. After the preceding controller is implemented on the Rijke tube, it is seen to eliminate the combustion instability. The effects of control on both the measured pressure spectrum and the time-domain oscillations are shown Fig. 3.33. Control gives a reduction of approximately 80 dB in the microphone pressure spectrum: This represents a reduction of four orders of magnitude. The controller obtains control rapidly, with oscillation amplitudes down to 10% of their unstable levels in fewer than 10 oscillations. Although the loudspeaker voltage is initially large in order to attain control, once control has been achieved very little actuator effort is needed to maintain it.

CONTROL STRATEGIES – ADAPTIVE CONTROL. The type of model-based controllers considered so far are robust to small changes, but are essentially designed for a single operating condition. Their redesign would be necessary should the operating conditions change significantly. Adaptive controllers, whose parameters are continually updated to track plant changes, offer an efficient means of achieving control across a range of operating conditions. Although some adaptive controllers, such as neural networks, gain information about the plant offline [137, 138], others treat the combustion system as a 'black box' and rely on an online system identification. The first adaptive controllers to be applied to combustion oscillations were least-mean-squares (LMS) controllers [139, 140]. These have the form of infinite-impulse-response (IIR) filters whose coefficients are updated according to the LMS algorithm [141]. Despite recent improvements to the online system identification within the algorithm [142], LMS controllers are unlikely to be sufficiently robust for use in practical applications as they rely on assumptions such as the primary noise source being slowly varying and do not offer any guarantees of global stability.

Possibly the most physics-based type of adaptive control is guided directly by the Rayleigh criterion [105, 143]. Open-loop testing is first used to determine the time delay between the actuator input signal and the heat release rate as a function of frequency. When control is activated, a real-time observer, similar to the one described in Subsection 3.2.3, is used to detect and track the frequencies, amplitudes, and phases of several combustor modes. This real-time detection combined with the open-loop information then allows fuel to be added unsteadily such that the heat release rate is exactly out of phase with the pressure fluctuations. Even though only

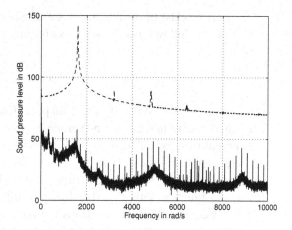

Figure 3.33. The effect of the model-based controller on the Rijke tube instability.

(a) The microphone pressure spectrum, — with control, - - - without control.

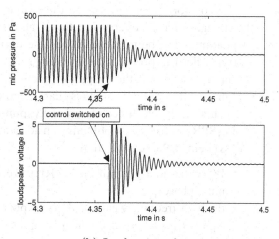

(b) In the time domain

a gain–time-delay controller is used, the observer ensures that any secondary peaks created are detected so that the control action can adapt to the new frequencies. This approach has been demonstrated experimentally on a large-amplitude instability in a small-scale gas rocket motor, in which the observer detected the most dominant mode, and has since been extended to a full-scale gas turbine combustor [144].

A particularly attractive approach to adaptive control, which avoids the need for either detailed prior measurements or online system identification, is to use *self-tuning regulators* (STRs). The remainder of this subsection discusses the use of STRs as adaptive controllers.

STR ADAPTIVE CONTROL THEORY. STRs are model-based adaptive controllers [145]. They use some general properties of the open-loop system to infer the controller structure needed for stability and then update the controller parameters by using a Lyapunov-based algorithm. A brief outline of STR background theory is now provided.

For the class of combustion systems whose OLTF has relative degree[3] n^* equal to 1 or 0, it follows from root-locus arguments that a first-order compensator of the form

$$K(s) = k_1 \frac{s + z_c}{s + p_c} \qquad (3.62)$$

can always stabilise the system if its parameters are chosen correctly. The adaptive control algorithm seeks to find stabilising values for the parameters of this first-order compensator. Note that real combustion systems have infinite dimension because of the presence of time delays. In practice, the requirement that $n^* = 0$ or 1 therefore translates to the requirement that the OLTF is well approximated by a rational transfer function with $n^* = 0$ or 1 over a sufficiently wide frequency range.

If we define $W_{cl}^*(s)$ as the transfer function of the resulting closed-loop stabilised system, the derivation of such an adaptive control algorithm relies on $W_{cl}^*(s)$ being strictly positive real (SPR). If $W_{cl}^*(s)$ is SPR, it is possible to form a positive-definite Lyapunov function which depends on the states (in a state-space sense) of $W_{cl}^*(s)$ and a control vector $\mathbf{k}$. Because the Lyapunov function is positive-definite, it can be thought of as an energy function. The time derivative of the Lyapunov function is guaranteed to be negative if the time derivative of the control vector $\mathbf{k}$ obeys a certain rule: This provides the updating rule for the controller parameters. When this updating rule is used, the Lyapunov function is guaranteed to decrease monotonically in time and the control vector $\mathbf{k}$ is guaranteed to converge to a stabilising value (which we call $\mathbf{k}^*$). The existence of such a Lyapunov function – which depends on $W_{cl}^*(s)$ being SPR – is crucial for obtaining an adaptation algorithm that guarantees stability. $W_{cl}^*(s)$ being SPR requires that

- $W_{cl}^*(s)$ has no right half-plane (RHP) zeros (i.e., the closed-loop system is minimum phase),
- the high-frequency gain of $W_{cl}^*(s)$ is positive, and
- $\Re[W_{cl}^*(s)] > 0$ for all s, which is equivalent to the phase of $W_{cl}^*(s)$ lying in the range $-\pi/2$ to $\pi/2$.

For a more rigorous definition of SPR transfer function, see Narendra and Annaswamy [145]. We now look at how each of these three requirements relates to the properties of combustion systems.

Attention is restricted to longitudinal combustion systems governed by 1D acoustic behaviour. For these, active control is typically achieved by fuel actuation at a single location and pressure sensing at a single location. The OLTF needed for control purposes is from the actuator signal V_c to the sensor pressure p_{ref}. A general model describing the constituent physical processes is shown in Fig. 3.34. It includes the effect of the actuator, the flame, and the acoustics.

Physically, changes in the actuator voltage V_c result in fuel mass-flow-rate fluctuations m_f'. These and the air mass-flow-rate fluctuations at the combustion zone, m_a', combine to change the equivalence ratio according to $\phi'/\overline{\phi} = m_f'/\overline{m}_f - m_a'/\overline{m}_a$. It is generally agreed that fluctuations in the heat release rate respond to both fluctuations in the equivalence ratio and in the air mass flow rate [146, 147]. Recent experimental

---

[3] The relative degree of a transfer function n^* is defined as the degree of the denominator minus the degree of the numerator.

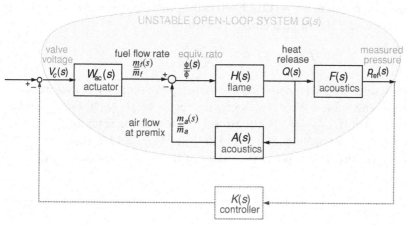

Figure 3.34. Block diagram for a general longitudinal combustion system.

work suggested that, when the flame is compact, heat release fluctuations are dominated by the response to equivalence ratio changes [148]. Assuming a compact flame, the heat release fluctuation Q' is modelled by the relationship $Q(s) = H(s)\phi(s)/\bar{\phi}$, where $H(s)$ is the flame model and $s = i\omega$ is the Laplace transform variable.

The response of the pressure measurement, p_{ref}, to the heat release fluctuation, Q', is modelled by an acoustic transfer function $F(s)$, and the response of the fractional air mass flow fluctuation at the combustion zone, $m'_a/\bar{m}_a$, to the heat release fluctuation Q' is modelled by $A(s)$. The overall OLTF $G(s)$ can then be expressed as

$$G(s) = \frac{p_{ref}(s)}{V_c(s)} = \frac{W_{ac}(s)H(s)F(s)}{1 + A(s)H(s)}. \tag{3.63}$$

Using this general model, we can interpret the three requirements for $W^*_{cl}(s)$ to be SPR as follows:

1. $W^*_{cl}(s)$ has no RHP zeros: The zeros of the open-loop plant $G(s)$ become the zeros of $W^*_{cl}(s)$. $G(s)$ has no RHP zeros if the actuator transfer function $W_{ac}(s)$, the FTF $H(s)$, and the acoustic transfer function $F(s)$ are all free of RHP zeros. It can be shown that the latter condition requires that the acoustic reflection coefficients at all combustor boundaries have magnitudes less than unity and that the temperature fluctuations generated by unsteady combustion are largely dissipated before reaching the end of the combustor [149].
2. The high-frequency gain of $W^*_{cl}(s)$ is positive: This requires that the sign of the high-frequency gain g_0 of the open-loop system is also positive, $\text{sgn}(g_0) > 0$. As long as the sign of g_0 is known, however, the algorithm can be modified to guarantee stability.
3. $\Re[W^*_{cl}(s)] > 0$ for all s: This corresponds to the phase of $W^*_{cl}(s)$ lying between $-\pi/2$ and $\pi/2$. Meeting this requirement means that the OLTF must have a relative degree n^* of 0 or 1 [because $n^*(W^*_{cl}) = n^*(G)$]. The phase change caused by any time delay must also be negligible. Furthermore, the zeros and poles must be 'close' in some sense; for example in systems dominated by second-order poles and zeros, the pole and zero pairs would need to interlace.

For a combustion system whose OLTF obeys the preceding properties, the phase compensator of Eq. (3.62) is implemented in the form

$$K(s) = k_1(t)\frac{s + z_c}{s + z_c + k_2(t)}, \qquad (3.64)$$

where z_c is prescribed and $k_1(t)$ and $k_2(t)$ are the adaptive parameters that we seek to tune to stabilising values. We define a control vector $\mathbf{k}(t) = [k_1(t)\, k_2(t)]^T$ and a data vector $\mathbf{d}(t) = [p_{\mathrm{ref}}(t)\, \frac{V_c(t)}{(s+z_c)}]^T$. The control signal sent to the actuator, $V_c(t)$, can then be written as the dot product of $\mathbf{k}(t)$ and $\mathbf{d}(t)$, i.e., $V_c(t) = \mathbf{k}(t) \cdot \mathbf{d}(t)$.

It can be shown that if we update the values of the control vector according to the rule

$$\dot{\mathbf{k}}(t) = p_{\mathrm{ref}}(t)\mathbf{d}(t), \qquad (3.65)$$

then the elements $k_1(t)$ and $k_2(t)$ are guaranteed to converge to stabilising values. The mathematical details of the control proof are not reproduced here (see [134, 145] for details). In practice the equation $\dot{\mathbf{k}}(t) = \Gamma p_{\mathrm{ref}}(t)\mathbf{d}(t)$ is used, where Γ is a diagonal matrix with entries γ_1 and γ_2. These are convergence coefficients, which determine the speed of the control action. If they are too large, i.e., if control attempts to force a response too rapidly, numerical problems may occur and the actuator bandwidth may become limiting. If they are too small, the controller may not be effective from within the limit cycle. Guidance on choosing suitable values is provided in [150].

EXPERIMENTAL IMPLEMENTATION OF THE STR. The STR algorithm was implemented experimentally on the same Rijke tube described in Subsection 3.2.4 [134]. To test whether it was able to adapt to changes in operating condition, the length L of the tube was varied by use of the 'trombone-like' arrangement to vary the unstable oscillation frequency. The results are compared with those for a fixed-parameter phase compensator, of the type designed in Subsection 3.2.4, in Fig. 3.35. It can be seen that, although the fixed parameter loses control after the length of the tube is varied, the adaptive controller maintains control.

The same STR was recently applied to a lean premixed prevapourised combustion rig operating at atmospheric conditions, in which modulation of the fuel supply was used for actuation [151]. The primary instability was reduced by up to 30 dB, and initial robustness studies confirmed that the controller retained control for a 20% change in frequency and a 23% change in air mass flow rate.

Because of their applicability to a wide range of combustion systems and the fact that stability can be guaranteed, STRs hold real promise in terms of future implementation in full-scale combustion systems. Recent work extended their use to combustors with higher relative degree and significant time delays [134, 152, 153], combustors exhibiting undesirable reflections from boundaries [149], and combustors with an annular shape exhibiting multiple instability modes [150]. Work on how to extend the adaptive algorithm to more complex combustion systems is ongoing.

FULL-SCALE IMPLEMENTATION OF ACTIVE CONTROL. Although many sophisticated feedback controllers have been applied to combustion models and laboratory-scale rigs, there have been very few reported full-scale demonstrations of feedback control. All used fuel modulation and very simple (and in many ways restrictive) controller designs.

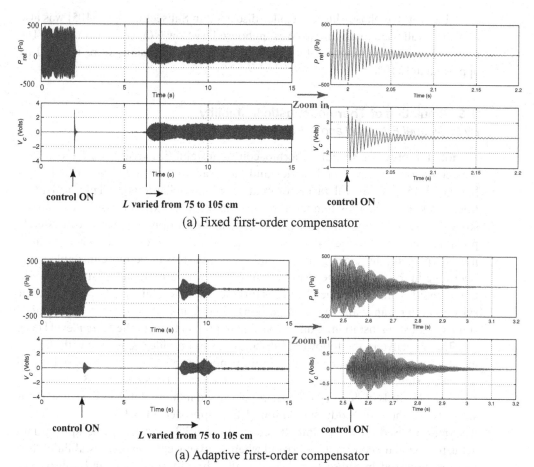

(a) Fixed first-order compensator

(a) Adaptive first-order compensator

Figure 3.35. Control of Rijke tube under varying tube length L (from [134]).

The first full-scale demonstration was in 1988 on the afterburner of a Rolls-Royce RB 199 military turbofan engine [118]. Actuation was achieved by spilling fuel from the engine, rather than adding it, using high-response electrohydraulic servovalves. The modulated fuel was approximately 5%–10% of the mean. Using a simple gain–time-delay controller, a 12-dB reduction in the dominant low- 'buzz' frequency was obtained.

A decade later, active control was performed on a Siemens heavy-duty industrial gas turbine [117, 154]. The combustor was annular, and the unstable modes were azimuthal. The pressure at several locations around the combustor circumference was measured, with a separate control system used at each location. Actuation was performed by modulating the fuel to the pilot flames by use of a high frequency solenoid value supplied by MOOG. Simple gain–phase-shift controllers were used, and the dominant frequency was reduced by 17 dB. It was found that the required controller parameters changed with operating condition.

Researchers at United Technologies Research Center then demonstrated active control on full-scale liquid-fuelled lean premixed combustors [124, 155]. With a solenoid valve to modulate the fuel supply, a 16-dB reduction of the dominant mode in a single combustor and a 6.5-dB reduction in a 67.5° sector cut from a full combustor annulus were obtained.

The adaptive phase-shift controller discussed in Subsection 3.2.4 [105] was also applied at full scale to a Siemens-Westinghouse Dry Low NO_x combustor [144]. The dominant mode was reduced by 15 dB, and the NO_x emissions were reduced by approximately 10%.

3.3 Simulation of Thermoacoustic Instability
By L. Gicquel F. Nicoud and T. Poinsot

As noted in previous sections, thermoacoustic instabilities arise from the coupling between acoustic waves and flames and can lead to high amplitude instabilities [29, 74, 75, 156]. In general, these instabilities induce oscillations of all physical quantities (pressure, velocities, temperature, etc.); in the most extreme cases, they can destroy the burner by inducing large-amplitude flame motion (flashback) or unsteady pressure (material fatigue). Because the equivalence ratio oscillates when instabilities are present, there is a general trend for combustors to be more unstable when operating in the lean regime. Also, because of new international constraints, pollutant emissions must be reduced and gas turbine manufacturers need to operate their systems under leaner and leaner conditions. Consequently there is a need to understand combustion instabilities and to be able to predict them at the *design* level [157].

The objective in the following subsections is to provide the reader with the relevant information regarding the description, modelling, and computation of thermo acoustic instabilities. The basic equations are first recalled in Subsection 3.3.1. Among the possible levels of description, two are discussed in more detail in the subsequent subsections, the large-eddy simulation (LES) approach in 3.3.2 and a 3D linear description based on the Helmholtz equation in 3.3.3. Because using appropriate acoustic boundary conditions is critical when analysing combustion instabilities, this issue is discussed in Subsection 3.3.4. Finally, the different tools and approaches discussed are used to study the thermoacoustic behavior of an industrial annular combustor in Subsection 3.3.5.

3.3.1 Basic Equations and Levels of Description

Three types of numerical or semi-analytical methods have been considered so far to predict and describe these instabilities:

1. LESs of all relevant scales of the reacting, turbulent, compressible flow where the instability develops. Many recent studies demonstrated the ability of this method to represent flame dynamics [158–164], as well as the interaction between reaction zone and acoustic waves [165–168]. However, even when simulations confirm that a combustor is unstable, LES calculations do not say why and how to control the instability. Besides, because of its intrinsic nature (full 3D resolution of the unsteady Navier–Stokes equations), LES remains very CPU demanding, even on today's computers (CPU stands for central processing unit).
2. Low-order methods in which the geometry of the combustor is modelled by a network of homogeneous (constant-density) 1D or 2D axisymmetric acoustic elements for which the acoustic problem can be solved analytically [102, 119, 169–172]. Jump relations are used to connect all these elements, enforcing pressure continuity and mass conservation and accounting for the dilatation induced by

an infinitely thin flame, if any. The acoustic quantities in each segment are related to the amplitudes of the forward and backward acoustic waves, which are determined such that all the jump relations and the boundary conditions are satisfied. This can be achieved for only a discrete set of frequencies ω, which are the roots of a dispersion relation in the complex plane. The main advantage of low-order methods is that they allow the representation of a complex system with only a few parameters, thus allowing an extensive use for predesign–optimization–control purposes. However, the geometrical details of the combustor cannot be accounted for, and only the first 'equivalent' longitudinal or orthoradial modes are sought.

3. As an intermediate step between LES and low-order methods, one may consider using a finite-element or finite-volume technique to solve for an equation (or a system of equations) describing the space–time evolution of small-amplitude perturbations. A set of linear transport equations for the perturbations of velocity, temperature, and density can be derived by linearizing the Navier–Stokes equations [86], in which the local unsteady heat release appears as a forcing term. The resulting system of linear partial differential equations for the fluctuating quantities can be solved, for example in the time domain [173]. Depending on the coupling between the flame and acoustics, especially the phase between the pressure and heat release fluctuations, some modes present in the initial field can be amplified and grow exponentially; after a while, the unsteady field is dominated by the most amplified mode, which can then be analyzed [173]. To facilitate the description of time-delayed boundary conditions and also to obtain more information about the damped or less-amplified mode, it is worth solving the set of linear equations in the frequency space, as proposed by [174] for the wave propagation through a complex baseline flow. If applied within the combustion instability framework, this would give rise to an eigenvalue problem, the eigenvalues being related to the (complex-valued) frequencies of the thermoacoustic modes. Combined with LES, this approach proved useful in providing understanding of the structure and nature of the instabilities observed in academic or industrial burners [167, 175–177].

Except when the thermoacoustic analysis relies on LES, viscous contributions are generally neglected together with the mixture inhomogeneities. The latter assumption amounts to considering a gas mixture in which all species share the same molar weight and heat capacity, which is acceptable for typical practical flames. A direct consequence is that the difference in heat capacities $r = c_p - c_v$ is constant even if c_p, c_v, and their ratio γ may depend on temperature.

Under the preceding assumptions, the mass, momentum and entropy equations are, respectively,

$$\frac{D\rho}{Dt} = -\rho \frac{\partial u_\ell}{\partial x_\ell}, \tag{3.66}$$

$$\rho \frac{Du_\ell}{Dt} = -\frac{\partial p}{\partial x_\ell}, \tag{3.67}$$

$$\frac{Ds}{Dt} = r \frac{\dot{\Omega}}{p}, \tag{3.68}$$

where $\dot{\Omega}$ is the heat release per unit volume. Together with the state equation and entropy expression

$$\frac{p}{\rho} = rT, \quad s - s_{st} = \int_{T_{st}}^{T} \frac{c_p(T')}{T'} dT' - r \ln\left(\frac{p}{p_{st}}\right), \tag{3.69}$$

these transport equations describe the spatiotemporal evolutions of all relevant physical flow quantities.

Although thermoacoustic instabilities can lead to high-amplitude fluctuations, it is meaningful to consider the linear regime to analyse the conditions under which these instabilities appear. Equations (3.66)–(3.69) can be linearised by considering a simple case of large-scale small-amplitude fluctuations, denoted by $'$, superimposed on a *zero-Mach-number* mean flow, denoted by an overbar, which depends only on space. The instantaneous pressure, density, temperature, entropy, and velocity fields can then be written as $p = \bar{p} + p'$, $\rho = \bar{\rho} + \rho'$, $T = \bar{T} + T'$, $s = \bar{s} + s'$, and $u_\ell = u'_\ell$, where the quantities $p'/\bar{p}$, $\rho'/\bar{\rho}$, $T'/\bar{T}$, $s'/\bar{s}$, and $\sqrt{u'_\ell u'_\ell}/\bar{c}$ are of the order of ϵ, where $\epsilon \ll 1$ and $\bar{c} = \sqrt{\gamma \bar{p}/\bar{\rho}}$ is the mean speed of sound. Note that the zero-Mach-number assumption implies that $\partial \bar{p}/\partial x_\ell = 0$, from Eq. (3.67), and $\bar{\dot{\Omega}} = 0$, from Eq. (3.68), the latter condition being acceptable because only the fluctuating quantities are of interest in the linear analysis. The same assumption also implies that the approximation $D/Dt \approx \partial/\partial t$ holds for any fluctuating quantity because $\overline{u_\ell} \simeq 0$, and the nonlinear convective terms are always of second order in ϵ. For simplicity, the temporal fluctuations of the heat capacities are often neglected. Injecting the preceding expansions for the instantaneous flow quantities into Eqs (3.66)–(3.69) and keeping only terms of the order of ϵ, one obtains the following set of linear equations for the fluctuating quantities ρ', u'_ℓ, s', and p':

$$\frac{\partial \rho'}{\partial t} + u'_\ell \frac{\partial \bar{\rho}}{\partial x_\ell} + \bar{\rho}\frac{\partial u'_\ell}{\partial x_\ell} = 0, \tag{3.70}$$

$$\bar{\rho}\frac{\partial u'_\ell}{\partial t} + \frac{\partial p'}{\partial x_\ell} = 0, \tag{3.71}$$

$$\frac{\partial s'}{\partial t} + u'_\ell \frac{\partial \bar{s}}{\partial x_\ell} = r\frac{\dot{\Omega}'}{p_0}. \tag{3.72}$$

The linearised state equation and entropy expression are

$$\frac{p'}{\bar{p}} - \frac{\rho'}{\bar{\rho}} - \frac{T'}{\bar{T}} = 0, \quad s' = c_p\frac{T'}{\bar{T}} - r\frac{p'}{\bar{p}}. \tag{3.73}$$

To close the set of Eqs. (3.70)–(3.73), a model must be used to express the unsteady heat release $\dot{\Omega}'$ in terms of the other fluctuating quantities.

FLAME RESPONSE. Modelling the unsteady behaviour of the flame is the most challenging part in the description of thermoacoustic instabilities [146]. Several models were proposed in the past to describe the response of conic or V-shaped laminar flames [178], accounting for nonlinear saturation effects [89, 179] and equivalence ratio fluctuations [180, 181]. Most models describe the global (integrated over space) heat released in the whole flame zone. For premixed flames, the most natural way to proceed is to relate this global quantity to the acoustic velocity in the cold-gas region upstream of the flame region. The idea behind this approach is that the heat release is mainly controlled by the fresh gas flow rate if the flame speed is specified. The most classical model follows seminal ideas by Crocco [182, 183] and is referred to as the $n - \tau$ model. This 1D formulation stipulates that the global heat release at time t is proportional to a time-lagged version of the acoustic velocity at a reference upstream position $\mathbf{x}_{\text{ref}}$, usually taken at the burner mouth:

$$\dot{\Omega}'_{\text{tot}} = \int_V \dot{\Omega}'(t) \, d\mathbf{x} = S_{\text{ref}} \frac{\gamma \overline{p}}{\gamma - 1} \, n \, u'(\mathbf{x}_{\text{ref}}, t - \tau). \tag{3.74}$$

In this expression, the left-hand-side term is the heat release fluctuations integrated over the flow domain V, S_{ref} is the cross-section area of the burner mouth, u' denotes the fluctuating velocity component in the direction x of the main flow that feeds the flame, the interaction index n controls the amplitude of the flame response to acoustic perturbations, and τ is the time delay between the acoustic perturbation and the response of the flame. This latter parameter controls the phase between the acoustic pressure and the unsteady heat release in the flame zone, and thus the value of the Rayleigh index is

$$\mathcal{R} = \int_t \int_V p' \, \dot{\Omega}' \, d\mathbf{x} \, dt. \tag{3.75}$$

According to the classical Rayleigh criterion, flame–acoustics coupling promotes the appearance of instabilities if $\mathcal{R} > 0$, showing the importance of the parameter τ in the description and prediction of thermoacoustic instabilities.

Models for the global response of the flame are justified only for acoustically compact flames, in which the typical length of the flame region L_f is small compared with the characteristic acoustic wavelength L_a. This condition is not always met. It is then natural to use a local flame model that relates the local unsteady heat release to a reference acoustic velocity in the injector mouth. The natural way to proceed is then to write

$$\frac{\dot{\Omega}'(\mathbf{x}, t)}{\dot{\Omega}_{\text{tot}}} = n_u(\mathbf{x}) \frac{u'_\ell[\mathbf{x}_{\text{ref}}, t - \tau_u(\mathbf{x})] \, n_{\text{ref},\ell}}{U_{\text{bulk}}}, \tag{3.76}$$

where $n_u(\mathbf{x})$ and $\tau_u(\mathbf{x})$ are fields of interaction index and time lag and $n_{\text{ref},\ell}$ are the components of a fixed unitary vector defining the direction of the reference velocity. The scaling by the total heat release $\dot{\Omega}_{\text{tot}}$ and the bulk velocity U_{bulk} have been used to make sure that $n_u(\mathbf{x})$ has no dimension. Obviously this modelling approach allows more degrees of freedom than any global model to represent the actual response of a typical industrial flame (two fields of parameters instead of two real numbers). However, a large amount of pointwise data is required for tuning such models, and for obvious technological reasons these data can hardly be obtained experimentally. As discussed in Subsection 3.3.2, the alternative is then to use compressible reacting LES to investigate the response of a turbulent flame submitted to acoustic perturbations.

Using the local flame model given in Eq. (3.76), the transport equation for s', Eq. (3.72), can be rewritten as

$$\frac{\partial s'}{\partial t} + u'_\ell \frac{\partial \overline{s}}{\partial x_\ell} = \frac{r}{\overline{p}} \frac{\dot{\Omega}_{\text{tot}}}{U_{\text{bulk}}} n_u(\mathbf{x}) \, u'_\ell[\mathbf{x}_{\text{ref}}, t - \tau_u(\mathbf{x})] \, n_{\text{ref},\ell}, \tag{3.77}$$

and the set of Eqs. (3.70)–(3.77) can be solved to determine the thermoacoustic properties of the system.

3.3.2 LES of Compressible Reacting Flows

LES [184, 185] is nowadays recognized as an intermediate approach to the more classical Reynolds-averaged Navier–Stokes (RANS) methodologies [185, 186]. Although conceptually very different, these two approaches aim at providing new systems of governing equations to mimic the characteristics of turbulent flows. Recent studies using LES have shown the potential of this approach for reacting flows (see reviews in [164] or [169]). LES is able to predict mixing [168, 187–190], stable flame behaviour [191–194], and flame–acoustic interaction [168, 195–197]. It is also used for flame transfer function evaluation [176, 198, 199] needed for Helmholtz solvers (see subsection 3.3.3). Although LES seems very promising for industrial applications, it remains computationally too intensive to integrate in the design cycle of the next generation of gas turbines. For example a typical single-sector LES computation, as subsequently presented, usually costs of the order of 50, 000 CPU hours. Helmholtz solvers, on the other hand, offer great flexibility and allow the prediction of combustion instabilities when new combustion chambers are designed. The compuntional cost with this approach and for the complete combustion chamber is more of the order of 200 CPU hours. It is also important to note that, although most academic setups used to study combustion instabilities [157, 169, 200, 201] are limited to single burners and are subjected mainly to longitudinal acoustic modes, real gas turbines exhibit mostly azimuthal modes [170, 202, 203] because of the annular shape of their chambers [157].

The governing equations for RANS and LES are respectively obtained by ensemble averaging [185, 186] and filtering the set of compressible Navier–Stokes equations. These operations yield unclosed terms that are to be modelled. In RANS simulations, the unclosed terms are representative of the physics taking place over the entire range of frequencies present in the ensemble of realizations used for averaging. In LES, the operator is a spatially localized time-independent filter of given size Δ, to be applied to a single realization of the studied flow. Resulting from this 'spatial average' is a separation between the large (greater than the filter size) and small (smaller than the filter size) scales. The unclosed terms are representative of the physics associated with the small structures (with high frequencies) present in the flow. Figure 3.36 illustrates the conceptual differences between RANS [Fig. 3.36(a)] and LES [Fig. 3.36(b)] when applied to a homogeneous isotropic turbulent field.

Because of the filtering approach, LES allows a dynamic representation of the large-scale motions whose contributions are critical in complex geometries. The LES predictions of complex turbulent flows are henceforth closer to the physics because large-scale phenomena such as large vortex shedding and acoustic waves are embedded in the set of governing equations [169].

For the reasons just presented, LES has a clear potential in predicting turbulent flows encountered in industrial applications, especially in the context of thermo-acoustic instabilities. In particular and in conjunction with Helmholtz solvers, LES can provide the estimation and validation of the model used to represent the thermo-acoustic coupling: i.e., the FTF.

THE LES SUBGRID SCALE (SGS) MODELS. LES for reacting flows involves the spatial Favre filtering operation that reduces for spatially, temporally invariant and localised

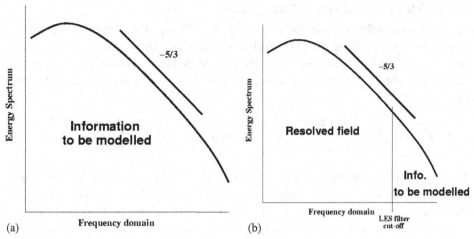

Figure 3.36. Conceptual representation of (a) RANS and (b) LES applied to a homogeneous isotropic turbulent field.

filter functions [204, 205] to

$$\widetilde{f(\mathbf{x}, t)} = \frac{1}{\overline{\rho(\mathbf{x}, t)}} \int_{-\infty}^{+\infty} \rho(\mathbf{x}', t) f(\mathbf{x}', t) G(\mathbf{x}' - \mathbf{x}) \, d\mathbf{x}', \tag{3.78}$$

where G denotes the filter function.

In the mathematical description of compressible turbulent flows with chemical reactions and species transport, the primary variables are the species densities $\rho_i(\mathbf{x}, t)$, the velocity vector $u_\ell(\mathbf{x}, t)$, the total energy $E(\mathbf{x}, t) \equiv e_s + 1/2\, u_\ell u_\ell$ and the fluid density $\rho(\mathbf{x}, t) = \sum_{i=1}^{N} \rho_i(\mathbf{x}, t)$.

The application of the filtering operation to the instantaneous set of compressible Navier–Stokes transport equations with chemical reactions yields the LES transport equations [169], which contain the so-called SGS quantities that need modelling [184, 206].

The SGS Velocity Stress Tensor

The unresolved SGS stress tensor $\overline{\tau_{ij}}^t$ is modelled with the Boussinesq assumption [185, 186, 207]:

$$\overline{\tau_{ij}}^t - \frac{1}{3} \overline{\tau_{\ell\ell}}^t \, \delta_{ij} = -2 \, \overline{\rho} \, \nu_t \, \widetilde{S}_{ij}, \quad \text{with } \widetilde{S}_{ij} = \frac{1}{2} \left(\frac{\partial \widetilde{u}_i}{\partial x_j} + \frac{\partial \widetilde{u}_j}{\partial x_i} \right) - \frac{1}{3} \frac{\partial \widetilde{u}_\ell}{\partial x_\ell} \, \delta_{ij}. \tag{3.79}$$

In Eq. (3.79), $\widetilde{S}_{ij}$ is the resolved strain-rate tensor and ν_t is the SGS turbulent viscosity. Most SGS turbulent viscosity models [207–210] take the generic form

$$\nu_{\text{SGS}} = C_m \Delta^2 \overline{\text{OP}}(\mathbf{x}, t), \tag{3.80}$$

where C_m is the constant of the model, Δ is the subgrid characteristic length scale (in practice the size of the mesh), and $\overline{\text{OP}}$ is an operator of space and time, homogeneous to a frequency, and defined from the resolved fields.

The SGS Species and Energy Flux Models

The SGS species flux $\overline{J_\ell^i}^t$ and the SGS energy flux $\overline{q_\ell}^t$ are, in most cases, respectively modelled by use of the species SGS turbulent diffusivity $D_t^i = \nu_t/Sc_t^i$, where Sc_t^i is the turbulent Schmidt number ($= 0.7$ for all i). The SGS thermal conductivity for energy flux is also obtained from ν_t by $\lambda_t = \overline{\rho}\,\nu_t\,C_p/Pr_t$, where Pr_t is a turbulent Prandtl number ($= 0.7$):

$$\overline{J_\ell^i}^t = -\overline{\rho}\left(D_t^i\,\frac{W_i}{W}\,\frac{\partial \widetilde{X}_i}{\partial x_\ell} - \widetilde{Y}_i\,V_\ell^c\right), \text{ with } \overline{q_\ell}^t = -\lambda_t\,\frac{\partial \widetilde{T}}{\partial x_\ell} + \sum_{i=1}^{N} \overline{J_\ell^i}^t\,\widetilde{h}_s^i. \qquad (3.81)$$

In Eq. (3.81), the mixture molecular weight W and the species molecular weight W_i can be combined with the species mass fraction to yield the expression for the molar fraction of species i: $X_i = Y_i W/W_i$. V_ℓ^c is the diffusion correction velocity resulting from the Hirschfelder and Curtiss approximation [169] and $\widetilde{T}$ is the Favre filtered temperature that satisfies the modified filtered state equation $\overline{p} = \overline{\rho}\,r\,\widetilde{T}$ [211, 212]. Finally, $\widetilde{h}_s^i$ stands for the enthalpy of species i. Note that the performances of the closures could be improved by use of the dynamic formulations described in [208, 211, 213–215].

The Dynamic Thickened-Flame (DTF) Model

The LESs of turbulent reacting flows imply the modelling of SGS combustion terms. One model employed in the context of thermoacoustic instabilites is the thickened-flame (TF) model [161]. Following the theory of laminar premixed flames [88], the flame speed S_L^0 and the flame thickness δ_L^0 may be expressed as,

$$S_L^0 \propto \sqrt{\hat{\alpha}\,A}, \qquad \delta_L^0 \propto \frac{\hat{\alpha}}{S_L^0} = \sqrt{\frac{\hat{\alpha}}{A}}, \qquad (3.82)$$

where $\hat{\alpha}$ is the thermal diffusivity and A is the pre-exponential constant of the reaction rate. Increasing the thermal diffusivity by a factor F, the flame speed is kept unchanged if the pre-exponential factor is decreased by the same factor [216]. This increases the flame thickness by factor F, which is easily resolvable on a coarser mesh. However, additional information needs to be supplied to reproduce the effect of the turbulence–chemistry interaction at the SGSs [165, 217, 218]. This is the intent of the so-called efficiency function $\mathcal{E}$ [161]. When thickening is applied everywhere in the flow, the model is limited to fully premixed combustion. If mixing is present, then thickening will strongly interfere with the physics by increasing the diffusion artificially everywhere in the flow. Thus diffusion flames are inappropriate configurations for such an approach. To compute partially premixed or non-premixed flames [169], a modified version of the TF model is used [168, 175, 218, 219]: the DTF model.

With the DTF model, the SGS fluxes are modified to become

$$\overline{J_\ell^i}^t = -(1-S)\,\overline{\rho}D_t^i\,\frac{W_i}{W}\,\frac{\partial \widetilde{X}_i}{\partial x_\ell} + \overline{\rho}\,\widetilde{Y}_i\,V_\ell^c, \text{ with } \overline{q_\ell}^t = -(1-S)\,\lambda_t\,\frac{\partial \widetilde{T}}{\partial x_\ell} + \sum_{i=1}^{N} \overline{J_\ell^i}^t\,\widetilde{h}_s^i,$$

$$(3.83)$$

where S is a sensor that detects reaction zones, i.e., derived from an Arrhenius type of law for example. The local thickening factor depends on the local mesh size:

Typically thickening must ensure that enough points are present in the flame zone. The thickening factor F is given by

$$F = 1 + (F_{\max} - 1)\, S, \quad F_{\max} = \frac{N_c}{\Delta_x}\delta_L^0, \tag{3.84}$$

where N_c is the number of grid points, typically 5–10, used to resolve the flame front and Δ_x is the local mesh spacing.

Although this approach is still being developed and further validations are needed, its ease of implementation and its success in prior applications [168, 175, 177, 219, 220] suggest its suitability for the problems of thermoacoustic instabilities such as those presented in this chapter.

NUMERICAL ISSUES IN LES SOLVERS. In many cases, turbulence results from the development, amplification, and saturation of unstable hydrodynamic modes of the main flow. Any numerical method used to compute such a flow must therefore be able to represent the growth of these modes: i.e., it must not be too dissipative. In high-Reynolds-number flows, the scale separation can be large [the integral to Kolmogorov length scale ratio is $\mathrm{Re}^{3/4}$, with $\mathrm{Re} = k^2/(\nu\varepsilon)$, where k is the turbulent kinetic energy, ν is the kinematic viscosity, and ε is the rate of turbulent kinetic energy dissipation], and it is worth minimising the number of grid points necessary to represent the smallest scales. Finally, the effective dissipation at the Kolmogorov scale must not be overestimated if the actual flow Reynolds number is to be accounted for [with $\mathrm{Re} = k^2/(\nu\varepsilon)$, any extra dissipation decreases Re]. Numerics, especially in the context of LES, faces an important constraint, i.e., numerical dissipation must be as small as possible for all the length scales present in the flow. Such a constraint is the reason why spectral methods [221] were considered until the early 1990s as the only appropriate methods for performing DNS or LES of turbulent flows. However, in the case of complex geometries or boundary conditions, spectral methods cannot be used and the simulations must be based on finite-volume, finite-element, or finite-difference methods. The three methods can be used for unsteady simulations as long as appropriate spatial and temporal time-stepping procedures are used.

As mentioned earlier, the numerical error must be controlled and minimised for all length scales present in the unsteady flow to be computed. This means that the accuracy of a numerical scheme cannot simply be reduced to its order of accuracy. As far as unsteady flow computations are concerned, it is necessary to perform a wavelength-based numerical analysis in which one considers a harmonic perturbation and compares how the discrete and the exact derivatives operate on this perturbation [222]. The effective-to-exact-wave-number ratio, κ'/κ, can be used to quantify the errors related to the numerical scheme. Considering the simple linear convection equation, this ratio can be interpreted as an error in the speed of propagation of a perturbation of wavelength κ. In a general case, κ'/κ can be written in the form $\kappa'/\kappa = E(\kappa\Delta x) = E_r(\kappa\Delta x) + i\, E_i(\kappa\Delta x)$, and the effective equation solved numerically reads

$$\frac{\partial f}{\partial t} + u_0\, E(\kappa\Delta x)\,\frac{\partial f}{\partial x} = 0. \tag{3.85}$$

Assuming that the initial condition is $f(x, t = 0) = \exp(i\kappa)$, the exact solution of Eq. (3.85), or equivalently the solution for the linear convection equation provided by the numerical scheme (with perfect time advancement), is simply,

$$f(x, t > 0) = \exp\left[\kappa \, E_i(\kappa \Delta x) \, u_0 \, t\right] \, \exp\left\{i\kappa \left[x - E_r(\kappa \Delta x) \, u_0 \, t\right]\right\}. \tag{3.86}$$

When $\kappa \Delta x$ tends to zero or the number of grid points per wavelength tends to infinity, $\kappa'/\kappa = E(\kappa \Delta x)$ tends to unity and the exact solution is recovered, i.e., $\exp\left[i\,\kappa\,(x - u_0\,t)\right]$. When the imaginary part of κ'/κ is not zero, $E_i(\kappa \Delta x) \neq 0$, the amplitude of the harmonic perturbation is not conserved; it is damped if $E_i(\kappa \Delta x) < 0$ and unbounded if $E_i(\kappa \Delta x) > 0$. The effective-to-exact-wavelength ratios of some classical finite-difference schemes are reported in Table. 3.2 and plotted in Fig. 3.37.

Centered finite-difference schemes have real-valued κ'/κ ratios and thus are nondissipative, whatever their order is. This property is not shared by the biased schemes, which all introduce dissipation. Note also that a property shared by all finite-difference schemes is that they cannot propagate wiggles accurately $[E(\pi) = 0]$.

The same analysis can be performed for second-derivative schemes for which κ'^2/κ^2 ratios are of interest. For LES, the effective energy dissipation issued by second-order derivatives at the small scales cannot be close to zero, whatever the SGS model is. Indeed, the energy flux from the largest to the smallest scales should be balanced to avoid an energy accumulation at the smallest resolved scales [223].

A key issue when a LES of turbulent flows is performed is the necessity of using virtually non-dissipative schemes to handle flow fields that contain a lot of energy at large wave numbers. Because numerical errors are large for the smallest scales, the risk for such a computation to run unstable is high. As far as incompressible Navier–Stokes equations are concerned, experience shows that the kinetic energy must be conserved if a stable and dissipation-free numerical method is sought. Indeed, such a property ensures that the sum of the square of the velocities cannot grow, even though nonlinear interactions between modes exist: A numerical scheme that conserves kinetic energy cannot lead to an unbounded growth of oscillations [224–227]. It should be noted that the concept of kinetic energy conservation is valid only for incompressible or low-Mach-number flows. Because the extension of this principle to compressible situations is quite difficult [228], one usually relies on numerical stabilization by means of artificial viscosity to stabilize the LES of such flows [229, 230].

To conclude on LES and the difficulties of the approach, modelling is clearly needed for the problem to be solved adequately from a purely mathematical and physical point of view. Numerics is, however, crucial and is also to be adequately addressed so as to properly qualify the modelling strategy at hand. Fundamental issues such as error propagation and cancellation are still not clear in LES codes. Parallelisation and associated alogorithms are another difficulty that emphasize the need for careful validation and developments of such tools.

FTF BASED ON LES – EVALUATION AND VALIDATION To predict and avoid combustion instabilities [169, 200, 201, 231], a well-known method is the identification of the combustion chamber response or FTF to acoustic waves introduced into the combustor by loudspeakers or rotating valves. This identification of the FTF is usually performed either numerically or experimentally. Two methods may be found in the literature to analyse this response: the purely acoustic (PA) approach [232–236] and

Table 3.2. *Classical finite-difference (FD) formula for the spatial first derivative and associated error.*

FD	Name	E_r	E_i
$[f_i - f_{i-1}]/\Delta x$	1st order upwind	$[\sin(\kappa\Delta x)]/\kappa\Delta x$	$[\cos(\kappa\Delta x) - 1]/\kappa\Delta x$
$[f_{i+1} - f_{i-1}]/2\Delta x$	2nd order centred	$[\sin(\kappa\Delta x)]/\kappa\Delta x$	0
$[3f_i - 4f_{i-1} + f_{i-2}]/2\Delta x$	2nd order upwind	$[\sin(\kappa\Delta x)][2 - \cos(\kappa\Delta x)]/\kappa\Delta x$	$-[\cos(2\kappa\Delta x) - 4\cos(\kappa\Delta x) + 3]/\kappa\Delta x$
$[-f_{i+2} + 8f_{i+1} - 8f_{i-1} + f_{i-2}]/12\Delta x$	4th order centred	$[\sin(\kappa\Delta x)][4 - \cos(\kappa\Delta x)]/3\kappa\Delta x$	0

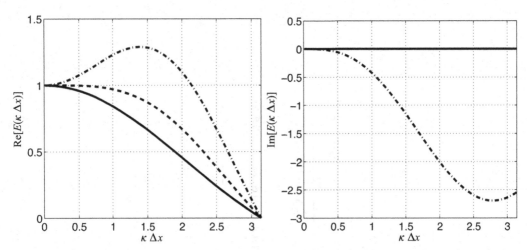

Figure 3.37. Effective-to-exact-wavelength ratios for the schemes displayed in Table. 3.2. The imaginary part (right plot) is zero for centered schemes. The first-order upwind and the second-order centred schemes share the same real part (left plot). These graphs can be interpreted as the effective-to-exact convection-velocity ratio, or as the effective-to-exact first-derivative ratio. Second-order centred, solid line; fourth-order centred, dashed line; second-order upwind, dot–dashed line.

the $n - \tau$ approach [169, 182, 183, 237–239]. In the PA approach, the burner is considered a 'black box' and a two-ports formulation (based on acoustic pressure and velocity perturbations) is used to construct a transfer matrix linking acoustic fluctuations on both sides of the burner. In the $n - \tau$ approach, pressure measurements are replaced with a global heat release measurement (usually based on optical methods). The heat release fluctuations are then related to the flow velocity modulations at the combustor inlet. Both PA and $n - \tau$ methods can be used experimentally or numerically. Numerical experiments comparing FTF results show that the $n - \tau$ approach often leads to an ill-defined problem in which the measured transfer function depends on acoustic impedances upstream and downstream of the combustor.

The essential drawback of the original $n - \tau$ model comes from the fact that it tries to correlate heat release perturbations to velocity perturbations only. With the FTF model presented in [240] (called 'extended $n - \tau$'), a consistent formulation can be used for any location of the reference point by introducing the effects of pressure perturbation on heat release: The FTF model is formulated with the local unsteady pressure and velocity measured upstream of the flame:

$$\frac{\dot{\Omega}'}{\dot{\Omega}_{\text{tot}}} = A_u \frac{u'(\mathbf{x}_a, t - \tau_u)}{\bar{\tilde{c}}} + A_p \frac{p'(\mathbf{x}_a, t - \tau_p)}{\bar{p}}, \qquad (3.87)$$

where the unsteady velocity, pressure, and heat release are scaled respectively by the sound speed $\bar{\tilde{c}}$, the mean pressure $\bar{p}$, and the mean integrated heat release $\dot{\Omega}_{\text{tot}} = \int_V \dot{\Omega}(\mathbf{x}) \, d\mathbf{x}$.[4] Equation (3.87) contains four unknowns, A_u, A_p, τ_u, and τ_p, which depend on the point where velocity and pressure fluctuations are measured.

---

[4] Assuming that all the fuel is burnt, $\dot{\Omega}_{\text{tot}}$ may be estimated with the fuel mass flow rate $\dot{m}_F$ and the heat of reaction Q:

$$\dot{\Omega}_{\text{tot}} = Q \, \dot{m}_F. \qquad (3.88)$$

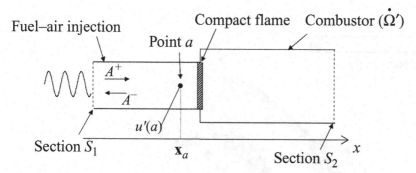

Figure 3.38. Schematic view of the laminar premixed flame configuration used to validate the PA and extended $n - \tau$ models.

These parameters may be determined when a new state is added to have exactly enough equationsas in the PA method.

A simple laminar premixed flame stabilized in a duct is shown in Fig. 3.38; it is possible to demonstrate that the extended n–τ model is fully compatible with the PA approach [240], as illustrated in Fig. 3.39, which compares two matrices constructed from the PA (lines with circles) and extended $n - \tau$ (solid line) models for the flame configuration shown in Fig. 3.38.

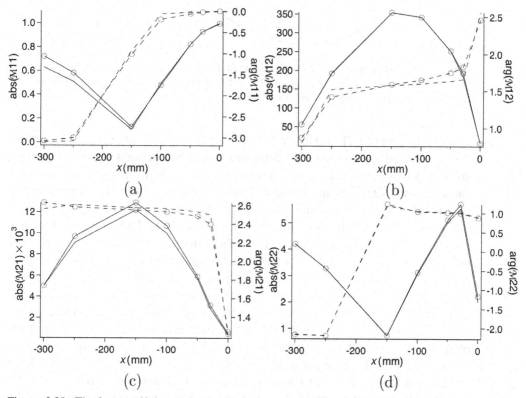

Figure 3.39. The four coefficients of the transfer matrix $\mathcal{M}$ of a laminar burner for different positions of the reference point $\mathbf{x_a}$. Solid line, absolute value; dashed line, phase (rad); line and symbols, PA; line, extended $n - \tau$ [240].

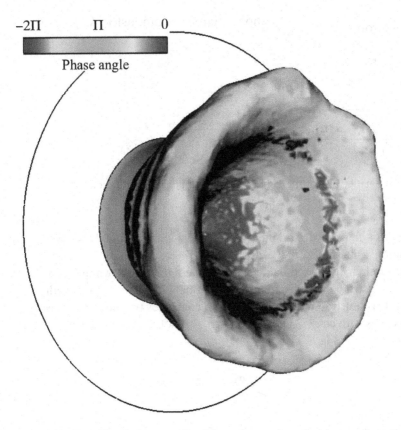

Figure 3.40. Forced turbulent premixed flame in a real gas turbine configuration: isosurface of heat release coloured by the local value of the delay as an estimate with LES and the classical $n - \tau$ model [198]. (See colour plate.)

The concept of FTF and its modeling being clearly defined in the context of the PA or extended $n - \tau$ formalisms [240], two options are available for its estimation: laboratory measurements or numerical simulations. In the former, turbulent closure and combustion models are introduced and validation of the FTF is needed. For real turbulent flow configurations, two approaches are conceptually able to produce estimates of the FTF: RANS (or more suitably unsteady RANS) and LES. A gas turbine configuration with turbulent flow is shown in Fig. 3.40, and Fig. 3.41 compares numerical predictions [198] and experimental measurements of the FTF by use of the original $n - \tau$ model [169, 241] in the configuration of Fig. 3.40. Experimental results are given by the phase comparison between the signal given by a photomultiplier (with an OH* frequency filter) that is directly linked to heat release, and a hot-wire-probe velocity signal at the inlet of the combustion chamber. Values of the LES phase $\phi_\omega = -\tau_\omega * \omega$ (where $\omega = 2\pi f$ is the the angular frequency, f is the frequency, and τ_ω is the time delay) are in good agreement with experimental values (Fig. 3.41) and confirm the potential of LES when compared with that of RANS.

The differences between RANS and LES show that the heat release fluctuations are not solely linked to the convection of an initial perturbation but can result from a complex interaction between acoustics and the flow in the chamber. That observation

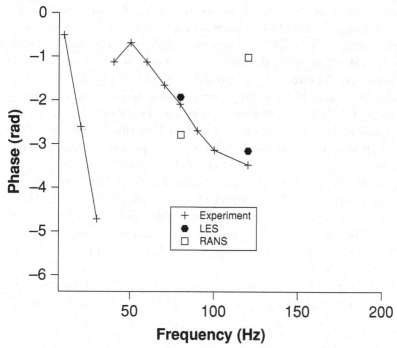

Figure 3.41. Comparison of LES, RANS, and experiment phases [198].

supports the need for a fully spatially and temporally dependent numerical approach for a proper numerical estimation of the FTF. As of today, only LES seems able to provide such a comprehensive framework, as illustrated in real applications.

3.3.3 3D Helmholtz Solver

Despite the potentials of LES, simpler and faster methods are often needed to design stable combustors. Solving the wave equation in reacting flow and identifying linearly unstable modes is such a technique and is found in Helmholtz solvers. These tools offer great flexibility and permit the prediction of combustion instabilities beforehand while the new combustion chambers are designed. As noted earlier, although most academic setups used to study combustion instabilities [157, 169, 200, 201] are limited to single burners and are subjected mainly to longitudinal acoustic modes, real gas turbines exhibit mostly azimuthal modes [170, 202, 203] because of the annular shape of their chambers [157]. LES is then very difficult and expensive to use whereas the Helmholtz solver can be applied economically. This section presents the required wave equation and its solution methodology using linear algebra techniques.

WAVE EQUATION. Taking the time derivative of Eq. (3.70), adding the divergence of Eq. (3.71), and using Eqs. (3.72) and (3.73) to eliminate ρ' yields the following wave equation for p':

$$\frac{\partial}{\partial x_\ell}\left(\frac{1}{\overline{\rho}}\frac{\partial p'}{\partial x_\ell}\right) - \frac{1}{\gamma\overline{p}}\frac{\partial^2 p'}{\partial t^2} = -\frac{\gamma-1}{\gamma\overline{p}}\frac{\partial \dot{\Omega}'}{\partial t} \tag{3.89}$$

when the Mach number of the mean flow is zero (i.e., $\overline{u}_\ell/\overline{c} = 0$). An order-of-magnitude analysis suggests that this assumption ($\overline{u_\ell} \simeq 0$) is valid when the characteristic Mach number $M = \sqrt{\overline{u_\ell u_\ell}}/\overline{c}$ of the mean flow is small compared with L_f/L_a, where L_f is the flame-zone thickness and L_a is the typical acoustic wavelength [242]. However, the effect of the approximation $M \simeq 0$ on the shape, frequency of oscillation, and stability of the thermoacoustic modes is far from being well understood [243–247]. Recent studies suggest that the validity domain of the zero-mean-flow assumption might be rather small [248]. Nevertheless, this somewhat restrictive assumption is necessary for deriving a wave equation for the thermoacoustic perturbations. This situation is different for classical aeroacoustics, in which combustion is not present and in which a wave equation for the perturbation potential can be derived if the baseline flow is assumed homentropic and irrotational [249]. Because assuming the mean flow to be homentropic is not realistic when dealing with combustion instabilities, assuming that the mean flow is at rest is the most convenient way to simplify the formalism, the alternative being to deal with the complete set of linearised Euler equations [174, 248].

Because Eq. (3.89) is linear, it is natural to introduce harmonic variations at frequency $f = \omega/(2\pi)$ for pressure, velocity, and local heat release perturbations:

$$p' = \Re\left[\hat{p}(\mathbf{x})\exp(-i\,\omega t)\right],$$

$$u'_\ell = \Re\left[\hat{u}_\ell(\mathbf{x})\exp(-i\,\omega t)\right],$$

$$\hat{\Omega}' = \Re[\hat{\Omega}(\mathbf{x})\exp(-i\,\omega t)]. \tag{3.90}$$

Introducing Eqs. (3.90) into Eq. (3.89) leads to the following Helmholtz equation:

$$\frac{\partial}{\partial x_\ell}\left(\frac{1}{\overline{\rho}}\frac{\partial \hat{p}}{\partial x_\ell}\right) + \frac{\omega^2}{\gamma\overline{p}}\hat{p} = i\,\omega\,\frac{\gamma-1}{\gamma p_0}\,\hat{\Omega}(\mathbf{x}), \tag{3.91}$$

where $\overline{\rho}$ and γ depend on the space variable $\mathbf{x}$ and the unknown quantities are the complex amplitude $\hat{p}(\mathbf{x})$ of the pressure oscillation at frequency f and angular frequency ω. In the frequency space, the zero-Mach-number assumption leads to $i\,\omega\,\hat{u}_\ell = (\partial\hat{p}/\partial x_\ell)/\overline{\rho}$ and the flame model, Eq. (3.76), translates into

$$\hat{\Omega}(\mathbf{x}) = \frac{\dot{\Omega}_{\text{tot}}}{i\,\omega\,\overline{\rho}(\mathbf{x}_{\text{ref}})\,U_{\text{bulk}}}\,n_u(\mathbf{x})\,e^{i\,\omega\,\tau_u(\mathbf{x})}\,\frac{\partial\hat{p}(\mathbf{x}_{\text{ref}})}{\partial x_k}n_{\text{ref},k}. \tag{3.92}$$

Introducing Eq. (3.92) into Eq. (3.91) leads to

$$\frac{\partial}{\partial x_\ell}\left(\frac{1}{\overline{\rho}}\frac{\partial\hat{p}}{\partial x_\ell}\right) + \frac{\omega^2}{\gamma\overline{p}}\hat{p} = \frac{\gamma-1}{\gamma\overline{p}}\frac{\dot{\Omega}_{\text{tot}}}{\overline{\rho}(\mathbf{x}_{\text{ref}})U_{\text{bulk}}}n_u(\mathbf{x})e^{i\,\omega\,\tau_u(\mathbf{x})}\frac{\partial\hat{p}(\mathbf{x}_{\text{ref}})}{\partial x_k}n_{\text{ref},k}. \tag{3.93}$$

The applications discussed in [175, 176, 250] as well as in Subsection 3.3.5 are based on the solution of Eq. (3.93) but the methodologies developed can be applied to a more general case in which the complex amplitude of the heat release is given by [242]

$$\hat{\Omega}(\mathbf{x}) = \hat{\mathcal{L}}_u\left[\frac{\partial\hat{p}(\mathbf{x})}{\partial x_\ell}\right] + \hat{\mathcal{L}}_p\left[\hat{p}(\mathbf{x})\right], \tag{3.94}$$

where $\hat{\mathcal{L}}_p$ and $\hat{\mathcal{L}}_u$ are two linear operators acting on $\hat{p}$ and its gradient, respectively. Of course the general formulation of Eq. (3.94) has the potential to include more physical effects than the local $n - \tau$ model described by Eqs. (3.76) and (3.92). Notably, it allows relating the unsteady heat release to the complete acoustic field at the reference position $\mathbf{x}_{\text{ref}}$ instead of the velocity field only, consistent with the matrix identification approach for flame modelling [251, 252] [see also Eq. (3.87)]. Although the effects of the acoustic pressure are often neglected in flame-transfer formulations, relating the unsteady heat release to the complete acoustic field (velocity and pressure) is highly desirable for cases in which the flame is not compact or when its distance with the injector mouth is not small compared with the acoustic wavelength [240].

BOUNDARY CONDITIONS. Denoting by $\mathbf{n}_{\text{BC}} = (n_{\text{BC},1}, n_{\text{BC},2}, n_{\text{BC},3})$ the outward unit normal vector to the boundary $\partial\Omega$ of the flow domain, three types of boundary conditions are usually used for acoustics:

- Zero pressure: This corresponds to fully reflecting outlets where the outside pressure is imposed strongly at the flow boundary, zeroing the pressure fluctuations:

$$\hat{p} = 0 \quad \text{on} \quad \text{boundary} \quad \partial\Omega_D, \tag{3.95}$$

where the subscript D in $\partial\Omega_D$ refers to the subset of $\partial\Omega$ where this Dirichlet boundary condition holds.

- Zero normal velocity, viz., $\hat{u}_k\, n_{\text{BC},k} = 0$: This corresponds to fully rigid walls or reflecting inlets where the velocity of the incoming flow is imposed, zeroing the velocity fluctuations. Under the zero-Mach-number assumption, Eq. (3.71) can be used to rewrite this condition as a Neumann condition for the acoustic pressure:

$$\frac{\partial\hat{p}}{\partial x_k}n_{\text{BC},k} = 0 \quad \text{on} \quad \text{boundary} \quad \partial\Omega_N, \tag{3.96}$$

where the subscript N in $\partial\Omega_N$ refers to the subset of $\partial\Omega$ where this Neumann boundary condition holds.

- Imposed reduced complex impedance $Z = \hat{p}/\overline{\rho c}\,\hat{u}_k\, n_{\text{BC},k}$. Under the zero-Mach-number assumption, this condition can be rewritten as a linear relationship between the acoustic pressure and its gradient in the normal direction to the boundary:

$$\overline{c}Z\frac{\partial\hat{p}}{\partial x_\ell}n_{\text{BC},\ell} - i\,\omega\,\hat{p} = 0 \quad \text{on} \quad \text{boundary} \quad \partial\Omega_Z, \tag{3.97}$$

where the subscript Z in $\partial\Omega_Z$ refers to the subset of $\partial\Omega$ where the reduced impedance is imposed.

Associated with the homogeneous boundary conditions (3.95), (3.96), and (3.97) on $\partial\Omega = \partial\Omega_D \bigcup \partial\Omega_N \bigcup \partial\Omega_Z$, Eq. (3.93) defines a nonlinear eigenvalue problem whose solutions provide the shape, frequency, and growing–damping rate of the relevant thermoacoustic modes.

Assuming that the sound speed $\overline{c}$ and the density $\overline{\rho}$ distributions over space are known, Eq. (3.93) can be solved with a Galerkin finite-element method to transform this equation into a nonlinear eigenvalue problem of size N (the number of nodes in the finite-element grid used to discretize the geometry, except those nodes belonging to $\partial\Omega_D$, where $\hat{p} = 0$ is known) of the form

$$[A][P] + \omega[B(\omega)][P] + \omega^2[C][P] = [D(\omega)][P], \qquad (3.98)$$

where $[P]$ is the column vector containing the nodal values of the eigenmode at frequency ω and $[A]$ and $[C]$ are square matrices depending on only the discretized geometry of the combustor and mean flow fields $\bar{c}$ and $\bar{p}$. Matrix $[B]$ contains information related to the boundary conditions and thus depends on ω because in general Z is frequency dependent. Matrix $[D]$ contains the unsteady contribution of the flame, i.e., $\dot{\Omega}'$, and usually depends nonlinearly on the mode frequency ω; see Eq. (3.92). Thus Eq. (3.98) defines a nonlinear eigenvalue problem that must be solved iteratively, the kth iteration consisting of solving the quadratic eigenvalue problem in ω_k defined as

$$\{[A] - [D(\omega_{k-1})]\}\,[P] + \omega_k[B(\omega_{k-1})][P] + \omega_k^2[C][P] = 0. \qquad (3.99)$$

A natural initialization is to set $[D](\omega_0) = 0$ so that the computation of the modes without acoustic–flame coupling is in fact the first step of the iteration loop. Usually only a few (typically fewer than five) iterations are enough to converge towards the complex frequency and associated mode.

Note that a quadratic problem must be solved at each iteration in Eq. (3.99). These problems are rather well known from a theoretical point of view; they can be efficiently solved numerically once converted into an equivalent linear problem of size $2 \times N$ [253], for example by making use of a parallel implementation of the Arnoldi method [254] available in the P-ARPACK library. Another option is to solve the quadratic eigenvalue problem directly without linearizing it; a specific algorithm must then be used instead of the Arnoldi approach. A good candidate is the Jacobi–Davidson method [255], which was recently applied successfully to combustion instability problems [256]. Another way to proceed is to define the kth iteration in the following way:

$$\{[A] - [D(\omega_{k-1})] + \omega_{k-1}[B(\omega_{k-1})]\}\,[P] + \omega_k^2[C][P] = 0, \qquad (3.100)$$

so that a linear eigenvalue problem must be solved at each subiteration and the classical Arnoldi iterative method [254] can be used. This latter formulation showed good potential for large-scale problems (N of the order of 10^6) arising from the thermoacoustic analysis of annular combustors [257]. More details can be found in [242].

ACCOUNTING FOR DISSIPATIVE EFFECTS. The linear formulation previously described is dissipation free because no damping terms were taken into account for its derivation (except for the acoustic radiation at boundaries, which can be modeled by a complex-valued impedance). However, damping effects should be included in some practical cases, for example when dealing with modern combustors for which perforated liners are increasingly used. Multiperforated plates (MPs) are widely used in combustion chambers of turbofan engines to cool the chambers walls exposed to high temperatures [258]. These plates consist of submillimeter apertures, across which the mean pressure jump forces a cold jet through the holes, from the casing into the combustion chamber. The microjets then coalesce to form a cooling film.

Because of the tiny diameter of the perforations, the holes cannot be meshed for numerical computations, and a model is required for the effect of perforated plates. This problem is encountered not only in CFD calculations [259, 260], but also in computing acoustic modes of a combustion chamber because MPs are known to have a damping effect on acoustics [99, 261–265]. As discussed in Subsection 3.2.2, when dealing with an array of circular apertures of diameter $2a$ and aperture spacing d, it is useful to introduce the Rayleigh conductivity $K = 2a(\eta + i\delta)$ [30] relating the 'smoothed' velocity normal to the plate u' to the acoustic pressure jump across the plate, $\Delta p' = p'(x = 0^-) - p'(x = 0^+)$ [see Eq. (3.44)]:

$$\bar{\rho}\frac{\partial u'}{\partial t} = \frac{2a(\eta + i\delta)}{d^2}\Delta p'. \qquad (3.101)$$

Using momentum equation (3.71), one obtains

$$\frac{\partial \hat{p}}{\partial x_\ell}n_\ell = \frac{K}{d^2}[\hat{p}(x = 0^+) - \hat{p}(x = 0^-)], \qquad (3.102)$$

which can be used as a Neumann boundary condition on both sides of a MP present in the computational domain. Once the geometrical properties of the plate (a, d) are selected together with the bias flow velocity $(\bar{u})$, analytical expressions for η and δ [261] (see also Fig. 3.17) allow using Eq. (3.102) to express the pressure gradient normal to the MP as a function of the angular frequency ω. A numerical procedure similar to the one used for accounting for complex-valued impedance (see Subsection 3.3.3) can then be used to solve the eigenvalue problem. This allows one to account for the acoustic damping related to the acoustic-to-vortical energy transfer while keeping an inviscid formulation based on the zero-Mach-number assumption [257].

3.3.4 Upstream–Downstream Acoustic Conditions

In practical applications, the combustion chamber where the zero-Mach-number thermoacoustic analysis is relevant is surrounded by decelerated–accelerated regions. Thus upstream and downstream boundary conditions must be prescribed to account for the acoustic impedance of the compressor and turbine stages. These complex-valued impedances can be assessed analytically under the so-called compact assumption, discussed next, or numerically, in the more general case.

ACOUSTIC IMPEDANCE UNDER THE COMPACT ASSUMPTION. In the low-frequency limit, the acoustic wavelength is much larger than the characteristic length of the upstream and downstream devices that surround the combustor. Using the mass, energy, and entropy conservations, Marble and Candel [266] established the relations linking the different perturbations in the case of planar waves travelling throughout quasi-1D devices (some analytical results can also be obtained in the case of circumferential modes in a choked nozzle [267] or for 2D baseline flows [268]). For example, one can show analyticaly that the reflexion coefficient of a compact choked nozzle equals

$$\frac{1 - (\gamma - 1)M/2}{1 + (\gamma - 1)M/2}, \qquad (3.103)$$

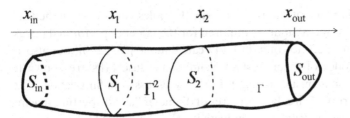

Figure 3.42. Quasi-1D flow domain for the computation of impedance.

where M is the Mach number of the flow entering the nozzle. More details regarding the analytical treatment can be found in the references previously cited as well as in [269].

ACOUSTIC IMPEDANCE OF NON-COMPACT ELEMENTS. A numerical approach can also be used to compute the acoustic impedance of diffusers or nozzles under the isentropic mean flow assumption [270]; it can be seen as a way to extend previous analytical results [266] to non-compact nozzles. The appropriate equations to be considered are the quasi-1D linearised Euler equations written in the frequency space and under the constant-mean-entropy assumption. Once discretised, these equations can be converted into an linear algebraic system:

$$[A][V] = [BT], \tag{3.104}$$

where $[V]$ is the discrete counterpart of the vector of acoustic unknowns $V = (\hat{p}, \hat{u})^T$, the matrix $[A]$ depends on both ω and the details of the spatial discretisation, and the right-hand-side term comes from possibly non-homogeneous boundary conditions.

For any quasi-1D flow domain with inlet and outlet sections $\mathcal{S}_{in}$ and $\mathcal{S}_{out}$, respectively (see Fig. 3.42), the following procedure is used to compute the equivalent acoustic impedance:

1. Fix the frequency ω.
2. Impose a non-zero forward-propagating acoustic wave at the inlet section $\mathcal{S}_{in}$. The corresponding boundary condition relates $\hat{p}$ and $\hat{u}$ at $x = x_{in}$, viz.,

$$2\mathcal{A}^+ \exp\left(i\,k^+ x_{in}\right) = \hat{p} + \overline{\rho}\overline{c}\,\hat{u},$$

where $\mathcal{A}^+ \exp\left(i\,kx\right)$ stands for the forward-propagating wave, $k = \omega/\overline{c}$ is the acoustic wave number, and $\mathcal{A}^+$ is the associated pre-exponential factors that are set to any non-zero value to ensure that the inlet condition is non-homogeneous.
3. Define the appropriate boundary condition to be prescribed at the outlet section depending on whether the mean flow is subsonic or supersonic.
4. Solve the corresponding linear system, Eq. (3.104).
5. Compute the acoustic equivalent impedance as $Z_{in} = \hat{p}/\overline{\rho}\overline{c}\,\hat{u}$, assessed at $x = x_{in}$.

Note that, when the mean flow is subsonic, there is a backward-propagating wave entering the domain through the outlet section $\mathcal{S}_{out}$ so that an outlet boundary condition is required. In this case, the preceding procedure turns out to provide a way to transform a supposedly known acoustic boundary condition at $\mathcal{S}_{out}$ to another condition at $\mathcal{S}_{in}$. For example, when the flow domain is a nozzle, this procedure allows us to displace an acoustic boundary condition at a high-speed section to an

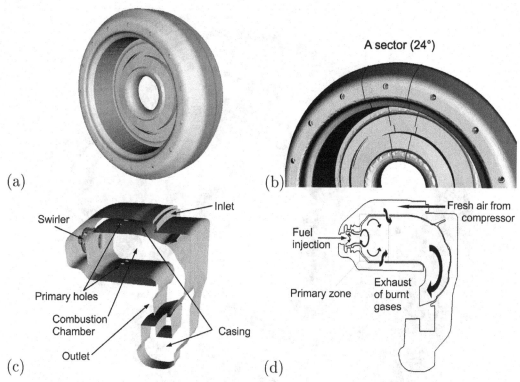

Figure 3.43. The annular helicopter combustor: (a) full annular view of the computational domain, (b) identification of one sector comprising the full annular combustor, (c) detailed view of one single sector, and (d) the expected flow distribution within the sector.

upstream, low-Mach-number position. In the case in which the nozzle is choked, no extra acoustic condition is required because no wave can enter the domain through the outlet section $\mathcal{S}_{\text{out}}$. In the particular case in which the outlet section coincides with the location of the throat, the proper acoustic impedance to impose at $\mathcal{S}_{\text{out}}$ is given by [157, 266],

$$Z_{\text{th}} = \frac{2\mathrm{d}\bar{u}/\mathrm{d}x - i\,\omega}{(\gamma - 1)\mathrm{d}\bar{u}/\mathrm{d}x - i\,\omega}, \tag{3.105}$$

and the preceding procedure allows us to convert this impedance condition valid at the sonic throat to another condition valid at an upstream, low-Mach-number location.

3.3.5 Application to an Annular Combustor

The target configuration chosen to illustrate the proposed LES–Helmholtz solver strategy corresponds to an annular helicopter combustion chamber equipped with 15 burners designed for a helicopter by Turbomeca, shown in Fig. 3.43. Each burner contains two co-annular counter-rotating swirlers. The fuel injectors are placed in the axes of the swirlers. To avoid uncertainties in boundary conditions the chamber's casing is also computed. The computational domain starts after the inlet diffuser and ends at the throat of the high-pressure stator. In this subsection, the flow is choked, allowing for an accurate acoustic representation of the outlet. The air and fuel inlets

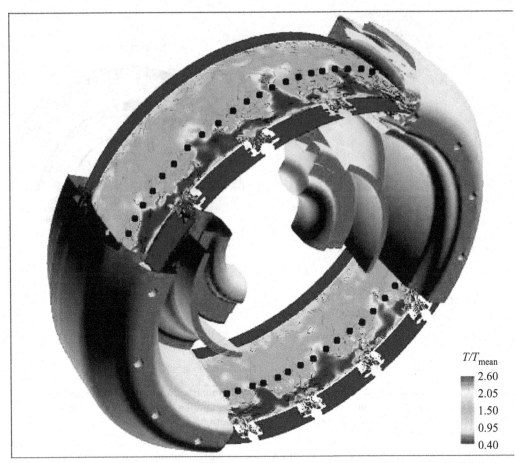

T/T_{mean}
2.60
2.05
1.50
0.95
0.40

Figure 3.44. 3D view of the computational domain; temperature field on a cylindrical plane passing through all the swirlers with velocity magnitude isocontours. Black dots denote typical problem locations for which diagnostics are subsequently provided. (See colour plate.)

use non-reflective boundary conditions [271]. The air flowing at 578 K in the casing feeds the combustion chamber through the swirlers, films, and dilution holes. To simplify the LES, fuel is supposed to be vaporised at the lips of the injector and no model is used to describe liquid-kerosene injection, dispersion, and vaporization.

All LESs presented here use the Smagorinsky approach [207] to model SGS stresses. Combustion is modeled with Arrhenius-type reaction rates: A reduced one-step scheme for JP10–air flames fitted to match the full scheme's behavior for equivalence ratios ranging from 0.4 to 1.5 [219, 272] is used. Five species explicitly solved are JP10, O_2, CO_2, H_2O, and N_2. Turbulence–flame interaction is modeled with the DTF model [161, 168, 218, 273] described earlier. A high-order spatial and temporal scheme (two-step Taylor-Galerkin Colin [274]) is used to propagate acoustic waves with precision. Computations are obtained for (a) the entire configuration (15 burners) and (b) a single-sector [see Fig. 3.43(b)] computational domain for FTF evaluations prior to (c) Helmholtz analysis.

MASSIVELY PARALLEL LES OF THE FULL ANNULAR CHAMBER. In the first computation [275, 276], shown in Fig. 3.44, the whole chamber is simulated from the diffuser

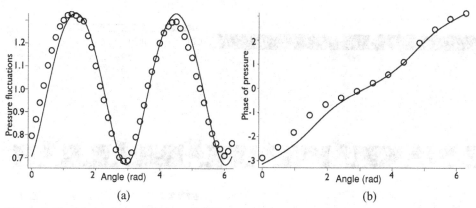

Figure 3.45. (a) Angular variation of $\frac{P_{rms}}{P_{mean}}$ on a ring passing through probes located in the lower part of the chamber casing and (b) pressure signal phase for probes located in front of all swirlers (Fig. 3.44). Note that sector 1 is used as reference: - model, o LES.

outlet to the high-pressure stator nozzle by a 9 009 065-node and 42 287 640-cell mesh.[5] The LES captures the self-excited instability and results (unsteady pressure rms and phase fields) show that it is characterised by two superimposed rotating modes with different amplitudes,[6] shown in Fig. 3.45. The turning modes.[7] are found to modulate the flow rate through the 15 burners, and the flames oscillate back and forth in front of each burner shown in Fig. 3.46, leading to local heat release fluctuations. Because of the rotating motion of the modes, all individual transfer functions of all the burners are the same, i.e., no mechanism of flame interactions between burners within the chamber is identified. Note that, although such computations are very CPU demanding and not realistically possible in an industrial context, they allow validation of the hypotheses introduced in the use of a hierarchy of computational modelling based on single-sector LES (b) and Helmholtz s olvers (c).

The flow-rate fluctuations in this computation are such that they have an impact on the flame response and the heat release perturbations, Fig. 3.47. Swirler flow-rate and pressure fluctuations are out of phase. The phase between global heat release and swirler flow-rate fluctuations satisfies a Rayleigh criterion for the 15 sectors.

It is possible to evaluate the response to the flow-rate oscillations by computing the transfer function between inlet velocity fluctuations and mean single-sector unsteady heat release [182, 183]. Figure 3.48 shows the modulus n and the phase τ of the classical $n - \tau$ model for each of the 15 burners. Amplitudes and delays are fairly constant, suggesting a common response for all burners. This has important implications for steps (b) and (c) of the proposed computational hierarchy.

---

[5] Resolution effects on LES of real configurations have been addressed in [272, 277].

[6] Paschereit et al. [203, 278] propose a nonlinear theoretical approach showing that standing-wave modes can be found at low oscillation amplitudes but that only one rotating mode is found for large-amplitude limit cycles.

[7] A simple model can be used to recover the amplitude and phase of the pressure signals found in the LES for the different burners by considering two counter-rotating acoustic modes P_+ and P_- such that

$$P_+ = A_+ e^{i \cdot k \cdot \theta - i\omega t}, \quad P_- = A_- e^{-i \cdot k \cdot \theta - i\omega t}. \tag{3.106}$$

When $A_- = 0.33 A_+$, the amplitude and phase of the resulting acoustic pressure $P_+ + P_-$ compare very well with the LES data (Fig. 3.45).

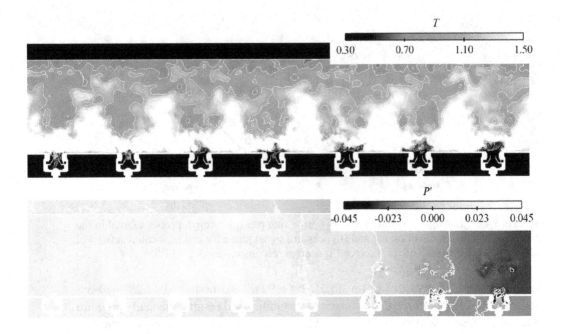

Figure 3.46. Detailed view of half of the burners. top, temperature field with temperature isocontour; bottom, pressure fluctuations with $p' = 0$ isoline.

SINGLE-SECTOR LES: FTF EVALUATION. Based on the previous observations, single-sector forced LES should be sufficient to retrieve the FTF necessary for the 3D Helmholtz solver computations. For the case of interest, a single-sector forced LES is obtained for an inlet acoustic modulation at 600 Hz. The local response amplitude coefficient, $n(\mathbf{x})$ at 600 Hz, is plotted in three different planes in Fig. 3.49. One remarks that the flame response is spread in the primary zone; it is neither homogenous nor symmetric. In the following discussion, the response amplitude coefficient $n(\mathbf{x})$ is kept in its local formulation in order to take into account the flame-response inhomogeneities. This flame response is assumed to be independent of the frequency

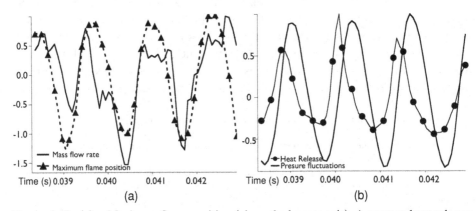

Figure 3.47. (a) –, Maximum flame position (along the burner axis), ▲ average heat release; (b) •, Heat release; – pressure fluctuation average. All results are obtained for sector 1.

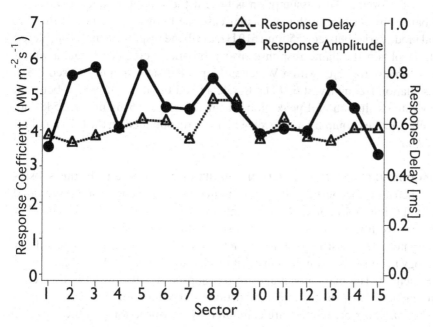

Figure 3.48. Sector-by-sector flame response to the self-exited mode: ○, response amplitude; •, response delay (ms).

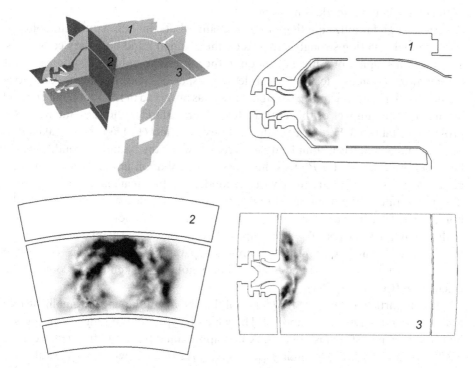

Figure 3.49. Inhomogeneities of the amplitude response coefficient $n(\mathbf{x})$.

of the acoustic forcing. This assumption is false if the forcing frequency varies in a broad band, but because in this particular case the known frequency of the first azimuthal mode varies between 550 and 610 Hz, the flame response can be considered constant. To obtain the flame response over a broader frequency range, a white-noise inlet forcing, together with a Whiener–Hopf inversion, should be used [251]. At the oscillation frequency of 600 Hz, the delay and the phase obtained between peak flow-rate oscillation and peak global heat release is 0.6761 ms or 2.548 rad, respectively, and they compare quite well with the data obtained on the full chamber; see Fig. 3.48.

THERMOACOUSTIC STABILITY PREDICTION – 3D HELMHOLTZ SOLVER. In this subsection, a detailed thermoacoustic study of the annular helicopter combustor is provided by use of a 3D Helmholtz solver. First, the influence of the geometry of the chamber in a non-reactive flow is investigated. It is shown that the casing, the swirler, and the primary holes have a strong influence on the acoustics of the azimuthal mode. A methodology to compute azimuthal instabilities in the context of Helmholtz solvers is then presented. The main advantages of this approach is its low CPU time cost. The coupling between acoustics and combustion is accounted for by a classical $n - \tau$ model. These n and τ parameters are computed by post-processing of the previous single-sector LES, and the corresponding fields are extended to a multi-injection annular combustion chamber under the independence sector assumption in annular combustor [250] (ISAAC). This assumption is validated thanks to the LES of the full annular helicopter combustion chamber previously discussed. Then the stability of the first azimuthal mode is investigated.

Because Helmholtz equations are essentially elliptic in their nature, the solution of the model equation strongly depends on the geometry and boundary conditions. This raises a simple but critical question in the context of industrial systems: What computational domain, technological devices, asperities, and boundary conditions are needed to properly capture the right physics as observed in LES or on the engine? To illustrate the impact of this critical step, several 3D Helmholtz simulations are provided in Table 3.3. For these results, combustion effects (FTFs) are not taken into account and geometrical complexity is increased while the boundary conditions are kept identical. The vector P gives the pressure distribution in the domain, the real part of ω gives the eigenfrequency of the mode, and the imaginary part of ω gives the growth rate or the damping of the eigenmode that is due to acoustic flux at the boundaries $\partial \Omega_Z$. In the calculation presented in Table 3.3 there is no acoustic flux at the boundaries because $Z = \infty$ is imposed everywhere. The combustion chamber (CC) and the casing (C) are first computed separately. Then they are connected with the swirler (CC+C+S) and finally primary holes are taken into account in the calculation (CC+C+S+PH).

In this particular case, the frequency of the first azimuthal mode remains within a narrow band between 500 and 700 Hz, which is typically the range of frequency involved in combustion instabilities. A first approximation of the frequency can be easily obtained with the formula $f_{approx} = \hat{c}_{mean}/(2\pi R_{mean})$, where $\hat{c}_{mean}$ stands for the mean sound speed in the domain and R_{mean} is the mean radius of the annular configuration. With $\hat{c}_{mean} = 750$ m/s, $f_{approx} = 680$ Hz. This approximation is in the

Table 3.3. *Influence of the geometry on the first azimuthal eigenmode calculated by Helmholtz solver (light grey denotes a pressure antinode and blackish grey a pressure node)*

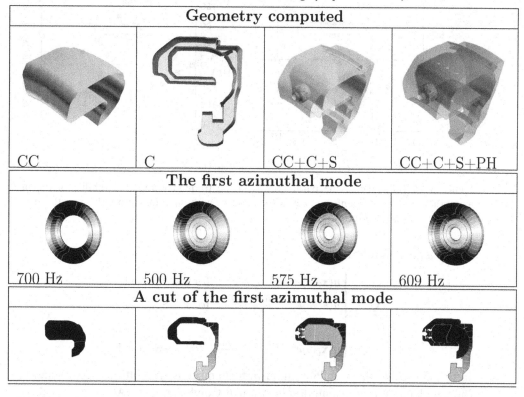

Geometry computed			
CC	C	CC+C+S	CC+C+S+PH
The first azimuthal mode			
700 Hz	500 Hz	575 Hz	609 Hz
A cut of the first azimuthal mode			

range of the results of the full calculation given in Table 3.3, as well as the frequency observed in the full annular LES. It is clear that the frequency of the first azimuthal mode is greater for CC+C+S than for CC and it is lower than for C. Note also that the computed modes are not purely azimuthal and that the azimuthal modes also have a longitudinal component. Adding the primary holes in the calculation changes this longitudinal component as well as the mode frequency. Those results emphasize the necessity of treating carefully all the geometrical details to model correctly the acoustics involved in combustion instabilities.

The coupling between acoustics and combustion is accounted for by a classical $n - \tau$ model. In its global formulation, it relates the fluctuations of the total heat release to the fluctuating velocity at a reference point, denoted as $\mathbf{x}_{ref}$. As underlined previously, this reference point must be chosen in the injection zone where acoustic fluctuations influence the flow rate or the equivalence ratio and it must be as close as possible to the combustion zone [240]. $\vec{u}_1(\mathbf{x}_{ref}) \cdot \vec{n}$ is the longitudinal velocity leaving the swirler. The time delay τ controls the stability of the configuration according to the Rayleigh criterion [29]. The local $n - \tau$ approach allows us to account for the inhomogeneities of the flame response. Obviously, in this full annular chamber, the coupling between the fluctuating heat release and the acoustic fluctuation cannot be described by a FTF involving only one point of reference. The local formulation is

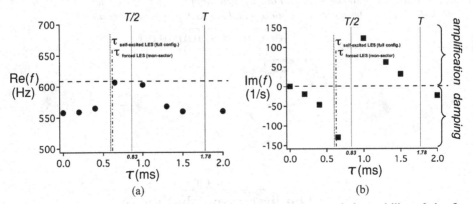

Figure 3.50. Influence of the parameter τ on the frequency and the stability of the first azimuthal mode: (a) eigenfrequency, • with and - - without combustion; (b) growth rate, ■ with and - - without combustion.

thus extended as follows:

$$\frac{\hat{\dot{\Omega}}'(\mathbf{x})}{\dot{\Omega}_{\mathrm{tot}}} = \begin{cases} n(\mathbf{x}, \omega)e^{i\omega\tau(\mathbf{x})}\frac{\vec{\hat{u}}'(\mathbf{x}_{\mathrm{ref}_1}) \cdot \vec{n}_1}{U_{\mathrm{bulk}}} & \text{for } \mathbf{x} \in \text{Sector 1} \\ n(\mathbf{x}, \omega)e^{i\omega\tau(\mathbf{x})}\frac{\vec{\hat{u}}'(\mathbf{x}_{\mathrm{ref}_2}) \cdot \vec{n}_2}{U_{\mathrm{bulk}}} & \text{for } \mathbf{x} \in \text{Sector 2} \\ \dots \\ n(\mathbf{x}, \omega)e^{i\omega\tau(\mathbf{x})}\frac{\vec{\hat{u}}'(\mathbf{x}_{\mathrm{ref}_{15}}) \cdot \vec{n}_{15}}{U_{\mathrm{bulk}}} & \text{for } \mathbf{x} \in \text{Sector 15.} \end{cases} \quad (3.107)$$

In this approach, the annular combustion chamber is split into 15 sectors and the Independence Sector Asumption in Annular Combustor (ISAAC) is assumed. The heat release fluctuations in a given sector are driven only by the fluctuating mass flow rates that are due to the velocity perturbations through its own swirler. The ISAAC assumption is implicitly used in most studies of annular combustors [119, 170, 203]. It is true only if flames issuing from neighbouring burners do not interact, a property that is known to be false in certain cases [279] but seems to be accceptable in this case, as presented previously in the massively parallel LES of the full helicopter combustion chamber, Fig. 3.48. Based on ISAAC, the FTF can be retrieved from the single-sector LES previously presented and used in the context of the full annular Helmholtz solver, as previously described.

Note that, in the following section, τ will be used in its global form and a sensitivity analysis of the combustor stability to this parameter is performed. The reaction index coefficient is used in its local form, $n_l(\mathbf{x})$, to take into account the space inhomogeneities of the flame response to a given excitation.

Under the ISAAC assumption, the local amplitude single-sector response coefficient $n_l(\mathbf{x})$ is duplicated in all sectors and test values for τ are chosen between 0 and 2 ms. Because of the geomtry, which is azimuthally periodic, the stability behaviour in this direction is also periodic in τ. Results are plotted in Fig. 3.50. When τ varies, the frequency of the azimuthal eigenmode varies between 555 and 610 Hz. When τ is in $[0, T/2]$, with T the period, the azimuthal mode is damped. The delay measured in the full LES and in the single-sector LES is ≈ 0.65 ms so that the burner is obviously operating very close to its stability limits. The same stability ranges are predicted analytically in a simple annular configuration with an infinitely thin flame [280].

3.3.6 Conclusions

The prediction of thermoacoustic instabilities in aeronautical gas turbine engines and more generally premixed and partially premixed burners is not an easy task. Currently, industrials are faced with such instabilities at the end of the design process. Recent developments in numercial applications allow us to partially apprehend such problems. In this book, a strategy based on the use of the fully unsteady LES approach and Helmoholtz solvers is proposed. Applications to a single sector and a full annular gas turbine helicopter combustion chamber proves the methodology to be applicable. Technological details such as the swirler geometry and primary and secondary holes are proven to be of importance in the determination of the eigenmodes and eigenfrequencies of the acoustics field obtained with the Helmholtz solver. With the help of the ISAAC hypothesis, the prediction of the azimuthal instabilities is also accessible and in very good agreement with full annular LES for the same configuration. Note, however, that current limitations, applicable to LES and Helnholtz solvers, rely on the proper determination of the acoustic impedances at the inlet and the outlet of the computational domains. Simple approximations are possible, but their extension to more complex systems is not so clear and further investigations seem needed. Finally, the estimation of the FTF, needed for the Helmholtz solver to determine the stability of the eigenmodes, remains a critical point. Although it can be retrieved from single-sector LES for a given forcing frequency, determination of the FTF for the entire frequency range of interest implies large computer costs.

REFERENCES

[1] A. Smithells and H. Ingle, The structure and chemistry of flames, *J. Chem. Soc.* **61**, 204–216 (1892).
[2] M. Hertzberg, Selective diffusional demixing: Occurrence and size of cellular flames, *Prog. Energy Combust. Sci.* **15**, 203–239 (1989).
[3] J. Buckmaster, P. Clavin, A. Linán, M. Matalon, N. Peters, G. Sivashinsky, and F. A. Wiiliams, Combustion theory and modelling, *Proc. Combust. Inst.* **30**, 1–19 (2005).
[4] M. Matalon, Flame dynamics, *Proc. Combust. Inst.* **32**, 37–82 (2009).
[5] D. Bradley, P. H. Gaskell, and X. J. Gu, Burning velocities, Markstein lengths, and flame quenching for spherical methane–air flames: A computational study, *Combust. Flame* **104**, 176–198 (1996).
[6] D. Bradley, M. Lawes, K. Liu, S. Verhelst, and R. Woolley, Laminar burning velocities of lean hydrogen–air mixtures at pressures up to 1.0 MPa, *Combust. Flame* **149**, 162–172 (2007).
[7] D. Bradley, P. H. Gaskell, and X. J. Gu, The mathematical modeling of liftoff and blowoff of turbulent non-premixed methane jet flames at high strain rates, *Proc. Comb. Inst.* **27**, 1199–1206 (1998).
[8] X. J. Gu, M. Z. Haq, M. Lawes, and R. Woolley, Laminar burning velocity and Markstein lengths of methane–air mixtures, *Combust. Flame* **121**, 41–58 (2000).
[9] G. Darrieus, Propagation d'un front de flamme: Assai de théorie des vitesses anomales de déflagration par developpement spontané de la turbulence, presented at the Sixth International Congress on Applied Mechanics, Paris, 1946.

[10] L. D. Landau and E. M. Lifshitz, *Fluid Mechanics* (Pergamon, Oxford, 1987).

[11] J. K. Bechtold and M. Matalon, Hydrodynamic and diffusion effects on the stability of spherically expanding flames, *Combust. Flame* **67**, 77–90 (1987).

[12] D. Bradley, Instabilities and flame speeds in large-scale premixed gaseous explosions, *Philos. Trans. R. Soc. London A* **357**, 3567–3581 (1999).

[13] P. Clavin, Dynamic behaviour of premixed flame fronts in laminar and turbulent flows, *Prog. Energy Combust. Sci.* **11**, 1–59 (1985).

[14] D. Bradley and C. M. Harper, The development of instabilities in laminar explosion flames, *Combust. Flame* **99**, 562–572 (1994).

[15] D. Bradley, Flame propagation in a tube: The legacy of Henri Guénoche, *Combust. Sci. Technol.* **158**, 15–33 (2000).

[16] G. Jomaas, C. K. Law, and J. K. Bechtold, On transition to cellularity in expanding spherical flames, *J. Fluid Mech.* **583**, 1–26 (2007).

[17] D. Bradley, R. A. Hicks, M. Lawes, C. G. W. Sheppard, and R. Woolley, The measurement of laminar burning velocities and Markstein numbers for iso-octane air and iso-octane-n heptane mixtures at elevated temperatures and pressures in an explosion bomb, *Combust. Flame* **115**, 126–144 (1998).

[18] D. Bradley, M. Lawes, and M. S. Mansour, Explosion bomb measurements of ethanol-air laminar gaseous flame characteristics at pressures up to 1.4 MPa, *Combust. Flame* **156**, 1462–1470 (2009).

[19] D. Bradley, C. G. W. Sheppard, R. Woolley, D. A. Greenhalgh, and R. D. Lockett, The development and structure of flame instabilities and cellularity at low Markstein numbers in explosions, *Combust. Flame* **122**, 195–209 (2000).

[20] L. Filyand, G. I. Sivashinsky, and M. L. Frankel, On self-acceleration of outward propagating wrinkled flames, *Physica D* **72**, 110–118 (1994).

[21] Yu. A. Gostintsev, A. G. Istratov, and Yu. V. Shulenin, Self-similar propagation of a free turbulent flame in mixed gas mixures, *Combust. Explos. Shock Waves* **24**, 563–569 (1988).

[22] D. Bradley, T. M. Cresswell, and J. S. Puttock, Flame acceleration due to flame-induced instabilities in large-scale explosions, *Combust Flame* **124**, 551–559 (2001).

[23] V. Karlin and G. Sivashinsky, Asymptotic modelling of self-acceleration of spherical flames, *Proc. Combust. Inst.* **31**, 1023–1030 (2007).

[24] I. R. Hurle, R. B. Price, T. M. Sugden, and A. Thomas, Sound emission from open turbulent premixed flames, *Proc. R. Soc. London A* **303**, 409–427 (1968).

[25] G. A. Batley, A. C. McIntosh, and J. Brindley, Baroclinic distortion of laminar flames, *Proc. R. Soc. London A* **452**, 199–221 (1996).

[26] G. H. Markstein, in *Non-Steady Flame Propagation*, AGARDograph 75 (Pergamon, Oxford, 1964).

[27] D. P. J. McCann, G. O. Thomas, and D. H. Edwards, Gas dynamics of vented explosions. Part 1: Experimental studies, *Combust. Flame* **59**, 233–250 (1985).

[28] H. Guénoche, in G. H. Markstein, (ed.), *Non-Steady Flame Propagation*, AGARDograph 75 (Pergamon, Oxford, 1964), Chap. E, pp. 107–181.

[29] Lord Rayleigh (John William Strutt), The explanation of certain acoustic phenomena, *Nature (London)* **18**, 319–321 (1878).

[30] Lord Rayleigh (John William Strutt), *The Theory of Sound* (Dover, New York, 1945), Vol. II.

[31] A. P. Dowling and S. Hubbard, Instability in lean premixed combustors, *Proc. Inst. Mech. Eng. A* **214**, 317–332 (2000).

[32] A. S. Al-Shahrany, D. Bradley, M. Lawes, K. Liu, and R. Woolley, Darrieus–Landau and thermoacoustic instabilities in closed vessel explosions, *Combust. Sci Technol.* **178**, 1771–1802 (2006).

[33] A. S. Al-Shahrany, D. Bradley, M. Lawes, and R. Woolley, Measurement of unstable burning velocities of iso-octane-air mixtures at high pressure and the derivation of laminar burning velocities, *Proc. Combust. Inst.* **30**, 225–232 (2005).

[34] W. E. Kaskan, An investigation of vibrating flames, *Proc. Combust. Inst.* **4**, 575–591 (1953).

[35] G. Searby, Acoustic instability in premixed flames, *Combust. Sci. Technol.* **81**, 221–231 (1992).

[36] Luc Bauwens, C. Regis Bauwens, and I. Wierzba, Oscillating flames: multiple scale analysis, *Proc. R. Soc. London A* **465**, 2089–2110 (2009).

[37] S. Kerampran, D. Desbordes, B. Veyssière, Study of the mechanisms of flame acceleration in a tube of constant cross section, *Combust. Sci. and Technol.* **158**, 71–91 (2000).

[38] V. Bychkov, Analytical scalings for flame interaction with sound waves, *Phys. Fluids* **11**, 3168–3173 (1999).

[39] R. C. Aldredge, Saffman–Taylor influence on flame propagation in thermo-acoustically excited flow, *Combust. Sci. Technol.* **177**, 53–73 (2005).

[40] D. Durox, T. Schuller, N. Noiray, A. L. Birbaud, and S. Candel, The Rayleigh criterion and acoustic energy balance in unconfined self-sustained oscillating flames, *Combust. Flame* **156**, 106–119 (2009).

[41] P. Cambray and G. Joulin, Length scales of wrinkling of weakly-forced, unstable premixed flames, *Combust. Sci. Technol.* **97**, 405–428 (1994).

[42] V. Bychkov, Importance of Darrieus–Landau instability for strongly corrugated turbulent flames, *Phys. Rev. E* **68**, 0066304-1066304-12 (2003).

[43] R. N. Paul and K. N. C. Bray, Study of premixed turbulent combustion including Landau–Darrieus instability effects, *Proc. Combust. Inst.* **26**, 259–266 (1996).

[44] D. Bradley, P. H. Gaskell, X. J. Gu, and A. Sedaghat, Premixed flamelet modelling: Factors influencing the turbulent heat release rate source term and the turbulent burning velocity, *Combust. Flame* **143**, 227–245 (2005).

[45] I. G. Shepherd and R. K. Cheng, The burning rate of premixed flames in moderate and intense turbulence, *Combust. Flame* **127**, 2066–2075 (2001).

[46] J. F. Driscoll, Turbulent premixed combustion: Flamelet structure and its effect on turbulent burning velocities, *Prog. Energy Combust. Sci.* **34**, 91–134 (2008).

[47] D. Bradley, M. Lawes, and M. S. Mansour, correlation of turbulent burning velocities of ethanol–air, measured in a fan-stirred bomb, up to 1.2 MPa, *Combust. Flame* **158**, 123–138 (2011).

[48] D. Bradley, A. K. C. Lau, and M. Lawes, Flame stretch rate as a determinant of turbulent burning velocity, *Philos. Trans. R. Soc. London A* **338**, 357–387 (1992).

[49] D. Bradley, M. Lawes, K. Liu, and R. Woolley, The quenching of premixed turbulent flames of iso-octane, methane and hydrogen at high pressures, *Proc. Combust. Inst.* **31**, 1393–1400 (2007).

[50] C. K. Law, D. L. Zhu, and G. Yu, Propagation and extinction of stretched premixed flames, *Proc. Combust. Inst.* **21**, 1419–1426 (1986).

[51] Y. Dong, A. T. Holley, M. G. Andac, F. N. Egolfopoulos, S. G. Davis, P. Middha, and H. Wang, Extinction of premixed H_2/air flames: Chemical kinetics and molecular diffusion effects, *Combust. Flame* **142**, 374–387 (2005).

[52] J. M. Donbar, J. F. Driscoll, and C. D. Carter, Strain rates measured along the wrinkled flame contour within turbulent non-premixed jet flames, *Combust. Flame* **125**, 1239–1257 (2001).

[53] S. R. N. De Zilwa, I. Emiris, J. H. Uhm, and J. H. Whitelaw, Combustion of premixed methane and air in ducts, *Proc. R. Soc. London A* **457**, 1915–1949 (2002).

[54] D. Bradley, P. H. Gaskell, X. J. Gu, M. Lawes, and M. J. Scott, Premixed turbulent flame instability and NO formation in a lean-burn swirl burner, *Combust. Flame* **115**, 515–538 (1998).

[55] D. Bradley, M. Lawes, and K. Liu, Turbulent flame speeds in ducts and the deflagration/detonation transition, *Combust. Flame* **154**, 96–108 (2008).

[56] D. Bradley, Hot spots' and gasoline engine knock, *J. Chem. Soc. Faraday Trans.* **92**, 2959–2964 (1996).

[57] N. Kawahara, E. Tomita, and Y. Sakata, Auto-ignited kernels during knocking combustion in a spark-ignition engine, *Proc. Combust. Inst.* **31**, 2999–3006 (2007).

[58] D. Bradley and G. T. Kalghatgi, Influence of autoignition delay time characteristics of different fuels on pressure waves and knock in reciprocating engines, *Combust. Flame* **156**, 2307–2318 (2009).

[59] D. Bradley and R. A. Head, Engine autoignition: The relationship between octane numbers and autoignition delay times, *Combust. Flame* **147**, 171–184 (2006).

[60] K. Fieweger, R. Blumenthal, and G. Adomeit, Self-ignition of S.I. engine model fuels: A shock tube investigation at high pressure, *Combust. Flame* **109**, 599–619 (1997).

[61] A. E. Lutz, R. J. Kee, J. A. Miller, H. A. Dwyer, and A. K. Oppenheim, Dynamic effects of autoignition centers for hydrogen and C1,2-hydrocarbon fuels, *Proc. Comb. Inst.* **22**, 1683–1693 (1988).

[62] D. Bradley, C. Morley, X. J. Gu, and D. R. Emerson, Amplified pressure waves during autoignition: Relevance to CAI engines, SAE paper 2002-01-2868 (2002).

[63] X. J. Gu, D. R. Emerson, and D. Bradley, Modes of reaction front propagation from hot spots, *Combust. Flame* **133**, 63–74 (2003).

[64] D. Bradley, X. J. Gu, and D. R. Emerson, Modes of reaction front propagation from hot spots in flammable gaseous mixtures, in D. Bradley, D. Drysdale, and V. Molkov (eds.), *Proceedings of The Fourth International Seminar on Fire and Explosion Hazards, FireSERT* (University of Ulster, 2004), pp. 201–208.

[65] J. Pan and C. G. W. Sheppard, A theoretical and experimental study of the modes of end gas autoignition leading to knock in SI engines, SAE paper 942060 (1994).

[66] G. T. Kalghatgi, D. Bradley, J. Andrae, and A. J. Harrison, A possible mechanism for superknock events, in *Internal Combustion Engines: Performance, Fuel Economy and Emissions* (Institution of Mechanical Engineers, London, 2009).

[67] J. Stokes, T. H. Lake, and R. J. Osborne, A gasoline engine concept for improved fuel economy – the lean boost system, SAE paper 2000-01-2902 (2000).

[68] P. Van Blarigan, Development of a hydrogen fueled internal combustion engine designed for single speed/power operation, SAE paper 961690 (1996).

[69] C. G. Bauer and T. W. Forest, Effect of hydrogen addition on the performance of methane-fueled vehicles. Part 1: Effect on SI engine performance, *Int. J. Hydrogen Energy* **26**, 55–70 (2001).

[70] S. M. Aceves and J. R. Smith, Hybrid and conventional hydrogen engine vehicles that meet EZEV emissions, SAE paper 970290 (1997).

[71] J. Nygren, J. Hult, M. Richter, M. Aldén, M. Christensen, A. Hultqvist, and B. Johansson, Three dimensional laser induced fluorescence of fuel distributions in an HCCI enguine, *Proc. Combust. Inst.* **29**, 679–685 (2002).

[72] J. E. Dec, W. Hwang, and M. Sjöberg, An investigation of thermal stratification in HCCI engines using chemiluminescent imaging, SAE paper 2006-01-1518 (2006).

[73] J. H. Chen, E. R. Hawkes, R. Sankaran, S. D. Mason, and H. G. Im, Direct numerical simulation of ignition front propagation in a constant volume with temperature inhomogeneities. Fundamental analysis and diagnostics, *Combust. Flame* **145**, 128–144 (2006).

[74] F. Culick, Combustion instabilities in liquid-fueled propulsion systems: An overview, presented at the *AGARD Conference on Combustion Instabilities in Liquid-Fueled Propeller Systems*, Seuille-Sur-Seine, France, 1998.

[75] A. Putnam, *Combustion Driven Oscillations in Industry* (Elsevier, New York, 1971).

[76] S. Candel, Combustion dynamics and control: Progress and challenges, *Proc. Combust. Inst.* **29**, 1–28 (2002).

[77] G. Richards and M. Janus, Characterization of oscillations during premix gas turbine combustion, *J. Eng. Gas Turbines Power* **120**, 294–302 (1998).

[78] G. Richards, D. L. Straub, and E. Robey, Passive control of combustion dynamics in stationary gas turbines, *J. Propul. Power* **19**, 795–810 (2003).

[79] R. Steele, L. Cowell, S. Cannon, and C. Smith, Passive control of combustion instability in lean premixed combustors, *J. Eng. Gas Turbines Power* **122**, 412–419 (2000).

[80] D. Gysling, G. Copeland, D. McCormick, and W. Proscia, Combustion system damping augmentation with Helmholtz resonators, *J. Eng. Gas Turbines Power* **122**, 269–274 (2000).

[81] V. Bellucci, P. Rohr, C. Paschereit, and F. Magni, On the use of Helmholtz resonators for damping acoustic pulsations in industrial gas turbines, *J. Eng. Gas Turbines Power* **126**, 271–275 (2000).

[82] I. Dupere and A. Dowling, The use of Helmholtz resonators in a practical combustor, *J. Eng. Gas Turbines Power* **127**, 268–275 (2005).

[83] N. Joshi, M. Epstein, S. Durlak, S. Marakovits, and P. Sabla, Development of a fuel air premixer for aero-derivative dry low emissions combustors, presented at the ASME *Conference IGTI ASME paper 94-GT-253* (1994).

[84] I. Hughes and A. Dowling, The absorption of sound by perforated linings, *J. Fluid Mech.* **218**, 299–335 (1990).

[85] J. Eldredge and A. Dowling, The absorption of axial acoustic waves by a perforated liner with bias flow, *J. Fluid Mech.* **485**, 307–335 (2003).

[86] B. Chu, On the energy transfer to small disturbances in fluid flow (part i), *Acta Mech.* **1**, 215–234 (1964).

[87] N. H. G. J. Bloxsidge, A.P. Dowling, and P. Langhorne, Active control of reheat buzz, *AIAA J.* **26**, 783–790 (1988).

[88] F. A. Williams, *Combustion Theory* (Benjamin Cummings, Menlo Park, CA, 1985).

[89] A. Dowling, Nonlinear self-excited oscillations of a ducted flame, *J. Fluid Mech.* **346**, 271–290 (1997).

[90] W. P. A. A. Peracchio, Nonlinear heat-release/acoustic model for thermoacoustic instability in lean premixed combustors, *J. Eng. Gas Turbines Power* **121**, 415–421 (1999).

[91] S. R. Stow and A. Dowling, A time-domain network model for nonlinear thermoacoustic oscillations, *J. Eng. Gas Turbines Power* **131**, 031502 1–10 (2009).

[92] T. Scarinci, C. Freeman, and I. Day, Passive control of combustion instability in a low emissions aeroderivative gas turbine, ASME paper ASME-2004-53767 (2004).

[93] M. S. Howe, Attenuation of sound in a low Mach number nozzle flow, *J. Fluid Mech.* **91**, 625–673 (1979).

[94] M. S. Howe, The influence of vortex shedding on the diffraction of sound by a perforated screen, *J. Fluid Mech.* **97**, 641–653 (1980).

[95] D. W. Bechert, Sound absorption caused by vorticity shedding demonstrated with a jet flow, *J. Sound Vib.* **70**, 389–405 (1980).

[96] P. M. Morse and K. Ingard, *Theoretical Acoustics* (McGraw-Hill, New York, 1968).

[97] A. Dowling and J. E. Ffowcs Williams, *Sound and Sources of Sound* (Ellis Horwood Chichester, England, 1983).

[98] I. J. D. Dupere and A. P. Dowling, The absorption of sound by Helmholtz resonators with and without a mean flow, presented at the Eighth AIAA/CEAS Aeroacoustics Conference, AIAA paper 2002-2590 (2002).

[99] A. Cummings, Acoustic nonlinearities and power losses at orifices, *AIAA J.* **22**, 786–792 (1984).

[100] J. C. J. Rupp and A. Spencer, Interaction between the acoustic pressure fluctuations and the unsteady flow field through circular holes, presented at the
ASME Turbo Exposition, ASME paper GT2009-59263 (2009).

[101] A. Dowling and S. Stow, Acoustic analysis of gas turbine combustors, *J. Propul.
Power* **19**, 751–764 (2003).

[102] S. R. Stow and A. P. Dowling, Modelling of circumferential modal coupling
due to Helmholtz resonators, ASME paper 2003-GT-38168 (2003).

[103] C. H. Wang and A. P. Dowling, Actively tuned passive control of combustion
instabilities, presented at the International Colloquium on Combustion and
Noise Control, Cranfield University, UK, 2003.

[104] D. Zhao and A. S. Morgans, Tuned passive control of combustion instabilities
using multiple Helmholtz resonators, *J. Sound Vib.* **320**, 744–757 (2009).

[105] Y. Neumeier and B. T. Zinn, Experimental demonstration of active control of
combustion instabilities using real time modes observation and secondary fuel
injection, *Proc. Combust. Inst.* **26**, 2811–2818 (1996).

[106] J. M. deBedout, M. A. Franchek, R. J. Bernhard, and L. Mongeau, Adaptive-
passive noise control with self-tuning Helmholtz resonators, *J. Sound Vib.* **202**,
109–123 (1997).

[107] H. Matsuhisa, B. Ren, and S. Sato, Semi-active control of duct noise by a
volume-variable resonator, JSME *Int. J.* **35**, 223–228 (1992).

[108] F. Lui, S. Horowitz, T. Nishida, L. Cattafesta, and M. Sheplak, A tunable
electromechanical Helmholtz resonator, presented at the AIAA/CEAS Aeroacoustics Conference, AIAA paper AIAA-2003-3145 (2003).

[109] D. Zhao, C. A'Barrow, A. S. Morgans, and J. Carrotte, Acoustic damping of a
Helmholtz resonator with an oscillating volume, *AIAA J.* **47**, 1672–1679 (2009).

[110] D. Zhao and A. S. Morgans, Tuned passive control of perforated liners, presented at the AIAA/CEAS Aeroacoustics Conference, AIAA paper AIAA-
2009-3406 (2009).

[111] A. P. Dowling and A. S. Morgans, Feedback control of combustion oscillations,
Annu. Rev. Fluid Mech. **37**, 151–182 (2005).

[112] K. R. McManus, T. Poinsot, and S. M. Candel, A review of active control of
combustion instabilities, *Prog. Energy Combust. Sci.* **19**, 1–29 (1993).

[113] S. Ducruix, T. Schuller, D. Durox, and S. Candel, Combustion dynamics and
instabilities: Elementary coupling and driving mechanisms, *J. Propul. Power*
19, 722–734 (2003).

[114] N. Docquier and S. Candel, Combustion control and sensors: A review, *Prog.
Energy Combust. Sci.* **28**, 107–150 (2002).

[115] M. Heckl, Active control of the noise from a Rijke tube, *J. Sound Vib.* **124**,
117–133 (1988).

[116] W. Lang, T. Poinsot, and S. Candel, Active control of combustion instability,
Combust. Flame **70**, 281–289 (1987).

[117] J. Seume, N. Vortmeyer, W. Krause, J. Hermann, C. Hantschk, P. Zangl, S. Gleiss,
D. Vortmeyer, and A. Orthmann, Application of active combustion instability
control to a heavy duty gas turbine, *J. Eng. Gas Turbines Power* **120**, 721–726
(1998).

[118] A. J. Moran, D. Steele, and A. P. Dowling, Active control and its applications,
in *Proceedings of the RTO AVT Symposium on Active Control Technology
for Enhanced Performance Operational Capabilities of Military Aircraft, Land
Vehicles and Sea Vehicles* (NATO, Bramschweig, Germany, 2000).

[119] A. S. Morgans and S. R. Stow, Model-based control of combustion instabilities
in annular combustors, *Combust. Flame* **150**, 380–399 (2007).

[120] P. J. Dines, Active control of flame noise, Ph.D. thesis (University of Cambridge,
Cambridge, 1983).

[121] J. Hermann, A. Orthmann, and S. Hoffmann, Active instability control of combustion oscillations in heavy duty gas turbines, in *Proceedings of the Sixth*

International Congress on Sound and Vibration, 1999), The Technical University of Denmark, Copenhagen, pp. 3341–3352.

[122] T. Poinsot, F. Bourienne, S. Candel, E. Esposito, and W. Lang, Suppression of combustion instabilities by active control, *J. Propul. Power* **5**, 14–20 (1989).

[123] J. Hermann, S. Gleis, and D. Vortmeyer, Active instability control (AIC) of spray combustors by modulation of the liquid fuel flow rate, *Combust. Sci. Technol.* **118**, 1–25 (1996).

[124] J. R. Hibshman, J. M. Cohen, A. Banaszuk, T. J. Anderson, and H. A. Alholm, Active control of combustion instability in liquid-fueled sector combustor, ASME paper ASME-99-GT-215 (1999).

[125] J. E. Ffowcs-Williams, Antisound, *Proc. R. Soc. London A* **395**, 63–88 (1984).

[126] P. J. Langhorne, A. P. Dowling, and N. Hooper, A practical active control system for combustion oscillations, *J. Propul. Power* **6**, 324–333 (1990).

[127] M. Fleifil, J. P. Hathout, A. M. Annaswamy, and A. F. Ghoniem, The origin of secondary peaks with active control of thermoacoustic instability, *Combust. Sci. Technol.* **133**, 227–265 (1998).

[128] J. M. Cohen and A. Banaszuk, Factors affecting the control of unstable combustors, *J. Propul. Power* **19**, 811–821 (2003).

[129] A. S. Morgans and A. P. Dowling, Model-based control of combustion instabilities, *J. Sound Vib.* **299**, 261–282 (2007).

[130] A. M. Annaswamy, M. Fleifil, J. W. Rumsey, R. Prasanth, J. P. Hathout, and A. F. Ghoniem, Thermoacoustic instability: Model-based optimal control designs and experimental validation, *IEEE Trans. Control Syst. Technol.* **8**, 905–918 (2000).

[131] D. U. Campos-Delgado, K. Zhou, D. Allgood, and S. Acharya, Active control of combustion instabilities using model-based controllers, *Combust. Sci. Technol.* **175**, 27–53 (2003).

[132] S. Murugappan, S. Acharya, D. C. Allgood, S. Park, A. M. Annaswamy, and A. F. Ghoniem, Optimal control of a swirl-stabilized spray combustor using system identification approach, *Combust. Sci. Technol.* **175**, 55–81 (2003).

[133] J. E. Tierno and J. C. Doyle, Multi mode active stabilization of a Rijke tube, in *Proceedings of the ASME Winter Annual Meeting: Active Control of Noise and Vibration* (American Society of Mechanical Engineers, New York, 1992), Vol. 38.

[134] S. Evesque, A. P. Dowling, and A. Annaswamy, Self-tuning regulators for combustion oscillations, *Proc. R. Soc. London A* **459**, 1709–1749 (2003).

[135] R. C. Dorf and R. H. Bishop, *Modern Control Systems*, 9th ed. (Prentice-Hall, Englewood Cliffs, NJ, 2001).

[136] G. F. Franklin, J. D. Powell, and A. Emami-Naeini, *Feedback Control of Dynamic Systems*, 4th ed. (Prentice-Hall, Englewood Cliffs, NJ, 2002).

[137] R. Blonbou, A. Laverdant, S. Zaleski, and P. Kuentzmann, Active control of combustion instabilities on a Rijke tube using neural networks, *Proc. Combust. Inst.* **28**, 747–755 (2000).

[138] E. Gutmark, K. J. Wilson, K. C. Schadow, B. E. Parker, R. L. Baron, and G. C. Smith, Dump combustor control using polynomial neural network, presented at the AIAA 31st Aerospace Sciences Meeting and Exhibit, AIAA paper AIAA-93-0117 (1993).

[139] G. Billoud, M. A. Galland, C. H. Huu, and S. Candel, Adaptive control of combustion instabilities, *Combust. Sci. Technol.* **81**, 257–283 (1992).

[140] A. Kemal and C. Bowman, Real time adaptive feedback control of combustion instability, in *Proceedings of the 26th Symposium (International) on Combustion* (Combustion Institute, Pittsburgh, Pennsylvania, 1996), pp. 2803–2809.

[141] B. Widrow and S. Stearns, *Adaptive Signal Processing* (Prentice-Hall, Englewood Cliffs, NJ, 1985).

[142] S. Evesque and A. P. Dowling, LMS algorithm for adaptive control of combustion oscillations, *Combust. Sci. Technol.* **164**, 65–93 (2001).

[143] S. S. Sattinger, Y. Neumeier, A. Nabi, B. T. Zinn, D. J. Amos, and D. D. Darling, Sub-scale demonstration of the active feedback control of gas-turbine combustion instabilities, *J. Eng. Gas Turbines Power* **122**, 262–268 (2000).

[144] C. E. Johnson, Y. Neumeier, M. Neumaier, B. T. Zinn, D. D. Darling, and S. S. Sattinger, Demonstration of active control of combustion instabilities on a full-scale gas turbine combustor, presented at the ASME Turbo Exposition, ASME paper 2001-GT-0519 (2001).

[145] K. Narendra and A. M. Annaswamy, *Stable Adaptive Systems* (Prentice-Hall, Englewood Cliffs, NJ, 1989).

[146] T. Lieuwen, Modeling premixed combustion–acoustic wave interactions: A review, *J. Propul. Power* **19**, 765–781 (2003).

[147] C. A. Armitage, A. J. Riley, R. S. Cant, A. P. Dowling, and S. R. Stow, Flame transfer function for swirled LPP combustion from experiments and CFD, presented at the ASME Turbo Exposition, ASME paper GT2004-53820 (2004).

[148] R. Balachandran, Experimental investigation of the response of turbulent premixed flames to acoustic oscillations, Ph.D. thesis (Engineering Department, Cambridge University, Cambridge, 2005).

[149] A. S. Morgans and A. M. Annaswamy, Adaptive control of combustion instabilities for combustion systems with right-half plane zeros, *Combust. Sci. Technol.* **180**, 1549–1571 (2008).

[150] S. J. Illingworth, A. S. Morgans, Adaptive feedback control of combustion instability in annular combustors, *Combust. Sci. Technol.* **182**, 143–164 (2010).

[151] A. J. Riley, S. Park, A. P. Dowling, and S. Evesque, Adaptive closed-loop control on an atmospheric gaseous lean-premixed combustor, presented at the ASME Turbo Exposition, ASME paper GT-2003-38418 (2003).

[152] S. I. Niculescu and A. M. Annaswamy, An adaptive Smith-controller for time-delay systems with relative degree $n^* \leq 2$, *Syst. Control Lett.* **49**, 347–358 (2003).

[153] S. Evesque, A. M. Annaswamy, S. Niculescu, and A. P. Dowling, Adaptive control of a class of time-delay systems, *J. Dyn. Syst. Meas. Control* **125**, 186–193 (2003).

[154] S. Hoffmann, G. Weber, H. Judith, J. Hermann, and A. Orthmann, Application of active combustion control to siemens heavy duty gas turbines, in presented at the RTO AVT Symposium on Gas Turbine Engine Combustion, Emissions and Alternative Fuels, Lisbon, Portugal, 1998.

[155] J. M. Cohen, N. M. Rey, C. A. Jacobson, and T. J. Anderson, Active control of combustion instability in a liquid-fueled low-Nox combustor, *J. Eng. Gas Turbines Power* **121**, 281–284 (1999).

[156] T. Poinsot and S. Candel, Interactions between acoustics and combustion, *Acoustics* **88** (1988).

[157] T. Lieuwen and V. Yang, Combustion instabilities in gas turbine engines. Operational experience, fundamental mechanisms and modeling, *Prog. Astronaut. Aeronaut.* **210** (2005). Available online at http://knovel.com/web/portal/browse/display?_Ext_KNOVEL_DISPLAY_bookid=2641&Vertical ID=0.

[158] P. E. Desjardins and S. H. Frankel, Two dimensional large eddy simulation of soot formation in the near field of a strongly radiating nonpremixed acetylene–air jet flame, *Combust. Flame* **119**, 121–133 (1999).

[159] T. Murota and M. Ohtsuka, Large-eddy simulations of combustion oscillation in premixed combustor, presented at the International Gas Turbine and Aeroengine Congress and Exposition, ASME paper 99-GT-274 (1999).

[160] D. Caraeni, C. Bergstrom, and L. Fuchs, Modeling of liquid fuel injection, evaporation and mixing in a gas turbine burner using large eddy simulation, *Flow Turbulence Combust.* **65**, 223–244 (2000).

[161] O. Colin, F. Ducros, D. Veynante, and T. Poinsot, A thickened flame model for large eddy simulations of turbulent premixed combustion, *Phys. Fluids* **12**, (2000) 1843–1863.

[162] C. D. Pierce and P. Moin, Progress-variable approach for large eddy simulation of non-premixed turbulent combustion, *J. Fluid Mech.* **504**, 73–97 (2004).

[163] L. Selle, G. Lartigue, T. Poinsot, R. Koch, K.-U. Schildmacher, W. Krebs, B. Prade, P. Kaufmann, and D. Veynante, Compressible large-eddy simulation of turbulent combustion in complex geometry on unstructured meshes, *Combust. Flame* **137**, 489–505 (2004).

[164] H. Pitsch, Large eddy simulation of turbulent combustion, *Annu. Rev. Fluid Mech.* **38**, 453–482 (2006).

[165] C. Angelberger, F. Egolfopoulos, and D. Veynante, Large eddy simulations of chemical and acoustic effects on combustion instabilities, *Flow Turbulence Combust.* **65**, 205–222 (2000).

[166] Y. Huang and V. Yang, Bifurcation of flame structure in a lean premixed swirl-stabilized combustor: Transition from stable to unstable flame, *Combust. Flame* **136**, 383–389 (2004).

[167] S. Roux, G. Lartigue, T. Poinsot, U. Meier, and C. Bérat, Studies of mean and unsteady flow in a swirled combustor using experiments, acoustic analysis and large eddy simulations, *Combust. Flame* **141**, 40–54 (2005).

[168] P. Schmitt, T. J. Poinsot, B. Schuermans, and K. Geigle, Large-eddy simulation and experimental study of heat transfer, nitric oxide emissions and combustion instability in a swirled turbulent high pressure burner, *J. Fluid Mech.* **570**, 17–46 (2007).

[169] T. Poinsot and D. Veynante, *Theoretical and Numerical Combustion*, 2nd ed. (Edwards, Ann Arbor, MI, 2005).

[170] S. R. Stow and A. P. Dowling, Thermoacoustic oscillations in an annular combustor, ASME paper, 2001-GT-37 (2001).

[171] S. Evesque and W. Polifke, Low-order acoustic modelling for annular combustors: Validation and inclusion of modal coupling, presented at the International Gas Turbine and Aeroengine Congress and Exposition, ASME paper GT-2002-30064 (2002).

[172] S. Evesque, W. Polifke, and C. Pankiewitz, Spinning and azimuthally standing acoustic modes in annular combustors, presented at the 9th AIAA/CEAS Aeroacoustics Conference, AIAA paper 2003-3182 (2003).

[173] C. Pankiewitz and T. Sattelmayer, Time domain simulation of combustion instabilities in annular combustors, *J. Eng. Gas Turbines Power* **125**, 677–685 (2003).

[174] P. Rao and P. Morris, Use of finite element methods in frequency domain aeroacoustics, *AIAA J.* **44**, 1643–1652 (2006).

[175] C. Martin, L. Benoit, Y. Sommerer, F. Nicoud, T. Poinsot, LES and acoustic analysis of combustion instability in a staged turbulent swirled combustor, *AIAA J.* **44**, 741–750 (2006).

[176] L. Selle, L. Benoit, T. Poinsot, F. Nicoud, and W. Krebs, Joint use of compressible large-eddy simulation and Helmoltz solvers for the analysis of rotating modes in an industrial swirled burner, *Combust. Flame* **145**, 194–205 (2006).

[177] A. Roux, L. Y. M. Gicquel, Y. Sommerer, and T. J. Poinsot, Large eddy simulation of mean and oscillating flow in a side-dump ramjet combustor, *Combust. Flame* **152**, 154–176 (2007).

[178] T. Schuller, D. Durox, and S. Candel, A unified model for the prediction of laminar flame transfer functions: Comparisons between conical and V-flames dynamics, *Combust. Flame* **134**, 21–34 (2003).

[179] N. Noiray, D. Durox, T. Schuller, and S. Candel, A unified framework for non-linear combustion instabilities analysis based on the flame describing function, *J. Fluid Mech.* **615**, 139–167 (2008).

[180] T. Lieuwen and B. T. Zinn, The role of equivalence ratio oscillations in driving combustion instabilities in low NO_x gas turbines, *Proc. Combust. Inst.* **27**, 1809–1816 (1998).

[181] T. Sattelmayer, Influence of the combustor aerodynamics on combustion instabilities from equivalence ratio fluctuations, presented at the International Gas Turbine and Aeroengine Congress and Exhibition, ASME paper 2000-GT-0032 (2000).

[182] L. Crocco, Aspects of combustion instability in liquid propellant rocket motors. Part i., *J. Am. Rocket Soc.* **21**, 163–178 (1951).

[183] L. Crocco, Aspects of combustion instability in liquid propellant rocket motors. Part ii., *J. Am. Rocket Soc.* **22**, 7–16 (1952).

[184] P. Sagaut, *Large Eddy Simulation for Incompressible Flows* (Springer, New York, 2002).

[185] S. B. Pope, *Turbulent Flows* (Cambridge University Press, Cambridge, 2000).

[186] P. Chassaing, *Turbulence en mécanique des fluides*, in *Analyse du phénomène en vue de sa modélisation à l'usage de l'ingénieur* (Cépaduès-éditions, Toulouse, France, 2000).

[187] C. Prière, L. Y. M. Gicquel, A. Kaufmann, W. Krebs, and T. Poinsot, Les of mixing enhancement: LES predictions of mixing enhancement for jets in cross-flows, *J. Turbulence* **5**, 1–30 (2004).

[188] C. D. Pierce and P. Moin, A dynamic model for subgrid scale variance and dissipation rate of a conserved scalar, *Phys. Fluids* **10**, 3041–3044 (1998).

[189] H. Pitsch and H. Steiner, Large eddy simulation of a turbulent piloted methane/air diffusion flame (Sandia flame D), *Phys. Fluids* **12**, 2541–2554 (2000).

[190] B. Vreman, B. Geurts, and H. Kuerten, Comparison of numerical schemes in large-eddy simulation of the temporal mixing layer, *Int. J. Numer. Meth. Fluids* **22**, 297–311 (1996).

[191] W. W. Kim and S. Menon, An unsteady incompressible Navier Stokes solver for LES of turbulent flows, *Int. J. Numes. Meth. Fluids* **31**, 983–1017 (1999).

[192] W. W. Kim, S. Menon, and H. C. Mongia, Large-eddy simulation of a gas turbine combustor, *Combust. Sci. Technol.* **143**, 25–62 (1999).

[193] H. Forkel and J. Janicka, Large-eddy simulation of a turbulent hydrogen diffusion flame, *Flow, Turbulence Combustion* **65**, 163–175 (2000).

[194] N. Branley and W. P. Jones, Large eddy simulation of a turbulent non-premixed flame, *Combust. Flame* **127**, 1914–1934 (2001).

[195] L. Selle, F. Nicoud, and T. Poinsot, The actual impedance of non-reflecting boundary conditions: implications for the computation of resonators, *AIAA J.* **42**, 958–964 (2004).

[196] G. Staffelbach, L. Y. M. Gicquel, and T. Poinsot, Highly parallel large eddy simulation of multiburner configurations in industrial gas turbines, in *Complex Effects in Large Eddy Simulation*. Springer, Berlin and Heidelberg. Lecture Notes in Computational Science and Engineering **56**, 325–336 (2007).

[197] G. Staffelbach, L. Gicquel, and T. Poinsot, Highly parallel large eddy simulations of multiburner configurations in industrial gas turbines, in, *Complex Effects in Large Eddy Simulation* Vol. 56 of Lecture Notes in Computational Science and Engineering Series (Springer, Berlin, 2006), pp. 326–336.

[198] A. Giauque, L. Selle, T. Poinsot, H. Buechner, P. Kaufmann, and W. Krebs, System identification of a large-scale swirled partially premixed combustor using LES and measurements, *J. Tarbulence* **6**(21), 1–20 (2005).

[199] A. Kaufmann, F. Nicoud, and T. Poinsot, Flow forcing techniques for numerical simulation of combustion instabilities, *Combust. Flame* **131**, 371–385 (2002).

[200] S. Candel, Combustion instabilities coupled by pressure waves and their active control, in *Proceedings of the 24th Symposium (International) on Combustion* (The Combustion Institute, Pittsburgh, 1992), pp. 1277–1296.

[201] D. G. Crighton, A. P. Dowling, J. E. F. Williams, M. Heckl, and F. Leppington, *Modern Methods in Analytical Acoustics*, Lecture Notes (Springer-Verlag, New York, 1992).

[202] W. Krebs, P. Flohr, B. Prade, and S. Hoffmann, Thermoacoustic stability chart for high intense gas turbine combustion systems, *Combust. Sci. Technol.* **174**, 99–128 (2002).

[203] B. Schuermans, C. Paschereit, and P. Monkiewitz, Non-linear combustion instabilities in annular gas-turbine combustors, AIAA paper 2006-0549 (2006).

[204] A. A. Aldama, Filtering techniques for tubulent flow simulation. Lecture Notes in Engineering, Vol. 64, Springer-Verlag, Berlin, (1990).

[205] B. Vreman, B. Geurts, and H. H. Kuerten, On the formulation of the dynamic mixed subgrid-scale model, *Phys. Fluids* **6**, 4057–4059 (1994).

[206] J. Ferziger, Large eddy simulation: An introduction and perspective, in: O. Métais and J. Ferziger (eds.), *New Tools in Turbulence Modelling*, Les Editions de Physique (Springer-Verlag, Berlin, 1997), pp. 29–47.

[207] J. Smagorinsky, General circulation experiments with the primitive equations: 1. The basic experiment., *Mon. Weather Rev.* **91**, 99–164 (1963).

[208] D. K. Lilly, A proposed modification of the Germano sub-grid closure method, *Phys. Fluids* **4**, 633–635 (1992).

[209] M. Germano, U. Piomelli, P. Moin, and W. Cabot, A dynamic subgrid-scale eddy viscosity model, *Phys. Fluids* **3**, 1760–1765 (1991).

[210] F. Nicoud and F. Ducros, Subgrid-scale stress modelling based on the square of the velocity gradient, *Flow, Turbulence Combust.* **62**, 183–200 (1999).

[211] P. Moin, K. D. Squires, W. Cabot, and S. Lee, A dynamic subgrid-scale model for compressible turbulence and scalar transport, *Phys. Fluids* A **3**, 2746–2757 (1991).

[212] F. Ducros, P. Comte, and M. Lesieur, Large-eddy simulation of transition to turbulence in a boundary layer developing spatially over a flat plate, *J. Fluid Mech.* **326**, 1–36 (1996).

[213] M. Germano, Turbulence: The filtering approach, *J. Fluid Mech.* **238**, 325–336 (1992).

[214] S. Ghosal and P. Moin, The basic equations for the large eddy simulation of turbulent flows in complex geometry, *J. Comput. Phys.* **118**, 24–37 (1995).

[215] C. Meneveau, T. Lund, and W. Cabot, A Lagrangian dynamic subgrid-scale model of turbulence, *J. Fluid Mech.* **319**, 353 (1996).

[216] T. D. Butler and P. J. O'Rourke, A numerical method for two-dimensional unsteady reacting flows, in *Proceedings of the 16th Symposium (International) on Combustion* (The Combustion Institute, Pittsburg, PA, 1977), pp. 1503–1515.

[217] C. Angelberger, D. Veynante, F. Egolfopoulos, and T. Poinsot, Large eddy simulations of combustion instabilities in premixed flames, in *Proceedings of the Cummer Program* (Center for Turbulence Research, NASA Ames/Stanford University, Stanford, CA, 1998), pp. 61–82.

[218] J.-P. Légier, T. Poinsot, and D. Veynante, Dynamically thickened flame LES model for premixed and non-premixed turbulent combustion, in *Proceedings of the Summer Program* (Center for Turbulence Research, NASA Ames/Stanford University, Stanford, CA, 2000), pp. 157–168.

[219] G. Boudier, L. Y. M. Gicquel, T. Poinsot, D. Bissières, and C. Bérat, Comparison of LES, RANS and experiments in an aeronautical gas turbine combustion chamber, *Proc. Combust. Inst.* **31**, 3075–3082 (2007).

[220] A. Sengissen, Simulation aux grandes échelles des instabilités de combustion: vers le couplage fluide/structure – th/cfd/06/12, Ph.D. thesis (Université de Montpellier, Montpellier, France, 2006).

[221] S. Orszag, Numerical methods for the simulation of turbulence, *Phys. Fluids* **12**, 250 (1969).

[222] R. Vichnevetsky and J. B. Bowles, *Fourier Analysis of Numerical Approxima-tions of Hyperbolic Equations*, SIAM Studies in Applied Mechanics (Society for Industrial and Applied Mathematics, Philadelphia, 1982).

[223] F. Nicoud, F. Ducros, and T. Schönfeld, Towards direct and large eddy simula-tions of compressible flows in complex geometries, in *Notes in Numerical Fluid Mechanics* R. Fredrich, P. Bontoux (eds.) **64**, 157–171 (1998).

[224] Y. Morinishi, T. Lund, O. Vasilyev, and P. Moin, Fully conservative higher order finite difference schemes for incompressible flow, *J. Comput. Phys.* **143**, 90–124 (1998).

[225] Y. Morinishi, O. Vasilyev, and T. Ogi, Fully conservative finite difference scheme in cylindrical coordinates for incompressible flow simulations, *J. Com-put. Phys.* **197**, 668–710 (2004).

[226] F. Nicoud, Conservative high-order finite difference schemes for low-Mach number flows, *J. Comput. Phys.* **158**, 71–97 (2000).

[227] K. Mahesh, G. Constantinescu, and P. Moin, A numerical method for large-eddy simulation in complex geometries, *J. Comput. Phys.* **197**, 215–240 (2004).

[228] A. E. Honein and P. Moin, Higher entropy conservation and numerical stability of compressible turbulence simulations, *J. Comput. Phys.* **210**, 531–545 (2004).

[229] A. Jameson, W. Schmidt, and E. Turkel, Numerical solution of the Euler equa-tions by finite volume methods using Runge–Kutta time stepping schemes, presented at the 14th Fluid and Plasma Dynamic Conference, Palo Alto, CA, 1981.

[230] O. Colin, Simulations aux grandes échelles de la combustion turbulente prémélangée dans les statoréacteurs, Ph.D. thesis (INP, Toulouse, France, 2000).

[231] T. Poinsot, A. Trouvé, D. Veynante, S. Candel, and E. Esposito, Vortex driven acoustically coupled combustion instabilities, *J. Fluid Mech.* **177**, 265–292 (1987).

[232] M. Abom, A note on the experimental determination of acoustical two-port matrices, *J. Sound Vib.* **155**, 185–188 (1991).

[233] W. Polifke and C. O. Paschereit, Determination of thermo-acoustic transfer ma-trices by experiment and computational fluid dynamics, *ERCOF-TAC Bulletin*, 1998, p. 38.

[234] C. O. Paschereit and W. Polifke, Characterization of lean premixed gas-turbine burners as acoustic multi-ports, in *Bulletin of the American Physical Soci-ety/Division of Fluid Dynamics* (American Physical Society, San Francisco, CA, 1997).

[235] C. O. Paschereit, P. Flohr, and B. Schuermans, Prediction of combustion os-cillations in gas turbine combustors, presented at the 39th AIAA Aerospace Sciences Meeting and Exhibit, AIAA paper 2001-0484 (2001).

[236] C. Pankiewitz, A. Fischer, C. Hirsch, and T. Sattelmayer, Computation of trans-fer matrices for gas turbine combustors including acoustics/flame interaction, presented at the 9th AIAA/CEAS Aeroacoustics Conference and Exhibit, AIAA paper AIAA-2003-3295 (2003).

[237] W. Krebs, B. Prade, S. Hoffmann, S. Lohrmann, and H. Buchner, Therm oacoustic flame response of swirl flames, ASME paper (2002).

[238] D. Bernier, S. Ducruix, F. Lacas, S. Candel, N. Robart, and T. Poinsot, Trans-fer function measurements in a model combustor: Application to adaptive instability control, *Combust. Sci. Technol.* **175**, 993–1013 (2003).

[239] W. S. Cheung, G. J. M. Sims, R. W. Copplestone, J. R. Tilston, C. W. Wilson, S. R. Stow, and A. P. Dowling, Measurement and analysis of flame transfer function in a sector combustor under high pressure conditions, ASME paper, 2003-GT-38219 (2003).

[240] K. Truffin and T. Poinsot, Comparison and extension of methods for acoustic identification of burners, *Combust. Flame* **142**, 388–400 (2005).

[241] L. Crocco, Research on combustion instability in liquid propellant rockets, in *Proceedings of the 12th Symposium (International) on Combustion* (The Combustion Institute, Pittsburgh, 1969), pp. 85–99.

[242] F. Nicoud, L. Benoit, C. Sensiau, and T. Poinsot, Acoustic modes in combustors with complex impedances and multidimensional active flames, *AIAA J.* **45**, 426–441 (2007).

[243] J. J. Keller, W. Egli, and J. Hellat, Thermally induced low-frequency oscillations, *J. Appl. Math. Phys.* **36**, 250–274 (1985).

[244] A. P. Dowling, The calculation of thermoacoustic oscillations, *J. Sound Vib.* **180**, 557–581 (1995).

[245] W. Polifke, C. Paschereit, and K. Doebbeling, Suppression of combustion instabilities through destructive interference of acoustic and entropy waves, presented at the 6th International Conference on Sound and Vibration, Copenhagen, Denmark, 1999.

[246] W. Polifke, C. Paschereit, and K. Doebbeling, Constructive and destructive interference of acoustic and entropy waves in a premixed combustor with a choked exit, *Int. J. Acoust. Vib.* **6**, 135–146 (2001).

[247] T. Sattelmayer, Influence of the combustor aerodynamics on combustion instabilities from equivalence ratio fluctuations, *J. Eng. Gas Turbines Power* **125**, 11–19 (2003).

[248] F. Nicoud and K. Wieczorek, About the zero Mach number assumption in the calculation of thermoacoustic instabilities, *Int. J. Spray Combust. Dyn.* **1**, 67–112 (2009).

[249] M. S. Howe, *Acoustics of Fluid–Structure Interaction* (Cambridge University Press, Cambridge, 1998).

[250] C. Sensiau, F. Nicoud, and T. Poinsot, A tool to study azimuthal and spinning modes in annular combustors, *Int. J. Aeroacoust.* **8**, 57–68 (2009).

[251] W. Polifke, A. Poncet, C. O. Paschereit, and K. Doebbeling, Reconstruction of acoustic transfer matrices by instationary computational fluid dynamics, *J. Sound Vib.* **245**, 483–510 (2001).

[252] C. O. Paschereit, W. Polifke, B. Schuermans, and O. Mattson, Measurement of transfer matrices and source terms of premixed flames, *J. Eng. Gas Turbine Power* **124**, 239–247 (2002).

[253] F. Tisseur and K. Meerbergen, The quadratic eigenvalue problem, *SIAM Rev.* **43**, 235–286 (2001).

[254] R. Lehoucq and D. Sorensen, Arpack: Solution of large scale eigenvalue problems with implicitly restarted Arnoldi methods, available at www.caam.rice.edu/software/arpack, User's guide (1997).

[255] G. Sleijpen, H. Van der Vorst, and M. van Gijzen, Quadratic eigenproblems are no problem, *SIAM News* **8**, 9–10, September (1996).

[256] C. Sensiau, F. Nicoud, M. van Gijzen, and J. van Leeuwen, A comparison of solvers for quadratic eigenvalue problems from combustion, *Int. J. Numer. Meth. Fluids* **56**, 1481–1487 (2008).

[257] E. Gullaud, S. Mendez, C. Sensiau, and F. Nicoud, Damping effect of perforated plates on the acoustics of annular combustors, presented at the 15th AIAA/CEAS AeroAcoustics Conference, AIAA paper AIAA-2009-3260 (2009).

[258] A. H. Lefebvre, *Gas Turbines Combustion* (Taylor & Francis, Washington, D.C., 1999).

[259] S. Mendez and F. Nicoud, Large-eddy simulation of a bi-periodic turbulent flow with effusion, *J. Fluid Mech.* **598**, 27–65 (2008). Available at http://www.cerfacs.fr/ cfdbib/repository/TR/_CFD/_06/_110.pdf.

[260] S. Mendez and F. Nicoud, Adiabatic homogeneous model for flow around a multiperforated plate, *AIAA J.* **46**, 2623–2633 (2008).

[261] M. S. Howe, On the theory of unsteady high Reynolds number flow through a circular aperture, *Proc. R. Soc. London A* **366**, 205–223 (1979).

[262] J. Dasse and S. Mendez, F. Nicoud, Les of the acoustic response of a perforated plate, presented at the 14th AIAA/CEAS AeroAcoustics Conference, AIAA paper AIAA-2008-3007 (2008).

[263] N. Tran, S. Ducruix, and T. Schuller, Damping combustion instabilities with perforates at the premixer inlet of a swirled burner, *Proc. Combust. Inst.* in press.

[264] X. Jing and X. Sun, Experimental investigations of perforated liners with bias flow, *J. Acoust. Soc. Am.* **106**, 2436–2441 (1999).

[265] S. H. Lee, J. G. Ih, and K. S. Peat, A model of acoustic impedance of perforated plates with bias flow considering the interaction effect, *J. Sound Vib.* **303**, 741–752 (2007).

[266] F. E. Marble and S. Candel, Acoustic disturbances from gas nonuniformities convected through a nozzle, *J. Sound Vib.* **55**, 225–243 (1977).

[267] S. Stow, A. P. Dowling, and T. Hynes, Reflection of circumferential modes in a choked nozzle, *J. Fluid Mech.* **467**, 215–239 (2002).

[268] N. A. Cumpsty and F. E. Marble, The interaction of entropy fluctuations with turbine blade rows; a mechanism of turbojet engine noise, *Proc. R. Soc. London A* **357**, 323–344 (1977).

[269] M. Leyko, F. Nicoud, S. Moreau, and T. Poinsot, Numerical and analytical investigation of the indirect noise in a nozzle, in *Proceedings of the Summer Program* (Center for Turbulence Research, NASA Ames, Stanford University, Stanford, CA, 2008), pp. 343–354.

[270] N. Lamarque and T. Poinsot, Boundary conditions for acoustic eigenmodes computation in gas turbine combustion chambers, *AIAA J.* **46**, 2282–2292 (2008).

[271] T. Poinsot and S. Lele, Boundary conditions for direct simulations of compressible viscous flows, *J. Comput. Phys.* **101**, 104–129 (1992).

[272] G. Boudier, L. Gicquel, and T. Poinsot, Effects of mesh resolution on large eddy simulation of reacting flows in complex geometry combustors, *Combust. Flame* **155**, 196–214 (2008).

[273] O. Colin and M. Rudgyard, Development of high-order Taylor–Galerkin schemes for unsteady calculations, *J. Comput. Phys.* **162**, 338–371 (2000).

[274] V. Moureau, G. Lartigue, Y. Sommerer, C. Angelberger, O. Colin, and T. Poinsot, High-order methods for DNS and LES of compressible multi-component reacting flows on fixed and moving grids, *J. Comput. Phys.* **202**, 710–736 (2005).

[275] G. Staffelbach, L. Gicquel, G. Boudier, and T. Poinsot, Large Eddy Simulation of self excited azimuthal modes in annular combustors, *Proc. Combust. Inst.* **32**, 2909–2916 (2009).

[276] G. Boudier, N. Lamarque, G. Staffelbach, L. Gicquel, and T. Poinsot, Thermo-acoustic stability of a helicopter gas turbine combustor using large-eddy simulations, *Int. J. Aeroacoust.* **8**, 69–94 (2009). Available at http://www.cerfacs.fr/ cfdbib/repository/TR_CFD_08_37.p%df.

[277] G. Boudier, G. Staffelbach, L. Gicquel, and T. Poinsot, Mesh dependency of turbulent reacting large-eddy simulations in a gas turbine combustion chamber, presented at the Quality and Reliability of LES Workshop, Leuven, Belgium, 2007.

[278] B. Schuermans, V. Bellucci, and C. Paschereit, Thermoacoustic modeling and control of multiburner combustion systems, presented at the International Gas Turbine and Aeroengine Congress and Exposition, ASME paper 2003-GT-38688 (2003).

[279] T. Poinsot, Analyse des instabilités de combustion de foyers turbulents prémélangés, Thèse d'etat (Université d'Orsay, France, 1987).

[280] W. Krebs, G. Walz, and S. Hoffmann, Thermoacoustic analysis of annular combustor, presented at the 5th AIAA Aeroacoustics Conference, AIAA paper 99-1971 (1999).

4 Lean Flames in Practice

4.1 Application of Lean Flames in Internal Combustion Engines
By Y. Urata and A. M. K. P. Taylor

> In the area of pollution, where there is strong social responsibility, it should be no special feat to burn materials completely. Success will often be a matter of funds and engineering design [1].

One of the two main 'drivers' for the adoption of lean flames in internal combustion (IC) engines has been the potential to reduce the emission of some pollutants. Since Bernard Lewis wrote these words in 1970, the same year as the enactment of the landmark extension to the Clean Air Act (CAA) – the so-called 'Muskie' Act – the design and technology behind IC engines for automobile application have seen many changes that indeed greatly reduced pollution. These include the widespread introduction of gasoline injection systems, three-way catalysts, the high-speed direct injection (DI) diesel engine, turbocharging, and high-pressure common-rail technology, to name a few at random. The related science has also progressed in ways amply described elsewhere in this volume: As to whether this has constituted 'no special feat', the reader will decide. Be that as it may, research into combustion in IC engines is now arguably going through one of its most exciting phases in the quest to 'burn materials completely', that is, with as little pollution as possible and with as much thermodynamic efficiency as possible. This part of the chapter describes the way in which engineering design can exploit combustion, predominantly lean combustion, in IC engines not only to reduce pollution, but also – by promoting fuel-conversion efficiency, the other main 'driver' – to husband the finite resource of petroleum-derived fuels [2]. The latter is important not only in the economic sense but also in the quest to reduce anthropogenic emissions of CO_2 to the atmosphere. Road transport is responsible for the release of between 20% and 26% of total CO_2 emissions in western economies and is the second largest greenhouse-gas- (GHG-) emitting sector in the European Union (EU), so that, although the EU as a whole has reduced its emissions of GHGs by just under 5% over the period from 1990 to 2004, the CO_2 emissions from road transport have increased by 26% [3, 4]. Thus even modest improvements to indicated fuel-conversion efficiency (subsequently we refer to fuel economy) can have important effects on GHG emissions, although these will be inevitably limited by the time constant associated with the rate of removal of engines with less efficient technology from the fleet of engines currently in use.

The improvement in fuel economy consequent on lean-burn combustion, based on the thermodynamics of fuel–air cycle analysis, is well known [5, 6], as is the reduction in NO_x emission that is due to the concomitant lower flame temperature. Both are desirable ends in principle, but the technical interest in lean-burn combustion is framed by legislative as well as by commercial requirements: Over the past four decades, up to three generations of lean-burn engines have been withdrawn from the market owing to increasingly stringent emissions legislation. Therefore the next subsection briefly describes past, current, and likely impending legislation that has been primarily related to emission standards but is increasingly concerned with limiting GHGs emissions, which are closely related to improving fuel economy. Subsection 4.1.2 presents three combustion concepts that make use of lean-burn combustion in IC engines: One is spark-ignited (SI) and the other two are compression-ignited (CI). The latter subsection sets the scene for a description of the role of experiments in Subsection 4.1.3: The final section summarises this contribution and outlines the technological and scientific challenges.

4.1.1 Legislation for Fuel Economy and for Emissions

Fuel Economy, the Clean Air Act and Greenhouse Gases

Good fuel economy – in principle one of the main advantages of lean-burn combustion – is presumably a desirable attribute for most purchasers of an IC engine, and in the United States CO_2 emissions did indeed decrease rapidly from 1975 to 1981 after the 1973–1974 oil embargo. However, emissions increased [7] from 1987 to 2004, decreasing thereafter – currently by 8%. This increase in the United States arose despite the existence of legislation since 1975 that the 'corporate average fuel economy (CAFE)' must be set at the maximum feasible level, subject to economic practicability and the effect of other standards on fuel economy as part of the 'Energy Policy Conservation Act'. In addition, the U.S. Congress established the 'Gas Guzzler Tax', provisions in the Energy Tax Act of 1978, to discourage the production and purchase of fuel-inefficient vehicles. The original purpose of CAFE was to reduce *energy* consumption: Another purpose may soon be added as a consequence of a U.S. Supreme Court ruling in 2007 that the CAA covers GHGs. Thereby, in early December 2009, the U.S. Environmental Protection Agency (EPA) was able to state that new motor-vehicle engines contribute to GHG air *pollution*, which endangers both public health and welfare [8]. At the time of writing [9], a proposed 'National Program' would require – by 2020 – vehicles to meet an estimated so-called *combined* average emissions level of 250 g of CO_2 per mile, equivalent to 35.5 miles per U.S. gallon or 6.6 L per 100 kms (note that the 2009 CAFE regulations stand at 27.5 and 23.1 miles per U.S. gallon for cars and light trucks, respectively), assuming that the automobile industry were to meet this CO_2 emission level solely through fuel-economy improvements. Thus, at least in the United States, it seems that likely future legislation trends may favour improvements in fuel-conversion efficiencies to which lean-burn combustion may be able to contribute. In practice, the importance of lean-burn combustion relative to other means of improving fuel consumption has yet to emerge: In the United States, for example, average light-duty vehicle weight has increased by about one quarter since 1980 whereas acceleration performance has become faster by about one third, and reductions in both these metrics could be a cost-effective method for the lowering of emissions. Power-train hybridisation, cylinder deactivation, and so-called 'stop–start' technologies are already making useful contributions to the lowering of fuel consumption. Although none of these approaches excludes the use of lean-burn combustion, all add – to a greater or a lesser extent – to the cost: Lean-burn combustion is an attractive candidate but its cost-effectiveness has yet to be full established.

In Europe, the need for commitment to measures to reduce CO_2 emissions from cars was recognized from the late 1980s [10, 11] and, as party to the UN Framework Convention on Climate Change in 1992, Europe committed itself to stabilizing CO_2 emissions by the year 2000 at 1990 levels. As a consequence, in 1999 [12], the Automobile Manufacturers' Associations entered into a negotiated self-commitment for a 25% reduction in CO_2 emission for the average of new passenger cars sold in the EU, i.e., to 140-g CO_2/km, to be achieved by 2008. The important aspect of this voluntary commitment, for the purposes of this chapter, was that manufacturers had to achieve the CO_2 target 'mainly' by technological developments. In practice, changes in fleet CO_2 emissions arose to some extent by the substitution of SI engines by CI engines rather than by any intrinsic improvement in the technology of either engine type. However, it became clear by 2007 that the revised EU objective of 120-g CO_2/km by 2012 would not be met. As a consequence, a legally binding regulation has come into force, (EC) No 443/2009, which establishes an average emissions requirement for new passenger cars of 130-g CO_2/km to be achieved by means of improvement in vehicle motor technology: For a gasoline vehicle, that represents 5.4 L per 100 km or 52 miles per Imperial gallon [13]. Thus, from a current average emissions level of CO_2 from new passenger cars in the EU of around 160-g CO_2/km, this represents a 19% reduction. It is the emission level currently attained by a two-seater, 720-kg vehicle, powered by a three-cylinder turbocharged engine [14]. From 2020 onwards, this regulation sets a target of 95-g CO_2/km as the average emission for the new car fleet [15].

Despite the disparity between the emissions targets on the two continents, the planned and existing legislation places an additional premium on finding ways to improve fuel economy in the near future. This is likely to be achieved through the adoption of several different approaches: Research and development on lean-burn combustion currently constitutes one of these approaches.

Emission Legislation: Why Not Zero Levels?

AMBIENT AIR QUALITY. By 1950, Haagen-Smit's landmark work at CalTech had identified ozone as a major component of smog that caused severe respiratory health difficulties; and by 1952 he had shown that vehicle emissions were a contributor to the generation of ozone, specifically nitrogen oxides (NO_x) and hydrocarbons (HCs) (in this text HC is used interchangeably with the term volatile organic compounds, VOCs), which are both ozone precursors. However, it was only after 1970 that governments worldwide enacted coherent legislation to regulate pollutant emissions from IC engines. In part, this two-decade delay arose from the need to set vehicle emission regulation within the broader context of other sources of pollution and the development of associated legislation for the setting of the maximum levels of pollutants in the *ambient air* (the CAA [16] in the United States and its European equivalent, originally [17] and, in its most recent form, [18]). The legislation, broadly speaking, has become increasingly and significantly stringent with each revision of the related acts and directives, such as the European "Clean Air for Europe (CAFE)" programme [19] (not to be confused with the same acronym for the U.S. corporate average fuel economy regulation). To provide the necessary perspective on the extent to which vehicle emission legislation influences the development of lean-burn combustion in IC engines, it is useful to recall the twin purposes of ambient air legislation, which has broad remit. In the United States, it is to protect human health and the environment: The related standards are as follows:

- 'Primary standards', setting limits to protect public health, including the health of sensitive populations such as people with asthma, children, and elderly people; this entails controlling concentrations of air pollutants that cause smog, haze, acid rain, and other problems and reducing emissions of toxic pollutants that are known to cause, or are suspected of causing, cancer or other serious health effects.
- 'Secondary standards', to set limits to protect public welfare, including protection against decreased visibility, damage to animals, crops, vegetation, and buildings from acidification and eutrophication.

European legislation similarly has 'the aim of reducing pollution to levels which minimise the harmful effects on human health and on the environment and improving information to the public on the risks involved' [20].

In the United States, six principal pollutants are identified, of which the first four subsequently listed are the more relevant to the emissions from IC engines (lead having been phased out of gasoline fuel and low-sulphur fuels being increasingly, though not universally, available):

- ground-level (i.e., tropospheric) ozone, O_3
- particle pollution
- nitrogen dioxide, NO_2
- carbon monoxide, CO
- lead, Pb
- sulphur dioxide, SO_2

For these there exist 'National Ambient Air Quality Standards' (NAAQS) [21]. Note that NO_2 is the component of NO_x of greatest interest and serves as the indicator for the entire NO_x family: Control measures that reduce NO_2 can generally be expected to reduce population exposures to all NO_x gases [22].

In view of what can arguably (see subsequent discussion) be called the overriding current health concern, it is perhaps surprising that legislation to control the adverse effects of fine particulate matter has appeared only in the past 15 years or so in the United States – and only in the past couple of years in Europe. By 1972, it was known that particle size has a multimode distribution, with the smaller peak being associated almost entirely with anthropogenic emissions. Nevertheless, as late as 1982, the U.S. EPA had to conclude [23, 24] – albeit in a heavily qualified and guarded review – that poor *visibility* was still the most quantifiable effect on public welfare. In the United States, it was only in 1987 that standards specifically for inhalable particles (as opposed to 'total suspended particles'), in the form of PM_{10}, appeared (see the following dicussion for the definition of PM_{10}). The existence of fine particles, $PM_{2.5}$, was recognised but, in 1987, these were the subject only of an 'advance notice' – and even this was issued as a 'secondary standard' NAAQS, once again on the grounds of visibility. The main reason was that, in 1987 the '... available epidemiology information (as opposed to "controlled exposure" tests), in which aerometry did not clearly separate the two (size categories), could not be used to support two PM indicators...' [25]. Particulate matter (hereafter PM) is currently regulated separately as airborne particles smaller than 10 and 2.5 μm in diameter (also known as PM_{10}, or inhalable coarse particles, and $PM_{2.5}$, or fine particles) since 1987 and 1997, respectively, in the United States: Equivalent European legislation for $PM_{2.5}$ was enacted only in 2008. Thus, as – and if – results from new epidemiological and controlled studies appear, a revision of current standards may be expected in future. As will be subsequently seen, legislation on particulate emissions is of particular importance to gasoline SI lean-burn combustion. (There are, of course, many more pollutants: In the United States, there is a list of 187 such 'hazardous air pollutants' [26] – a subset of which forms the 'air toxins from motor vehicles' [27] that extends to more than 100 substances. Although some (e.g., benzene) of these arise through evaporation or unburned fuel, others are consequent on incomplete combustion of compounds in the fuel, such as toluene, xylene, formaldehyde, acetaldehyde, and 1,3-butadiene. Model-based estimates suggest that such air toxics may account for up to half of all cancers attributed to outdoor sources of air toxics. Bachmann [25] tentatively discusses likely future action on these substances.

Similar air-quality standards exist in European legislation [18, 28]. Of particular relevance to the subject of this article is the 2005 European Strategy on air pollution up to 2020 described in [29], following the CAFE programme in 2001. In particular, it places the emphasis on the most harmful pollutants to public health, namely

- tropospheric ozone, and
- fine particles, i.e., $PM_{2.5}$. A similar emphasis on particulates is placed by [30], (cited by [31]) and by the Scientific Committee on Health and Environmental Risks (SCHER) of Directorate General of Health and Consumer Protection of the European Commission [32] that notes that there is increasing epidemiological evidence that $PM_{2.5}$ may be related to adverse health effects, especially in susceptible populations and vulnerable groups.

Both of these pollutants are relevant to the emissions from IC engines, and hence the associated legislation is important to the research and development of engines, including those based on lean burn, or so-called 'low-temperature' (see later) combustion.

Just how harmful are these pollutants? It has been estimated that, in the year 2000, there were some 21 000 cases of 'hastened death' that were due to ozone [29]. Of even greater concern, however, is that exposure to PM was estimated to reduce the average of 'statistical life expectancy' by approximately 8 months in the 'EU-25', equivalent to 348 000 premature mortalities per annum. Even by 2020, [33] states that the EU will still be a 'long way' from achieving the objectives of the 6th Environmental Action Programme [19]: With effective implementation of only the legislation current in 2005, there would still be 272 000 premature deaths in 2020 that would be due to PM pollution.

Once the pollutants of importance are identified, how are standards set? The NAAQS pollutants were originally regulated by developing health based or environmentally based criteria for setting permissible levels (these are guidelines that are drawn up on so-called science-based criteria: Hence the NAAQS six are also known as 'criteria pollutants'). This required the painstaking gathering of evidence of undesirable effects and is a 'risk-based' category of environmental protection. The alternative approach to emissions standards requires controls on 'best available technology' or 'maximum achievable control technology' (BAT and MACT, respectively) [25] and it is this which has held sway since the revision of the CAA in 1990. Bachmann [25] provides an interesting perspective on the development of standards in the United States.

A particular challenge is set by ozone and $PM_{2.5}$, because no safe levels have yet been identified for ozone or for an unambiguous threshold for the $PM_{2.5}$ dose: It appears to depend on the health-effect endpoint, populations, and vulnerability [34]. So what is the legislator to do? This challenge and its consequences are discussed in [25] in detail: '...The O_3 preambles also give EPA's first statement on the fundamental issue of why the agency believed the CAA requirements for safety and restrictions against considering costs did not require levels approaching zero:'

> The decision is made more difficult by the fact that the Clean Air Act ... does not permit him to take factors such as cost or attainability into account in setting the standard; it is to be a standard which will adequately protect public health. The Administrator recognizes, however, that controlling Ozone to very low levels is a task that will have significant impact on economic and social activity. It is thus important that the standard not be any more stringent than protection of public health demands.

'Subsequent articulations of this view would stress the actual wording of [the] standards must be "requisite" (*i.e.*, no more than necessary) to protect public health with an adequate margin of safety...'. As a consequence, for PM, Bachmann [25] also relates that the review of the standards was '... also among the most contentious, fraught with intrusive litigation, dueling expert panels, and a little intrigue...' and describes how the standards had been arrived at by 2007. In particular, he describes '...the focus on the more general issue so many academic reviews had raised over the years. That is, given a wide range of scientific opinion and no clear thresholds [for PM and ozone], what criteria do EPA use for making the ultimate normative choice on the standards...'. In the end, a judgement was made by the U.S. Supreme Court, which in 2001 issued a '... unanimous decision upholding EPA's position that the statutory requirement that NAAQS be "requisite" to protect public health with an adequate margin of safety sufficiently guided EPA's discretion, affirming EPA's approach of setting standards that are neither more nor less stringent than necessary...'. As further evidence becomes available, the standards are reassessed. For example, in early 2010, the standards for ozone and NO_2 were made more stringent, and those for fines were revised in 2006, the latter being indicative of the current, active concern with this particular pollutant of the ambient, which has implications for vehicle exhaust emission levels. In addition, concerns have been raised about the health impact of ultrafine particles (commonly defined as those below 0.1 μm) but, as yet, no legislation addresses these concerns.

In any event, it is clear that a significant reduction of these substances in the ambient will have beneficial effects in terms of public health and will also generate benefits for ecosystems. In the EU, the process by which objectives were established is summarised in [33]: Briefly, consideration was given to the cost and benefits of closing the gap, by 2020, between the projected baseline at 2020 in the absence of further measures and the 'maximum technically feasible reduction' (MTFR) by 2020. The MTFR would still result in 96 million EU-wide cumulative years of life lost by 2020 as opposed to 137 million for the baseline. Control costs were found to increase rapidly at about 75% between the baseline and the MTFR. Accordingly, compared with the situation in 2000, the strategy [33] set specific long-term objectives (for 2020):

- 47% reduction in loss of life expectancy as a result of exposure to particulate matter;
- 10% reduction in acute mortalities from exposure to ozone;
- reduction in excess acid deposition of 74% and 39% in forest areas and surface freshwater areas, respectively;
- 43% reduction in areas or ecosystems exposed to eutrophication.

To achieve these objectives, it was estimated that SO_2 emissions must decrease by 82%, NO_x emissions by 60%, VOCs by 51%, ammonia by 27%, and primary $PM_{2.5}$ (particles emitted directly into the air) by 59% compared with those in the year 2000. In terms of health, the number of premature deaths should fall from 370 000 that are due to $PM_{2.5}$ and ozone in 2000 to 230 000 in 2020 (compared with 293 000 in 2020 without the strategy).

In conclusion to this subsection, it should be noted that, despite the existence of air-quality standards for up to four decades, there remain several regions and areas in both the EU and the United States that do not attain these standards.

Vehicle Emissions

The attainment of 'levels of air quality that do not give rise to significant negative impacts on, and risks to, human health and the environment' [19], especially in urban areas and densely populated regions, implies legislation limiting the emission of atmospheric pollutants from motor vehicles. Of pollutants in the ambient, European road transport was responsible for 41% of total NO_x emission and 22% of total non-methane volatile organic compounds (NMVOCs) emissions in 2004 [31]. Road transport is held to be the dominant source of ozone precursors, contributing 36% in the European Environment Agency's (EEA) 31 [31]. The transport sector *as a whole* (i.e., including aviation, rail, and shipping, as well as road) contributed 29% of total $PM_{2.5}$ emissions in 2000 [33] (cited by [31]): Note that most diesel exhaust particles are considerably smaller than 2.5 μm. Similarly, in the United States, motor vehicles are responsible for nearly one half of smog-forming VOCs, more than half of the NO_x emissions, about half of the toxic air pollution, and 75% of CO emissions. This is despite the fact that emissions from new cars are more than 90% cleaner than in 1970 [16].

This apparent ineffectiveness of legislation has to be placed in perspective. In the United States, the total vehicle distance that people travel has increased (by 178% in the United States between 1970 and 2005) and the type of vehicle driven has changed (by 2000, 'light-duty trucks' accounted for half of all new passenger car sales). Thus for this reason, if no other, legislation continues to require a reduction in emissions. Similarly, European NO_x (and HC) emissions must be reduced because many member states will otherwise be unable to fulfil the requirements of the National Emissions Ceilings Directive (2001/81/EC) and the proposal for revision of the air-quality directives [34, 35]. In addition, new – and generally, although not always, lower – emission standards are required as research shows new cause for increasing concern for health: $PM_{2.5}$ is a case in point.

Standards for 'tailpipe-out' emission levels exist for CO, HCs, NO_x, and PM. *Hydrocarbons* are regulated as non-methane organic gases (NMOGs or VOCs) mainly because these are precursors for ozone. However, since about 2004 in the state of California, formaldehyde is subject to a separate pollutant category, the latter development being potentially important for the future as ethanol biofuel is being introduced. Although NO_2 is the regulated pollutant in European air-quality legislation, currently NO_x is the regulated emission for engines, with NO_2 being formed from the atmospheric oxidation of nitric oxide, NO, as well as by direct emission, depending on the conditions of combustion. Although about 90% of the NO_x mixture emitted by road vehicles was NO, in the future this emphasis may change [31] for three reasons, thereby contributing to the expectation that the regulated ambient level will be exceeded. First, diesel engines are an increasing fraction of the fleet, and these engines emit a greater proportion of NO_2. Second, some diesel particulate filters work on the principle of deliberately converting NO in the exhaust stream to NO_2 over an oxidation catalyst unit and using the NO_2 to oxidise the particulates held on the filter, thus regenerating the trap. Third, NO_2 increases in slow-moving traffic [36]. *Particulate* emission may be either primary or be added to, secondarily, by chemical reaction with NO_x (as well as ammonia and SO_2), and the increasing concern with the $PM_{2.5}$ contribution has already been noted. To ensure that emissions of ultra-fine particulate pollutants (0.1 μm and below) are controlled, it is proposed in the future to introduce a number-based approach to emissions of particulate pollutants in addition to the mass-based approach that is currently used. The number-based approach to emissions of particles should draw on the results of the United Nations Economic Commission for Europe (UN/ECE) Particulate Measurement Programme (PMP) and be consistent with the existing ambitious objectives for the environment. The standards '... would be set so that they broadly correlate with the mass standards of the current proposal...'. If appropriate, there should also be specification of the value of the admissible level of NO_2 component in the NO_x limit value. However, the intention is that the use of a particle-number standard is a means to ensure that developments in filter technology continue to focus on the removal of ultra-fine particles. Consequently, measuring the number of particulates instead of their mass could be considered a more effective means of regulation in the future.

The legislation, worldwide, is specified in great detail, as is to be expected given its commercial implications and the considerable public interest. Many references exist that describe the following items:

- Vehicle classifications (e.g., passenger cars, light-duty trucks, heavy-duty trucks),
- test procedures (which – currently at least – are different for Japan, Europe, and the United States, with California applying more stringent limits than apply federally),
- emission standards (which are currently common for diesel and gasoline-fuelled vehicles in the United States, but not so for Europe. Note also that the standards may be in units of either grams per kilometer or grams per kilowatt hour),
- emission categories, compliance, in-field monitoring, the increasingly important requirement for 'on-board diagnosis' systems [37, 38] and, particularly, for the important effects of transients [39].

Introductions more closely related to lean-burn engines are to be found in [13, 40–43] and, particularly for biofuels, [44]; [45] is a more detailed research monograph. The point of interest here, however, is the current and likely future general influence of this and similar legislation on the development of lean-burn engines. As will be seen later, this section focuses on three applications of lean-burn combustion, namely, SI engines and homogeneous charge-compression-ignition (HCCI) engines fuelled by gasoline or diesel, the latter particularly in the guise of 'low-temperature combustion' (LTC). Thus we consider the legislation for fuels and both SI and CI engines.

European Emission Standards – Passenger Cars and Light Commercial Vehicles

The standards will have undergone six major revisions by 2014 [46] since the original directive in force in 1992 (91/441/EEC). The latest requirements take effect in two stages, Euro 5 starting from 2009 and Euro 6 from 2014. For this section, the changes in the emission limits for stoichiometrically fuelled gasoline vehicles are not of interest, other than to note that the limits were lowered, partly to prevent the price differential relative to the more fuel – and hence CO_2 – efficient diesel engine from widening excessively.

Gasoline – lean burn SI. Until the introduction of the 'Euro 5a' emission stage [46] in late 2009, particulate emissions were unregulated for gasoline engines. However, as subsequently described at greater length, *lean-burn* DI (LBDI) engines can provide not only large improvements in fuel economy but also can, unfortunately, produce levels of particulates that are significantly higher than those of diesel engines fitted with filters. As a consequence, from late 2009 onwards, particulate emissions from this particular class of engine are regulated, initially by mass per unit distance travelled, at a level broadly comparable with limits in other countries. This is in view of the expectation that this technology might account for only 10% of market share by 2015. The PM limit for LBDI gasoline cars was carefully selected so as to enable the technology to develop without the need for any form of particulate filter to be installed [31]. This is arguably an important concession to the development of lean-burn combustion.

The legislation applies generally to SI gasoline-fuelled vehicles ('positively ignited', in the wording of the legislation – a distinction that is of significance as CI engines based on *gasoline* fuel are subsequently considered), so LBDI engines must also meet the same regulations for the emissions of CO, HC, and NO_x. For stoichiometrically fuelled engines, so-called three-way catalytic converters – for CO, HC and NO_x – are a well-developed, long-established, and effective technology. In contrast, for lean-burn engines, the three potential NO_x after-treatment technologies are regrettably not (yet, at least) comparable to the three-way catalyst in terms of cost benefit. The conversion efficiency of catalysts is currently inadequate; and lean NO_x traps and selective catalytic reduction with on-board urea are also not as well developed as three-way catalytic converters. The development of an economic, reliable, durable, and efficient lean NO_x after-treatment is an enabling technology for the LBDI engine, and the current generation of these technologies is at present available on only a few so-called premium vehicles. The science and technology underpinning the removal of NO_x upstream of emission to the atmosphere are critically important to the successful development and commercialisation of lean-burn combustion in SI engines. These subjects are, however, beyond the scope of this section, but an excellent introduction to the technology is given in [13]. Nevertheless it is worth noting that some stakeholders have commented, at the time the legislation was open for comment [31], that reducing the permitted emission levels for NO_x could be problematic for DI vehicles.

Diesel – CI. The main intention of the current Euro 5 regulation (which came into force in 2005) was to reduce the emission of PM from diesel cars by 80%, relative to Euro 4, to 5 mg/km. This large reduction was selected because more lenient emission limits were found to be insufficiently stringent to ensure a significant improvement in ambient air quality [47]. It has become possible to set such low limits in view of the relatively recent development of 'internal' and 'external' engine measures: The former include advanced fuel injection systems and advanced combustion control of the type described later in this section. The latter include diesel particulate filters (DPFs), although these imply, because of the associated pressure drop, a lowering of the efficiency of the diesel engine. It is worth noting that the emission limit is comparable with emissions from current engines fitted with DPFs and with emissions legislation in the United States and Japan. However, the new regulations are so stringent that it is difficult to measure such low levels of emission with current particulate measurement techniques and, as a consequence, an even stricter limit would have given rise to difficulties for both manufacturers and for monitoring [47, 48]. For example, under some typical test conditions, the mass change on a filter paper can be as low as 10 or 20 μg: This is comparable to the weight of a human fingerprint.

The subsequent regulation, Euro 6 (scheduled to enter into force in 2014), also takes account of the rapid developments in diesel-engine-related technologies (such as turbochargers and combustion control through advanced fuel injection equipment, to mention only two). The legislator's intent is mainly to reduce the emissions of NO_x from diesel cars by more than 50% for some categories of vehicle, to 80 mg/km, with the result that the regulation level will be similar to that for a gasoline engine. This development thus sees an end, for these categories, to the historically less stringent

Table 4.1. *Changes in health impacts associated with Euro 5 and 6 in 2020 [49]*

Parameter	Reduction due to Euro 5	Reduction due to Euro 5 and 6	Increase in % benefit	Unit	Pollutant
Acute mortality (all ages)	72	112	57	Premature deaths	O_3
Chronic mortality (30 yr+)	2 000	3 800	90	Premature deaths	PM
Chronic mortality (all ages)	20 600	35 900	74	Life years lost	PM
Restricted-activity days (RADs, 15–64 yr)	1 850 000	3 180 000	71	Days	PM

NO_x limit for diesel engines. However, it has been more difficult to control NO_x emissions from diesel than from (stoichiometrically fuelled) gasoline engines: This can be gauged from the fact that 'type-approval' values of NO_x and PM emissions for diesel vehicles have – for Euro 4 and 3 and in contrast to gasoline vehicles – been clustered towards the legislative limits, signifying that the emission limits are 'technology forcing'. The associated technology for such control downstream of the engine [e.g., selective catalytic reduction (SCR), using ammonia as the reducing agent; lean NO_x catalysis (LNC), using either HCs from the engine or exhaust fuel injection as the reducing agent; and storage catalysts, lean NO_x traps (LNT), with periodic regeneration] remains under development and is relatively costly. Note also that all three technologies imply a direct or indirect lowering of the efficiency of the engine by a few percentage points. The legislators concluded that there was a need for a balance to be struck relating to the cost-effectiveness of the emission limits corresponding to a cost estimate [49] for complying with Euro 5 and 6 relative to Euro 4, of the order of €590 (sales-weighted average cost per diesel vehicle). It is, however, acknowledged that this estimate may be too large by a factor of 2. Per tonne of NO_x, these measures are about 15% more costly than abatement from other, non-transport sectors. The balance in terms of cost-effectiveness needs to take into account, on the one hand, the affordability of Euro 6 generation vehicles both for continued renewal of the diesel fleet, so that Euro 6 standards gain market share as rapidly as possible, and to maintain the market share of diesel vehicles, which make a significant contribution to reduced CO_2 emissions. On the other hand, the argument for abating NO_x emissions from road vehicles is also strong, given that high transport NO_2 emission places at particular risk the relatively large population who live close to roads. The net result of Euro 5 and 6 measures is predicted to be a 26% reduction in NO_x emissions and a 5% reduction in HC emissions of those required in 2020 and an 11% overall reduction of $PM_{2.5}$ relative to Euro 4 alone. It is estimated that the reduction in emission limits this is due to Euro 6 relative to Euro 5 increases the health benefits by between 60 and 90%, as shown in Table 4.1.

European Emission Standards – Heavy-Duty Vehicles (Diesel CI)

The standards have undergone major revisions five times since the original directive in 1988 (88/77/EEC), the most recent being the so-called 'Euro V', which entered into force in late 2008. Even so, [33] identified that a review of the emission limits for heavy-duty vehicles beyond Euro V is needed to meet the air-quality targets for 2020. Thus there is a proposal for a new regulation, Euro VI, in 2014 [50], the main aspects of which are a 66% reduction in the mass of particulate emissions and a reduction of 80% in NO_x emissions. The new emission limits are comparable to the U.S. 2010 standards and '... reduce the risk of vehicles producing unnecessary levels of pollution ... without imposing excessive costs ... ' [34]. In other words, a balance is once again struck between higher environmental standards and the continued affordability of heavy-duty vehicles for the operators. However, it should be noted that, in the future, legislators may have to take into account the fact that there is strong evidence that heavy-duty diesel vehicles make a major contribution to NO_x emissions and to primary NO_2 emissions [51].

Emissions – Perspectives for Lean-Burn Combustion

The legislation for vehicle emissions sits within the broader context of the quality of the ambient air. As noted in [25] in the context of the United States,

> ... Certainly, conventional air pollution is not the national priority today that it was at the time of the passage of the 1970 CAA [Clean Air Act] Amendments. ... Yet, we still have a significant number of areas that do not meet the O_3 and $PM_{2.5}$ standards, ... and acid and nutrient deposition is still a concern for some ecosystems. We are well on the way to significant new improvements in all of these areas, but the end is not fully in sight. Moreover, whereas risks from air pollution are, on average, modest, it is still reasonable to expect that the cumulative effect may number in tens of thousands of early deaths nation-wide. ... mainstream experts believe the risks remain significant.

The situation is broadly similar in Europe, and, on both continents, it is likely that legislators will continue to review ambient air quality and the contribution of transport to it. Even the currently known evidence could justify further reduction in the emission levels, and it is likely that new epidemiological or controlled evidence will appear, particularly for particulates, that will strengthen this tendency. In addition, the impending change in regulation of emission of $PM_{2.5}$ from a mass-number to a particle-number basis may be accompanied by more information about research into the importance of particle size. It will be interesting to know the health effects, if any, of ultra-fine particles.

In terms of emissions, the stoichiometrically fuelled SI gasoline engine with a three-way catalytic converter has set, and continues to set, a high standard for IC engines with other combustion modes to meet. Legislation in Europe governing the CI diesel engine has allowed in the past either larger emission levels or extended periods of 'grace' for the emission levels to be lowered (a situation that currently includes the SI LBDI engine). This has been, in part at least, because of the lower CO_2 emissions from diesel engines. As technological solutions to produce lower diesel emissions have started to appear, however, recent legislation suggests that the compromise is struck in favour of a modest sacrifice of some of the fuel efficiency to improve the ambient air quality. There is thus little question that, whatever other merits the application of lean-burn combustion to IC engines may have in terms of fuel economy, these must be accompanied by low emissions, particularly of fine particulates, at an acceptable financial and fuel economy cost.

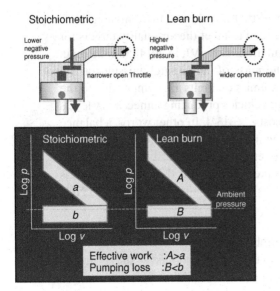

Figure 4.1. Comparison of schematic indicator diagrams for stoichiometric and lean-burn operation for a throttled four-stroke SI engine showing the reduced pumping losses and larger work output for the latter.

4.1.2 Lean-Burn Combustion Concepts for IC Engines

The technical applications of lean-burn combustion can be categorised according to the mode of combustion, either spark ignition or compression ignition, and these are subsequently considerd in turn.

Lean-Burn SI Combustion

CONCEPTS. One method of introducing the fuel, usually but not necessarily gasoline, into an SI engine is upstream of the inlet valves, through either (in the past) a carburettor or (currently) a 'port fuel injector' (PFI). The fuel mixes with the inducted air so that, at the time of ignition, a premixed gas surrounds the spark plug: The control of the fuel flow rate leads to control of the engine's output (load, speed). The physical distance between the point of introduction of the fuel and the spark plug, combined with the need to ensure a combustible mixture in the vicinity of the spark plug at ignition, has *generally* led to the fuel and air flow rates being metered to remain in approximately stoichiometric proportions. This thus implies the need to 'throttle' the mass flow rate of air through the engine at part load and leads to the p–V diagram for throttled operation shown in Fig. 4.1, with the well-known associated loss of pumping work. However, as briefly described later, it has been possible to produce designs that *stratify* the charge, and burn a portion of the fuel through lean combustion, and so extend the engine's lean stable operating limit by several air–fuel ratios (AFRs) from stoichiometric, say up to about 25:1. The first benefit of doing so is that the loss to pumping work is smaller for a given power output, which is of great technical importance because of the large amount of throttling needed by a conventional, unstratified lean-burn SI engine during urban drive cycles. The second benefit is thermodynamic and is shown in Fig. 4.2: The fuel-conversion efficiency as a function of the equivalence ratio, based on isochoric combustion and a fuel–air cycle analysis, increases with increased leaning of the mixture. This is because the effective value of the ratio of the specific heat increases over the expansion stroke

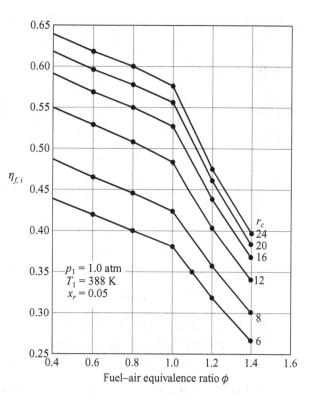

Figure 4.2. Fuel–air cycle results for indicated fuel-conversion efficiency as a function of equivalence ratio and compression ratio, r_c, (octene fule, inlet pressure p_1 is 1 atm, temperature T_1 is 388 K, and the residual gas fraction x_r is 5%). Cited by [52].

because of the dependence of the ratio of the specific heats of the burned gases both on the temperature and on the AFRs as shown in Fig. 4.3 [52]. The third benefit relates to emissions and is shown in Fig. 4.4 [53], which shows that NO_x (in fact, mostly nitric oxide, NO) falls with increasingly lean operation. This is because the temperature dependence of the reactions reduces the equilibrium concentration of NO as the flame temperature, which is consequent on increasingly leaner mixtures, falls [54]. HC emissions, in this figure, are shown initially to reduce as the *overall* equivalence ratio starts to become lean, but may rise again with increasingly lean operation as the reduced flammability of the mixture causes emissions from the flame quenching near cold walls, crevices, and, ultimately, misfires. As noted in [5], lean operation is the reason that some engines, operating at a lean overall equivalence ratio, could nearly meet early emission standards without the need for further emission control methods. In a stratified-charge engine, the equivalence ratio ranges from near stoichiometric close to the spark-plug gap to leaner than stoichiometric with increasing distance away from the gap. As a consequence, the emission characteristics of a stratified-charge lean-burn engine are a composite between the undesirable emissions that occur near stoichiometric mixtures and the more desirable characteristics in the lean region. There is thus a compromise to be struck between having reliable ignition (which requires a near-stoichiometric mixture to surround the gap) and the benefits of lean operation. Given the strong dependence of NO_x emission on temperature, there is an incentive to minimise the volume occupied by near-stoichiometric mixtures. However, we anticipate the following discussion in stating that, although the reduction in NO_x consequent on burning lean is welcome, for SI engines at least the reduction is insufficient to meet the limits currently in force. As

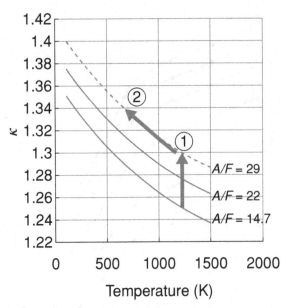

Figure 4.3. Variation of the ratio of the specific heat, $\kappa = c_p/c_v$, as a function of the temperature of the gas with the components of the gas as functions of the AFR as parameters.

(1) leaner combustion gas component
(2) lower combustion gas temperature

a consequence, some sort of NO_x after-treatment is required. This requirement is a significant, although not insuperable, technological burden for lean-burn combustion in SI engines, whether fuelled by port fuel injection or by DI, as described in the following paragraphs.

The other method is to inject the fuel directly into the combustion chamber, raising the possibility of dispensing with the throttle, thereby removing the pumping loss altogether, controlling the load of the engine instead solely by controlling the

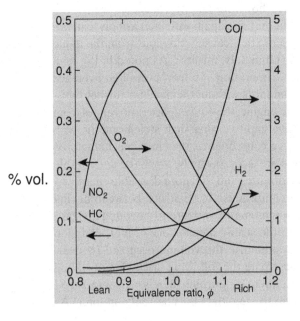

Figure 4.4. SI engine emissions for different fuel–air equivalence ratios. Reprinted from [53]. (Also see Fig. 1.1.)

fuel flow rate. In this approach, the injection timing is usually an important variable because this can determine whether an ignitable mixture is established in the vicinity of the spark plug near the top dead center (TDC) of compression. The possibility of achieving this goal over such a large implied variation of *overall* equivalence ratio arises largely because the physical distance between the injector and the spark plug is greatly reduced with DI, making it is possible to position a relatively small, yet ignitable, cloud of fuel and air around the spark plug. The approach essentially relies on retarding the timing of the start of injection: The leaner the required overall AFR, the nearer the start of injection is to compression TDC. DI also raises the possibility of increasing the cycle efficiency that is due to lean burn far beyond that attainable by a PFI design, in principle, by operating at overall AFRs beyond 65:1, or an overall equivalence ratio leaner than about 0.2 (see subsequent discussion). At low power, this implies operation with relatively late injection to prevent the fuel cloud from becoming too lean. Indeed, to have reliable ignition in this case, it is necessary to have a *spatially stratified* mixture, close to stoichiometric at the spark-plug gap (at the time of ignition) and leaner farther away, which is inflamed by the strongly growing flame kernel. As a consequence, the temperature rise in the gas is smaller than in a stoichiometrically fuelled engine, and hence a further benefit is that heat loss by conduction to the cylinder walls can be smaller. (Note that spatially uniform mixtures can also be used with DI: If the mixture is also stoichiometric this has the advantage that a three-way catalytic converter can be used, but there is no thermodynamic advantage. Alternatively, to gain some thermodynamic efficiency while retaining the convenience of generating spatially uniform mixture, the equivalence ratios must remain greater than about 0.7 to avoid misfires.)

Other advantages follow from DI that are not necessarily associated with lean combustion but that are important insofar as these strengthen the case for the adoption of equipment for DI – which is, as previously mentioned, the enabling technology for lean, and particularly stratified lean, burn combustion. The compression ratio of the engine can be increased because the likelihood of knock is decreased because of evaporative charge cooling, and because the 'end-gas' residence time can be reduced. Alternatively, the octane appetite of the engine is reduced at a given compression ratio. If injection takes place early, during the induction stroke (which, however, tends to result in homogeneous rather than stratified operation and hence tends to be limited to stoichiometric, high-load operation), the volumetric efficiency of the engine improves. Other advantages include [55] improved response to transient changes in load and speed, improved ignition and emissions during 'cold start' because of the lack of the accumulation of fuel on walls, as occurs with PFI, and higher tolerance of exhaust-gas recirculation (EGR) for reduction in NO_x and for dethrottling strategies. The control of emissions during the cold-start phase is known to be of importance in meeting emissions targets. If the fuel and air mixture is thought of as a 'cloud', isolated from the walls of the combustion chamber, then it may also be that wall quenching of the flame is reduced or eliminated altogether. Additionally, the large heat losses that would otherwise occur in the expansion stroke can be reduced. However, although the smaller heat transfer losses decrease the destruction of availability (exergy), this cannot all be realised as an increase in brake power [56]. Some is transformed into an increase in the exhaust-gas availability, and turbocharging is a favourable second-law process for recovery of this energy. As a consequence of the

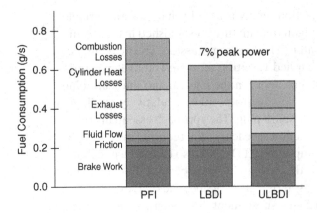

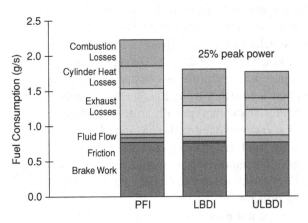

Figure 4.5. Comparison of fuel consumption for the PFI, LBDI, and ULBDI cases. 7% peak power for the top figure, and 25% peak power for the bottom figure [58].

preceding advantages, there are – in principle – substantial opportunities to reduce fuel consumption, although the magnitude of the improvement varies greatly with the location in the engine's performance map [57] and on the details of the engine technology.

To What Extent Do These Qualitative Arguments Result in an Improvement in Efficiency?

[58] provides an extensive theoretical analysis based on availability: the authors consider a 3.0-L stoichiometrically operated engine with port fuelled injection and a compression ratio of 10.5:1 and compare it with a mildly downsized (2.9-L) lean-burn ($1/\phi = \lambda = 1.7$) DI engine and operated at a compression ratio of 12:1. A third engine, a downsized (2.7-L), ultra-lean-burn ($\lambda = 5.0$), high-compression-ratio (16:1), DI engine is also considered (ULBDI). The peak brake work of the LBDI engine is about 40%, approximately 14 basis points higher than the PFI case; that of the ULBDI is 43%. Figure 4.5 summarises the results at 7% and 25% of full power on the basis that the brake work is the same. These values are representative of loads in which a large fraction of engine operation occurs in typical drive cycles. In both figures, the height of a bar gives the total fuel-consumption rate (in grams per second). Under lower-power (7%) conditions, the greater frictional losses and significantly lower exhaust losses in the progression from PFI to LBDI to ULBDI reflect the impact of increasing compression ratio from 10.5 to 12 to 16. The

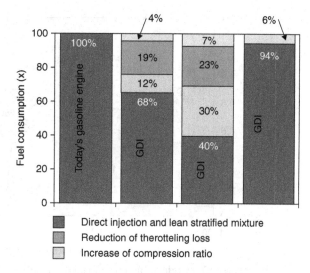

Figure 4.6. Theoretical fuel savings with gasoline DI. Cited by [57].

reduction in fluid flow losses (i.e., pumping losses) reflects the increasing degree of lean-burn operation. The higher λ of the ULBDI case ($\lambda = 5$ vs. 1.7 for LBDI) virtually eliminates pumping losses. Lean-burn operation also leads to lower cylinder heat losses by means of the reduced combustion temperatures. Whereas most of the losses decrease going from PFI to ULBDI, the combustion irreversibilities do not. This reflects the fact that lean-burn operation results in combustion at lower temperatures. Under the higher-power (25%) conditions, the qualitative benefits of LBDI as opposed to those of PFI are the same, reflecting the effects of lean operation and higher compression ratio. The LBDI and ULBDI cases are quite similar. For both cases, the AFR is slightly above $\lambda \approx 1.7$ (the LBDI lean limit), and consequently the major difference is the higher compression ratio. The slightly higher frictional losses, cylinder heat losses, and slightly lower exhaust losses for the ULBDI engine are all consistent with a higher compression ratio. Broadly similar conclusions are reached in [57], shown in Fig. 4.6, in considering a slightly different comparison, namely between a throttled stoichiometric PFI engine operating with a compression ratio of 11:1 and a DI engine with a compression ratio of 12.5:1 operating up to a $\lambda \approx 4$. Although these analyses show that large reductions in fuel consumption are possible, two points must be borne in mind. The first is that the improvements over a drive cycle will be more modest, because the drive cycle incorporates a wide range of speeds and loads. Thus [59] quotes [60], showing that – compared with a PFI without EGR – the advantage of a DI design is of the order of only a 15% fuel saving over the U.S. federal test procedure. In addition, [59] states that thermal management and re-generation of a lean NO_x, catalyst, engine transients, cylinder-to-cylinder variation, and injector aging can reduce the estimated cycle average benefit to 8%–12%.

The second point is that suitably optimised stratified combustion is increasingly difficult to achieve as the load demanded from the engine increases, for reasons that may include the excessive generation of NO_x and particulates (from excessively rich regions near the spark gap as well as from increased impingement of the fuel on surfaces including the piston crown), unburnt HCs from insufficiently premixed parts of the vapour cloud, and a degradation of fuel consumption [61]. Two of the latest production engines are based on so-called 'spray-guided' stratified-charge, DI

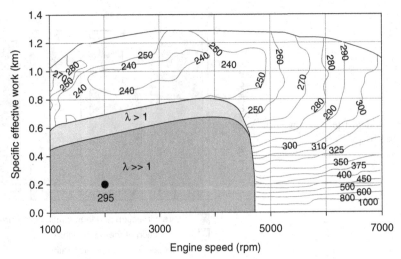

Figure 4.7. Mode of fuelling and contours of specific fuel consumption (g/kWh) on an engine map of a spray-guided DI combustion system. The excess air factor, λ, is the reciprocal of the equivalence ratio, ϕ [62].

design (see the following discussion) and [62, 63] report the engine maps together with the 'road-load' curve and the combustion mode superimposed on contours of specific fuel consumption (grams per kilowatt hour). These are reproduced in Figs. 4.7 and 4.8 and show two features of interest in the context of lean-burn combustion. The first is that the fuel consumption at the key point of 2000 rpm and a mean effective pressure of 2 bars, with stratified lean-burn operation, is reported as 295 and 290 g/kWh, respectively, which is claimed to be better than the best car diesel engines. The second is that the engines operate in lean-burn mode over a low-load, low-speed portion of the operation map only: The area, however, over which this lean mode obtains is reported to be a significant expansion over what was possible with

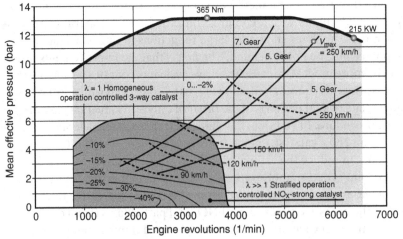

Figure 4.8. Engine map showing region of stratified mode and potential fuel-consumption improvement of a stratified DI engine compared with port injection. The excess air factor, λ, is the reciprocal of the equivalence ratio, ϕ [63].

previous generations of so-called wall-guided stratified-charge design (see following discussion). The importance of the latter observation is that fuel consumption is reduced due to stratified-charge DI not only over a large part of test drive cycles, but also in 'real-world' driving. One of the determinants for mode switching to homogeneous stoichiometric is governed by when the temperature of the exhaust gases exceeds the active temperature window of the NO_x storage converters, at about 500°C [63]. It is on this basis that it is claimed that SI stratified DI engines have the greatest single potential for a reduction in fuel consumption compared with all other concepts [63]. Once again, however, the reduction in NO_x emissions that is due to even such ultra-lean-burn conditions is insufficient to meet emissions legislation and elaborate, and costly, NO_x after-treatment measures are required for lean-burn operation. Both engines [63, 64] are reported to have an 'external' EGR system to reduce production of NO_x in stratified mode.

Because of these and other advantages, research and development into DI gasoline engines has a long history, dating from at least their use in aeroplane engines in the 1930s and in automobile racing engines since at least the 1950s. Stratified systems also have been researched since at least 1950. However, mass-production engines using DI and lean-burn combustion appeared only in the mid-1990s, in part because of the impetus of legislation relating to CO_2 emission and in part because important enabling technologies became available. These included

- advances in high-pressure common-rail system fuel injector technology (to be an accumulator and damper, initially at less than 12 MPa, and to be both reliable and have low parasitic loss), which enabled the generation of finely atomised fuel in the combustion chamber and to be able to inject the required fuel quantity into the cylinder; and
- development of control technology that enabled the selection of appropriate injection timing to achieve certain types of combustion and to track the rapid changes in load [55, 65–67].

It is useful to identify six processes associated with DI engines, whether stratified or not, namely fuel injection, spray atomization, droplet evaporation, charge cooling, mixture preparation, and the control of in-cylinder motion. DI SI engines are now classified by the last two processes: There are three methods by which fuel vapour from the evaporating spray is convected to the spark plug. A schematic of these methods, namely the wall-guided, air-guided, and spray-guided methods, is shown in Fig. 4.9 [66].

These methods lead to engine designs that can operate at an overall AFR of up to about 50 (or equivalence ratio, $\phi = 0.25$) by stratifying the charge while retaining a reliably ignitable mixture (that is, close to stoichiometric) in the vicinity of the spark plug. This is achieved, in principle, by retarding the start of injection during the compression stroke until it is quite close to the ignition point, by designing an injector with an appropriate spray pattern and with sufficiently finely atomised droplets to evaporate before ignition, and by control of the flow in the combustion chamber. The wall-guided concept generates bulk convection to transport the fuel spray or, better, its vapour, reliably to the vicinity of the spark plug at the time of ignition. Subsequently the mixture must allow stable propagation of the flame through the increasingly lean vapour cloud. Even from this brief description, the

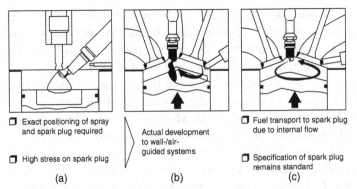

☐ Exact positioning of spray
   and spark plug required

☐ High stress on spark plug

Actual development
to wall-/air-
guided systems

☐ Fuel transport to spark plug
   due to internal flow

☐ Specification of spark plug
   remains standard

(a)   (b)   (c)

Figure 4.9. Classification of DI systems for SI engines: (a) spray-guided, (b) wall-guided, (c) flow- (or air-) guided. Cited by [66].

need is clear for precise management of the intake port geometry, of the in-cylinder flow, especially the profile of the piston crown, and its interaction with the spray, and hence of the spray's position and direction. These aspects of design are essential for a successful operation over the requisite wide range of speeds and loads. Broadly, the required 'endpoint' is to have [59] a slightly rich mixture at the spark-plug electrodes at ignition, a near-stoichiometric mixture surrounding the initial flame kernel to facilitate flame propagation, and steep gradients at the edge of the fuel cloud to minimise the amount of fuel that is below the flammability limit and that will give rise to unburnt HC emissions. As Drake and Haworth [59] point out, the management process must also take into account the effects of cycle-by-cycle variations to produce a robust design.

Fuel Premixing Upstream of the Inlet Valves

Prechamber design. Figure 4.10 shows an early four-stroke passenger-car SI engine that made use of lean-burn combustion for the purpose of lowering emissions: The design [68] was among the first 'stratified-charge engines' in regular production [53], at least outside the former Soviet Union (there, designs of engines based on so-called LAG combustion (meaning avalanche activated combustion) [69], an early

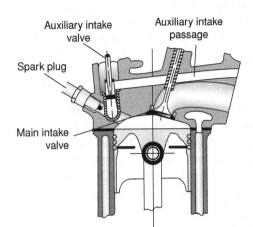

Auxiliary intake
valve

Auxiliary intake
passage

Spark plug

Main intake
valve

Figure 4.10. Prechamber stratified-charge engine with prechamber inlet valve and auxiliary carburettor (E. Taguchi, personal communication).

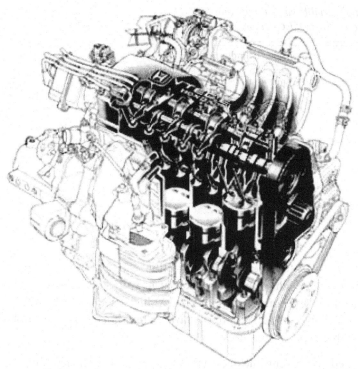

Figure 4.11. Four-cylinder, four-valve, PFI stratified-charge automobile engine with variable valve control system (VTEC-E) [70].

form of HCCI engine, as subsequently described below, were developed based on Nobel Laureate Semenov's work on chain branching in chemical reactions). The purpose of the design was to make use of the low-emission attributes of lean burn combustion to meet the legislative requirements responding to increased awareness of the noxious effects of pollutant emissions such as CO, NO_x, and unburnt HCs, most notably in the Clean Air Act Extension of 1970. However, the specific fuel consumption of the engine was also good for its time, while avoiding the need for any catalytic conversion. The design was an example of a 'divided chamber' or 'prechamber' that provided jet ignition to the lean mixture in the main combustion chamber. A more recent prechamber design was described in [5].

PFI design. As legislative requirements became more stringent, a four-stroke, four-valve, PFI, stratified-charge engine eventually superseded this prechamber design [70], shown in Fig. 4.11 and with the associated specifications in Table 4.2. Figure 4.12 shows the qualitative features of the flow field when the VTEC-E (variable valve timing and lift electronic control – economy) mechanism activates only one of the two intake valves with appreciable lift; the mechanism itself is shown schematically in Fig. 4.13. The variable valve timing and lift mechanism consists of one camshaft with two cam profiles: one, denoted as the primary cam, is permanently engaged and produces a full lift in the corresponding valve. The other, secondary, cam produces a maximum lift of only 0.65 mm, as shown in Fig. 4.14, thereby deactivating it. The

Table 4.2. *Specification of variable valve timing and lift electronic control–economy (VTEC-E) engine*

Cylinder configuration	In-line cylinder
Bore × stroke (mm)	75 × 84.5
Displacement (cc)	1493
Compression ratio	9.3:1
Valve train	SOHC[†] VTEC-E
Number of valves	4
Inlet-valve diameter	27.5
Exhaust-valve diameter	23.5
Fuel injection system	PGM*-FI
Ignition system	PGM-IG

[†] SOHC: single overhead camshaft.
* PGM: programmed fuel injection.

primary rocker arm has a roller follower, to reduce friction, with a built-in hydraulic piston. At low engine speeds, each of the two rocker arms works independently of the other according to the corresponding cam profile. At high speeds, the hydraulic piston engages the rocker arm of the secondary valve so that both valves are fully lifted. This engine has three specific features:

1. At high engine speeds, it prevents the decrease of the volumetric efficiency by using straight ports and activating both intake valves. At these speeds, the charge

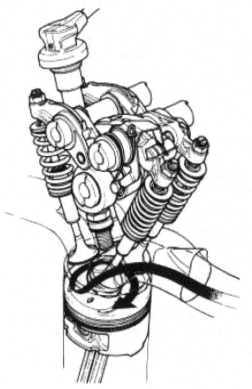

Figure 4.12. VTEC-E variable valve control system showing primary and secondary rocker arms in the foreground: The swirl component is generated at low speeds and loads [70].

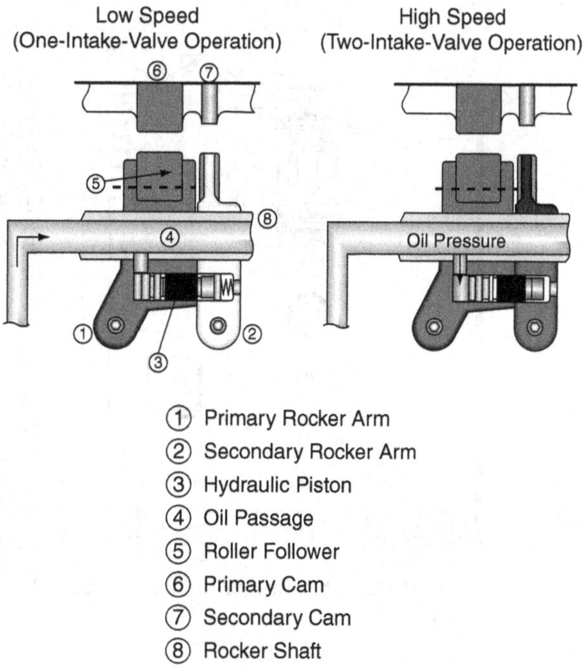

Low Speed
(One-Intake-Valve Operation)

High Speed
(Two-Intake-Valve Operation)

Oil Pressure

Figure 4.13. Mechanical and hydraulic actuation of the variable valve at low and high speeds [70].

① Primary Rocker Arm
② Secondary Rocker Arm
③ Hydraulic Piston
④ Oil Passage
⑤ Roller Follower
⑥ Primary Cam
⑦ Secondary Cam
⑧ Rocker Shaft

is stoichiometric and changes to lean burn at lower speeds. With few exceptions, such mode switching has been the hallmark of most lean burn combustion systems.

2. At low engine speeds, it generates mean swirl, as well as mean tumble, in the cylinder by means of a mechanism that deactivates one of the two intake valves, producing a complex overall mean flow pattern [71]. The swirl and tumble ratios were about 2.5 and 1.2, respectively;

3. At low engine speeds, it produces a lean-burn axially stratified charge by controlling the fuel injection timing in the intake stroke.

The second feature, and the generation of mean tumble in particular, is important in the following context. Lean mixtures have a slower burning rate than do

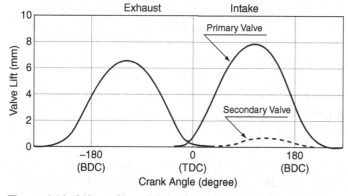

Figure 4.14. Lift profiles of the primary and secondary valves, the latter during low-speed operation [70]. BDC, before dead centre; TDC, top dead centre.

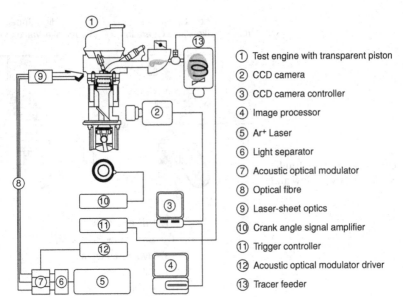

1. Test engine with transparent piston
2. CCD camera
3. CCD camera controller
4. Image processor
5. Ar+ Laser
6. Light separator
7. Acoustic optical modulator
8. Optical fibre
9. Laser-sheet optics
10. Crank angle signal amplifier
11. Trigger controller
12. Acoustic optical modulator driver
13. Tracer feeder

Figure 4.15. Schematic of the particle coded-pulse velocimeter [73]. CCD, charge-coupled device.

stoichiometric ones, and this can lower the fuel-conversion efficiency of the cycle as well as increase exhaust emissions. The burning rate can be increased by promoting the scale and magnitude of the turbulent velocity fluctuations during combustion, and [5] describes the use of squish-jet motion, produced by the geometry of a specially design piston crown, to achieve these ends. Here it is likely that the conventional description [72] of generating turbulence near and after TDC is the important mechanism. Arguments based on the conservation of angular momentum suggest that the velocities associated with the tumble vortex motion should increase by a factor related to the compression ratio of the engine. A particle coded-pulse velocimeter [73], shown in Fig. 4.15, made 3D measurements of velocity in an optically accessed engine. The measurements are shown in Fig. 4.16. The measurement plane was 16 mm below the spark plug, illuminating a section across the bore, and the engine speed was 1500 rpm: 450° CA (crank angle) corresponds to the intake flow whereas the other three figures show the flow during the compression stroke. The colour contour (see colour plate) shows the tumble (axial) component of velocity: red shows axial velocity components flowing from piston crown to the spark plug: Blue shows the opposite direction. The arrows show the swirl flow component. It is clear that the tumble component velocity magnitudes remain unchanged during compression. Thus the additional energy that is due to compression that is, in principle, put into the tumble vortex (as the density of the gases increase) is converted, by means of tumble vortex breakdown, into turbulent kinetic energy. Also, in agreement with expectation [72], the magnitude of the swirl component of velocity remains unaltered during compression.

The effect of injection timing on the NO_x emissions, on the co-variance in the indicated mean effective pressure (IMEP), and on the normalised AFR ratio at the spark plug (measured by a fast-acting sampling valve) as a function of the start of injection are shown in Fig. 4.17. The overall AFR was maintained at 23. The vertical

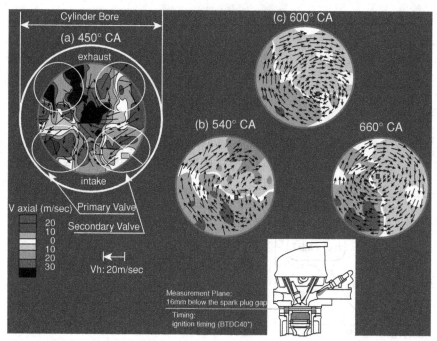

Figure 4.16. Three components of phase-averaged velocity at four crank-angle (CA) positions measured by the particle coded-pulse velocimeter. Note that the diameter of the field of view through the piston crown is less than the bore of the engine. Firing TDC is at 720° CA [73]. (See colour plate.)

band shows the optimum fuel injection timing during the intake stroke that enables quite stable combustion with a relatively low NO_x emission level. The questions thus arise as to how this rapid variation in the AFR arises with injection timing and whether further development could result in the engine running at even higher overall AFRs.

Although these measurements show how some aspects of the lean burn come about, the mechanism whereby injection upstream of open inlet valves leads to adequately stratified charge at the spark plug, after almost 360° of crankshaft revolution, is not clear. Extensive measurements by phase-Doppler anemometry of droplet size, velocity, and flux were carried out. These showed that gasoline injection against an open inlet valve controlled stratification through the amount of liquid phase entering the engine. The droplets evaporated at or near the cylinder liner, and gas-phase velocity measurements suggested that the resulting fuel-rich cloud was convected by the tumble vortex initially down towards the piston crown and subsequently back towards the cylinder head and into the pent-roof. Associated measurements of the fuel vapour within the cylinder measured by planar laser-induced fluorescence (PLIF) [74] confirmed this view of events. Figure 4.18 shows measurements at 1500 rpm, by PLIF, of the mixture distribution in the pent-roof, 6 mm below the spark plug at ignition [40° CA before top dead centre (BTDC)]. The light source was a XeCl (308-nm) laser sheet of 1-mm thickness, iso-octane was used as fuel and mixed with 10%vol/vol of 3-pentanone as the seed. A cooled charge-coupled device (CCD) camera with an image intensifier observed the fluorescence intensity: The mixture distribution strength is coded by colour, red showing the richer areas and

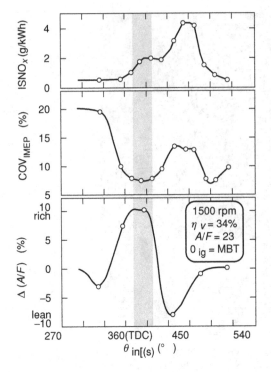

Figure 4.17. Influence of the timing of the start of injection on indicated specific NO$_x$ emissions, the covariance (COV) of the IMEP, and on the variation of normalised AFR. Intake TDC is at 360° CA [70]. MBT is minimum advance for best torque.

green showing the leaner ones (see colour plate). As the timing was changed from injection against a closed valve [Fig. 4.18(a), 0° CA] to against an opening valve during the induction stroke [Fig. 4.18(b), 20° CA after top dead centre (ATDC) and Fig. 4.18(c) 55° CA ATDC], a richer mixture appeared near the spark plug. Detailed

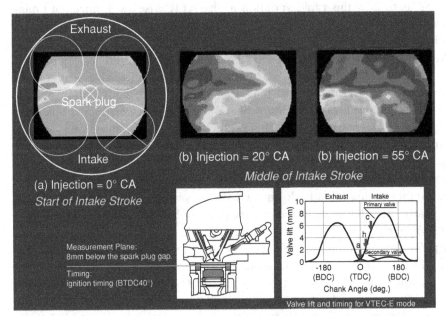

Figure 4.18. Contours of tracer PLIF measurements of fuel concentration near the spark-plug electrodes as functions of the start of injection timing (° CA afer TDC). Red is relatively richer, yellow is relatively leaner. Intake TDC is at 0° CA. (See colour plate.)

modifications to the design of the intake ports and to the fuel injection equipment were able to improve the limits of the design, for reasons described by Carabateas et al. [75], who performed measurements similar to those of [73]. To some extent, it remains surprising that a reasonably compact, relatively fuel-rich cloud can survive from just after TDC of inlet to TDC of compression. It may be that the suppression of turbulence by the swirl profile is an important near-wall mechanism [72].

As the AFR of the mixture is increased above stoichiometric, cycle-by-cycle variations in flame development and consequently in the IMEP become more prominent and eventually limit the range of lean-burn operation because these cause poor driveability. Therefore it is interesting to understand how, at least in principle, to extend the limits of the AFR at which engine designs operate. A critical stage of combustion in SI engines is during initial flame-kernel growth that corresponds, typically, to the duration of 0%–5% mass fraction burned (MFB). In particular, the correlation between the IMEP and the CA corresponding to 5% MFB ($\theta_{Xb5\%}$) has been found to be quite strong under low-load lean-burn conditions [76], although the figure of 5% is arbitrary in the sense that it was determined by experimental convenience: It is hard to measure lower values of MFB from the indicator diagram because of uncertainties. Imaging of the initial flame-kernel growth has proven particularly useful in investigating combustion behaviour at CA degrees closer to ignition timing than $\theta_{Xb5\%}$, corresponding to less than 0.1% MFB. The studies in [76–78] were concerned with investigation of such early stage flame-kernel activity by use of either stereoscopic or double CCD imaging per cycle. These images suggested that large variations affect the initial flame-kernel growth, either through stretch, wrinkling, inhibition of growth by wall proximity, or speed of convection towards, or away from, the electrodes. Surprisingly, and disappointingly, their findings suggested that the effects of flame convection, flame distortion, and heat losses to the electrodes were not the major contributors to cyclic variability. The differences between these quantities for very fast and very slow cycles were quite small. However, it was found that flames from fast cycles had occasionally shape properties similar to the ones of stoichiometric flames and a flame-kernel growth modelling approach also suggested that the level of cyclic variability in burning rates could be explained by large variations in the AFR. In subsequent work, [79] reported an experiment using chemiluminescence to evaluate the in-cylinder equivalence ratio of the reacting mixture and to further examine its contribution to the flame-growth speed and the cyclic variability in the CA by which 5% mass fraction was burned ($\theta_{Xb5\%}$). The results showed that the equivalence ratio exhibited large variations on a cycle-by-cycle basis and consistently produced negative correlation coefficients with $\theta_{Xb5\%}$, especially for lean set-operating conditions (AFR = 20–22). For injection against open inlet valves that yielded a stratified mixture, the degree of this correlation was in the range from about -0.4 to -0.8. Whether the level of cyclic variability in the fuel-droplet distribution patterns during the early intake stroke could account for the variations in the equivalence ratio was examined by Aleiferis et al. [80]. Their findings, based on light scattering from a planar laser sheet, showed consistent differences in the intensity scattered between poorer-than-average and better-than-average values of the IMEP. However, further work is needed to establish whether the larger scattered intensity associated with better-than-average cycles was due to a larger number of droplets or to larger droplets.

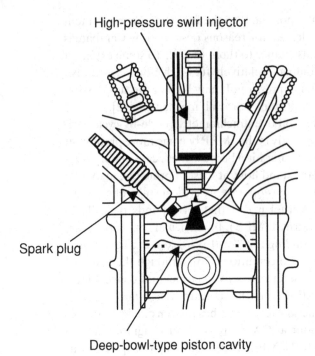

High-pressure swirl injector

Spark plug

Deep-bowl-type piston cavity

Figure 4.19. The i-VTEC I combustion chamber configuration [82].

Nevertheless, it is clear that PFI designs give rise to too much time between injection and ignition, owing to the distance between the injector and the spark plug, and too much turbulent mixing to hope that combustion at much leaner overall AFRs can be achieved. Fortunately, the ability to bring the injector and the spark plug closer together, and thereby burn at substantially leaner overall AFRs, began to be realised at commercial levels of robustness in the mid-1990s through DI of fuel into the combustion chamber, as described under the next heading.

Fuel Premixing Using DI

Wall-guided stratified DI concepts ('first generation'). Wall-guided designs were extensively developed and analysed by optical techniques, as shown by the comprehensive investigation [81] used to develop the first 'modern' commercial LBDI engine. We use the engine described in [82], one of the most recent designs to have used the stratified wall-guided approach, as an illustration of some of the design objectives that a stratified LBDI engine must achieve. The engine was based on the control of the burn rate by the VTEC mechanism (the same as the VTEC-E design previously described for the PFI) through the in-cylinder swirling motion of the gas. The engine is shown in Fig. 4.19, with the associated specifications in Table 4.3. In this design, a high-pressure atomiser was located in the centre of the combustion chamber with the fuel spray injected vertically downwards into the cylinder, towards a piston bowl, with the spark plug placed between the exhaust valves. As briefly described in the following discussion, optimisation was required of the mixture preparation and combustion through calibrations of the piston bowl, of the fuel-spray characteristics, and of the in-cylinder gas motion to enable stable combustion with an ultra-lean AFR. This centre-injection location eased the constraints on the timing of injection,

Table 4.3. *Specification of i-VTEC I engine*

Engine model code	K20B
Cylinder configuration	In-line 4-cylinder
Bore × stroke (mm)	86 × 86
Displacement (cc)	1998
Compression ratio	10:1
Valve train	DOHC[a] VTEC
Number of valves	4 per cylinder
Inlet-valve diameter	34
Exhaust-valve diameter	28
Fuel injection system	Programmed fuel injection
Fuel supply pressure (MPa)	8–10.5
Fuel	Regular (91 RON[*])
Max. power (kW/rpm)	115/6300
Max. torque (Nm/rpm)	188/4600

[a] Double overhead camshaft.
[*] Research octane number.

imposed by the injector and piston cavity geometry, found with the other choice for injector location, namely side injection. This made it possible to optimize injection timing for any operating condition and to achieve stratified combustion across a wide range of the operation map. However, in common with most engines of this design, stratified combustion was used on only a portion of one of the two VTEC-defined modes of the operation map; that is, only for the low loads and at low speeds. The general relationship between the brake-specific fuel consumption (BSFC) and the engine displacement is shown in Fig. 4.20. As the result of the stratified lean combustion, this engine achieved a 20% improvement in BSFC compared with that of engines of the same displacement under stoichiometric fuelling through the effects of reduced heat losses, increased specific heat ratio, and reduced pumping losses.

To meet the Japanese emissions regulations for 2005, during ultra-lean combustion, EGR was introduced. Figure 4.21 shows the influence of the EGR rate, at an engine speed of 1500 rpm and 118-kPa brake mean effective pressure (BMEP), on BSFC, engine-out NO_x emissions, and combustion stability: Note that the AFR could be as lean as 65:1 on this engine. The engine operating point that provided the

Figure 4.20. Comparison among BSFCs of PFI stoichiometric, PFI lean-burn and LBDI engines at 1500 rpm and a BMEP of 118 kPa. Modified from [82].

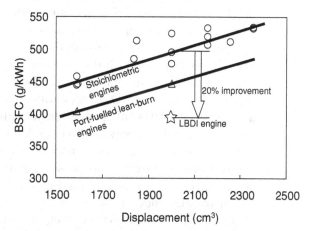

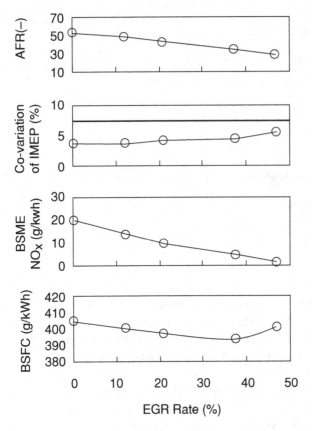

Figure 4.21. Influence of EGR on BSFC and engine-out emissions at 1500 rpm and a BMEP of 118 kPa. Modified from [82]. BSME is brake specific mean emission.

minimum BSFC had an EGR rate as high as 40% and yet it also has achieved stable combustion. In addition, this high EGR rate reduced engine-out NO_x emissions by 80% compared with the case without EGR, although the introduction of EGR slightly reduces thermodynamic efficiency because of its polytropic index and has, on occasion, been reported as increasing the emissions of hydrocarbons because of quenching. Together with an efficient lean NO_x catalytic converter, a vehicle with this LBDI engine achieved emissions at 25% of the level stipulated by the Japanese emissions regulations and was the first to obtain an 'ultra-low-emission vehicle' (ULEV) certification. To achieve this performance, higher conversion efficiencies of HCs during warm-up with stoichiometric operation, as well as during lean-burn operation, were realized through the precise control of AFR by use of a heated exhaust-gas oxygen (HEGO) sensor downstream of a three-way catalytic converter and a universal exhaust-gas oxygen (UEGO) sensor upstream from the converter. A high-efficiency lean NO_x catalyst with a conversion efficiency of about 98%–99% at around 300–350 °C of catalyst temperature was also installed. The vehicle was capable of a fuel economy of 15.0 km/L in the 10–15 Japanese driving cycles, which is 15% higher than the fuel-economy standard for the year 2010.

At high-power output, switching to two-intake-valve operation – with the intake stroke injection providing a homogenous stoichiometric mixture – increased the volumetric efficiency. Furthermore, the choice of injection geometry avoided wall wetting of the fuel on the cylinder wall, enabling reduction of unburned HCs (UHCs).

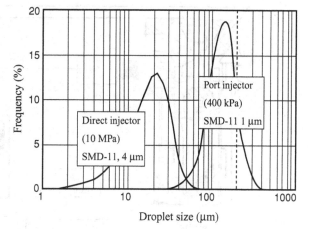

Figure 4.22. Comparison of SMDs between conventional port injector and high-pressure direct injector [82].

At low loads and speeds, to achieve ultra-lean stratified combustion, generating the adequately rich mixture clouds near the spark plug and achieving fast combustion were important. For mixture formation, it was better to restrain the in-cylinder flow to the extent possible. On the other hand, to realize rapid combustion, which improves thermal efficiency, a high turbulence intensity was required. To satisfy these opposing requirements and for the formation of stratification in the centre-injection combustion concept, the injector had to have a narrow spray angle, here set to 25°, and a spray that was capable of high penetration. The injection pressure was set at 10 MPa and the resulting Sauter mean diameter (SMD) of the fuel droplets was smaller than about 12 μm; see Fig. 4.22. Rapid combustion was obtained through the swirl generated by deactivation of one of the intake valves, using the VTEC mechanism.

To develop an efficient design of a combustion chamber, a commercial computational fluid dynamics (CFD) code (AVL "Fire" Version 7.1) was used, with experimental verification of its reliability of prediction of the mixture formation in the cylinder by comparison with experiment. Optical access through a glass piston crown provided measurements of the mixture preparation by use of PLIF, as described earlier. The instrument layout is shown in Fig. 4.23. The laser light sheet was introduced through the piston crown, and the mixture preparation in a vertical plane was observed under firing conditions. The comparison between PLIF experimental results and CFD estimations is shown in Fig. 4.24 for the particular conditions of an engine speed of 1500 rpm, an IMEP of 267 kPa, and an injection timing of 40° BTDC. A mesh was generated, with small amendments by hand, of between 50 000 cells at TDC and 90 000 cells at bottom dead centre (BDC) with an integration time step of 0.2° CA. Equations were solved in Reynolds-averaged form by use of the k–ε turbulence model, with boundary conditions at the valves provided by a duct acoustic simulation code (AVL BOOST). The wall temperature was fixed during the whole cycle. A richer-than-average mixture was found around the spark gap in both results, as well as on the opposite side of the cylinder bore from the spark plug, and the development of the mixture formation processes as functions of the CA was well matched. Figure 4.25 shows CFD results of the in-cylinder flow and the mixture preparation during stratified combustion. The penetration of the injected spray of the high-pressure fuel generated a toroidal vortex within the piston cavity,

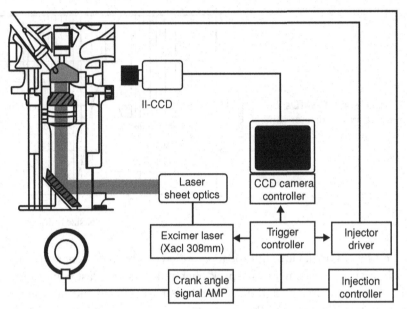

Figure 4.23. PLIF measurement system for visualisation of mixture formation [82].

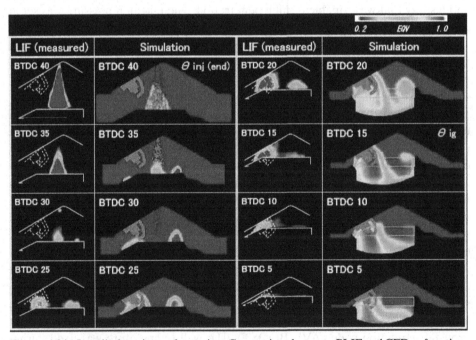

Figure 4.24. In-cylinder mixture formation: Comparison between PLIF and CFD as functions of the CA of the end of injection. Modified from [82]: Scale for equivalence ratio applies to CFD quantitatively, to experiment qualitatively. PLIF, planar laser-induced fluorescence. (See colour plate.)

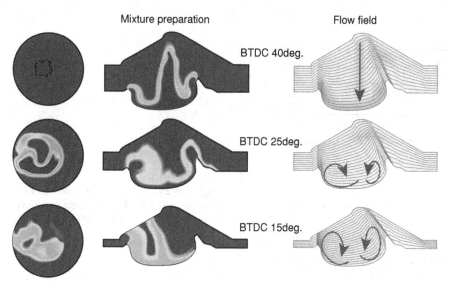

Figure 4.25. CFD predictions of in-cylinder mixture preparation [82]. (See colour plate.)

resulting in the required compact mixture cloud. The fuel spray injected was rapidly vapourised by impact with the piston wall, and the mixture was directed to the spark plug with the flow generated by the spray itself. At an ignition timing of 15° BTDC there was an ignitable mixture near the spark plug, and as a result even ultra-lean stratified combustion at an overall AFR of 65:1 could be reliably ignited.

Figure 4.26 shows combustion stability, engine-out HC, NO_x emissions and indicated specific fuel consumption (ISFC) as a function of the injection timing during stratified-charge operation at an engine speed of 1500 rpm and a BMEP of 118 kPa. The overall AFR was about 53 but nevertheless stable combustion across a wide range was achieved with an injection timing between 20° BTDC and 50° BTDC. The optimal injection timing for fuel economy, emissions, and combustion stability was found to be 40° BTDC. When the injection timing was advanced close to 60° BTDC, the mixture diffused, weakening the stratification and the co-variance of the indicated mean effective pressure (COV IMEP) became worse. On the other hand, if the injection timing was retarded to near 20° BTDC, the ISFC became worse. As the HC emissions increased in this scenario, there was probably an increase in late combustion and the delay in the combustion period caused the deterioration in the ISFC. The settings for injection timing were thus optimised for a good balance of emissions and ISFC across engine speeds and loads.

In other wall-guided stratified-charge DI engines, [59] reports on the emissions of soot from 'pool fires' on the piston surface, caused by the impact of the liquid fuel and its high-boiling-point components. Further consideration of such 'rich-burn' flames is outside the scope of this contribution, other than to note that smoke emissions are another restriction on the speed/load range for stratified engine operation. The ignition and combustion, as well as soot formation, of the premixed charge were examined through spectroscopic examination of the emission from the CN molecule, through OH* chemiluminescence, and using two-colour pyrometry respectively [83, 84]. The findings of [83, 84] are that optimum injection and ignition timing lead to an ensemble-averaged equivalence ratio at about $\phi = 1.5$, the bias to a rich average

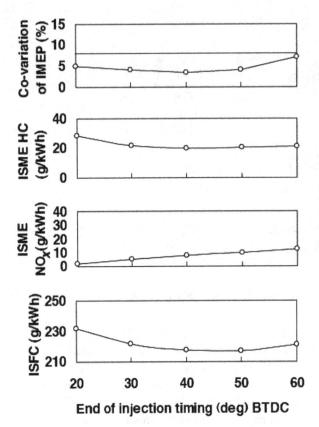

Figure 4.26. Influence of end-of-injection timing on ISFC and engine-out emissions at 1500 rpm and a BMEP of 118 kPa [82].

helping to obviate misfires (which occur only for $\phi < 0.5$) and also giving rise to tolerance of the large amounts of EGR. In addition, long-duration high-energy sparks are advantageous for reliable ignition of stratified mixtures. Combustion starts as a partially premixed flame that propagates through the flammable mixture. Sooting is observed in the hot post-flame gases as locally rich regions mix to burnable proportions, but this soot burns out rapidly because of its high temperature and rapid mixing with the surrounding hot lean regions.

Spray-Guided Stratified DI Concepts ('Second Generation'). The disadvantages of the wall- and air-guided designs are threefold from the point of view of realising lean-burn combustion [85]; the operating range with lean-burn combustion can be quite-limited, the efficiency potential is reduced by wall wetting and wall heat losses, and the CA at which 50% of the fuel mass has been burnt may sometimes be unfavourably phased.

The spray-guided system may be able to avoid these drawbacks. It involves having a spray discharging more or less along the axis of the cylinder with the electrodes of the spark plug placed close to the spray edge so as to be able to ignite the fuel vapour directly, without the need for further convective management by the in-cylinder flow. This design avoids – in principle at least – contact of the spray with any wall and thus avoids one source of UHC emission. The fuel cloud can consequently be smaller, resulting in faster combustion, with fewer cyclic variations in location or CA and with lower smoke emission. Spray guidance thus delivers [62] greater potential of the DI stratified concept, resulting in a wider operation map for stratified-charge combustion with improved fuel economy – provided that means can indeed be found to achieve these objectives successfully over a wide speed and load range of the engine's operation map. It is reported that, based on the new European drive cycle (NEDC) European drive cycle and compared with a wall-guided design, a spray-guided design had 4% better fuel economy, a factor of two lower UHC emissions and, as expected, lower smoke emissions [59].

From the preceding qualitative description, it is clear that spark location and timing relative to that of the spray – and the associated fuel vapour – is critical [86, 87]. The time available for generation of a vapour cloud conducive to lean burn combustion is short because the time available for injection and mixing is a few milliseconds. This implies the need for rapid fuel evaporation and hence droplets of the order of 20 μm in diameter, which in turn implies the need for development of a high-pressure (200-bars) gasoline fuel pump, and a design of spark plug open to the flow of mixture with a suitable – and spatially extended – spark discharge. Thus the relative orientation of the spark plug, the injector, and of the intake valves is an important parameter. Evaporation and subsequent mixing should also be rapid in order to avoid the generation of particulates, and hence this favours the development of fuel injection equipment operating at yet higher pressures than required for producing a fine spray. The UHC emissions can be reduced by reducing the time between the end of injection and the start of ignition because this results in less time for the fuel cloud to mix with the surrounding gases to form a too-lean mixture. It is beneficial to minimise – or preferably eliminate – the wetting of the spark plug by the liquid fuel (to avoid fouling), while nevertheless reliably establishing near the electrodes an ignitable vapour mixture at a thermodynamically appropriate ignition angle over a wide range of engine speeds – and hence wide range of flow speeds and also fluctuations of fuel vapour – while maintaining stratification of the remainder of the charge to derive the benefits associated with lean burn. It is therefore reasonable to inquire as to how, where, and when the vapour phase 'escapes' from the two-phase jet for this condition to be satisfied, given that the jet will be entraining the surrounding gases vigorously.

The fuel can be introduced into spray-guided engines by either high-pressure-swirl injectors, or solenoid-controlled multihole injectors, or piezoelectric 'outward-opening' injectors. Figure 4.27 compares these three designs at atmospheric and higher back-pressures. For repeatable formation of a lean cloud the cone angle should, as previously implied, ideally be independent of the back-pressure, and these particular multihole and piezoelectric injectors fulfil this requirement. It is also desirable that the fuel spray not penetrate too far, to avoid wall wetting, and the piezoinjector fulfils this requirement well. For the multihole injector, the rate of momentum transfer from the liquid to the surrounding gas is poorer: On the other hand, the multihole injector is likely to be less affected by the in-cylinder airflow relative to a pressure-swirl atomiser [88]. In addition, some multihole injectors induce a flow in the surrounding gas that can alter the so-called 'umbrella angle' of the injector [89]: However, other multihole designs can provide better spray targeting to avoid wall wetting. The rapid mechanical response of the piezoelectric injector permits multiple injections within the combustion cycle, and it is claimed [85] that this contributes to, and can extend, stratified-charge operation by preventing overly

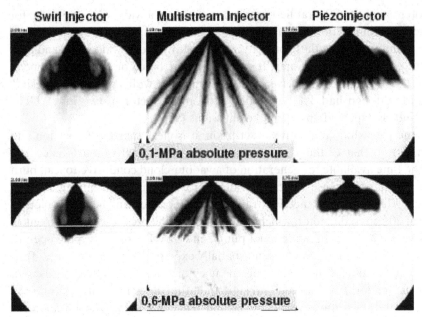

Figure 4.27. Spray images of a swirl, a multihole, and an outward-opening injector for atmospheric conditions and 0.6-MPa absolute back-pressure [85].

lean or rich areas, the former contributing to flame extinction and HC and CO emissions, the latter to soot formation. Furthermore, [67] reports that the greater the quantity of fuel to be injected, in other words the higher the load, the greater the effect of multiple injection on mixture formation. This observation thereby raises the possibility of extending lean-burn stratified operation to higher loads.

As a consequence of measurements by use of optical instrumentation, the fundamental processes forming the lean mixture are known, at least qualitatively. There are extensive parametric measurements of the velocity field accompanying the impulsively started spray issuing from, respectively, a hollow cone pressure-swirl injector and from an outward-opening piezoelectric injector [90–92]. For the pressure-swirl injector type, Fig. 4.28 shows the essence of the mechanism deduced from the measurements from a two-phase PIV instrument at two ambient pressures, the hollow cone spray collapsing at higher pressures. An inner-vortex structure and an outer-vortex structure are established by momentum exchange with the spray as it atomises, and this induced airflow is responsible for the structure of the spray as a function of the ambient pressure. At higher pressure, the entrainment at the outer surface of the cone is reduced because of the higher droplet density that entails an area of low pressure inside the spray, followed by the creation of a strong inner vortex. The latter drags the droplets to the nozzle axis, leading to a reduction in the spray cone angle. The temporal variations of spatial equivalence ratio were measured [93] and the effect of ambient air entrainment was compared for single and split injections, as shown in Fig. 4.29. The conclusion is that controlled split injection can contribute to the optimisation of combustion in DI spark ignition engines. The phenomenon of high concentration of liquid fuel that is due to droplets 'piling up' at the leading edge of such a spray is circumvented to some extent by use of appropriate split-pulse injections. Also, the existence of excessively lean mixture in

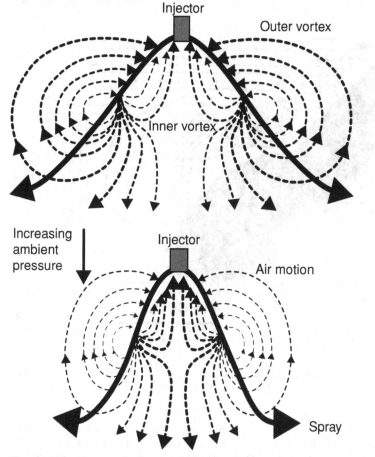

Figure 4.28. Pressure-swirl atomiser: mechanisms of spray–air interaction at different ambient pressures [90].

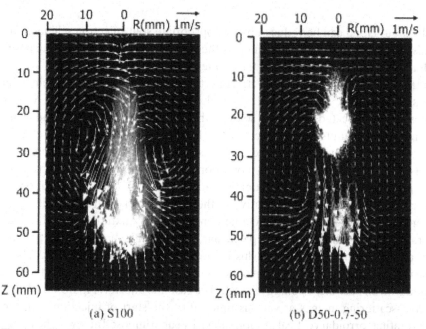

Figure 4.29. Pressure-swirl atomiser: Effect of ambient air entrainment on the mixture formations of (a) single injection and (b) split injections at 2.5 ms after start of injection. Injection pressure, 4.6 MPa; ambient pressure, 0.6 MPa; $T = 300$ K [93].

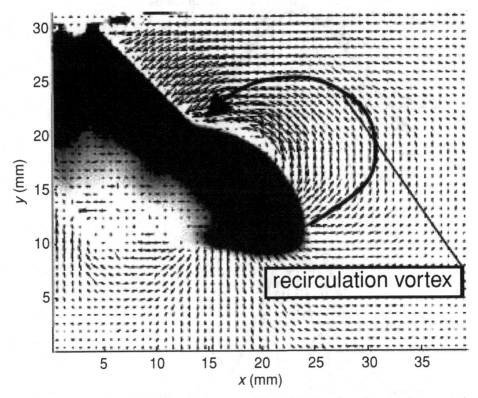

Figure 4.30. Piezoelectric injector: spray propagation and induced gas flow; back-pressure is 6.2 bars, injection pressure is 200 bars, 500 μs after start of injection. The maximum velocities in the picture correspond to 12 m/s [92]. The vector plots are blended with monochrome images of the spray, which is visualized with the Mie-scattering technique.

the fuel vapour cloud can be reduced by an appropriate split-injection scheme. The mechanism by which split injection changes the mixture and spray characteristics is through the spray-induced motion of the ambient air.

For the outward-opening piezoelectric injector, a broadly similar picture emerges [92], as shown in Figs. 4.30 and 4.31. As will be subsequently shown, the vortex structures can be expected to capture some of the evaporated fuel vapour and 'trap' it as a more-or-less uniform mixture: So the location of the vortex centres as a function of back-pressure, of injected fuel mass and, most important, of multiple injections – which improve operating conditions in stratified mode – is important. These aspects were investigated in detail, and it was found that the position of the two counter-rotating vortices relative to the nozzle is very stable. This leads to robust engine operation with predictable conditions for mixture formation over a wide range of operating points. In particular, multiple injections can control the position and strength of the vortex. This is shown clearly by Skogsberg et al. [94], who also investigated the development of the velocity and droplet-size characteristics associated with an outward-opening piezoelectric gasoline injector. Figure 4.32 shows planar Mie-scattering images at various times after the start of injection with the counter-rotating torroidal vortical motion shown by superimposed arrows. The upper parts of the fuel cloud remain in the vicinity of the spark plug well past the time

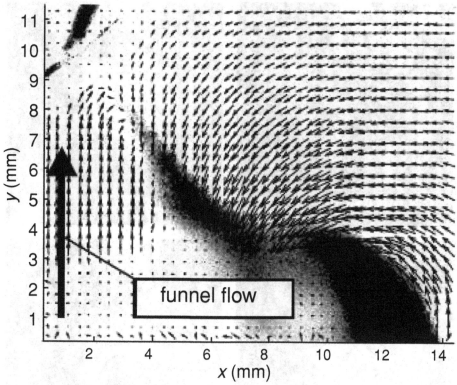

Figure 4.31. Piezoelectric injector: spray propagation and induced gas flow close to the injector; back pressure is 6.2 bars, injection pressure is 200 bars, 400 μs after start of injection. The maximum velocities in the picture correspond to 12 m/s [92]. The vector plots are blended with monochrome images of the spray, which is visualized with the Mie-scattering technique.

corresponding to the end of injection. This should be compared and contrasted with Fig. 4.33, which shows the corresponding image when the same mass of fuel is injected in two separate injections. A third vortex, introduced by the second injection, is formed at the outside edge of the spray. This results in the upper parts of the spray being pushed back towards the spark plug as a result of the interactions between the spray-induced vortices: A split-injection strategy can reduce the axial penetration, a phenomenon that may be useful in moderation to keep a suitable equivalence ratio near the spark plug – or a nuisance if the droplets strike the injector. Both of the currently commercially available stratified-charge engines make use of outward-opening piezoelectric injectors and both make use of up to three injections per cycle. A PLIF image of the fuel spray and vapour distribution for a stratified operating condition was reported in an optical engine [62]. It is claimed that the recirculation area of the vapour phase is formed at the spark-plug position, providing an ignitable mixture at the electrodes.

A great deal of research has been published for multihole nozzle sprays. Figure 4.34 [95] shows a planar visualisation of the liquid and vapour concentration of a six-hole 50° umbrella angle nozzle as a function of time after the start of injection in a quiescent chamber using the laser-induced exciplex fluorescence technique. The features of relevance here are that fresh air is entrained at point A-1

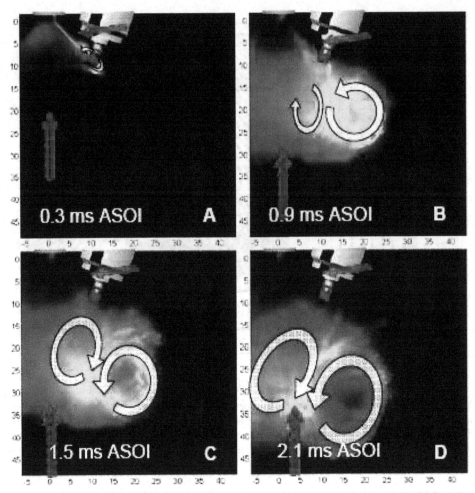

Figure 4.32. Piezoelectric injector: planar droplet–Mie scattering showing the liquid-fuel distribution at four time points from a single injection of 24-mg fuel. The injector is at the upper left; a spark plug, superimposed on the images, is at the top centre. ASOI denotes after start of injection [94].

as a result of the spray's motion. At point B-2, a vortex caused by the propagating spray transports vapour away from the liquid fuel: One can expect a similar, perhaps more efficacious, phenomenon to occur with an outward-opening piezoelectric injector. As the injection event progresses, the spray at point C-2 starts to move so as to effectively decrease the umbrella angle. By time D, when the injection has ended, the resulting combustible vapour cloud resides within the umbrella angle of the spray. Cycle-resolved simultaneous planar measurements of flow field and *quantitative* equivalence ratio in a engine under motored and fired conditions using *iso*-octane fuel were made recently [96]. Note that measurements and analysis suggest that the evaporation process is diffusion controlled and the effects of sequential evaporation are expected to be insignificant on fuel stratification [97]. This is part of a research programme seeking correlations between marginal engine operation and local instantaneous mixture and flow conditions [98]. Figure 4.35 shows the coupling of spray momentum with air–fuel mixing within a single cycle for a motored

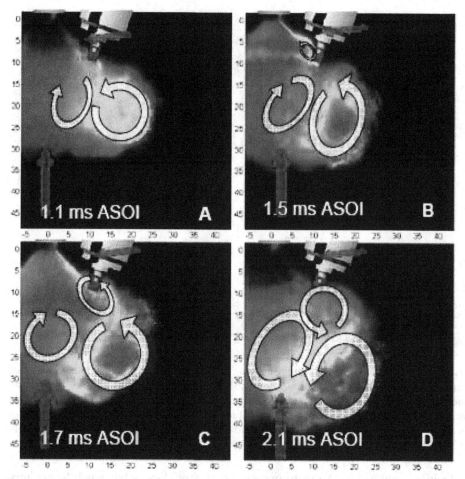

Figure 4.33. Piezoelectric injector: planar droplet–Mie scattering showing the liquid-fuel distribution at four time points from a split injection of 24 mg. ASOI denotes after start of injection [94].

engine operating at stratified conditions with one of the sprays deliberately targeted to strike the spark plug with the orientation shown: As might be expected from this figure, spark-plug orientation can affect both the ignition delay and the 10%–50% burn time [99]. Somewhat surprisingly, the majority of the liquid spray evaporated within 2° CA after the end of injection [100]. Also, the arc is strongly influenced by the gas motion induced by the spray and its vapour: Broadly, conditions that lead to shorter-duration arcs are unfavourable [101]. The first part of the fuel-vapour cloud has an equivalence ratio that is fuel rich but rapidly becomes fuel lean: This is also shown in Fig. 4.36, which is a scatter plot of spatial averages of equivalence ratio and velocity magnitude from the region close to the spark plug shown in the inset.

Although the overall ('delivered') equivalence ratio to a stratified-charge engine can be very lean, the optimum equivalence ratio at the time of ignition for reliable ignition ranges from about 0.6 [87] to 2 [59]. These values represent a compromise between conditions at advance, which have much richer conditions and high flow velocities, and conditions at retard, which have lower and more favourable

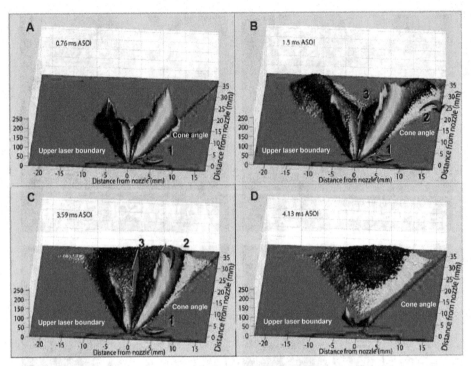

Figure 4.34. Multihole nozzle: direct imaging results using laser-induced exciplex fluorescence from a 50° 'umbrella' multihole nozzle, showing qualitative time-resolved results from measurements made in a chamber at 10 bars and 584 K. The results were processed so that the vapour concentration is shown by the magnitude of the z axis, and the liquid concentration is represented by the colour, red–high, blue–low in the z axis. ASOI is after start of injection [95]. (See colour plate.)

velocities – say below 15 m/s to prevent spark stretching – but much leaner mixture strength. The subsequent progress of combustion was gauged from spectrally filtered observation of the spatial and temporal development of emitted light. The discussion in [87] describes this for a particular engine as a two-stage process, as shown in Fig. 4.37, of the partially premixed combustion (visualised through OH* chemiluminescence) followed by a second stage of mixing controlled combustion of fuel-rich zones (visualised from the luminosity of hot soot, through black-body emission wavelengths). For advanced ignition (SA = 34° BTDC), when the equivalence ratio is much richer at the spark plug, the results show both premixed and rich combustion starting at the spark and propagating around the edges of the piston bowl. Retarded ignition (SA = 22° BTDC) results in much slower combustion than for the early ignition.

Lean-Burn, Compression-Ignition Combustion

CONCEPTS. To retain the advantages of unthrottled operation and to burn the fuel at yet leaner conditions than in DI SI engines, in the pursuit of further reduction in NO_x and smoke emissions, ideally to the point that no after-treatment is required at all, implies that the combustion in a premixed or partially premixed mode should not exceed a temperature of about 1800 K. Temperatures below 1800 K avoid the rapid rise in the generation of NO_x and both any rich equivalence ratios and

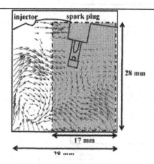

The area of measurement of fuel concentration measurements (the 17 mm × 28 mm region darkened region), and of velocity (the 28 mm × 28 mm viewing plane)

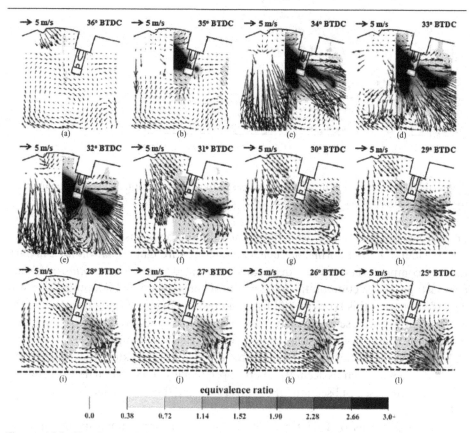

Figure 4.35. Simultaneous fuel concentration and velocity measurements inside a motored spray-guided DI engine at stratified conditions with a multihole injector. One of the sprays was deliberately aimed at the spark plug [96].

non-premixed flames are to be avoided as they give rise to particulates. It is also desirable to keep the temperature above about 1500 K to promote the conversion of CO to CO_2. These conditions imply operation near, or even below, the lean-mixture limit, which is about $\phi = 0.5$, depending on the amount of turbulence in the combustion chamber and on the compression ratio [52]. Alternatively, mixture strengths

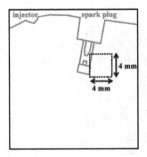

Spatial averages are based on the 4 mm × 4 mm region

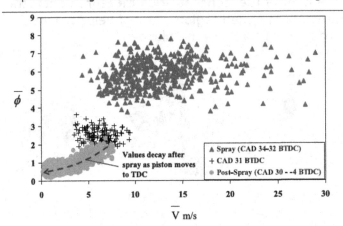

Figure 4.36. Scatter plot of the spatial averages of fuel concentration and velocity magnitude from 150 motored cycles. CAD denotes CA degrees [96].

up to stoichiometric can be used, but greatly diluted by ballast–bath gas: These are not considered here. In principle such lean mixtures are ignited, in whole or in part, through compression work and deliberately arranged to give rise to *quasi*-isochoric combustion: This section outlines the practical consequences of adopting this strategy. We note that this type of combustion in piston engines has many names in the literature. Here we use two widely, but not universally, adopted terms, namely homogeneous charge compression ignition (HCCI) and low-temperature combustion (LTC), the latter particularly for diesel-fuelled engines. The extent to which the former term, in particular, is accurate or appropriate has been debated and questioned in the literature (see Subsection 3.1.3, and [102]), and there is no need to repeat that debate here.

Although forms of HCCI have been known for some time, including some designs of two-stroke engines for model aircraft that run on appropriate blends of fuel, research on this topic began in the late 1970s and has since resulted in HCCI combustion becoming commercially available, albeit in small numbers, in a couple of two-stroke engines. In four-stroke diesel-fuelled engines, forms of LTC strategies reportedly found some commercial application since about 2000 [e.g., the 'modulated kinetics' (MK) [103] and the 'uniform bulky combustion' (UNIBUS) [104] approaches]. For gasoline-fuelled engines, many manufacturers of four-stroke engines have been researching this concept for some time [105, 106]. Other fuels, including natural gas and dimethyl ether (DME), have also been investigated [107, 108].

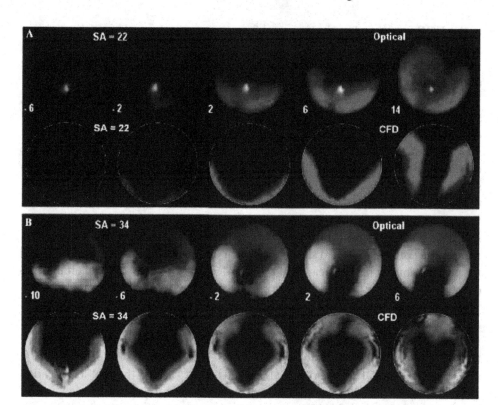

Figure 4.37. Effect of two values of ignition timing (denoted as SA = 22 and SA = 34, in units of degrees CA BTDC) on the progress of combustion in a spray-guided stratified-charge DI engine. Top row of each two-row set: visualisation through ensemble-averaged spectrally resolved imaging: OH* chemiluminescence (blue) and radiative emission from soot (red and green). Bottom row of each two-row set: CFD calculations of heat release rate (blue) and rich combustion products (yellow) [87]. (See colour plate.)

Overview of HCCI Engines. In an HCCI engine, ideally all of the fuel is premixed with the air before auto-ignition takes place, although the fuel mixture fraction (and temperature) may well be – deliberately or otherwise – spatially inhomogeneous to a greater or lesser extent. The premixing is achieved by use of a PFI or by injecting directly into the combustion chamber: In the latter case, it may be useful for the fuel injection equipment to be agile (in the sense that it can provide a considerable degree of flexibility in the number of injection pulses and the start and duration of each pulse). Depending on the fuel and the compression ratio, compression work provides some or all of the increase in the internal energy of the charge to promote auto-ignition. If additional energy is required, it is provided by a variety of techniques, including heating of the intake air [109], retention of residuals [110] or internal EGR [111]. The start of ignition at a thermodynamically appropriate point in the cycle implies some sort of control, usually through control of the initial temperature of the reactants and of its subsequent development during compression. This is because there is usually no 'external' event to promote ignition, such as a spark or fuel injection as in conventional SI or CI engines (however, spark-assisted HCCI is an area of active research [112–115] with conflicting reports on efficacy [59]. The use of a laser ignition source – based on the absorption of energy by moisture contained in the EGR at a wavelength of around 3 μm – was also recently investigated in a vessel in [116]). Investigations into means of control of the phasing of combustion over the range of speed and load for which HCCI/LTC can be achieved is thus one of the main avenues of research.

Although auto-ignition of the premixed charge can be virtually instantaneous, such isochoric combustion must be avoided. To avoid damage to the engine as the charge burns, and to limit the noise, vibration, and harshness from the engine, the rate of pressure rise must be limited to about 10 bars/° CA. With increasing load, this condition becomes increasingly hard to achieve and may eventually lead to the undesirable phenomenon of *engine knock* at an equivalence ratio as low as about 0.3 [117]. Computational estimates suggest that, with a truly spatially homogeneous charge, this rate of pressure rise would occur at a much lower equivalence ratio, and [118] shows, on the basis of measurements of chemiluminescence, that ignition fortunately starts at one or more isolated sites, rather than uniformly throughout the volume of the combustion chamber, because of more-or-less inevitable inhomogeneities in local conditions. This is corroborated by measurements of fuel distributions [119, 120]. It is argued that, in practice, it is the occurrence of either inevitable thermal stratification or incomplete mixing between the fresh charge and hot residuals [121] that gives rise to 'sequential' auto-ignition, starting with the hottest zone and proceeding thus throughout the charge to the coolest. This reduces the rate of rise in pressure at any given overall equivalence ratio. It is argued that it is mainly thermal stratification in the gases far from any thermal boundary layers that controls the maximum rate of rise in pressure [118], and [117] concludes that means to enhance 'naturally' occurring thermal stratification remain a challenge. An interesting question is whether any deflagration fronts coexist with the 'volumetric' combustion associated with auto-ignition. Some hold that there is evidence for this, at least under some circumstances, experimentally [122] and computationally [123]: Others hold that turbulence is likely to extinguish flames at such lean equivalence ratios. Recent debate is summarised in [59] and is also mentioned in Subsection 3.1.3.

It is known that retarding combustion after TDC is effective in allowing operation at equivalence ratios in excess of 0.3, while respecting the limit to the rate of pressure rise mentioned earlier by increasing the duration of burn. Dec and his co-workers [124–126] argue, on the basis of single-zone and multizone chemical kinetic modelling, that this is due neither the increase in the volume expansion rate associated with retarding after TDC nor to the lowering of chemical reaction rates associated with the lower combustion temperatures. Rather, it is due to an 'amplification' of the effects of thermal stratification: Expansion of the gases after TDC increases the induction time. However, increasing retard leads [127] initially to increased cycle-to-cycle variations, which are undesirable. These authors argue that, for a given combustion phasing, the variations are smaller for a fuel exhibiting two-stage, rather than single-stage, ignition (see subsequent discussion) by considering the higher rate of rise in temperature for a two-stage ignition fuel just before so-called 'hot ignition'. Further retarding leads to *misfire* and partial burning of the charge: For a two-stage ignition fuel, this leads to complete loss of ignition in subsequent cycles, and, for a single-stage fuel, this is followed by an advanced cycle with complete burning because of the presence of species in the partial burn cycle, which enhances the auto-ignition of the following cycle. With an increase in the load, that

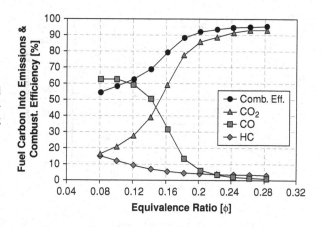

Figure 4.38. Combustion efficiency and emissions vs. equivalence ratio for well-premixed *iso*-octane and air; 1200 rpm; $p_{inj} = 135$ kPa, CR = 14 [117].

is, with an increase in ϕ, combustion phasing must be between the knock and the misfire limits. The angle between these decreases with an increase in ϕ and thus, currently at least, HCCI is limited to loads that are about half that typical of SI or CI operation. Thus it may be necessary to operate in two modes, with high power being provided in gasoline-fuelled engines by either conventional SI stoichiometric DI or, perhaps, by SI stratified DI. If this were the case, the maximum compression ratio of the engine would be limited by the SI mode of operation.

The operation of the HCCI engine is also limited at low load – for example operation at an equivalence ratio of, say, $\phi = 0.12$ at idle [117] – by poor combustion efficiency caused by high emissions of CO and UHCs. At these low loads, incomplete bulk gas reactions begin to dominate [128] because the temperature is too low (at higher-load operating points, these are from the near-wall regions and top land crevice). Figure 4.38 [117] shows that this tendency starts at $\phi \approx 0.2$ for the particular conditions of this figure. As the temperature falls below a level corresponding to a peak charge temperature of 1500 K, the OH concentration decreases rapidly and hence the main reaction for CO oxidation, namely CO + OH $\rightarrow$ CO$_2$ + H, is quenched by piston expansion. Figure 4.39 [129] suggests that fuel stratification by retarding injection can provide a solution but, at very retarded injection, NO$_x$ emission increases because regions of ϕ greater than about 0.6 arise.

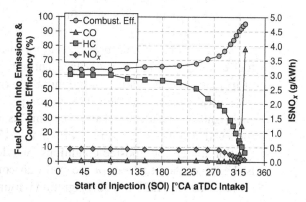

Figure 4.39. Combustion efficiency and emissions as functions of start of injection for an eight-hole injector. Low swirl (SR = 0.9) and $p_{inj} = 120$ bars [129].

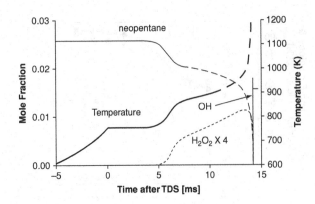

Figure 4.40. Computed temperature and concentrations of neopentane, H_2O_2, and OH, cited by [132].

Overview of Fuels, Including Single- and Two-Stage Auto-Ignition Characteristics. A study of the chemistry of auto-ignition is fundamental to the understanding of the operation of HCCI engines, and consequently much work has been published on this aspect (e.g., [130–132]). Figure 4.40 [132] shows the particular example of a fuel with two-stage ignition delay at a post-compression temperature of 757 K showing profiles of temperature and concentrations of hydrogen peroxide and OH. The first stage (associated with 'cool flames' and negative temperature coefficients in the variation of ignition delay times plotted against the reciprocal of temperature) shows a delay of about 5 ms to a temperature of about 880 K with second-stage ignition occurring at about 14 ms, when the temperature is between 900 and 1000 K. There is a modest temperature increase associated with the first ignition, which is the result of low-temperature HC oxidation. Although a complete calculation requires advanced reaction schemes, a readily appreciated description of the origin of two-stage combustion is provided by [133]. This reference analyses two-stage ignition for the particular example of *n*-heptane using largely an algebraic, rather than a computational, approach based on two reduced four-step mechanisms. The 'first stage ignition is related to a change from chain branching to chain breaking as a temperature threshold is crossed. The chain branching reactions build up ketohydroperoxides which dissociate to produce OH radicals. As the temperature threshold is crossed, this production ceases. The second stage is driven by the much slower production of OH radicals owing to the dissociation of H_2O_2. The OH radicals react with the fuel at nearly constant temperature until the latter is fully depleted'. A closed-form solution elegantly describes the first-stage ignition, the second-stage ignition, and the origin of the crucial 'temperature-crossover' point, as well as its magnitude. The crossover temperature is a weak function of pressure, at least over the range relevant to automotive boosting.

It is of considerable practical importance in HCCI engines as to whether a fuel exhibits a single- or two-stage ignition: The latter requires smaller amounts of increase in temperature of the initial charge before compression work is added or, equivalently, can operate at lower compression ratios. Typical fuels showing two-stage ignition include diesel and *n*-heptane as well as blends of the primary reference fuels with a sufficient content of *n*-heptane [134]. Typical fuels showing single-stage ignition under non-boosted conditions include gasoline, *iso*-octane, and ethanol. It is the molecular structure of the fuel that controls the type of ignition that is observed: Broadly [132], two-stage ignition is found in long, linear alkane fuel molecules

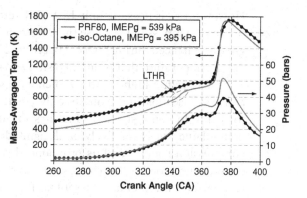

Figure 4.41. Temperature and pressure traces for *iso*-octane and PRF80, ($\phi = 0.40$) [135]. LTHR, low-temperature heat release.

and single-stage ignition in highly branched fuel molecules, containing primary C–H bonds that are difficult to abstract. Apart from the two major commercial fuels, gasoline and diesel, many other fuels have been used to fuel HCCI engines, including mixtures of two so-called primary reference fuels (PRFs), *iso*-octane and *n*-heptane. The advantage of PRF mixtures is that these allow a controlled parametric investigation into HCCI characteristics based on reasonably well-known and understood chemistry.

Figure 4.41 [135] shows the difference in auto-ignition, for a lean mixture, between a fuel with single-stage (*iso*-octane) and one with two-stage (PRF 80) ignition, the latter showing a low-temperature heat release (LTHR) at about 340° CA. The practical importance is that, because the two-stage fuel allows a lower intake temperature, a higher charge mass can be inducted into the cylinder, and hence a higher power output is delivered. The fuel with the two-stage ignition also has smaller cycle-to-cycle variations and can withstand greater retarding of ignition, as a consequence of LTHR damping out random variations in intake temperature and of the enhancement of intermediate-temperature chemistry that is due to the LTHR, respectively [117]. Therefore higher loads can be delivered by fuels with some two-stage ignition than with single-stage ignition: As discussed in [136], the limiting loads – depending on the extent of the LTHR – can be related to either to the available oxygen because of the amount of EGR required to retard combustion phasing or to the production of small amounts of NO_x that advance combustion phasing to runaway knock.

It is of fundamental as well as of practical importance (e.g., for engine control) to know how changes in ϕ shift the combustion phasing. An intriguing experiment [137] shows how to remove the thermal effects of varying ϕ (i.e., changes in wall temperature and residual temperature). For two-stage fuels (but not single-stage fuels), ϕ shifts the combustion phasing. These two-stage ignition fuels become sensitive because the intensity of the low-temperature auto-ignition reactions is proportional to the local fuel concentration. In turn, this makes the timing of the 'hot' main combustion event sensitive to the local ϕ. Therefore, introducing a range of ϕ in the combustion chamber leads to a staged combustion event with lower pressure-rise rates.

For a research-grade gasoline, the effect of intake pressure boosting is to enhance auto-ignition, as might be expected, so that, for example, intake temperature has to be reduced and cooled EGR has to be increasingly added [138]. The maximum gross IMEP (over the compression and expansion strokes only) increases with boost.

However, this is possible only by highly retarding combustion to avoid knock, and this in turn is possible because the 'intermediate-temperature heat release' becomes greatly enhanced during boosted operation. This latter is then able to counteract the fall in temperature that is due to piston expansion during highly retarded ignition.

Fuels with a high volatility, such as gasoline, form a mixture relatively easily, and, because this fuel exhibits only single-stage ignition, it is relatively difficult to ignite. In contrast, diesel fuel has the opposite characteristics: It is relatively non-volatile but is relatively easily ignited because of its cetane number through compression ignition and its pronounced two-stage ignition and combustion. As a consequence, combustion can easily become overadvanced. Although HCCI engines can be run on diesel and diesel-like fuels, this combination of characteristics has led researchers to consider an alternative, and perhaps more promising, approach to secure low-emission, high-efficiency combustion. Here we refer to this as low-temperature combustion (LTC) and consider it in greater detail under a separate subheading.

Brief Introduction to Control of HCCI Through Combustion Measures. With a compression ratio of between 11.5:1 and 15:1, a homogeneous air–fuel mixture with 'regular' gasoline will not realize stable compression ignition through a rise in the temperature of the air–fuel mixture caused by compression work from the piston alone, even with mixing from the inevitably hot residuals. The temperature of the mixture must be brought to about 1050 K to achieve ignition near TDC. Intake air heaters are a convenient way to aid ignition in laboratory research work but are not feasible for commercial vehicles because of the large amount of power that these consume. A practical alternative may be to recirculate *hot* exhaust gases (exhaust gas recirculation – EGR) either internally or externally. The addition of EGR to the fresh charge alters the composition and thermodynamic properties of the contents of the cylinder. The mole fraction of exhaust gases that is required decreases with increasing load as these become hotter and as the surface temperature of the cylinder wall increases. Eventually it is possible to raise the temperature to such a level that unacceptably high levels of NO_x begin to be generated. An alternative strategy might then be to operate with the fresh charge at stoichiometry (rather than ultra-lean), to dilute the charge strongly with EGR to bring down the rate of rise of pressure to acceptable limits, and to remove any NO_x emissions by a conventional three-way catalyst.

The use of exhaust gases as diluents has at least four effects [139, 140]:

- *Charge heating effect.* This describes the turbulent heat transfer between the exhaust and residual gases and the fresh charge.
- *Dilution, or O_2 concentration reduction, effect: retarding.* The retarding effect is stronger for two-stage ignition fuels than for single-stage ignition fuels.
- *Heat capacity, or thermodynamic cooling, effect: retarding.* Triatomic molecules (such as CO_2 and H_2O) have a higher value of c_v than does air and the retarding effect is stronger for two-stage, as compared with single-stage, ignition fuels.
- *Chemical effect, which is due to*
  - *enhancement of autoignition* due to the presence of H_2O; this effect is stronger for two-stage ignition fuels, and

- *Enhancement/suppression of auto-ignition* due to the presence of trace species. Auto-ignition is enhanced for single-stage ignition fuels but, for PRF 80, a two-stage ignition fuel with 80% *iso*-octane, ignition was retarded.

Zhao et al. [139] found that ignition is dominated by the charge heating effect but that combustion duration is dominated by the dilution and heat capacity effect. The maximum rate of apparent heat release was equally affected by the charge heating effect and by the combined dilution and heat capacity effect. From this, they conclude that, for high-octane fuels, such as gasoline, alcohols, and natural gas, it is advantageous to retain burned gases at as high a temperature as possible to promote auto-ignition. For diesel fuel and DME, cooled burned gas is preferable in order to increase the ignition delay period and thereby enhance the premixing.

Apart from the use of EGR in HCCI engines to control ignition timing (as hot EGR, cooled EGR, or internal EGR), high levels of cooled EGR have also been held to extend the high load limit through control of the rate of rise of pressure. This has been attributed to a reduction in the rate of apparent heat release that is due to lower combustion temperatures and reduced oxygen concentration. However, Dec et al. [141] recently pointed out that EGR (or high levels of retained residuals) also affects the phasing of combustion and that this might equally be the cause of the effects observed. They highlight, in addition to effects on the rate of pressure rise, inconsistent findings in the literature concerning the effects of EGR on NO_x emissions and thermal efficiency. Although EGR certainly *is* associated with extending the high load limit and with inducing low peak combustion temperatures and associated NO_x formation, these attributes are possibly the result of ignition retarding – which can be produced by means other than the introduction of EGR (for example, by adjusting the intake temperature). Their conclusion, from experiments, is the surprising finding that EGR *increases* NO_x levels at constant gross IMEP (that is, IMEP excluding pumping work) and constant combustion phasing: This is because, under these constraints, the introduction of EGR necessitates increased fuelling to compensate for the reduction in thermal efficiency consequent on the lower ratio of specific heats of the EGR gases. The increased fuel increases the peak combustion temperature. However, again under the same constraints, EGR did produce a small reduction in the rate of pressure rise, which in turn permitted a modest increase in the gross IMEP. However, the authors point out that the same increase could have been obtained without EGR by retarding the timing by only $1°$ CA at a constant rate of pressure rise. Their overall conclusion is that EGR has drawbacks but is necessary for the control of ignition timing (and can be used to permit the operation of a three-way catalytic converter). In addition, in stratified or poorly mixed out charge, EGR can limit NO_x emissions by limiting local peak temperatures.

Two-Stroke HCCI Gasoline Engines

HCCI research can be said to have started with the two-stroke engine, conferring – at part load – improvements in stability, fuel efficiency, exhaust emissions, noise, and vibration, and it remains an area of active research. Duret [142] surveyed recent research and shows the advantage of lean-burn HCCI in improving the phasing of combustion and increasing the duration to mitigate excessive rates of pressure rise. This engine type is also an example of early commercialisation of the HCCI concept.

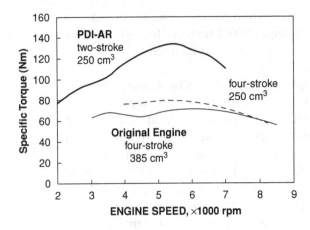

Figure 4.42. Comparison of the specific torque output between a conventional four-stroke engine and a two-stroke lean-burn 'active radical' (PDI-AR) engine [143].

We briefly mention recent further research [143] in the context of the 'downsizing' the potential of an HCCI two-stroke engine relative to a comparable, conventional SI four-stroke engine: This particular example runs closer to stoichiometric. Figure 4.42 shows the specific torque of the PDI-AR (pneumatic DI activated-radical – i.e. HCCI – engine) in comparison with two other four-stroke engines. The higher specific torque is a feature of any two-stroke engine: What is remarkable is that HCCI combustion can be maintained up to about 7000 rpm, which is considerably higher than has been possible for a four-stroke HCCI engine. The combination of DI and HCCI combustion means that the usual fuel 'short circuiting' is avoided and that NO_x emissions are low. As is usual with many HCCI engines, UHCs and CO emissions require an oxidation catalyst. Nevertheless, the fuel economy of this concept, over the 'ECE 40' drive cycle, is 68% better than the original two-stroke engine and 23% better than the comparable four-stroke.

Four-Stroke HCCI Gasoline Engines

In summary, reference [117] cites four limitations to HCCI operation:

- poor low-load combustion efficiency,
- limited high-load ability that is due to limitations of either rapid pressure rise or high NO_x emission or misfire,
- poor understanding of the effect of fuel composition on operating limits and design criteria,
- need to control combustion phasing over the load–speed range and through transients.

Here, we present some results [144] related to the first two points, namely how to expand the upper and lower boundaries of the HCCI operational range by controlling the compression ignition timing: that depends on, among other variables, the temperature, pressure, and the fuel concentration of the air–fuel mixture in the cylinder. As previously mentioned, for gasoline-fuelled engines, it is necessary to promote ignition, even at a 15:1 compression ratio, through the addition of energy to the intake charge, and one practical way to do this is by increasing the amount of EGR in the engine. There are several ways in which this can be done by using variable valve timing, as shown in Fig. 4.43: comparative analysis showed that

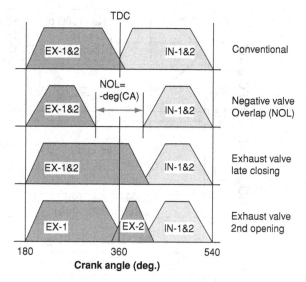

Figure 4.43. Valve timing events for the control of the amount of EGR using an electromagnetic valve train for a four-valve HCCI engine. EX- denotes exhaust valve, IN- denotes inlet valve [144].

negative valve overlap (NOL) was the only method that could achieve HCCI operation without further external heating of the intake charge. Figure 4.44 shows that the operational map of the engine with NOL, and PFI is inadequate, at the upper range, to the demands of the engine load and speed range required for the Japanese 10*15 mode test as well as, at the lower range, the frictional mean effective pressure of the engine. The upper limit of the operational map was set by the knocking limit, and the lower was set by the coefficient of variation of the IMEP ('COV IMEP'). As previously discussed, these tend to converge with increasing speed. The upper limit was successfully extended by use of DI, rather than PFI, with a multihole nozzle (with performance better overall than that provided by a pintle-type nozzle) while maintaining excellent fuel economy, Fig. 4.45. This was, however, at the expense of increased NO_x emission (Fig. 4.45), suggesting that, as previously discussed, the premixing is inhomogeneous. Another disadvantage shown in Fig. 4.45 is that DI in combination with internal EGR results in an increased coefficient of variation of the IMEP (relative to PFI with external heating of the intake air) because of cycle-by-cycle variations in the internal EGR temperature and in the generation

Figure 4.44. Operational range of HCCI engine with DI during negative valve overlap (N.O.L.) and compression ignition, also with DI. The compression ratio ϵ is 15:1. Pmf denotes the frictional mean effective pressure of the engine. 10–15 mode shows the operational range required of the engine to cover this drive cycle [144].

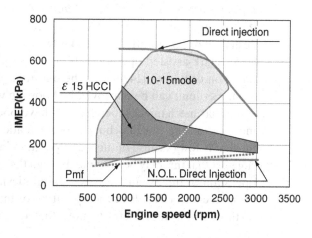

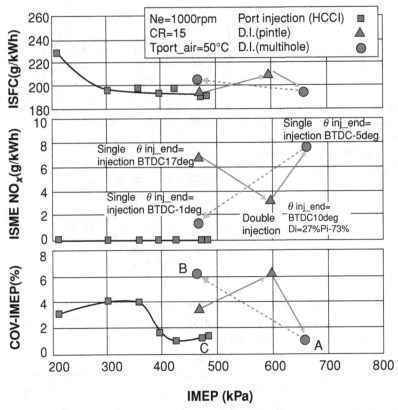

Figure 4.45. Comparison of fuel consumption (ISFC), emission of indicated specific NO_x and of the covariance of IMEP as a function of IMEP using various injection strategies: port injection, DI with a pintle nozzle, and DI using a multihole nozzle [144]. Point A, 4% internal EGR and DI through multihole injectors ($\theta_{injection\_end}$ = BTDC $-5°$ CA); point B, 20% internal EGR and DI through multihole injectors ($\theta_{injection\_end}$ = BTDC $-17°$ CA); point C, 30% internal EGR and no DI (HCCI with port fuel injection).

of the air–fuel mixture. The lower load range of HCCI could be extended by the injection of small amounts of fuel during NOL, although the result was somewhat marginal.

Further Extension of the Upper Limit: 'Boosting' Strategies. That the upper limit of the operating map of an HCCI engine can be improved by a wide variety of engine designs and configurations is well known (e.g., the use of a variable compression ratio [145], spark-assisted operation [146], and supercharging [144]) and research in this direction continues [147–149]. The question of interest here is the extent to which the upper limit can be improved by boosting while retaining lean-burn combustion (note that mechanical superchargers have poor type efficiency: Turbochargers are more promising but matching is problematic because the exhaust-gas enthalpy is inevitably low because of the lean operation in HCCI [150]). The work of [151–153] suggests that this may be possible through the use of pressure waves generated by the operation of the exhaust valves on a multicylinder engine. Although a description of the technical details of the operation of the boosting effect is outside the scope of this section, it is useful to note that recently it has proven possible to obtain

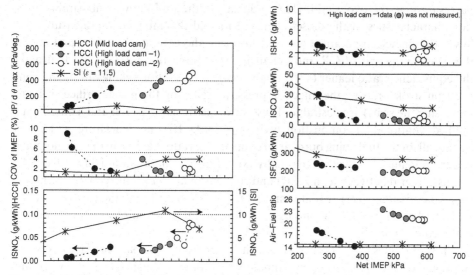

Figure 4.46. Comparison of the performance between HCCI boosted by blowdown super-charging and SI operation [152].

relatively high IMEP (650-kPa) results with lean-burn combustion (Fig. 4.46) [116]. This was achieved by generating large thermal stratification within the cylinder and by controlling the mixing between the hot EGR and the intake charge.

Four-Stroke Diesel HCCI; LTC

It is convenient to consider LTC strategies on a Kamimoto–Bae (i.e., temperature equivalence ratio) map (Fig. 4.47), which shows CO and UHC yields based on isothermal, isobaric homogeneous reactor simulations, together with two 'islands' where NO_x and soot (particulates) are formed [154]: see [154] for further details and a brief discussion on the applicability to diesel combustion. If 'conventional' diesel combustion is assumed adiabatic, it starts along a locus with auto-ignition at a ϕ of about 4 and goes to completion in a diffusion flame at $\phi = 1$, which unfortunately

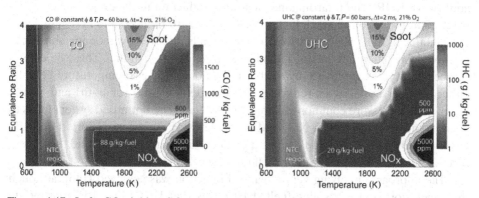

Figure 4.47. Left, CO yield at 2.0 ms obtained from a homogeneous reactor simulation of an n-heptane–air mixture, using the LLNL detailed n-heptane mechanism. The 88-g/kg-fuel isopleth corresponds to 2% of the fuel energy in CO [154]. Right, UHC yield for the same conditions as in the figure on the left. (See colour plate.)

passes *through* the two islands [117] and leads to the well-known undesirable emission characteristics of the diesel engine. To avoid this intersection, measures that lower the flame temperature (such as lowering the compression ratio and mixing the fresh charge with EGR and enhancing precombustion mixing) and that make the equivalence ratio leaner (such as advancing the injection timing during the compression stroke, raising the injection pressure, reducing the nozzle orifice size, and increasing the swirl) are desirable. These strategies result in cooler and leaner flames than conventional CI combustion, giving rise to the LTC nomenclature, with some, but not all, of the fuel being burnt at an equivalence ratio that is lean of stoichiometric and the remainder of the charge usually extending to richer than stoichiometric. LTC is thus a hybrid between lean HCCI and conventional non-premixed CI, which aims to keep emissions of particulates and NO_x low while retaining the control of combustion phasing through the timing of the (main) injection pulse, as in conventional diesel combustion.

The strategies for LTC operation with diesel fuel can be classified [117] as one of two kinds: either injection at approximately TDC or very advanced injection. The aim (for both approaches) is to allow sufficient time for mixing between the fuel vapour and the air before ignition (which implies that it is usually desirable to complete injection before ignition). We begin by describing the strategies and conclude this subsection with a review of the evidence for the extent of lean combustion.

The approach involving injection about TDC retains a closer coupling between the injection and the combustion events and therefore retains the hope of closer control of the phasing of ignition; but this also limits the time available to promote mixing. This approach makes use of largely 'conventional' injection equipment and cylinder bowl geometry and is sometimes known as 'modulated kinetics' (MK) [103]. Injection is retarded to ATDC, which provides some time for the mixing by delaying the auto-ignition that is due to the expansion – and hence reducing the temperature and density of the gases in the power stroke – and introducing large amounts of EGR, to lower the temperature and to reduce the oxygen concentration and further delay auto-ignition. Note that, as pointed out in [155], the increase in ignition delay that is due to EGR is offset by the lower in-cylinder ambient concentrations of oxygen, so that the equivalence ratios at ignition are likely to be unchanged by the addition of EGR. The retarding may lead to a reduction in the cycle efficiency but with the introduction of swirl, heat transfer losses are reduced and cycle efficiency is maintained [103]. The extension to higher loads involves overcoming shortened ignition delay times and the need to inject a larger quantity of fuel. These objectives can be achieved, respectively, by further reduction of the gas temperature field, by either cooling the EGR or reducing the compression ratio (or both) and by the use of a high-pressure common-rail injection system. The optimised system leads to a simultaneous reduction in NO_x and particulates. However, as reported in [117], levels of CO and HC emissions can be high under certain conditions, and these were recently measured by planar induced fluorescence techniques [154].

The approach involving very advanced injection was achieved [156], although the conclusion, as for gasoline-fuelled HCCI, is that HCCI diesel is appropriate up to medium loads, beyond which the rate of pressure rise is too abrupt and that the control of all the variables that can affect HCCI–diesel auto-ignition presents a considerable challenge [117]. It clearly allows greater time for some evaporation before ignition: However, it is now necessary to ensure adequate mixing and to avoid fuel impinging onto the cylinder wall. This led to the development of the design of a narrow-angle direct-injector concept [157] with an appropriate ("omega") piston bowl shape that was optimised for both low and medium loads, together with EGR to control the ignition delay, and able to switch the mode at high loads to conventional diesel combustion. An alternative way in which to reduce the spray penetration at low in-cylinder densities is to use multiple injections [104].

The control of the ignition timing is determined by the – necessarily relatively long – ignition delay: To avoid complete reliance on the inevitably delicate control of the ignition time through the low-temperature chemistry of the mixture, a hybrid, dual-injection scheme, known as UNIBUS [158], was developed, which involves advanced injection ($-50°$ ATDC) of a portion of the fuel followed by the remainder being injected, as in conventional CI designs, around TDC to act as a trigger for the ignition, and subsequent combustion, of the whole charge. Thus, the argument runs, much of the fuel is burnt in a lean environment, avoiding soot formation. As with the approach of the preceding paragraph, large amounts of EGR are introduced to dilute the combustion gases, thereby reducing flame temperatures. Experiments [159] have now led to a conceptual model for this scheme – at least for the case of an optically accessed heavy-duty engine (i.e., 139.7-mm bore, 97.8-mm bowl width operated with an equivalent of 60% EGR and operation at the threshold of soot formation). Comparable measurements for MK-like combustion were also presented in [160], but no conceptual model has yet been presented. Figure 4.48 shows the UNIBUS conceptual model, which differs significantly from the corresponding model of conventional diesel combustion [161] shown in Fig. 4.49. During the first injection, the liquid phase has disappeared by about $5°$ ASOI (after the start of injection) and a cool flame (identified by chemiluminescence) is present throughout the length of the jet: This is not observed in conventional diesel combustion. High-temperature second-stage ignition reactions start downstream at $13°$ ASOI, and broad distributions of OH radicals start to appear in the same region by $15°$ ASOI, indicating a mixture that is leaner and more uniform than for conventional diesel combustion. In the latter, OH is formed in a thin ribbon-like region around the periphery of the jet. Isolated islands of soot arise, near the edge of the bowl, which is mostly oxidised by $22°$ ASOI. These thus, presumably, identify fuel-rich regions – implying that most of the jet is therefore below sooting limits: For conventional diesel combustion, the mixture in the interior of the jet is quite rich ($\phi > 2$), leading to soot formation throughout the jet cross section. The reactions accompanying the second injection are closer to conventional diesel combustion: Cool flames (again identified by chemiluminescence) begin by $5°$ ASOI rapidly followed by the appearance of second-stage ignition and a putative non-premixed flame may arise on the perimeter of the jet. By $9°$ ASOI, soot is being produced more or less throughout the downstream region of the jet: The upstream part seems to be devoid of reaction until about $20°$ ASOI. After this time, soot remains downstream and weak broadband fluorescence, perhaps either 'from unburned fuel or other combustion intermediates', remains near the injector long after the end of combustion. This suggests that this region may be the origin of the HC emissions observed with LTC.

The essential determinant for the existence of lean-burn combustion is 'positive ignition dwell', that is, ignition starts after the end of injection [155]. This, in

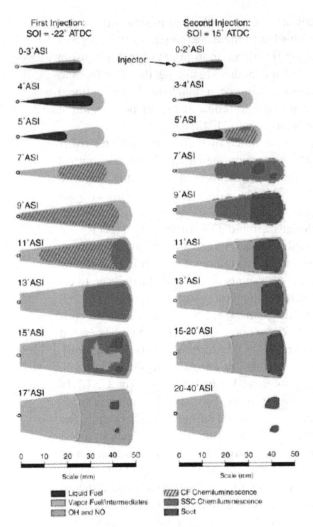

Figure 4.48. Schematic representation of the spray, combustion and emissions processes for the low-temperature, double-injection (UNIBUS-like) condition (first injection in left column and second injection in right column) [159]. SOI is start of injection, ASI is after start of injection (See colour plate.)

turn, arises at low load when the fuel jet does not establish a quasi-steady period. In experiments similar to those previously described [162], images show that ignition chemiluminescence begins between 20 and 40 mm downstream of the nozzle with the existence of formaldehyde during first-stage ignition: Related equivalence ratio measurements extend from fuel lean to fuel rich. Furthermore, in [155] it is suggested that the subsequent presence of significant OH throughout the jet's cross section implies that much of the downstream jet must have ϕ in the neighbourhood of unity. Further evidence for this comes from the measurement of the equivalence ratio *after the end of injection*, and thus relevant to the conditions considered here, by Rayleigh scattering [163]. The conditions are broadly comparable but were measured in a constant-volume chamber. The measurements show that much of the jet 30 mm downstream of the injector is near $\phi = 1$ with mixtures leaner than that closer to the injector. Finally, in [164], formaldehyde was measured by PLIF, and it was found that it formed strongly and remained in the upstream region of the jet. An associated chemical kinetic simulation suggests that formaldehyde would indeed persist if mixtures were too lean for complete combustion, say $\phi < 0.5$.

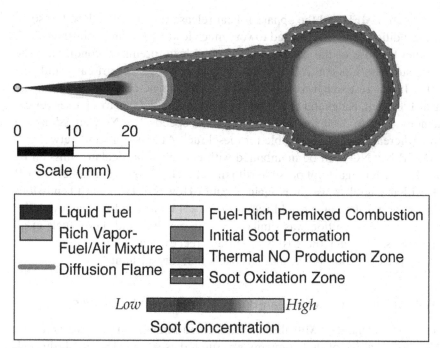

Figure 4.49. A schematic of the conceptual model of conventional diesel combustion with additional features (fuel-rich premixed flame, soot formation, soot oxidation, and NO formation zones) that are indicated by the data but have not yet been verified (J. E. Dec, personal communication and [161]). (See colour plate.)

The result is, however, a surprising one for two reasons [155]. First, the centreline decay of the equivalence ratio in a quasi-steady jet, without EGR in the ambient, is proportional to the reciprocal of distance from the nozzle. Thus expectation might have been that the equivalence ratios should be far in excess of unity in the region up to about 30 mm; and this expectation would have been further reinforced by the presence of EGR in the ambient, because mixtures in such a jet should be richer still because of the reduced oxygen concentration because of EGR in the ambient. Nevertheless, shortly after the end of injection, the jet has equivalence ratios near unity in the same region. Second, it is likely that the amount of entrained ambient gas is similar to that of a conventional jet, because at a given downstream location the cross-sectional area is comparable. However, shortly after the end of injection, the equivalence ratio becomes much lower, which implies that the mixing processes after the end of injection must be more rapid than those occurring during injection. In conclusion, in a quasi-impulsively stopped jet, the richest parts are far downstream and not upstream as in a quasi-steady jet. Thus the lean equivalence ratios near the injector after the end of injection may lead to the UHC emissions because of incomplete combustion for conditions with long ignition delays.

Gasoline as a CI Fuel. A recent study [165] used high-octane-number fuels, such as gasoline and ethanol, in the 'partially premixed combustion' mode in CI engines. The premixing is promoted because such fuels are both resistant to auto-ignition, hence increasing the opportunity for mixing, and highly volatile, with the result that the emissions of particulates were low. It was found possible to control the

CA at which 50% MFB and the apparent heat release rate could be low. For some operating conditions, the gasoline led to very much lower NO_x emissions than those of diesel fuel because of the better attainment of lean, premixed conditions. The rate of pressure rise was controlled by use of double injection, part early and part near TDC. Initial research was conducted in a 2-L single-cylinder diesel engine running at 1200 rpm. Kalghatgi et al. [166] recently extended this to a passenger-car displacement at up to 3000 rpm. At low loads and speeds, low NO_x emissions are attainable whereas this is not possible for diesel fuel. At 2000 rpm and between 10 and 12 IMEP, low NO_x can be maintained with negligible particulate emission by use of EGR, which is again not possible with diesel fuel. At 3000-rpm 13-bars IMEP, low NO_x with negligible smoke was again possible. However, HC and CO emissions were higher for gasoline and would require removal by an oxidising catalyst. The gravimetric fuel consumption was similar for both gasoline and diesel fuels at all operating points. Although there is, as yet, little direct evidence of the extent of lean combustion in these experiments, it seems likely that it is substantial.

4.1.3 Role of Experiments for Lean-Burn Combustion in IC Engines

After the announcement of Mitsubishi's first-generation gasoline DI engine in the mid-1990s, an executive of that company was quoted as saying 'The break-through came five years ago when, by using laser technology and high-speed cameras, engineers were able to study exactly what goes on at the moment of ignition' (cited in [59]). Related publications suggest that the development of this engine does indeed seem to have benefited [81] from the development of laser instrumentation that made planar, rather than point, measurements. Such techniques, in their infancy in the 1990s, have advanced greatly since then, and recent reviews of the contribution that can be made by optical instrumentation were published in [167] for gasoline engines, [168] for diesel engines, and [169] for HCCI engines. Instruments, although not all providing 'planar' information, are available for the measurement of droplet size, fuel concentration (by tracer LIF) and fuel–air ratio (FAR), of the concentration of each phase in a spray (by exciplex LIF), of the presence of cool flames and second-stage ignition, of NO concentration and soot particle size and volume fraction, of thermometry for the soot and gas phases [the latter by Rayleigh, LIF, or absorption thermometry, or coherent anti-Stokes Raman scattering (CARS), or microparticle thermographic phosphors) and of the concentration of HCs, H_2O, CO, and CO_2 by laser absorption. In addition, the measurement of flow velocity, by point or planar techniques, is widespread, and soot size and concentration are increasingly reported by use of laser-induced incandescence (LII) [170].

Some of the instruments in the preceding paragraph have been available, in one form or another, for well over a decade and are well known and widely used. The evolution of instruments continues; here we briefly mention some recent developments that are relevant to research into lean-burn combustion in engines. We start with the measurement of the atomisation of the liquid fuel. Here planar, as opposed to point (e.g., phase-Doppler velocimetry), measurements of the liquid phase have become available for surface impingement [171] and for simultaneous size and velocity measurements by interferometric laser imaging for droplet sizing (ILIDS) [172]: Where the spray is too dense for these techniques, or the droplets are not spherical, measurements by combined Mie and LIF [173] may be possible. In very dense sprays, it may be that so-called 'ballistic imaging' can provide some of the information required [174]. Although many of these techniques can, and have, been used in engines, optical access to the spray in the engine may be too limited in some circumstances, or it may be desirable to measure the spray in isolation from the engine to investigate the fundamental physics of atomisation, evaporation, and mixing – but at realistic pressures and temperatures. For these two reasons, basic research on sprays is being performed in high-pressure, high-temperature cells, or rapid-compression machines (e.g., [175–177]), which give the best possible optical access to measure spray formation, evaporation, entrainment, rates of mixture formation, ignition, and soot formation. Spray break-up length can be measured in such environments by the technique described in [178].

A particularly interesting technique is the simultaneous visualisation of the fuel in the liquid and vapour phases by use of spectral separation of the exciplex LIF the liquid phase from the gas-phase fluorescence: Exciplex fluorescence results from two tracers that combine in the liquid phase only to form an <u>exci</u>ted com<u>lp</u>ex). Recently, fully quantitative planar measurements have been reported in evaporating fuel sprays under conditions of pressure and temperature relevant to engines [179], based on full characterisation of the temperature dependence of the particular exciplex system used.

The measurement of the mixing of fuel vapour is usually idealised by the familiar technique of using a non-fluorescing fuel, which does not fluoresce by itself, with the addition of a fluorescing tracer (the technique called planar laser-induced fluorescence, PLIF), which has an evaporation rate, matched as far as possible, to that of the fuel [180]. This technique is being extended in a number of ways. Recently, publications have begun to appear that attempt to distinguish between various fractions of a gasoline fuel, e.g. [181], and *quantitative*, high-temporal-resolution PLIF [182] has been developed. Measurements of the 3D distribution of vapour have been technically possible for some time [183], although the technique will remain expensive. The fuel vapour from a diesel spray is similarly measured by a non-fluorescing fuel 'doped' with a fluorescent tracer: Here, the tracers have a higher boiling point than those used for gasoline. For diesel sprays, under certain conditions, Rayleigh scattering has also been used to measure fuel-vapour density: Experimentally, laser flare must be kept as low as possible with this delicate technique. So-called filtered Rayleigh scattering avoids flare because the only wavelength detected is that which is Doppler shifted by Brownian motion, the elastically scattered light from stationary surfaces being filtered out [184, 185].

To measure the fuel-vapour concentration, its temperature, or the equivalence ratio, recent developments use the general dependence of the PLIF tracer signal on both temperature and oxygen partial pressure by making two simultaneous measurements: See [186] for simultaneous measurements of temperature and residual gas concentration based on acetone laser-induced fluorescence. When toluene is used for temperature measurements of the gas phase, this approach has the great convenience of needing but a single excitation wavelength [187]. An alternative technique for the measurement of fuel vapour is based on the relative extinction of two wavelengths by the vapour phase and by the liquid droplets [188] although, because this is a line-of-sight technique, it relies on an assumption of axisymmetry to produce a planar measurement by the so-called 'onion–peeling' deconvolution algorithm. It has nevertheless produced useful insights, as previously mentioned [93]. A different technique for the measurement of temperature is by the addition of micrometre-sized phosphor powder to the intake air [189]. The cooling caused by the evaporation of directly injected fuel has been measured by rotational CARS (R-CARS), in preference to vibrational CARS, because it offers a more accurate measurement at low temperatures [190].

The future is likely to see increasing use of several instruments applied simultaneously, such as PLIF for the detection of formaldehyde as a flame-front marker, OH* for the interface between burned and unburned gases, and tracer PLIF for fuel. An early example is [191], which measured the local fuel concentration near the flame front and combustion, by means of tracer PLIF and OH-PLIF, in a DI engine. Other examples include the interaction among fuel concentration, plasma discharge, and the flame front [192, 193]. Kim et al. [154] demonstrated the PLIF of CO, which will also lend itself to combined measurement with other techniques.

In lean-burn SI engines, measurement of the FAR close to the time of ignition is important: The equivalence ratio is a stochastic variable, and there is need to know its distribution, at least, which is closely related to the subsequent growth rate of the initial flame kernel. This can be done either by measurement of the CN and CH radicals [83], which are formed by the spark discharge, or by measurement of the CH/OH ratio from the chemiluminescence of the subsequent flame as it grows [79]. The growth rate of the initial flame kernel and the conditions leading to misfire are of critical importance to stratified lean-burn engines. Recent developments [194] in high-speed imaging and automated image analysis (which permits a conditional analysis based on IMEP) permit 'binarised' visualisation based on seeding the reactants with micrometre-sized droplets and illuminating the field with a planar sheet of light: As the reaction proceeds, the droplets are burned and a 'negative' image of the flame can be seen. These have led to the conclusion that random, rare misfires (which are nevertheless important for pollutant emission levels) and poor burn cycles are due to local flow fluctuations that fail to move the kernel downwards into the main fuel cloud contained in the piston bowl. [100] and [195] report a series of measurements in a spray-guided DI engine that show the effect of spray momentum on the plasma motion, coupled with the role of local fuel concentration on ignition stability.

In the context of LTC CI engines, the occurrence of the cool flame was measured [196, 197] by excitation with frequency-tripled Nd:YAG (neodymium:yttrium aluminum garnet) lasers and the second-stage ignition by the disappearance of formaldehyde and its replacement with OH. The detection of the second stage of ignition in HCCI engines was made [198] by OH* chemiluminescence to observe the effects of thermal stratification: however, OH* chemiluminescence is a line-of sight technique [199]. Nevertheless, it was this technique that led to the conclusion that it is mainly thermal stratification within the bulk gases that controls the maximum rate of pressure rise and the unexpected conclusion that the effect of thermal stratification between the bulk gases and the boundary layer has a lesser role in reducing the maximum rate.

Broadening the scope of the discussion, there are two challenges that instrumentation may have to meet in future. The first is that there are differences in the behaviour and performance between optical and all-metal engines that can arise because of effects ranging from the difference in wall temperatures to the flow pulsations through the valves that are due to pressure waves originating in the emptying and filling of other cylinders. Thus some hold that, for some applications at least, it is preferable to measure an optically accessed cylinder that replaces one of the cylinders in a metal multicylinder engine. This is usually less convenient than making measurements in a single-cylinder, dedicated, optically accessed engine. In other cases, it is either necessary or adequate to measure in all-metal engines using, for example, access through an endoscope or through modifications to spark plugs [200]. The second challenge is that the role of cycle-to-cycle variability and misfiring is gaining importance, particularly with operation near conditions of marginal ignition and initial flame-kernel growth. This can arise as the combustion process becomes dependent on, for example, the details of the motion and mixing of the first or second fuel-vapour cloud in a spray-guided SI engine or on the details of the interaction between the high-speed spray and the convective motions in the engine. Fortunately, high-repetition-rate lasers and high-framing-rate cameras have recently become available, and it is now possible to measure and record several consecutive cycles or make CA-resolved planar measurements. An excellent example is described in [201].

4.1.4 Concluding Remarks

As early as the turn of the 19th century, the damage done to the *environment* as a result of the pollution (because of the burning of coal) caused the *New York Tribune* to fulminate:

> One of these days when the mischief is fully done, when our once pellucid and crystalline atmosphere is transformed into Chicago reek, and Pittsburgh smoke and London fog, men will begin to realize what they have lost, and will hold conventions, and pass resolutions, and enact laws, and spend great sums of money for the undoing of the mischief and the restoration of our atmosphere to its original state. ([202], cited by [25])

By the eve of the 1970 'Muskie Act' the damage to *health* had become the greater concern to a National Air Pollution Control Administration (NAPCA) epidemiologist:

> What we do know is that people get killed by air pollution, and I don't see any excuse for there being enough air pollution to kill people. Do you? ([203], cited by [25])

- This part of the chapter described the way in which engineering design can exploit predominantly lean combustion, not necessarily involving a deflagration front, in IC engines not only to reduce pollution, but also – by promoting fuel-conversion efficiency – to husband the finite resource of petroleum-derived fuels.
- The improvement in fuel economy consequent on lean-burn combustion, based on the thermodynamics of fuel–air cycle analysis, is well known, as is the reduction in NO_x emission that is due to the concomitant lower flame temperature.
- In the 1970s and part of the 1980s, lean-burn SI combustion in throttled IC engines was indeed able to deliver what were, at the time, great advances in fuel economy and reduction in emissions.

- However, the technical interest in lean-burn combustion is framed by increasingly demanding legislative constraints as well as by commercial requirements. European legislative constraints set a fuel-economy target of 95-g CO_2/km from 2020 and, in particular, particulate emissions that are now so low as to render their measurement difficult. Even so, it is estimated that emissions will cause many adverse health-related effects, including death. Lean-burn combustion has a role to play in achieving these constraints – as do other technologies and approaches outside the scope of this book.
- For spark ignition IC engines, lean-burn combustion has been allied with throttle-less direct injection technology to improve greatly part-load fuel economy relative to a throttled PFI stoichiometric burn equivalent. The important virtues of the latest designs, albeit produced in limited numbers of premium vehicles, are the achievement of
  - lean-burn combustion over a wide range of loads and speeds, and
  - low, but nevertheless close to legislative limits, emission of particulate matter.
  These desirable attributes are achieved through the use of 'spray-guided' designs exploiting the advantages of high-pressure common-rail piezoinjection. The engine-out NO_x emissions are low because of the use of lean burn, but still need substantial after-treatment. Unfortunately, both of these technologies are expensive, currently at least. However, much is becoming known, and it is possible to be optimistic that the costs and emission levels associated with LBDI SI engines will benefit from current research and development.
- In the context of diesel-fuelled CI engines, the development of low temperature combustion (LTC) holds the promise of *simultaneous*, large reductions in the emissions of particulate matter and NO_x. This is achieved through the lean combustion of some, if not all, of the diesel fuel, together with copious amounts of EGR. Once again, high-pressure common-rail fuel injection equipment has played an important role in the development of this approach. Much remains to be done, but recently conceptual models of the process have begun to emerge, and these will undoubtedly help further research and development of LTC. Equally exciting is the suggestion of using gasoline, or a fuel with evaporation and octane properties similar to those of gasoline, in the context of CI engines. This promises even further reduction in PM emissions.
- HCCI engines hold the promise of yet greater reduction in NO_x and particulate matter than LTC, possibly to the extent that neither pollutant would need after-treatment, while maintaining high fuel economy. This would be in the form of very lean equivalence ratios. The advantages and desirability of such combustion need no further elaboration. Nevertheless, the challenges in achieving sufficiently high IMEP while avoiding excessive rates of pressure rise remain, with research in this field making progress as described in the main body of the section.
- The role of experiments – particularly using optical instrumentation – in contributing to the development of lean-burn combustion in IC engines has been, and will continue to be, substantial.
- It has not been possible in a section of this length to discuss and illustrate the extensive research with, and use of, computational fluid dynamics, other than to show – very briefly and somewhat tangentially – its application in the

design and development process. At a fundamental level, there will continue to be interaction between experiments and the development of more accurate computational models of the processes associated with DI engines (namely fuel injection, spray atomization, droplet evaporation, charge cooling, mixture preparation, and the control of in-cylinder motion). This will undoubtedly prove increasingly beneficial in the future in both basic and applied research and in commercial development.

With reference to the first quotation in this final section, made over a century ago, humans *did* begin to realise what had been lost by the mid-20th century and did indeed start to pass resolutions and enact laws – if not to restore, then at least to mitigate some of the 'mischief'. Although 'great sums of money' have been spent, it is claimed that even greater sums of money have been saved by the introduction of clean air legislation (see, for example, [25]). Lean-burn (SI) combustion made a contribution to the restoration of our atmosphere, as well as to fuel economy, in the 1970s and has started to make a contribution once more. Lean-burn combustion is showing further promise for the medium term, in the guise of LTC, and for the long term, as HCCI. The extent to which lean-burn combustion will play a part in improving fuel economy and in reducing pollution, as opposed to competing technologies, remains to be seen. But it is certain that its evaluation will come about only through continued, and concerted, research and development.

Acknowledgements

We gratefully acknowledge the assistance of colleagues in the preparation of this contribution: Pavlos Aleiferis, Giles Bruneaux, Alexandros Charalambides, John Dec, Ingemar Denbratt, Mike Drake, Todd Fansler, Neil Fraser, Doug Greenhalgh, Yannis Hardalupas, Julian Kashdan, Tim Lake, Paul Miles, Richard Pearson, Khizer Tufail, Graham Wigley, Ernst Winklhofer, and Hua Zhao.

4.2 Application of Lean Flames in Aero Gas Turbines
By B. Jones

The combustion technology required for an aircraft engine depends on its size. The flight length for which an engine is to be used, the proportion of operating cost accounted for by fuel in different sectors of the engine market, and high pressure (HP) rotor-blade size drive the trend to increase pressure ratio with thrust.

Engines can, for convenience, be grouped in thrust ranges, as shown in Table 4.4. Within a thrust category there will always be a range of pressure ratios because manufacturers produce families based on one engine platform. An engine may be developed or adapted to cover a thrust range exceeding 1.5:1. Associated with pressure ratio is the combustor inlet temperature (see Fig. 4.50), and both these parameters strongly affect the combustor design.

CYCLE TEMPERATURE LIMITS. In Fig. 4.50, polytropic efficiency is assumed to increase from 0.89 at 10 overall pressure ratio (OPR) to 0.91 above 40 OPR. Compressor-disk material will limit the maximum permissible combustor inlet temperature to $1\,000\,K$

Table 4.4. *Engine thrust categories*

Thrust (kn)	OPR–A.D. 2000	OPR–A.D. 2000
180–360+	35–43	45–55
90–180	25–35	40
45–90	20–30	35
< 45	15	20

OPR, overall pressure ratio.

for the foreseeable future. NO_x emission will limit the turbine entry temperature (TET) to about 1850 K in civil aircraft. The TET includes the cooling effect of the HP turbine nozzle guide vane (NGV) cooling air admitted before the gas enters the HP turbine. The combustor mean outlet temperature is at least 100 K higher. This higher temperature is usually referred to as the combustor outlet temperature, which is not uniform. The HP turbine requires a radial gas temperature profile matched to blade stress and to avoid overheating the shrouds, which are hard to cool, and there will be both regular and random circumferential variations in temperature. For a TET of 1850 K there will be hot streaks in the gas entering the nozzle guide vane at temperatures above 2050 K arising from these non-uniformities.

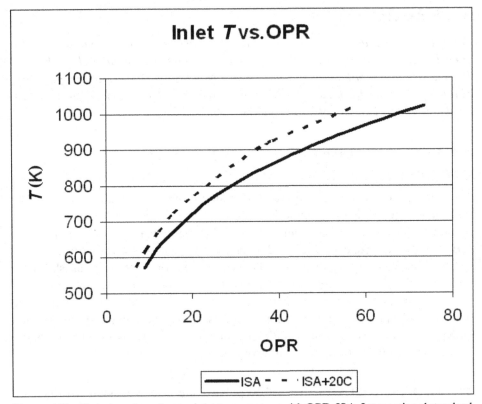

Figure 4.50. Variation of combustor inlet temperature with OPR. ISA: International standard atmosphere.

DIFFERENCES BETWEEN GROUND-BASED AND AVIATION ENGINES. Ground-based engines can be divided into two subdivisions – heavyweight engines designed specifically for power generation or gas pumping, and aeroderivatives, which naturally have more in common with their progenitors. Many ground-based engines burn natural gas – a relatively easy fuel to deal with. Diesel oil is much more difficult.

Ground-based gas turbine engines are not required to respond rapidly to a demand for power. An aeroengine must be able to accelerate from 15% to 95% thrust in 5s and is subject to numerous load changes during each flight. Its combustion system is susceptible to ingestion of large amounts of water when flying through heavy rain, and it must be capable of restarting the engine at high altitudes.

In aircraft, engine operability is a safety issue. In circumstances in which all engines are subject to abnormal conditions, such as during landing in a tropical storm, the passengers' lives depend on absolute reliability of the combustor. In ground-based installations, operability is an economic issue. These differences affect the achievement of very low pollution.

HEAVYWEIGHT ENGINES. These often form part of a co-generation or combined cycle plant in which exhaust heat is used for district heating or to produce steam. The cost of increasing pressure ratio, which brings with it complex cooling systems and expensive materials and manufacturing methods, does not justify the benefit of improved gas generator cycle efficiency. Energy lost in the gas generator is recovered downstream. Such engines usually operate at about 15:1 OPR and the combustors are subject to modest inlet temperature. Size and weight do not constrain combustor design, so the combustion system can be optimised for emissions control. Heavyweight engines operate at constant speed to maintain generator output frequency, and the air mass flow through the engine is restricted at low power by throttling the engine inlet. Over most of the power range they operate at constant HP TET and the range of the combustor FAR is relatively narrow.

AERODERIVATIVE ENGINES. These engines are based on the gas generators of large, high-bypass-ratio engines. The fan being removed, they lose the fan precompression of the gas generator flow, and the TET is reduced to achieve the long life required. They serve as gas generators to drive a separate power turbine. The pressure ratio is usually in the range 20 to 35. They are favoured in applications in which their low weight (oil-drilling platforms, high-speed ships, power generation or gas pumping in remote regions) or rapid starting for peak looping is of sufficient advantage to offset their higher cost.

All ground-based gas turbines are subject to more stringent emissions regulation than aeroengines. Regulations vary between countries – and even between states. California and Florida have extremely stringent regulations. Regulations governing aircraft are universal. Sometimes local variances can be met by employment of water or steam injection in ground-based systems, but in California there are restrictions on the use of water! Water injection has been considered during take-off in aircraft applications (see Subsection 4.2.8).

The differences just summarised result in combustor designs for ground-based engines that bear no resemblance to those employed in aeroengines. It is not that one

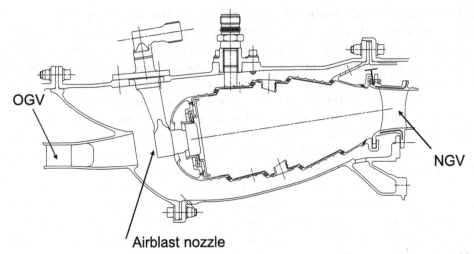

OGV

NGV

Airblast nozzle

Figure 4.51. A typical modern aero gas turbine combustor. OGV, compressor outlet guide vane. (Courtesy of Rolls–Royce Deutschland.)

group of designers has gained a great lead over the other. The duty of the combustor is totally different.

4.2.1 Background to the Design of Current Aero Gas Turbine Combustors

The technology on which modern aero gas turbine combustors is based was developed during the late 1970s in response to the introduction of emissions regulation and entered service from 1985. Although the major manufacturers began in 1972 from different starting points, determined by their past experience and the constraints imposed by their chosen engine configurations, designs converged. Over a period of 30 years, airflow through the airblast fuel injectors has doubled, but the injector design concepts have not changed, inlet diffusers have become much more compact, and the materials in use today are similar to those used 30 years ago. Thermal barrier coating improved during the 1980s to the point at which it could be depended on not to spall. During the 1990s, significant improvements in wall-cooling technology were introduced, notably tiled combustor walls and shallow-angled effusion-cooled cast walls, made possible by low-cost laser drilling. The saving in cooling air allowed greater control of stoichiometry within the reaction zones. Figure 4.51 is a representative design optimised for an engine of moderate size, pressure ratio, and cost.

Emissions reduction was achieved by incremental development, not by the introduction of new concepts. Extensive research carried out since 1972, reviewed in Subsection 4.2.6, is only now yielding alternative airworthy concepts that should enter service from 2010. It is sobering that, over the past 30 years, although combustion inefficiency at low power was reduced by an order of magnitude, the 50% reduction in NO_x achieved at fixed engine conditions was largely eroded by the increase in pressure ratio of 50%, which occurred over the same period. Clearly continued incremental development is not going to eradicate NO_x.

Table 4.5. *Comparison of fuel–air mixing parameters*

Parameter	Diesel	Gas turbine
Reaction-zone equivalence ratio	0.7	0.7
Combustion air mass flow (kg/s)	0.17	68
Fuel flow (kg/s)	0.008	3.18
Single injector flow rate (kg/s)	**0.008**	**0.159**
Approximate burning volume/inj L	**0.42**	**1.2**
Mixture mass flow per litre (kg/L)	**0.43**	**3.0**
Residence time (ms)	3	3
Injector pressure drop (bars)	**1000+**	**20**
Mean pressure (BMEP) (bars)	(18.5)	43

4.2.2 Scoping the Low-Emissions Combustor Design Problem

It is highly desirable that the aviation industry should contribute to the reduction in global emissions. A great deal has been achieved in the reduction of both UHCs and CO at low power and in eradicating visible smoke. In 2001 ACARE (Advisory Council for Aeronautical Research in Europe) declared the goal to reduce NO_x by 80% and to halve GHG emissions by 2020 C.E. [204]. The reductions are to be achieved in stages defined in a series of Strategic Research Agenda, the latest of which is the 2008 addendum. All these documents are freely available to download from the ACARE website. These reductions include improvements in engine efficiency, airframe weight, and drag reduction and improved traffic management, to avoid delays in landing, for example. The reduction in NO_x can best be achieved by lean combustion, requiring dispersal of the fuel very rapidly through up to 65% of the total combustor airflow – ideally before reaction starts.

To gain an idea of the magnitude of this challenge, we can compare the mixing problem with that in a diesel engine. We will compare a Volkswagen V6 2.5-L turbo diesel developing 132 kW at 4000 rpm with an 18-deg sector of a 300-kN thrust gas turbine having 20 fuel injectors. Although combustion is intermittent in the diesel engine, by considering all cylinders operating in sequence, we can think of the fuel flow as steady, treating the process as if it were taking place continuously in one cylinder. It is a rough-and-ready comparison, and Table 4.5 shows that it is not completely straightforward, but it brings out one point clearly. The quantity of fuel and air to be mixed in each fuel injector in a 300-kN thrust gas turbine is 20 times that of a 2.5-L high-speed diesel engine at full load, but the diesel fuel injector pressure drop is 50 times greater. The time available for mixing and combustion is similar.

Several factors hinder the application of lean premixed combustion in aircraft engines. In approximate order of influence these are as follows:

1. Airworthiness or safety – this obviously cannot be compromised and is the subject of rigorous international regulation.
2. Operability – which interacts strongly wtih safety and includes fast response to change in power demand, ability to cope with ingestion of large amounts of water during descent in tropical storms, and relight at altitudes above 8000 m.
3. Fuel employed – aviation kerosene, the specification of which, though tightly controlled in some respects, like viscosity and aromatic content, allows significant

variation in its constituents. Auto-ignition delay, always difficult to measure, will vary with fuel composition.

4. Size and weight.
5. OPR, which in large engines is likely to reach between 50:1 and 55:1 by 2020.

Emissions regulations, though strongly influencing design practice, are a consequence of what is feasible, given the preceding constraints. As we are here considering the application of new technology, we must take into account the changes likely in engine cycle during the probable technology development time scale. Technology that has not yet reached the stage of engine demonstration is unlikely to enter service within 10 years. Lean burn direct injection (LBDI) technology (not prevaporised premixed), capable of delivering an intermediate NO_x reduction of around 40%, has been under development for 10 years and has been demonstrated in engines. It is undergoing further development and the first combustor of this type, the General Electric (GE) TAPS (twin annular premixing swirler) combustor, will enter service in 2010.

4.2.3 Emissions Requirements

There are two issues to be considered: local air quality around airports, in which there is a high concentration both of air traffic and other sources of pollution, and the impact of emissions at high altitude. At present only the first of these is regulated. Emissions at altitude remain the subject of a discussion that began in the 1980s. Emissions regulations are applied worldwide and form part of the airworthiness certification procedure. Some European countries also apply a landing surcharge based on emissions.

LOCAL AIR QUALITY. Regulation is concerned with activity within a radius of 30 miles (48 km) from an airport and below an altitude of 3000 ft (914 m)[205]. So that engines can be evaluated on a sea-level testbed, where accurate measurement is possible, engine operation over a typical take-off and landing cycle within the sensitive zone is characterised by four thrust settings. The basis of the calculation is presented in Fig. 4.52.

NO_x, CO, and UHCs are measured at four thrust settings. These measured values weighted by time are summed and then normalised by dividing by engine sea-level static (SLS) thrust. The four thrust settings and the corresponding times are given in Fig. 4.52. Smoke is also measured, and the concentration is limited to ensure invisible exhaust.

To account for variation from engine to engine, an allowance is made in the emissions testing procedure. To avoid having to test a large number of engines to establish an average, a certificate may be obtained from a small number of engines, but the compliance limit is correspondingly tightened. For example, to pass on the basis of a single engine test, measured NO_x must be 14% below the regulatory limit.

No engine operates at these four conditions in practice. Most engines idle at below 7% thrust, and during approach the engine power cycles over a fairly wide range to maintain the required flight path. The power required during take-off and climb

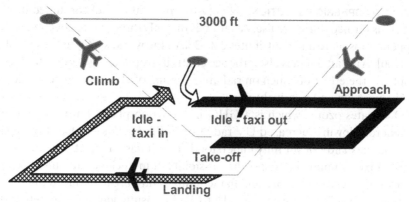

Operating mode	Power setting	Time in mode
1. Taxi / Idle	7% take-off thrust	26.0 minutes
2. Take-off	100% Std. day take-off thrust	0.7 minutes
3. Climb	85% take-off thrust	2.2 minutes
4. Approach	30% take-off thrust	4.0 minutes

Figure 4.52. The landing and take-off cycle and emissions regulations. (Courtesy of Rolls-Royce plc)

depends on aircraft payload and runway length. To conserve engine life, pilots use as little power as can safely be employed. NO_x output depends on the ambient temperature and humidity. The simplification is needed to enable a comparable rating of engines. The combustion system must be designed for real operation, however, not optimised for operation at these four power ratings. Fuel staging describes a method of dividing the combustor volume into separately fuelled zones, the boundaries of which may be defined by solid walls or by control of the airflow pattern. Distribution of fuel between zones is thrust dependent. Engine control with fuel staging is much more difficult in practice than the system that would be theoretically required for certification testing. Thrust must vary smoothly with throttle position over the entire power range and without lag beyond that imposed by the inertia of the turbomachinery. Certification emissions testing merely requires the engine to operate stably at four thrust settings.

The regulatory limits imply that combustion inefficiency that is due to CO must be less than 0.5% and less than 0.3% for UHCs at 7% thrust. Although, as already described, the regulations consider the four operating conditions, today's combustors are 100% efficient during all but ground idle operation. This will not necessarily be true when new lean-burn combustors enter service. The limit for NO_x at take-off is a function of engine thrust and pressure ratio, and is expressed in grams per 1000 kg of thrust per landing and take-off (LTO) cycle. Corrected to 15% O_2 for a large engine, the NO_x concentration at take-off today would be around 400 vppm – 8 times the limit for ground-based engines in most parts of the world, and 20 times the limit in California and Florida. Limits for land-based engines may be halved again in the near future. These figures are simplified to permit a comparison with much more stringent ground-based engine limits. The comparison reflects the limitations on combustor design imposed by airworthiness requirements.

GLOBAL ATMOSPHERIC POLLUTION. Ninety-five percent of fuel consumed in a long-haul flight is burned during cruise. Aviation is the only direct source of emissions into the upper atmosphere and contributes 2%–3% of the world's CO_2 to the greenhouse effect [206]. Aircraft engines also produce contrails by releasing water into the upper atmosphere, the effect of which on radiative forcing of climate change is, according to [206], uncertain and relatively small.

NO_x creates ozone, which is a GHG in the troposphere, but reduces it in the stratosphere, allowing increased UV radiation at the Earth's surface. The impact of NO_x is complex and not yet fully understood. That it has an effect constitutes a risk. Lean, premixed combustion raises the possibility of inefficiency at cruise because of CO emission, because pressure and flame temperature are so reduced that the residence time in the combustor of around 6 to 7 ms is insufficient for complete CO oxidation. This is economically unacceptable, so could not be permitted for that reason.

There are no regulations limiting emissions at altitude. CO_2 and H_2o vapour emissions are functions of the engine cycle, not of the combustor design, and there are powerful economic reasons to reduce fuel burn. The impact of NO_x and particulates is uncertain, though probably adverse. Legislation is hindered by the difficulty of demonstrating compliance. Combustor conditions vary greatly during cruise, as the aircraft weight and optimum flight altitude change during the flight. Cruise cannot be simulated on a sea-level testbed because of the fivefold difference in inlet pressure to the engine and emissions cannot be measured in flight. To operate a large engine in an altitude test facility is very expensive, and only one facility exists that is capable of doing this. Emission of NO_x at cruise is not necessarily coupled strongly with NO_x emission in LTO. The trend to increase the bypass ratio to reduce noise, for instance, greatly reduces NO_x measured on a sea-level testbed through reduced fuel burn, but the increased nacelle drag and engine weight negate this benefit at cruise.

FUTURE THINKING. There seems little doubt that NO_x regulations will become more stringent and the limits under discussion are challenging. Figure 4.53 shows the lowest projected future limits compared with their current production engine performance.

This diagram, with filled symbols representing current Rolls-Royce engines, assumes the same form as current legislation, and the current limits are included. However, the form of and limits in future legislation could change and some possibilities are listed here:

- The LTO parameter (Dp/Foo) may be replaced with parameters taking account of whole aircraft performance (emissions/passenger kilometre).
- Emissions during climb and cruise may be limited, calculating emissions by combining aircraft and engine performance data with sea-level measurement.
- Studies are ongoing into emissions reduction by better traffic management and ground operation of aircraft.
- Emissions charges and fuel tax are under consideration.
- New regulations could be applied to certification of new aircraft or engines as early as 2014.
- The question of cruise emission control for supersonic transports is still under discussion. There is at present no plan to build a supersonic transport. Supersonic

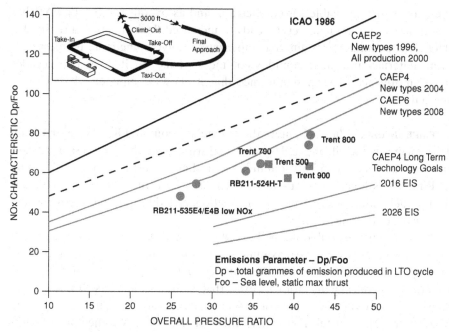

Figure 4.53. Current and future NO$_x$ limits (g/kN). The long-term goals are technology objectives. Legislation will be moderated if these cannot be met. (Courtesy of Rolls-Royce plc)

business jets are under discussion, but their flying hours being less than 10% of civil transports, and, the aircraft being small, emissions are more a matter of public relations than of real environmental concern. The number of supersonic business jets in service will be limited by their high cost of ownership.

4.2.4 Engine Design Trend and Effect of Engine Cycle on Emissions

Increasing fuel prices stimulate demand for more fuel-efficient engines and thus the OPR continues to increase. Increasing traffic density and the drive to reduce cost are leading to increased airframe size and weight and thus to a demand for more powerful engines. Airframe noise during landing increases with aircraft weight, and engine noise also tends to increase with thrust. To compensate, the engine manufacturers are driven to reduce exhaust velocity by increasing the bypass ratio. The engine core size decreases and, as the same total energy must be generated, the smaller core must work harder. Combustor inlet pressure, inlet temperature, and temperature rise increase. NO$_x$ increases by 1% for every Kelvin increase in compressor delivery temperature. A third of this is allowed for in the regulations through the dependence of the NO$_x$ limit on pressure ratio (see Fig. 4.50).

Altitude relight requires that the combustor volume scale wheares core flow, whereas other dimensions in the flow path scale by cross-sectional area. The combustor wall area per unit throughput thus increases and this further increases the proportion of the air used for wall-cooling flow beyond that required to compensate for the higher gas temperatures.

Less air is thus available for combustion, and as overall combustor FAR has increased, this trend in engine cycle leads to increased flame temperature, smoke, and NO_x. The manufacturers aim to compensate by developing more temperature-resistant materials and more complex cooling systems, but it takes two decades to make an advance of such significance that it allows component design to be revolutionised.

FUEL TEMPERATURE. The fuel enters the engine at about 400 K, having first been used to cool the engine oil. The fuel injector feed arm crosses the compressor gas path to inject the fuel on the combustor annulus centreline – a distance of up to 150 mm. As OPR increases, heat transfer to the fuel increases. If the fuel temperature exceeds 460 K, the fuel undergoes thermal decomposition, called lacquering or coking, and injector passages become blocked.

Injectors today are double skinned to reduce heat transfer, but there is a margin of only a few degrees beyond the maximum fuel temperature attained. NO_x reduction technology, which either increases the number of fuel injectors or increases the wetted area by increasing injector size, exacerbates this problem. As NO_x reduction is achieved by improving the dispersion of fuel in a large amount of air, it is impossible to avoid the problem of thermal management of the fuel. Fuel injector airflow is likely to increase more than fivefold during the next 10 years. Wetted area of the fuel passages within the injector will more than double. Fuel staging in which fuel is turned on and off within circuits in the injectors may result in stagnant fuel remaining in hot passages unless avoided by efficient purging and careful design. There is only a 4% pressure drop available to drive an air purge system. Weight precludes the option of carrying bottled gas for this purpose.

DEVIATION OF TURBOMACHINERY FROM SPECIFICATION. It is possible that the compressors or turbines may not fully meet their specification at entry into service. A shortfall in efficiency of the turbomachinery relative to specification has three effects on the combustor:

- An increased fuel burn, arising from reduced engine efficiency, directly increases NO_x output.
- The reaction zone is richer than designed, increasing smoke formation and NO_x production. This cannot be remedied without an expensive change to the fuel injector, which must be reversed when the turbomachinery problem is rectified.
- Compressor delivery temperature, increased by inefficiency, increases NO_x.

The more uniform the mixture of fuel and air prior to reaction, the greater the impact of these effects on NO_x. A 1% loss of compressor efficiency would produce an increase in flame temperature of about 10 K. The corresponding increase in NO_x from a lean premixed combustor would be about 11%–12%. A minor shortfall in engine performance against specification can be managed. If it renders the engine uncertifiable through its effect on NO_x, it becomes a far more serious liability.

4.2.5 History of Emissions Research to A.D. 2000

To gain some idea of the likely pace of future combustion technology development, it will be useful to review its history. In the early 1970s, GE and Pratt and Whitney

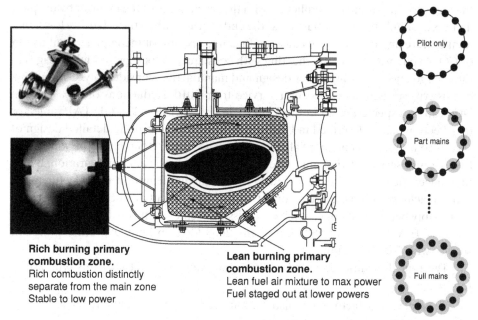

Rich burning primary combustion zone.
Rich combustion distinctly separate from the main zone
Stable to low power

Lean burning primary combustion zone.
Lean fuel air mixture to max power
Fuel staged out at lower powers

Pilot only

Part mains

Full mains

Figure 4.54. Lean DI combustor, showing on the right, circumferential staging (in which the grey circles represent active main injectors), centre, on the left the flame pattern, and on the top left, a lean, DI fuel injector beside an injector from a current Trent engine. (Courtesy of Rolls-Royce plc.)

(P&W) (United Technologies) started research into fuel-staged combustors under a NASA-funded contract. Fifteen fully annular combustors were evaluated, and two, the P&W Vorbix, an axially staged design, and GE's parallel double-annular combustor emerged as practical concepts. In Europe Rolls-Royce (R-R) developed and demonstrated a double-annular combustor. The engine demonstrations were not entirely successful. Judging from published literature, P&W and R-R stopped work on these concepts but GE continued, until in 1990 they launched the GE90 with a double-annular combustor. The technology was also incorporated into the CFM56. Today, 37 years after research started, the performance of these combustors in service does not justify their increased cost and complexity. Superficially, the International Civil Aviation Organisation (ICAO) data may suggest otherwise, but the reason for the low NO_x of the GE90 is its low specific fuel consumption (SFC) on a sea-level testbed, resulting from its high bypass ratio, rather than because of its combustor. Dp/Foo includes the SFC through division by SLS thrust at take-off. The performance of the GE90 double-annular combustor has been matched by incremental development of conventional combustors.

LEAN DIRECT INJECTION. The failure of double-row staged combustors to deliver the sought-for improvement in performance led to a change of direction in the late 1990s. GE and R-R developed lean DI combustors based on large, airblast atomisers passing more than 5 times the airflow of previous injectors. These devices require a central pilot zone that provides flame stability for ignition and a fuel-rich zone to achieve high efficiency at low power. Figure 4.54 illustrates an arrangement for combustion of this type, which is provided by the courtesy of Rolls-Royce, although other possibilities exist. The pilot flame needs protection from reaction quenching by the high airflow around it. Conflicting with this, preheating of the main air by the pilot flame is desirable to aid combustion at the end of cruise, when the relatively low flame temperature and pressure can cause inefficiency because of incomplete combustion of CO. Avoidance of fuel overheating with consequent blockage of fuel passages in these large injectors complicates design and manufacture considerably. Combustors of this type are scheduled to enter service from 2010. Reductions of 30%–40% in NO_x can be expected, both in the LTO cycle and at cruise. The GE TAPS injector (see Section 5.2) is described in U.S. patent. 7,464,553B2, but the detailed design of such a component is commercially sensitive. The GE TAPS combustor will be the first to enter service. Twin annular refers not to the combustor configuration, but to the concentric swirlers in the injector. The premixing length in the injector is too short to achieve complete prevaporisation and mixing, but any degree of premixing is probably better than none, and NO_x emissions achieved at certification, published on the GE website, were 58% below the 2008 CAEP6 limit. All other pollutant species were extremely low.

The four difficulties in successfully engineering an injector of this type are as follows:

1. To achieve satisfactory fuel distribution over the entire flight envelope.
2. To avoid overheating the fuel.
3. To avoid CO emission at end of cruise.
4. To avoid combustion instability.

A solution successfully addressing all of the preceding issues is a great achievement.

A fuel control strategy to ease the operability problems of lean combustors is circumferential fuel staging, whereby the main fuel circuit is split, enabling some main injectors to be turned off under conditions in which the main flame becomes too lean to completely consume the CO. This complicates the fuel system, and, through the consequent non-uniformity of the circumferential temperature profile, may reduce turbine efficiency and introduce stress and distortion into the turbine casing. These problems may not prevent application but are likely to add to the cost and time scale of development.

ULTRA-LOW NO_x COMBUSTION. At the end of the 1980s three significant programmes were launched to develop low-NO_x combustors. By far the biggest of these was the NASA Supersonic Transport programme, with the declared goal of achieving less than 5 EI NO_x at supersonic cruise. This programme continued for 10 years at an approximate cost of U.S. \$200 M.

In Europe, a series of research programmes was conducted, aimed at subsonic application and involving about one quarter that of the U.S. investment. In both the United States and Europe, lean premixed prevapourised (LPP) and rich–quench–lean (RQL) combustion and multipoint fuel injection were investigated. Other national and company programmes were conducted in parallel with these collaborative programmes.

The Japanese government funded a much smaller collaborative emissions reduction programme over the same time scale for supersonic transport application, involving a partnership between Kawasaki Heavy Industries (KHI) and R-R, which formed part of the a much larger multinational engine technology programme, HYPR.

Although all of these programmes successfully demonstrated very low NO_x in high-pressure combustion rigs, none produced anything close to an airworthy concept. Indeed, airworthiness issues were not addressed. The author knows of no feature in a current production engine derived from these massive efforts.

INCREMENTAL IMPROVEMENT OF CONVENTIONAL COMBUSTORS. Over the same period of about 37 years, visible smoke has been eliminated, combustion inefficiency at idle has been reduced by a factor of 10, and NO_x has been halved. However, increases in engine pressure ratio over the same period, from around 28 to 43, have eroded the net reduction in NO_x in service to around 20%. P&W developed [207] a combustor that they call TALON, which controls NO_x by careful admission of air and fuel to achieve good mixing, followed by rapid dilution with reduced residence time in the secondary burning zone. This is an effective and low-risk approach to NO_x reduction. This so-called RQL approach has been the subject of research for three decades. The difficulty with such a concept is to avoid smoke at high power and CO at low power, because the secondary burning zone is the part of the combustor in which these species are consumed. In reducing the formation time for NO_x the consumption time for these species is also reduced. P&W appear to have achieved substantial improvement by careful management of air admission and mixing. Indeed their claims match those made for lean Direct Injection (DI). It is neither clear whether the technology is at the same stage of development nor whether the claimed performance is enhanced because associated with the P&W prop–fan cycle, which has very low SLS SFC. The stage of development is important, as the optimisation of a concept to meet operational requirements in service usually results in some loss of emissions capability (see Subsection 4.2.7). The GEnx engine is certified for flight and should enter service during 2010.

COMBUSTION INSTABILITY. All attempts to reduce NO_x have increased the propensity of the combustor to generate self-excited pressure oscillations, even in conventional diffusion flame combustors. An amplitude above 15 kPa peak to peak is considered hazardous, and a higher pressure amplitude results in rapid failure of fuel pipes, oil pipes, or the combustion liner. The problem has not so far proved insoluble, but as azimuthal modes occur round the annulus they do not appear until relatively late in the development of a new combustor concept. By the time HP annular testing is being conducted, changes in injector configuration become very expensive and time consuming. Full sets are required rather than the one to four injectors needed for sector testing. Design and manufacture of a set typically take from 6 to 9 months. Cost of experimental components of this complexity is probably in the region of £100k per set, and test costs exceed that figure. Combustion instability, though audible, is not a significant contributor to engine noise. It is often encountered in production engines during acceleration. If it were audible under continuous load, when noise is measured, then the amplitude under acceleration would be considerably higher, and rapid mechanical failure would result.

4.2.6 Operability

The following aspects of engine operation are combustor dependent. In this subsection, variable geometry in the gas path is not included in the discussion. The associated increase in complexity and concern about reliability renders it impractical. It is discussed in Subsection 4.2.9

1. Thrust variation from full power to deceleration to flight idle during descent must be accommodated. Overall combustor AFR at take-off is around 35 and could in future reduce to 33. To achieve low NO_x at full power, the reaction-zone equivalence ratio must not exceed 0.65 (23 AFR) so about 65% of the combustor airflow must participate in combustion at full load. Efficiency must be greater than 99% at idle, when combustor overall AFR is around 100. To achieve this, the equivalence ratio in the reaction zone must be at least 0.6. A threefold change in overall AFR is required with scarcely any change in reaction-zone stoichiometry, so fuel staging is essential.
2. AFR during descent is between 100 and 150, depending on the engine design. A turn-down ratio[1] without extinction of greater than 3 is required. Lean extinction for premixed reactants at high pressure and temperature, relevant for gas turbine operation, occurs at 40 AFR. At low temperature and pressure it may occur at 30 AFR. Allowing a margin for safety, 35 AFR is probably the maximum allowable in the reaction zone at high power. The available turn-down ratio for a premixed combustor while operating at high inlet temperature is thus about 1.5, sufficient to cover the range of operation from take-off to end of cruise, but not the descent phase.
3. Altitude relight. AFR is adjustable, but combustor inlet pressure and temperature are 0.2 bar and 250 K. A pilot zone is needed to provide adequate reaction time for ignition and acceleration to flight idle.
4. Water ingestion. During descent in a tropical storm, water ingestion may be 5 times the fuel flow. It is good practice to allow a 100% weak extinction margin over the descent phase of the flight envelope to allow for this. This rule has proved adequate with combustors in which the reaction zone admits about a third of the combustor airflow. Doubling this figure may have an adverse effect on water ingestion performance, but no evidence of this has yet been reported. It may be that the water is diverted around the outside of the combustor and has little effect, but this cannot be assumed and needs to be demonstrated. The GE TAPS combustor has satisfactorily passed the water ingestion test required for flight certification.

The implication is that any combustor that employs lean burn with very good fuel–air mixing will need a pilot zone that confers operability similar to combustors of today. This zone will compromise NO_x emissions at high power.

Implications of Fuel Staging

Fuel staging has five consequences:

1. Thermal management of the lean-burn stage raises concern over injector blockage. There are three factors involved – the large wetted surface area, the difficulty of completely purging fuel from the fuel passages when the main injectors are not in use, and, if purging is not used, maintaining stagnant fuel below 460 K. The

---

[1] Turn-down ratio is the ratio of the FAR at maximum power to the FAR at minimum power.

pilot fuel is often used as a coolant, adding complexity to an already complicated assembly, and of course, adding to heat pickup by the pilot fuel.

2. Main injector fuel purging and reignition has an impact on emergency acceleration time, which is an airworthiness issue. To avoid compressor surge, staging must take place smoothly. An initial fuel spike must be avoided. There is less than a 1-s margin between the acceleration time achieved by high-bypass-ratio engines and the airworthiness requirement. The implications are that the fuel manifolds cannot be purged, for to refill them very quickly without creating a fuel surge when the liquid reached the nozzles would be very difficult.

3. During approach, frequent, large changes in thrust will give rise to cyclic staging at low altitude with the aircraft flaps in a high-drag configuration. Staging and any purging process must be completely reliable. The situation is more hazardous in heavy rain and is exacerbated by the fact that all engines undergo thrust changes simultaneously. Extinction could occur simultaneously in all engines, a situation from which recovery would be impossible. It is probably necessary to stage injectors in several groups to avoid excessively depleting pilot flow to furnish the main injectors with adequate fuel to light. This complicates the fuel control system, increasing the likelihood of component failure. It is important to remember that thrust changes must be smooth and predictable. The pilot must have virtually perfect control of engine behaviour.

4. The TET profile may change substantially during staging, resulting in deterioration of hot end components.

5. Fuel manifolds are exposed to high ambient air temperature and radiation from the HP engine casing, which can reach a temperature of 600 °C. Employing fuel control valves remotely from the engine core creates a fuel heating problem in the main manifold. The use of a single manifold in which fuel flows at all flight conditions requires the fuel staging valves to be located in a hostile environment. Concentric manifolds could be used so that the pilot fuel cools the main, but complexity and cost make that option unattractive.

These are not insoluble problems, but they increase the scale of the necessary technology development programmes. Understanding these complex, critical issues helps to explain why progress is seemingly slow.

4.2.7 Performance Compromise after Concept Demonstration

Research programmes necessarily concentrate on those aspects of combustor performance in which the stakeholders are interested – currently emissions reduction – and use test rigs that do not perfectly replicate the engine operating conditions, geometry, and compressor delivery flow velocity profile, wakes, and turbulence. Often the engine operating pressure and mass flow are higher than those of available rig test facilities. The core compressor power input for a large aeroengine is more than 100 MW. After the successful achievement of emissions targets in, say, a single-sector high-pressure combustion test rig, the combustor will be redesigned in annular form, with representative diffuser geometry, and at the scale required by the engine project. Optimisation of the combustor to achieve altitude relight, to eliminate combustion instability, to improve cooling, to adjust outlet temperature distribution to suit the

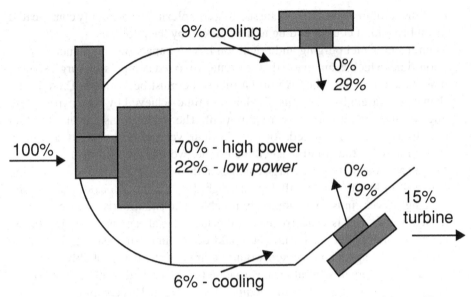

Figure 4.55. Typical flow modulation required in a variable-geometry combustor. (Courtesy of Rolls–Royce plc.)

turbine, to reduce cost and weight, to incorporate thermal management into the fuel injector, and finally to operate the combustor in the engine environment, give rise to loss of emissions performance. It is reasonable to suppose that the final NO_x will be 50% higher than that predicted from sector rig tests, and sometimes it could double.

4.2.8 Lean-Burn Options

VARIABLE GEOMETRY. Theoretically, variation of the airflow distribution in the combustor would solve the problem of operability, and this was demonstrated experimentally by R-R in the Japanese government funded hypr programme during the 1990s. [208]. Altitude relight was achieved at 15-Km altitude, adequate idle efficiency was achieved, and NO_x at cruise measured in a single-burner rig was well below the target EI of 5. Geometry was not adjusted during test, but the combustor was reconfigured and then tested at different conditions. When taken to the next stage – a three-burner sector – severe combustion instability occurred. Design studies performed in that programme failed to find a practical method of varying geometry in an engine. Figure 4.55 shows the range of air flows required. The main points to note are these:

1. The effective area change in the fuel injector is so great that dilution flow must also be changed to avoid an unacceptable increase in combustor pressure loss.
2. The combustor is annular. Modulation of the inner dilution jets by mechanical means appears impractical.
3. Any mechanical mechanism for flow modulation must withstand vibration and operate reliably at temperatures up to 650 °C.
4. Working clearances between moving parts allow leakage, impairing accuracy of flow control.

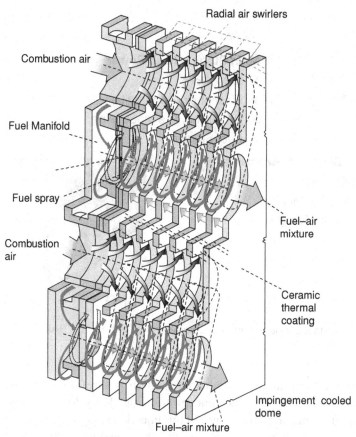

Figure 4.56. Macrolaminate injection system invented and developed by Parker Hannifin Corporation. (Courtesy of Parker Hannifin Corp.)

5. Fluidic devices such as vortex amplifiers waste half the available wall pressure loss to achieve flow stability, and to achieve substantial flow variation with the available wall pressure loss of 3%, they have to be far too big to fit within the available space.

MULTIPLE INJECTORS. GE, UTC, and NASA with Parker Hannifin [209] investigated the idea of using up to 1000 small injectors combined with fuel staging to achieve rapid, uniform mixing of fuel and air, with short residence time. This approach is used successfully for burning gas in the GE LM6000 engine.

The obstacles to be overcome for aeroengine application are

- The heat transfer to fuel, leading to injector coking; removal of the injectors for cleaning is impractical (probably true for most ultra-low-NO_x systems);
- The depth of the combustor renders diffuser design and accommodation within the bypass duct difficult and adds weight.

Parker Hannifin Corporation developed Macrolaminate injectors, in which a large array of injector elements is chemically etched into sheets, subsequently laminated by diffusion bonding or brazing (see Fig. 4.56). Both air and fuel passages can thus

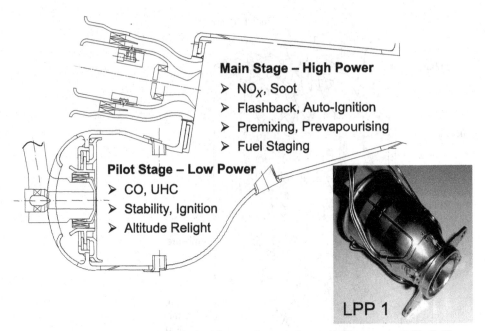

Figure 4.57. Radially fuel-staged double-annular combustor. (Courtesy of Rolls-Royce Deutschland)

be formed. Heat transfer from air to fuel must be limited by careful design. No application is known to the author, but the concept offers flexibility, low cost, and uniform distribution of fuel and air. Performance of this injection system was demonstrated in high-pressure research combustion rigs, and the emissions performance shows promise [210]. Very low NO_x was demonstrated at pressures up to 28 atm and with an inlet temperature of 870 K. There was no evidence of flashback into the swirl modules. High efficiency was also achieved at cruise conditions, so the concept appears well suited to fuel the main zone of a staged combustor.

An aspect of performance not discussed in the available literature, to the knowledge of the author, is that of altitude relight performance. The author believes that this would prevent application to the pilot zone of an aero gas turbine. The achievement of ignition at above 8000-m altitude requires a substantial recirculation volume in the primary reaction zone. The length of this zone in a combustor with a single ring of injectors is typically half the combustor depth, being determined by the length of the recirculation generated by the fuel injector and surrounding air swirlers. The recirculation length is proportional to the diameter of the injector. If, for example, the scale of the fuel injection device is reduced to one quarter, then the length of the recirculation-zone will also be reduced by 75% and with it the recirculation-zone volume. Altitude relight could not be achieved.

The concept lends itself to an axially staged arrangement with a separate, conventionally fuelled pilot zone similar in layout to that shown in Fig. 4.57 or to an arrangement in which the Macrolaminate nozzles are grouped around separate pilot injectors in a single annulus.

WATER INJECTION. Direct water injection into the combustor is a practical and powerful way to reduce NO_x at take-off. Substantial NO_x reduction can be achieved with a water-to-fuel ratio of around 1:1. If intimately mixed, a reduction in flame temperature of around 300 K can be achieved, and the volume of steam produced also reduces the oxygen and nitrogen concentrations in the combustion gas by about 10%, reducing reaction rate by 20%. NO_x is approximately halved by a reduction in flame temperature of 100 K, so to a first approximation a 10-fold reduction in NO_x can be expected. Reductions of this order have been demonstrated in engine tests.

A modern large aircraft with take-off thrust of 1000 kN will consume around 600 kg fuel during the first minute from the start of take-off. It would thus require this weight of water and would require a water tank and pumps, adding weight that would be carried throughout the flight. This increases fuel consumption, further adding to take-off weight and emissions.

Freezing of the water has to be avoided in extreme weather conditions, and concern has been expressed about the possibility of the wrong liquid being put into the wrong tank during fuelling. This may be very unlikely and would be hindered by baulking features on delivery nozzles. The author is not qualified to judge the risk, but the consequences of a mistake would clearly be very serious.

The idea has come up for discussion periodically but always the disadvantages have been considered too great. Increasing concern about NO_x could shift the balance.

LPP FUEL-STAGED COMBUSTION. LPP combustion has achieved very low NO_x in high-pressure test rigs. Fuel staging could take the form of a central pilot zone within each premixing duct outlet or a separate ring of conventional fuel injectors in a double-annular configuration. An example of this type is shown in Fig. 4.57. Of these, the former appears preferable. The pilot flame can be used to stabilise the LPP flame and to aid smooth light across during staging. The practical difficulties with LPP combustion are the risks of auto-ignition and flashback at high power, poor turn-down ratio, high CO at cruise, combustion instability, and the problem of dispersing the fuel uniformly in a very short time.

AUTO-IGNITION. There are very large differences in auto-ignition delay times for aviation fuels published over the last 40 years, arising from differences in the design of the test apparatus, the difficulty of the experiments, and the stochastic nature of the process. The various types of apparatus employed – shock tubes, piston compression devices, static bombs into which fuel is injected into preheated, pressurised air, and steady-flow combustion tunnels provide data over different temperature and pressure ranges. The most difficult conditions for obtaining data appear to be between 850 and 1100 K with pressures above 20 bars – just the range of conditions at which data are needed for LPP system design.

The steady-flow research rigs are closest in geometry to gas turbine premixing ducts. Effort is made to achieve uniform fuel distribution and air velocity across the duct, and of course there is no constraint on upstream settling length or on the complexity and size of the fuel injection apparatus.

If the auto-ignition delay time is affected by the variation in rig design, how much more is it likely to be affected by differences in premix duct design in gas turbines? They are constrained to fit within a length of less than 100 mm, are fed with a non-uniform velocity profile from a compressor immediately upstream, they may have to employ swirl to distribute the fuel, with wakes from swirl vanes, and the fuel distribution may vary greatly with engine power because of the variation of momentum ratio between air and fuel.

To quantify this problem we compare results published by leaders in this field. Some excellent work done by Spadeccini and TeVelde [211] of UTC was published by NASA in 1980. The value of this paper is enhanced by the publication of the measured – not merely correlated – data. Although tests were conducted over the range 700–1000 K, almost no auto-ignition events could be obtained at above 900 K, and there are no data at above 20 bars with a temperature higher than 800 K.

A wide range of delay time pressure dependence is reported. Spadaccini and TeVelde [211] report a P^{-2} dependence. Others, like Hayashi of JAXA report [212] P^{-1}, and results collected from various sources by Coats [213] of the UK Royal Aircraft Establishment (RAE) also show P^{-1} dependence. These observations illustrate the difficulty of formulating design rules. As most of the high-temperature data were obtained at 20 bars or less, and with large uncertainty on the pressure exponent, one cannot estimate auto-ignition delay time at 50 bars reliably. There are ratios greater than 2 in ignition delay time measured in different test facilities at the same conditions. Coats [213] included data obtained up to 60 bars, but the data reported in that survey from the mid-1980s indicate significantly longer delay times than the more recent publications. What safety margin should we apply in application of the experimental data to a real design?

For the moment let us not compound the problem by extrapolation. There are a lot of data at around 800 K and 20 bars corresponding to altitude cruise for a large engine and max take-off for an engine of below 65-kN thrust. Application of the Spadeccini and TeVelde correlation [211] gives a range of delay time from 0.75 to 1.25 ms, Hayashi et al. [212] reports 1.1 ms, and data collected by Coates [213] give a value of around 5 ms.

Bearing in mind that some compromise of performance may be expected in a premix duct designed for an engine, a safety factor of 50% relative to the shortest time measured in a laboratory experiment appears reasonable. This implies a duct residence time of 0.5 ms at 20 bars and 800 K. Assuming a flow velocity of 100 m/s in the premixing duct, the available mixing length is 50 mm. The author's experience with experimental, lean premixed combustors suggests that this estimate is conservative, but conservatism is necessary for the sake of safety. Auto-ignition delay experiments are conducted for short periods, and the experiment ends as soon as ignition occurs. A gas turbine operates for thousands of hours, heat from the flame can soak back into the premix duct, raising the boundary-layer temperature, compressor surge can reverse the flow in the duct temporarily – if flashback occurs then detection and shutdown are necessary or the flame must blow out of the duct. Compressor surge can be caused by bird ingestion during take-off. It is not a rare event in aircraft safety terms, in which a probability of 10^{-8} is considered remote.

It is clear from this preliminary assessment that LPP combustion may be impractical at higher pressures and temperatures than those previously assumed. For a large engine during take-off and climb, DI would be necessary, but to achieve reasonably low NO_x, even this must be a fuel-staged system. It does not appear practical to incorporate three kinds of injection system in one engine – a diffusion burner for operation at and below flight idle, an LPP duct injector for cruise, and a LBDT burner using the same air as the LPP burner for take-off and climb. The author believes that application of LPP should be limited to engines of below 20 OPR and thus to regional aircraft of 50 seats or fewer.

The engine manufacturer has to consider the implications of thrust growth, mentioned in the introduction of this section. It would be very costly to change the combustion system because a step in thrust growth rendered an existing LPP combustor unsafe to use. For example, the Rolls-Royce BR700 series of engines use the same combustor and range in thrust from 66 to 93 kN and in pressure ratio from 24 to 32, the higher pressure ratio adding about 75 K to the combustor entry temperature.

SUPERSONIC CIVIL TRANSPORTS – A SPECIAL CASE. Although no such aircraft are planned at present, they have been discussed repeatedly, discussions that have stimulated substantial research programmes, such as the NASA High Speed Research initiative of the 1990s and the Japanese Government programmes HYPR (1989–1999) and ESPR (2000–2004), sponsored by NEDO.

Supersonic transports must be treated as a special case for several reasons:

- They fly at high altitude, within the lower part of the ozone layer, which is sensitive to NO_x concentration.
- They consume around 50% more fuel per passenger kilometre than a subsonic aircraft.
- During cruise, ram precompression raises the combustor inlet temperature to over 850 K, considerably higher than in a subsonic aircraft. To limit cruise temperature, the SLS pressure ratio is low around 10:1 and because the atmospheric pressure at 15 km is only 0.115 bar, the combustor pressure at cruise will be around 10 bars. It is therefore probable that engines of this type, in which high pressure and high temperature do not occur simultaneously, could utilise LPP combustion.

PRACTICAL LIMITATIONS GOVERNING THE DESIGN OF A PREMIXING DUCT. For the purpose of illustration through which some understanding of the design problem may be gained, let us consider a simple, annular premixing duct supplied with fuel through radial jets. A swirling fuel prefilmer like that used in many airblast injectors, sandwiched between airstreams, could also be used, but radial dispersion across the airstream is difficult to achieve in a short distance. Cross-stream injection achieves efficient atomisation and mixing when optimised. Some excellent research was performed by Rachner et al. [214] on this subject.

The difficulty in the application of cross-stream injection is that jet penetration decreases with engine power. Whereas the air velocity is almost independent of power and thus air momentum varies with pressure, fuel momentum varies approximately with the square of the engine pressure. Jets optimised for cruise will over penetrate at take-off. However, the following issues have to be taken into account:

1. **Mixing in the duct:** The high velocity in the duct allows little pressure loss for the generation of turbulence. The dynamic head in the duct proposed for illustration is 3% of the total pressure. The available combustor pressure loss is about 4.5%. Any device used to produce turbulence must not create wakes in which fuel could auto-ignite.

2. **Premixing duct flow:** An engine suitable for a 50-seat regional jet such as the GE CF34-3 delivers 40.9-kN SLS thrust, at 21 OPR, with bypass ratio of 6.2. SLS SFC is 0.34 kg/kg/h, so fuel flow at full power is about 0.4 kg/s, 90% of which would pass through the premixing ducts, 10% through pilot nozzles. Combustor air mass flow would be about 17 kg/s, and total premixing duct flow would be 9.5 kg/s, giving an equivalence ratio of 0.65.

3. **Duct flow passage:** The design constraints are as follows:

    a. Experience shows that fuel jets less than 0.5-mm-diameter block easily.

    b. For a reasonable fuel pressure drop over the power range, the total number of jets must be limited, and they have to be shared among the premix ducts. The larger the number of ducts, the more widely spaced the jets are. There comes a point when circumferential uniformity of the fuel distribution is compromised.

    c. There must be enough space in the centre of the premix duct to accommodate a pilot injector, allowing for double skinning of the fuel passages for thermal insulation.

    d. The flow passage must contract in area by at least 10% between fuel injection plane and outlet to prevent flashback and to suppress wakes or separations in which auto-ignition could occur.

    An example calculation is as follows.

Dynamic head	3% P_{comb}
Premixing duct effective area	89 cm^2
Number of premix ducts	12
Fuel jet diameter	0.5 mm
Fuel jet C_d	0.75
Fuel ΔP	8 bars
Fuel flow	0.36 kg/s
Fuel flow per jet	0.0048 kg/s
Number of jets	72
Number of jets per duct	6
O.D. of duct	40.0 mm
I.D. of duct	27.4 mm
Radial thickness of insulation between pilot and main passages	5 mm
Max O.D. of pilot	17.4 mm
Min O.D. of pilot to pass 10% flow	13 mm
Duct height	7.2 mm
Jet pitch	14.3 mm

Jet penetration reaches almost to the full duct height, based on the correlation in [214], and is thus satisfactory.

4. **Discussion.** A jet pitch of twice the duct height is larger than one would like, but to reduce the diameter of the duct, so that height increases and pitch reduces, leaves insufficient space for a pilot and results in inadequate jet penetration. At high-altitude cruise, the jet penetration is reduced. In a business jet application, in which the cruise altitude might be 13 km, fuel pressure drop could fall to

less than 1 bar and the fuel jets would then penetrate to only about 60% of the duct height. Regional jets usually cruise at below 9000 m. Because business jets fly very few hours per year, they do not contribute significantly to the global emissions problem. Deterioration in cruise NO_x should not be a concern. The radial FAR profile would be biased to the inside at cruise, which might help combustion efficiency.

The preceding simple calculation indicates that a practical premix duct geometry can be formulated for an engine of modest pressure ratio. LPP combustion applied to engines of around 20 OPR should reduce LTO and cruise NO_x by up to 70% compared with today's values. The gain is compromised by the fact that the premix duct cannot be used at idle and during descent, and by the need for a diffusion pilot over the entire flight envelope.

COMBUSTION INSTABILITY. Because the mixture entering the combustor is uniform and lean, small perturbations in AFR affect the rate of heat release throughout the burning zone in the same sense. In a DI combustor, some zones are richer and some are leaner than stoichiometric, so the response to a change of AFR in these zones is of opposite sense. Pressure perturbations generated in the combustor modulate the airflow in the premixing duct, but not the fuel flow; hence there is a strong coupling between heat release and pressure. This may end up with serious consequences if the local conditions allow the amplitude of the pressure wave to grow in time. This includes flashback in the premixing duct, damage to the combustor, and damage to sensitive external components such as fuel and oil pipes, which is due to vibration, leading to fracture. Every lean-burn system with which the author has been involved during the past 20 years has shown a propensity to such cyclic instability. An important aspect of the problem is that there are several possible modes of instability. Some occur only in combustion rigs, some in engines, and these instabilities often depend on the inlet and outlet acoustic boundary conditions. Engine combustors are choked at outlet, and rig combustors operate with an outlet Mach number of around 0.3 and exhaust into a large volume. A rig combustor is usually fed with air from a large plenum. An engine combustor is closely coupled to the compressor. Thus instabilities may appear and seriously hinder rig development, which would not occur in an engine, and instabilities may appear for the first time when the combustor is engine tested. The risk must be assessed and included in the engine demonstration and product development programme. CFD tools can help the product development programme by answering many 'what if . . . ' questions. A good example of this is in the previous chapter on thermoacoustic instability. One should, however, understand that the quality and reliability of a CFD solution very much depends on the ability of various submodels to capture the relevant physical processes correctly and accurately.

4.2.9 Conclusions

The introduction of lean DI combustion systems can be expected to reduce NO_x by about 40%. For this technology to substantially permeate the entire world fleet will take at least 25 years. The only way in which NO_x can be further reduced is by the application of lean, premixed, prevapourised combustion. This may be feasible in engines of up to 20:1 pressure ratio, including supersonic civil aircraft, should these ever again be built. It would be wise to gain service experience in such engines before attempting to apply the technology at higher pressure ratios.

All lean-burn combustion systems require staged fuelling and thus add complication, weight, and cost and tend to reduce reliability. Reliability of engines for large aircraft is not only a safety concern but a concern for dealing with very large numbers of stranded passengers at airports and the rescheduling of their flights. One may think that these are problems to be dealt with by airlines, but inevitably such issues have large economic implications for engine manufacturers.

Although future increases in pressure ratio and temperatures will exacerbate the emissions problem, it is likely that compressor-disk material will limit the OPR to around 55 for several years. NO_x cannot be controlled if the TET exceeds about 1850 K. Because the optimum TET is linked to pressure ratio and bypass ratio, emissions control may limit engine cycles within the next 10 years.

REQUIRED EXPERIMENTAL RESEARCH EMPHASIS. It should be clear from the foregoing discussion that emissions reduction in aircraft gas turbines requires the consideration of all aspects of combustor performance and fuel injector design. Issues like altitude relight and thermal management are perceived by the research community to be the concern of engine designers and manufacturers, and thus to belong to work around technology readiness level 5.[2] The author believes that these issues need to be evaluated at least theoretically at technology readiness level 3 or 4, so that research may be directed to address the areas of greatest risk or uncertainty.

With lean-burn systems, the injector is of paramount importance, as it controls the entire combustion process. A single research project may focus on a very small area of the subject, but the researcher should understand how his or her work relates to the total problem. Collaboration among universities, research institutions, and industry is essential if the aim is to make improvements in environmental impact. The improvement will come from applications in engines, and it is in the transition from the simplified research rig to the engineering of a reliable product that all may be lost. Simplification and compromise are necessary in research, but there is a point at which simplification renders the experiment worthless.

For example, the investigation of the condition of fuel leaving a premix duct by the application of phase-Doppler anemometry on a cold, atmospheric rig, using water instead of fuel, would be a complete waste of time. Droplet mass and fuel–air momentum ratio are nowhere near engine values, droplet trajectory bears no relation to a practical case, there is no evaporation, and most of the atomised liquid runs down the walls of the duct and is inefficiently reatomised off the outlet inner and outer rims. If there is swirl in the duct, the liquid will be centrifuged onto the outer wall. Yet to conduct such an experiment with kerosene at conditions for which auto-ignition is possible is unsafe. Good experiments that are within the capability of university research departments and that offer adequate access for instrumentation can be designed through conversation between engine combustion specialists and

---

[2] TRL, NASA defined technology Readiness Levels from 1 to 9 [215], and their definitions are now widely accepted. The concept is very useful for controlling financial risk.

academic researchers. Access to industrial test facilities may enable researchers to bring their expertise to bear in a more realistic and safe environment.

Although it is not necessary to incorporate thermal management into research fuel injectors, they should be so designed that insulating air gaps are simulated by solid metal. This is also true for CFD studies. An additional total thickness of about 4 mm is required to allow for heat shields on both sides of a fuel passage.

Better understanding of fuel–air mixing and means to disperse fuel rapidly are needed. The effects of pressure and temperature, although expensive to replicate, cannot be ignored. Both have a strong impact on liquid atomisation, dispersion, and evaporation.

There is a need for experiments to validate and support the development of a better CFD code. Of particular importance is the description of the fuel atomisation process. There are two difficulties here:

1. The process is likely to be strongly influenced by the injector configuration. There are many variables in the injector geometry, and although airflow pattern is not affected by scale, fuel jets and drops do not scale in the same way.
2. We are primarily interested in the interaction of fuel and air at high power. The very high fuel flows in gas turbines produce sprays so dense that they are not amenable to measurement with techniques and technologies currently available.

Finally there is an obstacle to collaboration arising from the commercial sensitivity of this technology. Industry will not allow publication of research it has co-funded that investigates geometries close to the envisaged product. To do so would threaten competitiveness. For the same reason, industry will not divulge details of its future designs to aid research that it is not funding. Thus, in practice, either the science-based community must sacrifice its right to publish freely or industry will hold it at arms' length, and the effective application of precious skills to the problem of improving gas turbines will be limited.

It may appear that the message is 'It is all too difficult, don't attempt it!' That is not the view of this author, but 20 years spent in managing collaborative research of this kind has yielded some lessons. The message is 'Proceed with care as the problem is complex and challenging, yet crucial to understand!'

CFD. One of the patriarchs of gas turbine combustion, Professor A.H. Lefebvre, would assist his students to grasp the place of CFD in this field by bringing into his lecture room a T shirt, on the back of which, he told them, had been printed the major contributions to gas turbine combustion made by CFD. He would then hold it up to the audience. It was completely white!

Since the turn of the century that situation has changed. The massive computing power of parallel processors combined with the patient efforts of many highly competent people over the past 40 years are at last bearing fruit. The aerodynamic flow field generated by complex geometries such as lean-burn fuel injectors can now be calculated reliably. A great deal of work has been done to improve both combustion models and chemistry–turbulence interaction models. LES provides a greatly improved model of unsteadiness and turbulence, but requires careful meshing, and the cost and time expended in its application is a serious problem for industry. Scientifically, the subgrid modelling is challenging. At present it is common practice to use both LES and URANS to model unsteady phenomena. Research is ongoing in many universities to improve these models.

The importance of appropriate boundary conditions cannot be overemphasised. As it is comparatively easy to measure the aerodynamic flow field external to the combustor, but very difficult inside, and as CFD offers greatest benefit when applied to the complex, internal flow field, it is good practice to determine the boundary conditions experimentally as close to the flame space as possible. The most serious deficiency of CFD is in the definition of spray boundary conditions and modelling, and, as mentioned in the previous section, this is also very difficult experimentally. The empirical correlation [216] now used to estimate the SMD was generated from fuel injectors far different from those now under development for emissions control. Advanced, two-phase flow methods to predict primary break-up of the fuel spray are under development, but it will be some time before we can confidently apply these techniques.

The way we use CFD in combustion is to anchor it to existing experimental data and then use it for predictions cautiously. Availability of experimental data is extremely important as it enables the selection of the most appropriate model (there are about 10 turbulence models to choose from!)

Good work has been done in the representation of evaporating two-phase flow and fuel-spray development in cross-stream injectors [214, 217]. Another paper [218] is concerned with the break-up of a diesel injector spray and provides a good example of advanced two-phase flow modelling. The sprays under consideration are extremely dense, so experimental validation is not possible. Thus, aided by experimental validation, CFD is ready to be applied to the design of premixing ducts. Although not completely reliable, it is far better than guessing.

We are not able to predict CO emissions reliably. The empirical design rules for conventional combustors are adequate, but with lean burn, CO formation and consumption is a problem. Several effects contribute to this difficulty (spray boundary conditions, finite-rate chemistry, complexity of the partially and fully premixed combustion modelling, etc). Premixed combustion modelling is more difficult than diffusion-flame modelling, which makes application of CFD to lean burn more of a challenge. Some recent progress on this front is discussed in Chapter 2 of this book.

Interaction between pilot and main zones is important for both emissions control and flame stability. It is likely that unsteady flow effects are important here, and it may be asking too much of CFD to handle the problem. However the parallel use of CFD with experiment can be of enormous help in interpreting experimental results and guiding the aerothermal designer in defining modifications that will improve performance. The usefulness of CFD depends on the reliability with which the flow field and fuel distribution, among many other factors, can be modelled.

One of the advantages of CFD over experiment is that it is not limited by operating conditions: The problem is that neither can it be validated at high pressure and temperature. It can only be shown to work by experience in application. Here we must be wary of our own selectivity. When CFD yields a result contrary to observation of engine behaviour, there is a tendency either to adjust something that cannot be reliably estimated, for example SMD, select a different turbulence model to achieve agreement, or simply ignore and move on. When CFD shows a trend that matches engine behaviour, the results are naturally trumpeted from the rooftops. We all love to achieve something, and publishing our achievements is the only way of financing further work. But this normal behaviour is biased. The bias may be unavoidable, but let us at least not deceive ourselves. There is no such thing as a bad result in research, if competently conducted. We learn as much from our failures as our successes – but only if we give our failures equal weight.

CFD provides a useful approximation. We are often unable to estimate the level of confidence in our CFD predictions, and there is therefore a natural reluctance to rely on CFD to support initial design and component development. It is in these applications where the potential cost and time savings are greatest, and there is a strong expectation that continued improvement of the models together with experience in their application will yield significant benefits. Finally, with reference to time scale for improvement, for the past 30 years, the breakthrough that would transform the situation has always been from 2 to 5 years away. This has something to do with funding horizons and the patience of stakeholders. It is encountered in every field of research. The appropriate response is a wry smile!

4.3 Application of Lean Flames in Stationary Gas Turbines
By B. Jones

Stationary gas turbines, including those used on board ships as auxiliary power generators or main propulsion units, vary greatly in both size and design. Simple cycle machines range in power from 100 kW to a little over 300 MW, and combined cycle machines produce up to 450 MW.

The absence of airworthiness constraints described in the introductory subsection of Section 4.2 opens wider opportunities for innovation in the design of combustion systems for stationary engines. Land-based machines are subject to stringent emissions regulation because they often operate continuously at full load close to populated areas. The aeroengine industry has struggled for the past 35 years to make significant reductions in NO_x emissions, the advances in combustion technology being largely negated by increases in pressure ratio and turbine inlet temperature. In the industrial sector NO_x has been reduced by 90% to 95% by the application of lean combustion.

By integrating the gas turbine into a combined cycle in which exhaust heat is used to produce additional power from a steam turbine, cycle efficiency can be increased from that of an aeroengine – around 40% – to almost 60%. The efficient use of heat that would otherwise be wasted influences the choice of gas generator cycle. There is little to be gained from increased pressure ratio or very high TET, both of which greatly increase cost. Pressure ratios of stationary gas turbines are thus usually around 15:1, although there are engines that operate at more than double this pressure. Although aeroderivative engine combustors can be designed without regard to airworthiness, they have to be compatible with the mechanical design of the original engine core. Efforts to radically reduce emissions have, in the past two decades, led to a multiplicity of new designs that bear little resemblance to combustors used towards the close of the 20th century. The constraints of size, weight, and flight safety that so constrain the design of aeroengine combustors has led to convergence between competing designs to something approaching a single,

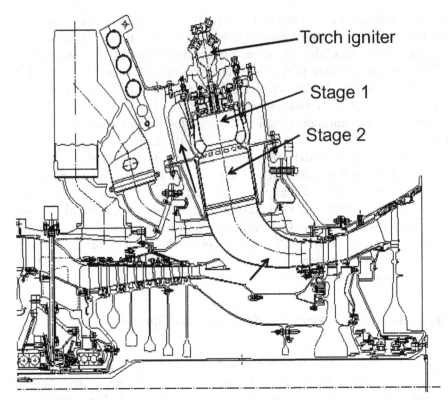

Figure 4.58. R-R industrial RB211 two-stage gas-burning LPP combustor. (Courtesy of Rolls-Royce plc.)

optimum solution. The absence of these constraints has led to remarkable diversity of designs in stationary engines.

4.3.1 Common Combustor Configurations

1. A ring of tubular combustors that may be arranged within an annular casing or in separate pressure casings mounted radially around the engine (Figs. 4.58 and 4.59). Where the tubular combustors are arranged axially, they are often angled to allow the fuel injectors to be outboard of and upstream of the compressor outlet, to reduce shaft length. Both configurations create serious aerodynamic problems in relation to the diffusion and distribution of the compressor outlet flow.
2. A single annular combustor fed by one or more rings of fuel injectors; see Fig. 4.60.
3. A single tubular combustor may be mounted radially on the side of the pressure casing in small engines or situated outside the engine gas path in a silo adjacent to the engine in large engines. This larger combustor may be fed by many fuel injectors, as shown in Fig. 4.61.
4. Interstage reheat is employed in Alstom's GT24 and GT26 engines, shown in Fig. 4.62. The incorporation of an interstage combustor to reheat the turbine gas increases the specific power output of the machine without raising the maximum cycle temperature. Alstom developed a premixing fuel injection system for this combustor that actually reduces the NO_x from the primary combustor [219].

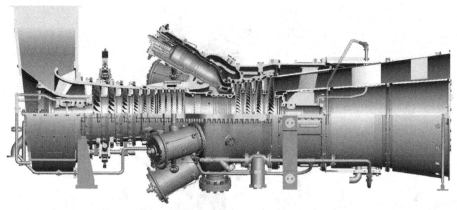

Figure 4.59. KHI M7A-03 8-MW gas turbine with tubular dry low-NO_x (DLN) combustion system. (Courtesy of Kawasaki Heavy Industries, Nishi Akashi, Japan)

The choice of configuration is driven by the engine operating conditions, mechanical design considerations, cost, and probably by the experience of the design team. In gas turbine design, innovation brings with it a period of trouble and high cost. The financial return often comes many years later. If a company has a successful design they will first consider scaling that. If the engine is of a high pressure ratio, the combustor casing will be highly stressed and the configuration will be selected with that in mind. This will also place a heavy demand on cooling, so the combustor wall area will need to be minimised. A silo or annular combustor will be preferred. A company with limited test facilities will want to avoid the use of an annular combustor if possible because the cost of rig testing is enormous compared with testing one of a set of tubular combustors.

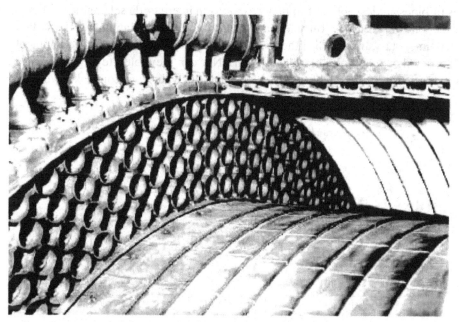

Figure 4.60. Alstom annular DLN combustor showing multiple premixed fuel injectors. (Courtesy of Alstom Power)

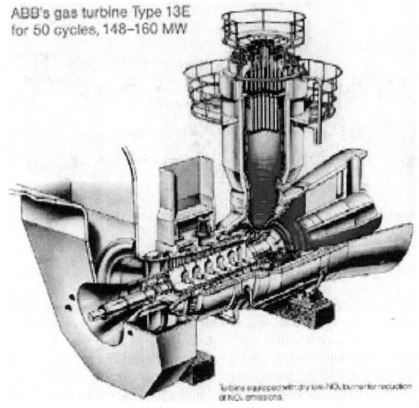

Figure 4.61. Large engine silo combustor with multiple fuel injectors.

4.3.2 Fuels

Gas turbines are less tolerant to fuel variations than either diesel engines or boilers for the following reasons. Turbine blades cannot tolerate contaminants like sodium, vanadium, potassium, or ash, either because of high-temperature corrosion or blockage of small cooling holes by solid deposits. Nor can they tolerate large changes in the calorific value of the fuel supplied. Low-calorific-value gas from coke ovens and

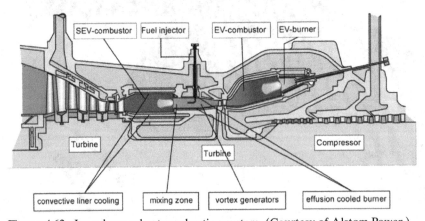

Figure 4.62. Lean-burn reheat combustion system. (Courtesy of Alstom Power.)

blast furnaces [coke-oven gas (COG), blast furnace gas (BFG), and Linz–Donawitz gas (LDG)] has a low burning velocity, approximately proportional to its heating value by volume, which may be less than 10% that of natural gas. Syngas has a much lower heating value than natural gas but a high burning velocity, as it can contain more than 40% hydrogen. These low-calorific-value gases require the use of special injectors to accommodate an increased volume flow rate. The dilution of the fuel by nitrogen and CO_2 reduces reaction rate, and combustion efficiency will be reduced at low power. Due consideration needs to be given to the power needed to raise the pressure of the greater mass of fuel and the size of the injectors required to accommodate it. The fuel supply pressure must be around 15 bars higher than the combustor pressure to accommodate the losses across the fuel control system and fuel injectors. Low-calorific-value gas is often mixed with natural gas to contain these problems. It will be apparent that there is a limit to the application of lean flames with these fuels. Lean combustion is already close to weak extinction and cannot tolerate a large reduction in flame speed or flame temperature. LPP systems cannot tolerate a significant increase in flame speed associated with the presence of hydrogen because of flashback.

Most engines used for power generation burn natural gas, and a significant proportion of these engines have the ability to burn both gas and liquid fuels. Sometimes there is a requirement to operate with gas mixtures, including propane, liquefied natural gas (LNG), or syngas to take advantage of low-cost alternative fuels. Gas turbines operating on gaseous fuels cannot tolerate liquid HCs in the fuel, and hence fuel preheating is commonly employed in cold climates or where gas contains heavier fractions. There is also widespread demand for the ability to burn light fuel oil as a back-up should the gas supply fail. Diesel fuel is the only fuel available on board ship.

Lean-burn combustions systems are more susceptible to changes in gas-fuel properties. Whereas a diffusion burning combustor can typically tolerate a ±20% variation in Wobbe index, LPP combustors can accept only a ±10% variation of this index. The Wobbe index, commonly used to denote changes in gas-fuel properties, is defined as the ratio of higher heating value in millijoules per normal cubic meter (Nm^3, i.e., at standard temperature and pressure) to the square root of the specific gravity of the fuel gas. The significance of the Wobbe index is that, for a given power output and injector effective flow area, a fuel with a high Wobbe index will require a lower injector pressure drop, increasing the likelihood of combustion instability. Depending on the injector design, it may also influence the degree of premixing achieved. The Wobbe index for natural gas varies with the source but is typically around 50 (1250 in BTU/ft^3 used in the United States), whereas propane has a Wobbe index of 80 (2000 in BTU/ft^3). Auto-ignition properties of fuels also need to be considered and are discussed in Subsection 4.3.9.

4.3.3 Water Injection

The simplest way to control NO_x is to reduce maximum flame temperature by injecting either water or steam. The ratio of water to fuel is typically around 1, and this ratio can go up to 1.6 with steam. There is an increase in power of up to 5% because more fuel can be burned without exceeding the HP turbine temperature limit, increasing the enthalpy, density, and specific heat of the flow through the turbine. The disadvantages, which drive the continued research of dry low-NO_x (DLN) solutions, are as follows:

1. The highly purified water required is not always available in sufficient quantities and it is costly. A 100-MW engine might consume more than 25 tons/h. There are places like California where there is high demand for power, there is little water, and there are stringent emissions regulations necessitated by the climate (see Subsection 4.3.4). Cycle efficiency is reduced by up to 4%.
2. CO emission increases, especially at part load, and may exceed local air-quality requirements.
3. Water and steam injection reduce component life and increase maintenance costs. Prandtl number and Reynolds number increase, thus increasing the heat transfer coefficient in the combustion gas. The injection of water or steam probably also modifies the combustor outlet temperature profile.
4. Continuous monitoring of exhaust emissions is required, whereas for DLN_x systems only an annual verification test is required. However, the DLN systems require continuous monitoring of combustion instability.

Offset against the costs associated with water or steam injection is that it avoids the problems that commonly occur in lean-burn systems, especially when combustion of diesel fuel is required. These are auto-ignition and flashback, combustion instability, and fuel system complexity. Maintenance costs and engine downtime are also higher with engines fitted with LPP DLN combustors. According to [220, p. 426], maintenance costs for engines using DLN combustors are 40% higher than for engines using diffusion burning. What is probably more serious in some installations are unscheduled shutdowns, usually caused by the onset of combustion instability. This problem is receiving a lot of attention and has been with us for over 20 years. The solution may not be imminent; however, we trust that it will be solved in due course. The subject of combustion instability is treated in depth in Chapter 3 of this book.

4.3.4 Emissions Regulations

There is no single document that defines regulations for stationary gas turbines. This subject is complex, and only those who market gas turbines or act as advisors to the industry can justify the time required to understand it and keep up with its development. The author of this section is indebted to the Industrial Power Division of Rolls-Royce plc in Montreal for providing advice and the references needed for this text. The documents are complex, and the author's research has not been exhaustive. The aim of this section is only to explain the targets to be met by the combustion engineer.

The EPA website, www.epa/nscep/, provides a free service, allowing any document published by the EPA to be downloaded as pdf files. This source is so abundant that finding something can be difficult, even with the help of the advanced search facility on the site.

The minimum U.S. federal standard for NO_x and SO_2 control is defined in [221], which also explains the rationale behind the legislation in great detail. The emissions controlled are NO_x and SO_2, which come from sulphur in the fuel. CO and particulate emissions are regulated separately but not across the entire U.S. continent. The federal regulations require gas turbines of greater than 110 MW to emit less than 15 ppmv NO_x and gas turbines of between 25 and 110 MW to emit less than 25 ppmv. Regulations vary from one country to another, state to state in the United States, and in California even district to district. California and Florida apply the most stringent regulations (see, for example, [222, 223]). The combination of geographic and climatic conditions in California results in the production of highly toxic ozone at ground level by the action of UV radiation on NO_x. In Florida it is because of the large orange-growing industry. Small concentrations of NO_x discolour the skins of oranges. Permitted concentrations of NO_x depend also on power plant efficiency and size. Local regulations are updated in response to the opening of new plant of a type not previously covered by legislation in the region and in response to the introduction of new technology. The use of best available control technology (BACT) and best available retrofit control technology (BARCT) is demanded by some U.S. states. This drives competition and of course requires amendment of legislation, making it difficult to keep up to date. For example, [223] was amended seven times between 1994, when it was first issued, and 2009. Failure to match BACT can exclude a company from certain sectors of the market. Exceptions are permitted to allow for off-design performance during start-up, which for some types of plant can take 3 to 4 hours, and for malfunction, but these exceptions must be recorded and reported. Sometimes heavy fines are incurred.

The requirements for the Sacramento metropolitan area [222] are summarised in the following list, as an example, and NO_x limits are corrected to 15% O_2, dry basis:

1. **Gaseous fuel firing**
   - 42 ppmv for units $\geq$ 0.3 MW and < 2.9 MW
   - 42 ppmv for units $\geq$ 2.9 MW and operated < 877 h/year
   - 25 ppmv for units $\geq$ 2.9 MW and < 10 MW and operated $\geq$ 877 h/year
   - 15 ppmv for units $\geq$ 10 MW, operated $\geq$ 877 h/year, without SCR
   - 9 ppmv for units $\geq$ 10 MW, operated $\geq$ 877 h/year, with SCR
2. **Liquid Fuel Firing**
   - 65 ppmv for units $\geq$ 0.3 MW and < 10 MW
   - 65 ppmv for units $\geq$ 10 MW and operated < 877 h/year
   - 42 ppmv for units $\geq$ 10 MW, operated $\geq$ 877 h/year, without SCR
   - 25 ppmv for units $\geq$ 10 MW, operated $\geq$ 877 h/year, with SCR

In [222] these limits are compared with the limits for other districts in California. They are typical, but the regulations for the San Joaquin Valley are the most stringent, requiring 5-ppmv NO_x for a new plant using certain gas turbine types. This would usually require an exhaust clean-up of some kind. Regulations are extremely detailed, sometimes specifying different limits for specific engine models. The San Joaquin Valley Authority imposes fines for non-compliance of up to U.S.$75 000 per ton of NO_x, increasing to U.S.$100 000 per ton [223] in 2014.

CO regulations also vary from place to place; 50 ppmv is the most common value applied, but 25 ppmv is required in the San Joaquin Valley for some installations.

The regulatory limits applicable to land-based gas turbines within the EU are specified in [224]. The rules apply to plants of greater than 50-MW heat input. NO_x for a gas-fired plant must be below 24 ppmv (15% O_2), 53 ppmv for liquid fuel. As in the United States, local air-quality rules may be more stringent than the EU regulations. Local air-quality rules often take into account the number of hours operated annually. There are complex variations related to the age of the power plant that require study. Emissions regulations sometimes specify pollution in mg/Nm³. 50-mg/Nm³ NO_x is equivalent to 24.3 ppmv corrected to 15% O_2. CO is not yet controlled by the EC, but is expected to be included in the near future.

Another variable is the lower limit of the power range over which emission regulations are applied. This may be anywhere from start-up to 70% power, with varying degrees of alleviation for warm-up and variation in ambient temperature. The rules are different for different types of plants because of the large differences in warm-up time and use.

4.3.5 Available NO$_x$ Control Technologies

Lean DI, unaided by steam or water, cannot achieve the required NO_x and CO control. There is too much excess air in the exhaust of a gas turbine for a three-way catalyst to work, and the exhaust temperatures from gas turbines preclude the use of most alternative options of employing catalysts. Selective catalytic reduction of NO_x using ammonia injected into the exhaust upstream of a catalyst can achieve NO_x below 5 ppm. Ammonia is, however, a hazardous substance, and its accidental release in a populated area would have serious consequences. Lean premixed combustion is the most widely used and most practical remedy. Chapter 7 of [223] provides a thorough assessment of chemical means of exhaust clean-up for a gas turbine plant.

I. Gas-burning systems: LPP combustors are now offered as standard equipments for most stationary gas turbine engines using gaseous fuels.

1. **Heavyweight engines**
   Asea–Brown–Boveri (ABB) demonstrated their first LPP combustor in a GT13 150-MW service engine in 1984. The combustor was of the silo type with multiple fuel injectors, and six engines entered service. NO_x below 100 ppm was achieved with low CO, a commendable achievement at that time [219]. In 1993 Alstom, which had bought ABB, introduced the EV burner, which is a conical premixing burner, into service [219], which proved so successful that it is now used in all Alstom engines (Fig. 4.63). The burner is able to burn fuels ranging from syngas, which has a high hydrogen content yielding a high flame speed and susceptiblity to flashback, to number two diesel oil. NO_x below 25 ppmv (corrected to 15% O_2) at full power and CO of 4 ppmv is claimed.

   Alstom introduced a variant of the EV burner, the SEV, for use in their interturbine combustors for the GT24 and 26 engines [219]. This fuel–air mixing device is modified to avoid auto-ignition at the high inlet temperature to which it is exposed in this application. The fuel is admitted with compressor delivery air to achieve a degree of premixing at lower temperature and to increase the thermal mass to delay fuel heating to its auto-ignition temperature. The injector also

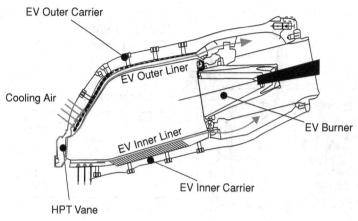

Figure 4.63. Alstom EnVironmental Burner in combustor cooled by external convection. (Courtesy of Alstom Power)

uses turbulators to enhance mixing. There are many other designs extant. [225] is an excellent 118-page report by the Electric Power Research Institute, California, which describes General Electric, Siemens Westinghouse, Mitsubishi, and Alstom heavyweight DLN combustion systems and how to optimise their fuel controls to minimise emissions and combustion instability. Dual fuel designs for large engines rely on a degree of water or steam injection to limit NO_x emission.

2. **Aeroderivative engines**
   Rolls-Royce Trent and RB211 derivatives employ several radially mounted tubular combustors, within each of which the fuel is admitted in two or three axially displaced stages; see Fig. 4.58. In addition there is a centrally mounted primary pilot injector to improve low power efficiency, lean stability, and combustion noise. This design style arose from a desire to achieve three objectives:
   a. to fit the combustor into the available space between compressor outlet and turbine inlet,
   b. to simplify the fuel system and engine control system by limiting the number of fuel stages to two or three (there are three stages in the industrial Trent engine), and
   c. to provide sufficient residence time to consume CO, unavoidably produced at some operating conditions because of the relatively small number of fuelling stages.

The advantage of axial staging over parallel staging is that the primary stage sufficiently preheats the gas entering the secondary stage to ensure efficient combustion at very lean secondary AFRs. The disadvantages are the combustor length required and the difficult aerodynamic problem of achieving a uniform air feed to the fuel–air premixing ducts. A detailed discussion on the staged mixture injection and combustion is provided in Subsection 5.1.1.

GE adopted a different strategy for their LM6000 engine, which is a derivative of their CF6 aeroengine. The LM6000 combustor is a short, axial, annular combustor that fits into the same length as the aerocombustor, although it is almost

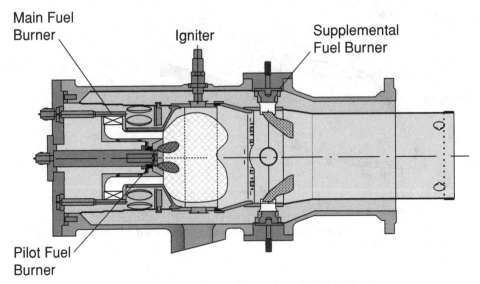

Figure 4.64. KHI DLN research combustor for engines of 8–18 MW output.

three times the depth of is its aeroengine counterpart. It contains 75 LPP fuel injection modules in three concentric rows. Because of the very short burning length available for CO consumption, the combustion temperature must be held within very narrow limits – 1825±25 K – to achieve desired emissions. Above 1850 K, NO_x is produced and below 1800 K there is insufficient time to complete the relatively slow oxidation of CO to CO_2. To achieve the narow temperature limits, 10 parallel fuel injection stages are employed. The compressor outlet diffuser has multiple passages to achieve a high degree of diffusion in a very short length.

3. **Small engines – below 25 MW**
Although legislation controlling emissions from these engines is not yet universal, considerable effort has been made to meet the requirements of the most sensitive regions of the world. Several companies have experimented with catalysts and ceramic material in this field, which is less challenging than that of the larger engines. The operating conditions are similar but the smaller size of the components reduces the manufacturing difficulties.

Because of their small size, there is more opportunity for application of the results of research performed by institutions with test facilities of limited capacity. For example, KHI developed tubular DLN combustors, shown in Fig. 4.64, which achieve 15-ppmv NO_x in engines, with a test facility capable of operating only up to 0.3 MPa, 725 K air inlet temperature [226]. The mass flow required to feed just one of six tubular combustors at less than 20% of engine pressure is only 3% of the full engine flow. This combustor is axially staged with a central pilot to aid stability and efficiency at part load. 15-ppmv NO_x is claimed from 50% to full load, together with less than 50-ppmv CO [226]. Such a rig could be operated using compressor bleed from an an existing engine, rather than investing in an expensive plant.

Air staging involving variable geometry, impractical in aeroengines, may be possible in stationary gas turbines. Air is sometimes bypassed to enrich combustion to reduce CO at low power, but here we are referring to redistribution of the airflow within the combustor to control the local equivalence ratio. A search will show that this subject has been researched widely because it is theoretically very attractive. Its implementation is rendered difficult by the temperature at which moving parts must operate reliably, and much of the research has been done by testing variable-geometry components locked in various positions by welding, to limit leakage. Fretting or seizure is obviously a problem where several thousand hours of operation are required. Much more attention needs to be given to the mechanical aspects of this problem, rather than expending effort in the relatively easy task of demonstrating that low emissions can be achieved by this approach.

II. Liquid-burning systems: Much of what was said in Subsection 4.2.9 relating to NO_x control for aircraft gas turbines is applicable here. The relaxation of space and weight constraints may allow fluidic control, vortex amplifiers, etc., to be employed to modulate air flow. Half the pressure loss across these devices, however, is taken for control stability and is not therefore available for mixing. Auto-ignition, discussed in Subsection 4.3.9, is a far greater hazard than with gaseous fuels.

An LPP system cannot operate with liquid fuel if the inlet air temperature is below the fuel boiling point. This is not a single temperature but a range over which different constituents in the fuel evaporate. At air inlet temperatures below 500 K, it is likely that a significant amount of number 2 diesel fuel will fail to evaporate and may be deposited on the walls of the premixing duct, where it can form deposits and create sites for auto-ignition and flashback. A diffusion pilot flame is thus necessary. If this is turned off at full load, it needs to be purged to prevent blockage.

4.3.6 Lean Blowoff

There is a limiting FAR below which combustion cannot be sustained. With the natural gas fuel, this is around 5% by volume – an equivalence ratio of about 0.5. For liquid fuels it is around 0.4 equivalence ratio. To achieve very low NO_x it is necessary to operate close to lean blowout, and therefore if the gas flow is to be reduced to reduce power, then some injectors must be turned off. This is called fuel staging. This problem makes it difficult to sustain combustion during a generator emergency load reduction. Pilot diffusion burners ease the problem greatly, but there is a significant NO_x penalty. The lean blowoff limit is insensitive to pressure but becomes richer as inlet temperature falls. As the lean blowoff limit is approached the flame becomes less stable and combustion instabilities, discussed in the next subsection, can occur. This problem is also eased by using pilot diffusion flame. Combustion instability is not, however, confined to operation close to the lean limit.

4.3.7 Combustion Instability

Combustion instability is critically important in DLN combustion systems. For decades the users of gas turbines have been able to depend on their running for long periods with little attention beyond scheduled maintenance. Owners of

DLN-equipped systems find that their engines require continuous monitoring and regular retuning. The OEM (original equipment manufacturer) often provides this monitoring service, but it is expensive. Although DLN systems have been in service for 25 years, problems of this kind continue.

Combustion instability is a cyclic process driven by coupling between the rate of heat release and pressure. These two parameters couple easily in a gas turbine, in that gas volume increases with temperature but turbine nozzle capacity increases only with the square root of temperature. Thus an increase in heat release rate reduces mass flow through the NGV, and the pressure rises in the combustor. This reduces the mass flow from the compressor. Cyclic coupling occurs for the following reasons:

1. When the fuel pressure drop across the injector is high, the fuel flow rate is not sensitive to combustor pressure. As the air mass flow falls, the mixture becomes richer, further increasing firing temperature.
2. When the fuel pressure drop across the injector is low, fuel flow is modulated by variations in combustor pressure. Time delays related to the transport of fuel into the burning zone can result in a phase relationship between heat release and pressure favourable to instability.

Lean premixed combustion amplifies this process. The FAR in the reaction zone is uniform and leaner than stoichiometric. Thus heat release throughout the mixture increases with FAR. When the fuel is injected directly into the combustor, some zones are richer and some are leaner than stoichiometric. As FAR increases, the fuel-rich volume releases less heat, damping to some extent the opposite effect in the fuel-lean volume. It must be borne in mind that, at high pressure and temperature, with the high turbulence generated in a gas turbine combustor, almost all the heat release has taken place before most of the dilution air is admitted. In a DI diffusion-flame combustor, there is little heat release in regions of flame where the equivalence ratio is less than 0.8. (Relatively slow consumption of remaining CO and carbon will continue, but with insufficient heat release rate to generate instability.)

Combustion pressure oscillations are likely to grow in amplitude when the phase relationship between heat release and pressure are within $90°$ of each other. This yields a large positive value for the left-hand side of relation (3.39), which is Rayleigh's criterion. Whether this is the case depends on reaction rates, mixture transport times in the reaction zone, and the ability of these to couple with one of the several pressure oscillation modes available. These are sometimes acoustic resonances, or the combustion-related processes may couple with convection of spatiotemporal fluctuations of temperature along the combustor, which can generate acoustic waves at the turbine nozzle. These waves are sometimes referred to as entropy waves. The many possible ways in which the aerodynamic, chemical, and acoustic processes can interact means that combustion instability is extremely likely to occur at some point in the engine-operating envelope. Frequencies may range from around 50 Hz to above 5 kHz. Frequencies above about 1 kHz are referred to as screech, frequencies below 150 Hz may be referred to as humming, rumble or buzz, frequencies between these two may be referred to as howl, and there is a low-frequency phenomenon known as chugging at around 10 Hz, which is associated with operation close to the lean extinction limit. This is not an acoustic phenomenon, and it is better thought of as intermittent combustion.

In reading about this subject and in consulting combustion specialists, one may encounter diverse opinions relating to the causes and description of the same problem. This arises from the complexity of the phenomena and the current state of knowledge. Such opinions may sometimes be expressed with an unjustified dogmatism, creating the impression that a problem is well understood and that the proposed action to mitigate it is certain to succeed. It is not in the interest of most advisors to give the impression that they are unsure about a subject in which they are supposed to be expert! The costs of being mistaken are, however, so high, that there is a case for obtaining a second opinion. Differing opinions may not help to solve the problem, but could lead to a more realistic assessment of programme risk and risk-mitigation strategies.

The author strongly recommends a study of [225] by those concerned with the practical aspects of this problem in a gas turbine plant, especially if they are about to work with DLN combustors for the first time.

In lean premixed combustors, a sustained amplitude above 30 kPa peak-to-peak is very likely to cause damage. Half this amplitude is usually recommended as a safety limit. Screech can cause damage at less than 10 kPa peak-to-peak, however. If the frequency of instability matches the resonant frequency of some component on the engine, then lower amplitudes can cause failure. Not only does the oscillation subject the combustor to stress, but it can excite mechanical vibration in components such as oil and fuel pipes, leading to fracture. If the amplitude exceeds about 4% of combustor pressure it will reverse the flow in the premixing duct, causing flashback. Flashback can also be caused by lower amplitudes.

Helmholtz dampers (see Subsection 3.2.2) are often employed in industrial engines to reduce the amplitude of combustion instability. These consist of a volume connected by a tube to the combustor. The frequency at which they absorb energy is a function of the volume of the damper and the length and cross-sectional area of the connecting tube or neck. The resonant frquency of this damper is

$$\omega = \hat{c}\sqrt{\frac{A}{VL}}, \tag{4.1}$$

where ω is frequency in rad/s, $\hat{c}$ is speed of sound, A is the area of neck, L is the length of neck, and V is the volume of the resonator cavity.

The absorption frequency bandwidth may be increased at the expense of the attenuation efficiency by increasing the cross-sectional area of the neck. Often, several absorbers are needed to absorb energy at different frequencies. They are fairly large, having a volume in the region of 10% that of the combustor. They must be connected to the combustor rather than to the outer air-feed annulus to be effective and may need a small air purge to prevent overheating by ingress of combustion products.

A novel means of preventing coupling between pressure and FAR is employed in the premixing ducts of the R-R industrial Trent engines. Natural gas is admitted through the end of the duct, and the air is admitted through perforations in the wall. The wavelength of instability is long compared with the length of the duct, so the air jets are all in phase. If the transit time of the fuel through the perforated section is at least twice the period of the oscillation then the fluctuations in FAR are greatly attenuated [227].

Combustion instability is such a common problem in LPP combustion systems that industrial DLN engines are continuously monitored and may have to be shut down and adjusted if instability reaches a certain amplitude [225]. This has a big financial impact if it shuts down a power station engine during a period of peak demand.

4.3.8 Flashback

Flashback occurs when the flame speed exceeds the velocity of the unburnt mixture of fuel and air. Premixing ducts are not cooled, so it is potentially hazardous. As flame speed is only a few meters per second, whereas flow velocity may be as high as 100 m/s, it is seemingly not difficult to avoid. It is important to avoid wakes from swirl vanes or fuel injectors and thick boundary layers that persist along the length of the premixing duct, and this is best achieved by contracting the passage to accelerate the flow towards the outlet. If too much swirl is imparted to the air to encourage mixing, then the flow on the centreline of the duct may reverse, allowing flashback along the core of the duct. Combustion oscillation adds to these problems. The discussion on "Practical limitations governing the design of a premixing duct" is applicable here.

4.3.9 Auto-Ignition

This subject is discussed in Subsection 3.1.3 of Chapter 3. Here the discussion is extended to include the effects of various fuels employed. Data provided in that section may be used for number 2 light diesel distillate. The auto-ignition delay time for diesel fuel will be lower than that for kerosene, but not to a degree that is significant when compared with the uncertainty in the kerosene data. Auto-ignition delay time is difficult to measure, and results depend on the method of measurement used. It is therefore necessary to be conservative in applying published data. The writer would always design for half the expected ignition delay time based on the shortest delay time measured in published research. (There is a need to exercise some judgement concerning the quality of published data.) The reason for this conservatism is that in engines there are aerodynamic phenomena that can be controlled far less easily than in a research test facility; notably large-scale turbulence and combustion-driven pressure oscillations in the premixing ducts.

Auto-ignition can be caused by some of the features that can cause flashback. It is not always easy to diagnose which of the two has occurred. Wakes, boundary layers, or small recirculation zones where unburnt fuel–air mixtures can reside can provide a site for auto-ignition. Also it can be triggered, as can flashback, from hot surfaces in the premix duct arising from radiation and conduction back from the flame zone. It is far less likely in industrial gas-burning engines, both because natural gas has a higher auto-ignition temperature than kerosene and because the compressor delivery pressure and temperature are lower than in aeroengines.

There is very little published auto-ignition data for methane or natural gas in the temperature range of interest. Most experiments are conducted at temperatures above 1100 K. On account of the heavier components in natural gas, the auto-ignition delay time is around half that of methane – but this is only a rule of thumb. There is considerable variation in the constituents of natural gas around the world. In engines without regenerative heat exchangers and with pressure ratios below 25:1, auto-ignition is extremely unlikely to occur with natural gas, provided that reasonable care is taken in the aerodynamic design of the premixing ducts. At pressure ratios above 30:1, although not a formidable problem, special care should be taken to streamline passages containing fuel–air mixtures. Where inlet air is preheated by a regenerative heat exchanger, temperatures above 850 K are potentially hazardous whereas at pressures above 30 bars without such a preheat, an auto-ignition delay of natural gas may be encountered. The author knows of no reliable published data on which to base design rules. That is not to say that the leading manufacturers do not have reliable data of their own.

Another possible source of pre-ignition is PM from compressor tip rubs. This is not a common cause, but cannot be ruled out from investigation of a problem. The cause of pre-ignition is rarely obvious – partly because so much damage results that evidence of the cause is destroyed.

4.3.10 External Aerodynamics

FAR deviations from the mean of more than 2% have a significant effect on NO_x in combustion systems designed to achieve values below 15 ppmv. There are inevitable variations from combustor to combustor through manufacturing tolerances and likewise in the fuel injectors. With the passing of time, deposits can build up in fuel passages, which increase variability. Variations in the air-feed pressure add to the total variation. Whereas aircraft gas turbine emissions are verified only once before leaving the factory, a stationary plant is checked annually.

The lengths of many LPP combustion systems are too great to fit between the compressor outlet and turbine inlet, so the compressor delivery air must be turned through a large angle to feed the LPP fuel injectors. The GE LM6000 DLN combustor avoids this problem, but its great depth requires a complex, multipassage compressor outlet diffuser. Both radially and axially disposed tubular combustors present the aerodynamicist with a challenge. To achieve the very low NO_x required and to avoid flashback, a high degree of uniformity in the air distribution into the premixing ducts is required. Where the airflow turns through a large angle and passes between tubular combustors, it is obviously difficult to achieve this. The problem is solved by careful development, supported by CFD and experiment, and may require some compromise in acceptance of a higher parasitic pressure loss with an associated loss of cycle efficiency [228].

4.3.11 Combustion Research for Industrial Gas Turbines

To conduct combustion research at full scale and full pressure and temperature is formidably expensive, even for a 50-MW aeroengine. For a 200-MW industrial engine the air compression and heating plant required would be colossal. Yet auto-ignition can be evaluated only at full pressure, temperature, and residence time. Some aerodynamic phenomena mentioned in the previous section require testing in an environment of realistic geometry, and some acoustic modes that drive combustion instability occur in only the full annulus of an engine system. The aeroengine industry has used extensive component rig testing since the inception of the gas turbine. This is justified by the high cost per unit power, the smaller size of the combustors, airworthiness requirements, and low weight that results in the need to design with the minimum stress safety factors. The large number of units produced – typically over 1000 for a single model and occasionally 10 times that number – makes extensive testing economically feasible. It has been customary for industrial engines to be developed under special arrangements with kindly disposed customers on the customer's site.

Aeroderivative engines are downrated by removal of the fan, which reduces the cycle pressure ratio by 30%, and to achieve the very long life at full load required in stationary applications. (An aeroengine spends only a small percentage of its life at above 85% load). Consequently the adaptation of aeroengines to industrial application has not, until relatively recently, required extensive research.

The necessity to drastically reduce emissions has transformed this situation. Ultra-low-NO_x combustion systems cannot be designed and tested in the production engine, although a demonstrator engine is often employed in the development of a generic design. Extensive research is required, followed by validation testing at high pressure and temperature. Combustion research in the industrial gas turbine field is now more competitive and technologically advanced than in the aeroengine field. As explained in Section 4.2, innovation in aeroengine combustion technology is limited by airworthiness considerations. Emissions control cannot take precedence over safety. In the industrial field the conflict is a lot less severe.

Generic research cannot take account of engine geometry. It is necessary to recognise the significance of this in evaluating results. As mentioned in Subsection 4.3.10, there are many sources of non-uniformity in the air and fuel supply to an engine combustor that are not present in a combustion rig. Emissions performance achieved in a research rig cannot therefore be replicated in an engine.

Two topics are of great importance: combustion instability, which has already been discussed at length, and auto-ignition of natural gas. As noted earlier, there is little reliable data describing the auto-ignition characteristics of natural gas at the temperatures and pressures encountered in modern gas turbines. Of particular interest are temperatures between 800 and 900 K at pressures between 15 and 40 bars. Variation in the chemical constituents of natural gas must be included. In devising laboratory research programmes with an eye on industrial application, physical scale and the turbulence structure in airflows is important. Attempting to achieve greater realism will increase the variability in results. The author believes that this must be accepted. Highly repeatable data from a completely non-representative environment is almost useless, as the engine designer does not know what safety margin to allow. Experimental data with realistic variability provide statistical information that is of practical use.

4.3.12 Future Trends and Research Emphasis

NO_x limits will probably become increasingly stringent, and in a few years' time the generally accepted limit will probably fall from 25 ppmv to 15 or even 10 ppmv. These values have already been achieved by DLN systems in some applications. It will be apparent from the foregoing discussion that the greatest obstacle to the achievement of lower NO_x is combustion instability. This is a very complex phenomenon, and there is much to be done both to better understand it and to devise methods for its control. Active control of fuel admission is a promising field of study for liquid fuel. This topic is discussed at length in Section 3.2. The greatest difficulty at present lies in the perfecting of valves that can modulate flow at the required frequency. The inertia of liquid fuel opposes every effort to rapidly change the flow rate, and reflected waves from geometric features in the pipe work tend to attenuate input variations. The closer the control valve is placed to the fuel nozzle, the better, but then the hostile environment creates problems in the valves. Because of the compressibility of gas it is difficult to see how active control of the gas flow can be transmitted to the injector final orifice. There is scope for the development of improved fuel–air mixing devices.

The discussion in Subsection 4.2.9 should be read in conjunction with this section, and the discussion in that section on the topic of CFD applies equally in the industrial gas turbine field.

REFERENCES

[1] B. Lewis, Outlook in combustion research, *Combust. Flame* **14**, 1– (1970).

[2] R. F. Sawyer, Science based policy for addressing energy and environmental problems, *Proc. Combust. Inst.* **32**, 45–56 (2009).

[3] COM(2007) 19 final: Communication from the Commission to the Council and the European Parliment, "Results of the review of the Community Strategy to reduce CO_2 emissions from passenger cars and light-commercial vehicles" (2007).

[4] A. M. K. P. Taylor, Science review of internal combustion engines, *Energy Policy* **36**, 4657–4667 (2008).

[5] R. L. Evans, Lean-burn spark-ignited internal combustion engines, in D. Dunn-Rankin (ed.), *Lean Combustion: Technology and Control* (Academic, New York, 2008).

[6] R. W. Schefer, C. White, and J. Keller, Lean hydrogen combustion, in D. Dunn-Rankin (ed.), *Lean Combustion: Technology and Control* (Academic, New York, 2008).

[7] Light-duty automotive technology, carbon dioxide emissions, and fuel economy trends: 1975 through 2009. Executive summary. Compliance and innovative strategies division and Transportation and Climate division, EPA 420-S-09-001 (Office of Transportation and Air Quality, U.S. Environmental Protection Agency, November 2009); http://www.epa.gov/otaq/cert/mpg/fetrends/420s09001.pdf.

[8] Federal Register (www.regulations.gov) under Docket ID EPA-HQ-OAR-2009-0171. Also http://www.epa.gov/climatechange/endangerment.html.

[9] National Highway Traffic Safety Administration, Docket NHTSA-2009-0059, Notice of intent to prepare an environmental impact statement for new corporate average fuel economy standards (U.S. Department of Transportation 2009).

[10] Council Directive 89/458/EEC of 18 July 1989, amending with regard to European emission standards for cars below 1.4 litres, Directive 70/220/EEC on the approximation of the laws of the Member States relating to measures to be taken against air pollution by emissions from motor vehicles, Official Journal L 226, 3.8.1989, pp. 1–3.

[11] Council Directive 91/441/EEC of 26 June 1991 amending Directive 70/220/EEC on the approximation of the laws of the Member States relating to measures to be taken against air pollution by emissions from motor vehicles, Official Journal L 242, 30/08/1991, pp. 0001–0106.

[12] 1999/125/EC: Commission Recommendation of 5 February 1999 on the reduction of CO_2 emissions from passenger cars, Official Journal L 040, 13/02/1999, pp. 0049–0050.

[13] T. Lake, J. Stokes, R. Osborne, and R. Murphy, The lean boost combustion system for improved fuel economy, in H. Zhao (ed.), *Advanced direct injection combustion engine technologies and development, Gasoline and gas engines* (Woodhead, Oxford, 2009), Vol. 1.

[14] P. Leduc, B. Dubar, A. Ranini, and G. Monnier Downsizing of gasoline engine: An efficient way to reduce CO_2 emissions, *Oil Gas Sci. Technol. Rev. IFP* **58**, 115–127 (2003).

[15] Regulation (EC) 443/2009 of the European Parliment and of the Council of 23 April 2009, setting emission performance standards for new passenger cars as part of the Community's integrated approach to reduce CO_2 emissions from light-duty vehicles, Official Journal L 140, 5.6.2009, p. 2.

[16] The plain English guide to the Clean Air Act, publication EPA-456/K-07-001, U.S. Environmental Protection Agency, Office of Air Quality Planning and Standards, Research Triangle Park, NC (2007). http://www.epa.gov/air/peg/understand.html.

[17] Council Directive 70/220/EEC of 20 March 1970 on the approximation of the laws of the Member States relating to measures to be taken against air pollution by gases from positive-ignition engines of motor vehicles, Official Journal L 076, 06/04/1970, pp. 0001–0022.

[18] Directive 2008/50/EC of the European Parliament and of the Council of 21 May 2008 on ambient air quality and cleaner air for Europe, Official Journal L 152, 11.6.2008, pp. 1–44.

[19] Decision 1600/2002/EC of the European Parliament and of the Council of 22 July 2002 laying down the Sixth Community Environment Action Programme, Official Journal L 242, 10/09/2002, pp. 0001–0015.

[20] A summary of Directive 2008/50/EC of the European Parliament and of the Council of 21 May 2008 on ambient air quality and cleaner air for Europe. http://europa.eu/legislation_summaries/environment/air_pollution/ev0002_en.htm.

[21] Title 40 of the United States Code of Federal Regulations: Protection of Environment. Part 50-National Primary And Secondary Ambient Air Quality Standards. Summarised at http://www.epa.gov/ttn/naaqs.

[22] Primary National Ambient Air Quality Standards for Nitrogen Dioxide; Final Rule. Environmental Protection Agency 40 CFR, Parts 50 and 58, Federal Register, Vol. 75, No. 26, Tuesday, February 9, 2010, Rules and Regulations. Summarised at http://www.epa.gov/air/nitrogenoxides/pdfs/20100124 presentation.pdf.

[23] J. Bachmann, L. Zaragoza, J. Cohen, and J. Haines, Review Of The National Ambient Air Quality Standards For Particulate Matter, Assessment Of Scientific and Technical Information, OAQPS Staff Paper, Strategies and Standards Division, Office of Air Quality Planning and Standards, U.S. Environmental Protection Agency, Research Triangle Park, N. C. 27711, EPA-450/5-82-001. Available at http://nepis.epa.gov.

[24] J. Bachmann, Air today, yesterday and tomorrow: Part 2. Presentations on air quality management and health effects (2008). Available at http://www.epa.gov/air/caa/.

[25] J. Bachmann, Will the circle be unbroken: A history of the U.S. National Ambient Air Quality Standards, *J. Air Waste Manage. Assoc.* **57**, 652–697 (2007).

[26] http://www.epa.gov.ttn/atw/pollsour.html.

[27] Environmental fact sheet, Air toxics from Motor Vehicles, EPA400-F-92-004. Available at http://www.epa.gov/oms/f02004.pdf.

[28] http://ec.europa.eu/environment/air/quality/standards.htm. Summarising Air Quality Standards in Directive 2008/50/EC of the European Parliament and

of the Council of 21 May 2008 on ambient air quality and cleaner air for Europe.

[29] COM 446, Final Communication From The Commission To The Council And The European Parliament: Thematic Strategy on Air Pollution (2005).

[30] Health aspects of air pollution results from the project "Systematic review of health aspects of air pollution in Europe". Publications, WHO Regional Office for Europe, Scherfigsvej 8, DK-2100 Copenhagen O, Denmark (2004). Available at http://www.euro.who.int/document/e83080.pdf.

[31] SEC 1745, Commission Staff Working Document. Annex to the proposal for a regulation of the European Parliament and of the Council on type approval of motor vehicles with respect to emissions and on access to vehicle repair information, amending Directive 72/306/EEC and Directive ../../EC [sic]. Impact Assessment (2005).

[32] Scientific Committee on Health and Environmental Risks (SCHER): Opinion on "New evidence of air pollution effects on human health and the environment", 18 March 2005. European Commission Health & Consumer Protection Directorate-General Directorate C – Public Health and Risk Assessment C7 – Risk assessment. Available at http://ec.europa.eu/health/ph_risk/committees/04_scher/docs/scher_o_009.pdf.

[33] SEC 1133, Commission Staff Working Paper. Annex to The Communication on Thematic Strategy on Air Pollution and The Directive on "Ambient Air Quality and Cleaner Air for Europe". Impact Assessment (2005).

[34] SEC 1718, Commission Staff Working Document. Annex to the proposal for a regulation of the European Parliament and of the Council on the approximation of the laws of the Member States with respect to emissions from on-road heavy duty vehicles and on access to vehicle repair information. Impact Assessment (2007).

[35] COM 447, final Proposal for a Directive of the European Parliment and of the Council on ambient air quality and cleaner air for Europe (2005).

[36] Nitrogen dioxide in the United Kingdom – Chapter 2. NOx emissions and emission inventories, The Air Quality Expert Group, Defra Publications, Admail 6000, London SW1A 2XX. This document is also available on the AQEG website: http://www.defra.gov.uk/environment/airquality/aqeg. Published by the Department for Environment, Food and Rural Affairs (2004).

[37] W. W. Pulkrabek, *Engineering Fundamentals of the Internal Combustion Engine*, 2nd ed. (Pearson Education International, Upper Saddle River, NJ, 2004).

[38] K.-H. Dietsche and M. Klingebiel (eds.), *Bosch Automotive Handbook*, 7th ed. Robert Bosch GmbH Plochingen, Wiley, Chichester (2007).

[39] C. D. Rakopoulos and E. G. Giakoumis, *Diesel Engine Transient Operation – Principles of Operation and Simulation Analysis* (Springer-Verlag, London, 2009).

[40] U. Spicher, Exhaust emissions and pollutant reduction, in R. van Basshuysen (ed.), *Gasoline Engine with Direct Injection* (Vieweg+Teubner GWV Fachverlag GmbH, Wiesbaden, Germany, 2009).

[41] P. Eastwood, Exhaust gas aftertreatment for light-duty diesel engines, in H. Zhao (ed.), *Advanced Direct Injection Combustion Engine Technologies and Development*, Vol. 2, *Diesel Engines* (Woodhead, Oxford, 2010).

[42] A. Cairns, H. Blaxhill, and N. Fraser, Exhaust gas recirculation boosted direct injection gasoline engines, in H. Zhao (ed.), *Advanced Direct Injection Combustion Engine Technologies and Development*, Vol. 1, *Gasoline and Gas Engines* (Woodhead, Oxford, 2009).

[43] J. W. G. Turner and R. J. Pearson, The turbocharged direct injection spark-ignition engine, in H. Zhao (ed.), *Advanced Direct Injection Combustion Engine Technologies and Development*, Vol. 1, *Gasoline and Gas Engines* (Woodhead, Oxford, 2009).

[44] J. D. Pagliuso and M. E. S. Martins, Biofuels for spark ignition engines, in H. Zhao (ed.), *Advanced Direct Injection Combustion Engine Technologies and Development*, Vol. 1, *Gasoline and Gas Engines* (Woodhead, Oxford, 2009).

[45] P. Eastwood, *Particulate Emissions from Vehicles* (Wiley-Professional Engineering, New York, 2008); also SAE book R-389.

[46] Commission Regulation (EC) 692/2008 of 18 July 2008 implementing and amending Regulation (EC) 715/2007 of the European Parliament and of the Council on type-approval of motor vehicles with respect to emissions from light passenger and commercial vehicles (Euro 5 and Euro 6) and on access to vehicle repair and maintenance information.

[47] COM 683 final. Proposal for a Regulation of the European Parliament and of the Council on type-approval of motor vehicles with respect to emissions and on access to vehicle repair information, amending Directive 72/306/EEC and Directive (ibid.).

[48] M. Adachi, Advanced systems and trends for powertrain emission measurements, presented at the 7th International Conference on Modelling and Diagnostics for Advanced Engine Systems (COMODIA 2008), Sapporo, Japan, 2008.

[49] Commission Staff Working Document. Impact Assessment for Euro 6 emission limits for light duty vehicles (2006).

[50] COM 851 final. Proposal for a regulation of the European Parliament and of the Council on type-approval of motor vehicles and engines with respect to emissions from heavy duty vehicles (Euro VI) and on access to vehicle repair and maintenance information (2007).

[51] Nitrogen dioxide in the United Kingdom – Summary. Product code PB 9144. The Air Quality Expert Group, Defra Publications, Admail 6000, London SW1A 2XX. This document is also available on the AQEG website: http://www.defra.gov.uk/environment/airquality/aqeg. Published by the Department for Environment, Food and Rural Affairs (2004). Available at http://www.defra.gov.uk/environment/airquality/aqeg.

[52] J. B. Heywood, *Internal Combustion Engine Fundamentals* (McGraw-Hill, New York, 1989).

[53] R. Stone, *Introduction to Internal Combustion Engines*, 3rd ed. (Palgrave, Basingstoke, UK 1999).

[54] J. F. Griffiths and J. A. Barnard, *Flame and Combustion*, 3rd ed. (Blackie Academic and Professional, London, 1995).

[55] W. Hentschel, Optical diagnostics for combustion process development of direct-injection gasoline engines, *Proc. Combust. Inst.* **28**, 1119–1135 (2000).

[56] C. D. Rakopoulos and E. G. Giakoumis, Second-law analyses applied to internal combustion engines operation, *Prog. Energy Combust. Sci.* **32**, 2–47 (2006).

[57] U. Spicher, Fuel consumption, in R. van Basshuysen (ed.), *Gasoline Engine with Direct Injection* (Vieweg+Teubner GWV Fachverlag GmbH, Wiesebaden, Germany, 2009).

[58] J. T. Farrell, J. G. Stevens, and W. Weissman, A second law analysis of high efficiency low emission gasoline engine concepts, SAE paper 2006-01-0491 (2006).

[59] M. C. Drake and D. C. Haworth, Advanced gasoline engine development using optical diagnostics and numerical modelling, *Proc. Combust. Inst.* **31**, 99–124 (2007).

[60] A. C. Alkidas and S. H. El Tahry, Contributors to the fuel economy advantage of DISI engines over PFI engines, SAE paper 2003-01-3101 (2003).

[61] T. H. Lake, J. Stokes, P. A. Whitaker, and J. V. Crump, Comparison of direct injection gasoline combustion systems, SAE paper 980154 (1998).

[62] C. Schwarz, E. Schünemann, B. Durst, J. Fischer, and A. Witt, Potentials of the spray-guided BMW DI combustion system, SAE paper 2006-01-1265 (2006).

[63] A. Waltner, P. Lueckert, U. Schaupp, E. Rau, R. Kemmler, and R. Weller, Future technology of the spark-ignition engine: Spray-guided direct injection with piezo injector, presented at the 27th International Vienna Motor Symposium (2006). See also P. Lueckert, A. Waltner, E. Rau, G. Vent, U. Schaupp, The new V6 gasoline engine with direct injection by Mercedes-Benz. MTZ (Motortechnische Zeitschrift) **11**, 1–6 (2006).

[64] P. Langen, T. Melcher, S. Missy, C. Schwarz, E. Schünemann, New BMW six- and four-cylinder petrol engines with high precision injection and stratified combustion, 28, presented at the Internationales Wiener Motorensymposium (2007) (Fortschritt-Berichten VDI, Austrian Society of Automotive Engineers, Vienna 2007).

[65] Y. Takagi, A new era in spark-ignition engines featuring high-pressure direct injection, *Proc. Combust. Inst.* **27**, 2055–2068 (1998).

[66] F. Zhao, M.-C. Lai, and D. L. Harrington, Automotive spark-ignited direct-injection gasoline engines, *Prog. Energy Combust. Sci.* **25**, 437–562 (1999).

[67] U. Spicher, J. P. Haentsche, and T. Heidenreich, Injection systems and system overview, in R. van Basshuysen (ed.), *Gasoline Engine With Direct Injection* (Vieweg+Teubner GWV Fachverlag GmbH, Wiesebaden, Germany, 2009).

[68] T. Date and S. Yagi, Research and development of the Honda CVCC engine, SAE paper 740605 (1974).

[69] A. K. Oppenheim, *Combustion in Piston Engines: Technology, Evolution, Diagnosis and Control* (Springer-Verlag, Berlin, 2004).

[70] K. Horie, K. Nishizawa, T. Ogawa, S. Akazaki, and K. Miura, The development of a high fuel economy and high performance four-valve lean burn engine, SAE paper 920455 (1992).

[71] P. G. Aleiferis, I. Prassas, and A. M. K. P. Taylor, Novel optical instrumentation meeting the challenges for energy and the environment, in Y. H. Mori and K. Ohnishi (eds.), *Energy and Environment: technological Challenges for the Future* (Springer-Verlag, Tokyo, 2001).

[72] J. L. Lumley, *Engines: An Introduction* (Cambridge University Press, Cambridge, 1999).

[73] Y. Hardalupas, A. M. K. P. Taylor, J. H. Whitelaw, K. Ishii, H. Miyano, and Y. Urata, Influence of injection timing on in-cylinder fuel distribution in a Honda VTEC-E engine, SAE paper 950507 (1995).

[74] M. Berckmueller, N. P. Tait, R. D. Lockett, and D. A. Greenhalgh, In cylinder crank angle resolved imaging of fuel concentration in a firing spark ignition engine using planar LIFS, *Proc. Combust. Inst.* **25**, 151–156 (1994).

[75] N. E. Carabateas, A. M. K. P. Taylor, J. H. Whitelaw, K. Ishii, K. Yoshida, and M. Matsuki, The effect of injector and intake port design on in-cylinder fuel droplet distribution, airflow and lean burn performance for a Honda VTEC-E engine, SAE paper 961923 (1996).

[76] P. G. Aleiferis, A. M. K. P. Taylor, J. H. Whitelaw, K. Ishii, and Y. Urata, Cyclic variations of initial flame kernel growth in a Honda VTEC-E lean-burn spark-ignition engine, SAE paper 2000-01-1207 (2000).

[77] P. G. Aleiferis, A. M. K. P. Taylor, K. Ishii, and Y. Urata, The nature of early flame development in a lean burn stratified-charge spark-ignition engine, *Combust. Flame* **136**, 283–302 (2004).

[78] P. G. Aleiferis, A. M. K. P. Taylor, K. Ishii, and Y. Urata, The relative effects of fuel concentration, residual-gas fraction, gas motion, spark energy and heat losses to the electrodes on flame kernel development in a lean-burn spark-ignition engine. *Proc. Inst. Mech. Eng. D J. Auto. Eng.* **218**, 411–425 (2004).

[79] P. G. Aleiferis, Y. Hardalupas, A. M. K. P. Taylor, K. Ishii, and Y. Urata, Flame chemiluminescence studies of cyclic combustion variations and air-to-fuel ratio

of the reacting mixture in a lean-burn stratified-charge spark-ignition engine, *Combust. Flame* **136**, 72–90 (2004).

[80] P. G. Aleiferis, Y. Hardalupas, A. M. K. P. Taylor, K. Ishii, and Y. Urata, Cyclic variations of fuel-droplet distribution during the early intake stroke of a lean-burn stratified-charge spark-ignition engine, *Exp. Fluids* **39**, 789–798 (2005).

[81] K. Kuwahara and H. Ando, Diagnostics of in-cylinder flow, mixing and combustion in gasoline engines, *Meas. Sci. Technol.* **11**, R95–R111 (2000).

[82] K. Horie, Y. Kohda, K. Ogawa, and H. Goto, Development of a high fuel economy and low emission four-valve direct injection engine with a center-injection system, SAE paper 2004-01-2941 (2004).

[83] T. D. Fansler, M. C. Drake, B. E. Stojkovic, and M. E. Rosalik, Local fuel concentration, ignition and combustion in a stratified charge spark ignited direct injection engine: Spectroscopic, imaging and pressure-based measurements. *Int. J. Eng. Res.* **4**, 61–86 (2003).

[84] B. D. Stojkovic, F. D. Fansler, M. C. Drake, and V. Sick, High-speed imaging of OH* and soot temperature and concentration in a stratified-charge direct-injection gasoline engine, *Proc. Combust. Inst.* **30**, 2657–2665 (2005).

[85] E. Achleitner, H. Bäcker, and A. Funaioli, Direct injection systems for Otto engines, SAE paper 2007-01-1416 (2007).

[86] U. Spicher, T. Heidenreich, and B. Xander, Mixture formation and combustion processes, in R. van Basshuysen (ed.), *Gasoline Engine with Direct Injection* (Vieweg+Teubner GWV Fachverlag GmbH, Wiesebaden, Germany, 2009).

[87] M. C. Drake, T. D. Fansler, and A. M. Lippert, Stratified-charge combustion: modelling and imaging of a spray-guided direct-injection spark-ignition engine, *Proc. Combust. Inst.* **30**, 2683–2691 (2005).

[88] T. Honda, M. Kawamoto, H. Katashiba, M. Sumida, N. Fukutomi, and K. Kawajiri, A study of mixture formation and combustion for spray-guided DISI, SAE paper 2004-01-0046 (2004).

[89] P. Dahlander, M. Skogsberg, and I. G. Denbratt, Effects of injector parameters on mixture formation for multi-hole nozzles in a spray-guided gasoline DI engine, SAE paper 2005-01-0097 (2005).

[90] G. Rottenkolber, J. Gindele, J. Raposo, K. Dullenkopf, W. Hentschel, S. Wittig, U. Spicher, and W. Merzkirch, Spray analysis of a gasoline direct injector by means of two-phase PIV, *Exp. Fluids* **32**, 710–721 (2002).

[91] K. D. Driscoll, V. Sick, and C. Gray, Simultaneous air/fuel-phase PIV measurements in a dense fuel spray, *Exp. Fluids* **35**, 112–115 (2003).

[92] W. Sauter, J. Pfeil, A. Velji, U. Spicher, N. Laudenbach, F. Altenschmidt, and U. Schaupp, Application of particle image velocimetry for investigation of spray characteristics of an outward opening nozzle for gasoline direct injection, SAE paper 2006-01-3377 (2006).

[93] T. Li, K. Nishida, and H. Hiroyasu, Effect of split injection on stratified charge formation of direct injection spark ignition engines, *Int J. Engine Res.* **8**, 205–218 (2007).

[94] M. Skogsberg, P. Dahlander, and I. G. Denbratt, Spray shape and atomization quality of an outward-opening piezo gasoline DI injector, SAE paper 2007-01-1409 (2007).

[95] M. Skogsberg, P. Dahlander, R. Lindgren, and I. Denbratt, Effects of injector parameters on mixture formation for multi-hole nozzles in a spray-guided gasoline DI engine, SAE paper 2005-01-0097 (2005).

[96] B. Peterson and V. Sick, Simultaneous flow field and fuel concentration imaging at 4.8 kHz in an operating engine, *Appl. Phys B.* **97**, 887–895 (2009).

[97] R. Zhang and V. Sick, Multi-component fuel imaging in a spray-guided spark-ignition direct-injection engine, SAE paper 2007-01-1826 (2007).

[98] J. D. Smith and V. Sick, Crank-angle resolved imaging of fuel distribution, ignition and combustion in a direct-injection spark-ignition engine, SAE paper 2005-01-3753 (2005).

[99] J. D. Smith and V. Sick, A multi-variable high-speed imaging study of ignition instabilities in a spray-guided direct-injected spark-ignition engine, SAE paper 2006-01-1264 (2006).

[100] J. D. Smith and V. Sick, Real-time imaging of fuel injection, evaporation and ignition in a direct-injected spark-ignition engine, presented at ILASS Americas, 19th Annual Conference on Liquid Atomisation and Spray Systems, Toronto, Canada, 2006.

[101] J D. Smith and V. Sick, Factors influencing spark behavior in a spray-guided direct-injected engine, SAE paper 2006-01-3376 (2006).

[102] H. Zhao, Motivation, definition and history of HCCI/CAI engines, in H. Zhao (ed.), *HCCI and CAI Engines for the Automotive Industry* (Woodhead, Cambridge, 2007).

[103] S. Kimura, Low temperature and premixed combustion concept with late injection, in H. Zhao (ed.), *HCCI and CAI Engines for the Automotive Industry* (Woodhead, Cambridge, 2007).

[104] I. Denbratt, Advanced concepts for future light-duty diesel engines, in H. Zhao (ed.), *Advanced Direct Injection Combustion Engine Technologies and Development*, Vol. 2, *Diesel Engines* (Woodhead, Oxford, 2010).

[105] F. Zhao, T. W. Asmus, D. N. Assanis, J. E. Dec, J. A. Eng, and P. M. Najt (eds.), Homogeneous charge compression ignition (HCCI) engines: Key research and development issues, SAE PT-94 (2003).

[106] W. Steiger and C. Kohnen, New combustion systems based on a new fuel specification, 27th International Vienna Motor Symposium 2006, VDI Research Reports, Series 12, No. 622, Vols. 1 and 2 (2006).

[107] N. Iida, Natural gas HCCI engines, in H. Zhao (ed.), *HCCI and CAI Engines for the Automotive Industry* (Woodhead, Cambridge, 2007).

[108] N. Iida, HCCI engines with other fuels, in H. Zhao (ed.), *HCCI and CAI Engines for the Automotive Industry* (Woodhead, Cambridge, 2007).

[109] P G. Aleiferis, A. G. Charalambides, Y. Hardalupas, and Y. Urata, Modelling and experiments of HCCI engine combustion with charge stratification and internal EGR, SAE paper 2005-01-3725 (2005).

[110] H. Zhao, Four-stroke CAI engines with residual gas trapping, in H. Zhao (ed.), *HCCI and CAI Engines for the Automotive Industry* (Woodhead, Cambridge, 2007).

[111] A. Fuerhapter, Four-stroke CAI engines with internal exhaust gas recirculation (EGR), in H. Zhao (ed.), *HCCI and CAI Engines for the Automotive Industry* (Woodhead, Cambridge, 2007).

[112] S. M. Begg, D. J. Mason, and M. R. Heikal, Spark ignition and spark-assisted controlled auto-ignition in an optical gasoline engine, SAE paper 2009-32-0072 (2009).

[113] H. Persson, B. Johansson, A. Remaon, The effect of swirl on spark-assisted compression ignition (SACI), SAE paper 2007-01-1856 (2007).

[114] P. G. Aleiferis, A. G. Charalambides, Y. Hardalupas, A. M. K. P. Taylor, and Y. Urata, Autoignition initiation and development of n-heptane HCCI combustion assisted by inlet air heating, internal EGR or spark discharge: an optical investigation, SAE paper 2006-01-3273 (2006).

[115] H. Yun, N. Wermuth, and P. Najt, Extending the high load operating limit of a naturally-aspirated gasoline HCCI combustion engine, SAE paper 2010-01-0847 (2010).

[116] D. K. Srivastava, M. Weinrotter, H. Kofler, A. K. Agarwal, and E. Wintner, Laser-assisted homogeneous charge ignition in a constant volume combustion chamber, *Opt. Lasers Eng.* **47**, 680–685 (2009).

[117] J. E. Dec, Advanced compression-ignition engines – understanding the in-cylinder processes, *Proc. Combust. Inst.* **32**, 2727–2742 (2009).

[118] J. E. Dec, W. Hwang, and M. Sjöeberg, An investigation of thermal stratification in HCCI engines using chemiluminescence imaging, SAE paper 2006-01-1518 (2006).

[119] A. Hultqvist, M. Christensen, B. Johansson, M. Richter, J. Nygren, H. Hult, and M. Aldén, The HCCI combustion process in a single cycle – High-speed fuel tracer LIF and chemiluminescence imaging, SAE paper 2002-01-0424 (2002).

[120] P. G. Aleiferis, A. G. Charalambides, Y. Hardalupas, A. M. K. P. Taylor, and Y. Urata, Axial fuel stratification of a homogeneous charge compression igni-tion (HCCI) engine, *Int. J. Vehicle Des.* **44**, 41–61 (2007).

[121] P. G. Aleiferis, A. G. Charalambides, Y. Hardalupas, A. M. K. P. Taylor, and Y. Urata, The effect of axial charge stratification and exhaust gases on com-bustion 'development' in a homogeneous charge compression ignition engine, *Proc. Inst. Mech. Eng. D J. Auto. Eng.* **222**, 2171–2183 (2008).

[122] A. Hultqvist, M. Christensen, B. Johansson, M. Richter, J. Nygren, J. Hult, and M. Aldén, The HCCI combustion process in a single cycle – high-speed fuel tracer LIF and chemiluminescence Imaging, SAE paper 2002-01-0424 (2002).

[123] D. J. Cook, H. Pitsch, J. H. Chen, and E. R. Hawkes, Flamelet-based modeling of auto-ignition with thermal inhomogeneities for application to HCCI engines, *Proc. Combust. Inst.* **31**, 2903–2911 (2007).

[124] M. Sjöberg and J. E. Dec, Effects of engine speed, fueling rate, and combustion phasing on the thermal stratification required to limit HCCI knocking intensity, SAE paper 2005-01-2125 (2005).

[125] M. Sjöberg, J. E. Dec, and N. P. Cernansky, Potential of thermal stratification and combustion retard for reducing pressure-rise rates in HCCI engines based on multi-zone modelling and experiments, SAE paper 2005-01-0113 (2005).

[126] M. Sjöberg and J. E. Dec, Influence of fuel autoignition reactivity on the high-load limits of HCCI engines, SAE paper 2008-01-0054 (2008).

[127] M. Sjöberg and J. E. Dec, A. Babajimopoulos, and D. Assanis, Comparing enhanced natural thermal stratification against retarded combustion phasing for smoothing of HCCI heat-release rates, SAE paper 2004-01-2994 (2004)

[128] M. Sjöberg and J. E. Dec, An investigation into lowest acceptable combustion temperatures for hydrocarbon fuels in HCCI engines, *Proc. Combust. Inst.* **30**, 2719–2726 (2005).

[129] W. Hwang, J. E. Dec, and M. Sjöeberg, Fuel stratification for low-load HCCI combustion: Performance & fuel-PLIF measurements, SAE paper 2007-01-4130.

[130] S. Tanaka, F. Ayala, and J. C. Keck, A reduced chemical kinetic model for HCCI combustion of primary reference fuels in a rapid compression machine, *Combust. Flame* **133**, 467–481 (2003).

[131] S. Tanaka, F. Ayala, J. C. Keck, J. B. Heywood, Two-stage ignition in HCCI combustion and HCCI control by fuels and additives, *Combust. Flame* **132**, 219–239 (2003).

[132] C. K. Westbrook, Chemical kinetics of hydrocarbon ignition in practical com-bustion systems, *Proc. Combust. Inst.* **28**, 1563–1577 (2000).

[133] N. Peters, G. Paczko, R. Seiser, and K. Seshadri, Temperature cross-over and non-thermal runaway at two-stage ignition of n-heptane, *Combust. Flame* **128**, 38–59 (2002).

[134] W. Hwang, J. E. Dec, and M. Sjöberg, Spectroscopic and chemical-kinetic analysis of the phases of HCCI autoignition and combustion for single- and two-stage ignition fuels, *Combust. Flame* **154**, 387–409 (2008).

[135] M. Sjöberg and J. E. Dec, Comparing late-cycle autoignition stability for single- and two-stage ignition fuels in HCCI engines, *Proc. Combust. Inst.* **31**, 2895–2902 (2007).

[136] M. Sjöberg and J. E. Dec, Influence of fuel autoignition reactivity on the high-load limits of HCCI engines, SAE paper 2008-01-0054 (2008).

[137] J. E. Dec and M. Sjöberg, Isolating the effects of fuel chemistry on combustion phasing in an HCCI engine and the potential of fuel stratification for ignition control, SAE paper 2004-01-0557 (2004).

[138] J. E. Dec and Y. Yang, Boosted HCCI for high power without engine knock and with ultra-low NO_x emissions – Using conventional gasoline, SAE paper 2010-01-1086 (2010) (also, Int. J. Engines, **3**, 750–767, 2010).

[139] H. Zhao, Z. Peng, and N. Ladommatos, Understanding of controlled autoignition combustion in a four-stroke gasoline engine, Proc. *Inst. Mech. Eng. D J. Auto. Eng.* **215**, 1297–1310 (2001).

[140] M. Sjöberg, J. E. Dec, and W. Hwang, Thermodynamic and chemical effects of EGR and its constituents on HCCI autoignition, SAE paper 2007-01-0207 (2007).

[141] J. E. Dec, M. Sjöberg, and W. Hwang, Isolating the effects of EGR on HCCI heat-release rates and NO_x emissions, SAE paper 2009-01-2665 (2009).

[142] P. Duret, Two-stroke CAI engines, in H. Zhao (ed.), *HCCI and CAI Engines for the Automotive Industry* (Woodhead, Cambridge, 2007).

[143] Y. Ishibashi, An application study of the pneumatic direct injection activated radical combustion two-stroke engine to scooter, SAE paper 2004-01-1870 (2004).

[144] Y. Urata, M. Awasaka, J. Takanashi, Y. Kakinuma, T. Hakozaki, and A. Umemoto, A study of gasoline-fuelled HCCI engine equipped with an electromagnetic valve train, SAE paper 2004-01-1898 (2004).

[145] G. Haraldsson, P. Tunestål, B. Johansson, and J. Hyvarnen, HCCI combustion phasing with closed-loop combustion control using variable compression ratio in a multi-cylinder engine, SAE paper 2003-01-1830 (2003).

[146] T. Urushihara, K. Yamaguchi, K. Yoshizawa, and T. Itoh, A study of a gasoline-fueled compression ignition engine-expansion of HCCI operation range using SI combustion as a trigger of compression ignition, SAE paper 2005-01-0180 (2005).

[147] D. Yap, M. Wyszynski, A. Megaritis, and H. Xu, Effect of inlet valve timing on boosted gasoline HCCI with residual gas trapping, SAE paper 2005-01-2136 (2005).

[148] A. Cairns and H. R. Blaxill, Lean boost and external exhaust gas recirculation for high load controlled autoignition, SAE paper 2005-01-3744 (2005).

[149] M. Martins and H. Zhao, 4-Stroke, multi-cylinder gasoline engine with controlled auto-ignition (CAI) combustion: A comparison between naturally aspirated and turbocharged operation, SAE paper 2008-36-0305.

[150] S. Mamalis, V. Nair, P. Andruskiewicz, D. Assanis, A. Babajimopoulos, N. Wermuth, and P. Najt, Comparison of different boosting strategies for homogeneous charge compression ignition engines – A modeling study, SAE paper 2010-01-0571 (2010).

[151] J. Takanashi, M. Awasaka, T. Kakinuma, M. Takazawa, and Y. Urata, A study of a gasoline HCCI engine equipped with an electromagnetic VVT mechanism – increasing the higher load operational range with the inter-cylinder EGR boost system, FISITA paper F2006P360; also JSAE paper no. 20068238 (2006).

[152] T. Kuboyama, Y. Moriyoshi, K. Hatamura, T. Yamada, and J. Takanashi, An experimental study of a gasoline HCCI engine using the blow-down super charging system, SAE paper 2009-01-0496 (2009).

[153] T. Kuboyama, Y. Moriyoshi, K. Hatamura, M. Suzuki, J. Takanashi, T. Yamada, and S. Gotoh, Effect of fuel and thermal stratifications on the operational range of an HCCI gasoline engine using the blow-down super charge system, SAE Paper 2010-01-0845 (2010).

[154] D. Kim, I. Ekoto, W. F. Colban, P. C. Miles, In-cylinder CO and UHC imaging in a light-duty diesel engine during PPCI low-temperature combustion, SAE paper 2008-01-1602 (2008).

[155] M. P. B. Musculus, L. M. Pickett, In-cylinder spray, mixing, combustion and pollutant-formation processes in conventional and low-temperature combustion diesel engines, in: H. Zhao (ed.), *Advanced Direct Injection Combustion Engine Technologies and Development, Vol 2, Diesel Engines* (Woodhead, Oxford, 2010).

[156] A. Helmantel and I. Denbratt, HCCI operation of a passenger car DI diesel engine with an adjustable valve train, SAE paper 2006-01-0029 (2006).

[157] B. Gatellier, Narrow angle direct injection (NADI) concept for HCCI diesel combustion, in H. Zhao (ed.), *HCCI and CAI Engines for the Automotive Industry* (Woodhead, Cambridge, 2007).

[158] L. Hildingsson, H. Persson, B. Johansson, R. C. Collin, J. Nygren, M. Richter, M. Aldén, R. Hasegawa, and H. Yanagihara, Optical diagnostics of HCCI and UNIBUS using 2-D PLIF of OH and formaldehyde, SAE paper 2005-01-0175 (2005).

[159] S. Singh, R. D. Reitz, M. P. B. Musculus, and T. Lachaux, Simultaneous optical diagnostic imaging of low-temperature, double-injection combustion in a heavy- duty DI diesel engine, *Combust. Sci. Technol.* **179**, 2381–2414 (2007).

[160] T. Lachaux, M. P. B. Musculus, S. Singh, and R. D. Reitz, Optical diagnostics of late-injection low-temperature combustion in a heavy-duty diesel engine, *J. Eng. Gas Turbines Power* **130** (2008). pp. 032808-1-9.

[161] J. E. Dec, A conceptual model of D.I. diesel combustion based on laser sheet imaging. SAE Paper 970873, *SAE Trans.* **106**, 1319–1348 (1997).

[162] C. L. Genzale, R. D. Reitz, and M. P. B. Musculus, Effects of piston bowl geometry on mixture development and late-injection low-temperature combustion in a heavy-duty diesel engine, SAE paper 2008-01-1330 (2008).

[163] M. P. B. Musculus, T. Lachaux, L. M. Pickett, and C. A. Idicheria, End-of-injection over-mixing and unburned hydrocarbon emissions in low-temperature-combustion diesel engines, SAE paper 2007-01-0907 (2007).

[164] T. Lachaux and M. P. B. Musculus, In-cylinder unburned hydrocarbon visualization during low-temperature compression-ignition engine combustion using formaldehyde PLIF, *Proc. Combust. Inst.* **31**, 2921–2929 (2007).

[165] G. T. Kalghatgi, P. Risberg, and H-E. Ångström, Partially pre-mixed auto-ignition of gasoline to attain low smoke and low NO_x at high load in a compression ignition engine and comparison with a diesel fuel, SAE paper 2007-01-0006 (2007).

[166] G. Kalghatgi, L. Hildingsson, and B. Johansson, Low NO_x and low smoke operation of a diesel engine using gasoline-like fuels in *Proceeding of the ASME Internal Combustion Engine Division* (American Society of Mechanical Engineering, New York,2009), ICES 2009-76034.

[167] V. Sick, Optical diagnstics for direct injection gasoline engine rsearch and development, in H. Zhao (ed.), *Advanced Direct Injection Combustion Engine Technologies and Development.* Vol. 1, *Gasoline and Gas Engines* (Woodhead, Oxford, 2009).

[168] C. Schulz, Optical diagnostics in diesel combustion engines, in H. Zhao (ed.), *Advanced Direct Injection Combustion Engine Technologies and Development*, Vol. 2, *Diesel Engines* (Woodhead, Oxford, 2010).

[169] M. Richter, Overview of advanced optical techniques and their applications to HCCI/CAI engines, in H. Zhao (ed.), *HCCI and CAI engines for the Automotive Industry* (Woodhead, Cambridge, 2007).

[170] L. de Francqueville, G. Bruneaux, and B. Thirouard, Soot volume fraction measurements in a gasoline direct injection engine by combined laser induced incandescence and laser extinction method, SAE paper 2010-01-0346 (2010).

[171] M. C. Drake, T. D. Fansler, A. S. Solomon, and A. A. Szekely, Piston fuel films as a source of smoke and hydrocarbon emissions from a wall-controlled spark-ignited direct-injection engine, SAE paper 2003-01-0547 (2003).

[172] Y. Hardalupas, S. Sahu, A. M. K. P. Taylor, and K. Zarogoulidis, Simultaneous planar measurement of droplet velocity and size with gas phase velocities in a spray by combined ILIDS and PIV techniques, *Exp. Fluids*, **49**, 417–434 (2010).

[173] L. Zimmer, R. Domann, Y. Hardalupas, and Y. Ikeda, Simultaneous laser induced fluorescence and Mie scattering for droplet cluster measurements, *AIAA J.* **41**, 2170–2178 (2003).

[174] S. Idlahcen, L. Mees, C. Roze, T. Girasole, and J.-B. Blaisot, Time gate, optical layout, and wavelength effects on ballistic imaging, *J. Opt. Soc. Ame. A* **26**, 1995–2004 (2009).

[175] G. Bruneaux and D. Maligne, Study of the mixing and combustion processes of consecutive short double diesel injections, SAE paper 2009-01-1352 (2009).

[176] R. Baert, B. Somers, P. Frijters, C. Luijten, W. de Boer, Design and operation of a high pressure, high temperature bomb for HD diesel spray diagnostics: guidelines and results, SAE paper 2009-01-0649 (2009).

[177] C. Crua, D. A. Kennaird, and M. R. Heikal, Laser-induced incandescence study of diesel soot formation in a rapid compression machine at elevated pressures, *Combust. Flame* **135**, 475–488 (2003).

[178] G. Charalampous, Y. Hardalupas, and A. M. K. P. Taylor, A novel technique for measurements of the intact liquid jet core in a coaxial air-blast atomizer, *AIAA J.* **47**, 2605–2615 (2009).

[179] T. D. Fansler, M. C. Drake, B. Gajdeczko, I. Düewel, W. Koban, F. P. Zimmermann, and C. Schulz, Quantitative liquid and vapor distribution measurements in evaporating fuel sprays using laser-induced exciplex fluorescence, *Meas. Sci. Technol.* **20**, 125–401 (2009).

[180] C. Schulz and V. Sick, Tracer-LIF diagnostics: Quantitative measurement of fuel concentration, temperature and fuel/air ratio in practical combustion systems *Prog. Energy Combust. Sci.* **31**, 75–121 (2005).

[181] R. Zhang and V. Sick, Multi-component fuel imaging in a spray-guided, spark-ignition, direct-injection engine, SAE paper 2007-01-1826 (2007).

[182] J. D. Smith and V. Sick, Quantitative, dynamic fuel distribution measurements in combustion-related devices using laser-induced fluorescence imaging of bi-acetyl in iso-octane, *Proc. Combust. Inst.* **31**, 747–755 (2007).

[183] J. Nygren, J. Hult, M. Richter, M. Aldén, M. Christensen, A. Hultqvist, and B. Johansson, Three-dimensional laser induced fluorescence of fuel distributions in an HCCI engine, *Proc. Combust. Inst.* **29**, 679–685 (2002).

[184] D. Hoffman, K.-U. Muench, and A. Leipertz, Two-dimensional temperature determination in sooting flames by filtered Rayleigh scattering, *Opt. Lett.* **21**, 525–527 (1996).

[185] A. P. Yalin and R. B. Miles, Ultraviolet filtered Rayleigh scattering temperature measurements with a mercury filter, *Opt. Lett.* **24**, 590–592 (1999).

[186] M. G. Loeffler, K. Kroeckel, P. Koch, F. Beyrau, A. Leipertz, S. Grasreiner, and A. Heinisch, Simultaneous quantitative measurements of temperature and residual gas fields inside a fired SI-engine using acetone laser-induced fluorescence, SAE paper 2009-01-0656 (2009).

[187] M. Luong, R. Zhang, C. Schulz, and V. Sick, Toluene laser-induced fluorescence for in-cylinder temperature imaging in internal combustion engines, Appl. Phys. B **91**, 669–675 (2008).

[188] M. Yamakawa, D. Takaki, Y. Zhang, K. Nishida, and T. Li, Quantitative measurement of liquid and vapor phase concentration distributions in a D.I. gasoline spray by the laser absorption scattering (LAS) technique, SAE paper 2002-01-1644.

[189] R. Hasegawa, I. Sakata, H. Yanagihara, G. Saerner, M. Richter, M. Aldén, and B. Johansson, Two-dimensional temperature measurements in engine combustion using phosphor thermometry, SAE paper 2007-01-1883 (2007).

[190] S. Hildenbrand, S. Staudacher, D. Brueggemann, F. Beyrau, M. C. Weikl, T. Seeger, and A. Leipertz, Numerical and experimental study of the vaporization cooling in gasoline direct injection sprays, *Proc. Combust. Inst.* **31**, 3067–3073 (2007).

[191] D. A. Rothamer and J. B. Ghandhi, Determination of flame-front equivalence ratio during stratified combustion, SAE paper 2003-01-0069 (2003).

[192] J. D. Smith and V. Sick, Crank-angle resolved imaging of fuel distribution, ignition and combustion in a direct-injection spark-ignition engine, *SAE Trans. J. Eng.* **114**, 1575–1585 (2005).

[193] J. D. Smith and V. Sick, High-speed fuel tracer fluorescence and OH radical chemiluminescence imaging in a spark-ignition direct-injection engine, *Appl. Opt.* **44**, 6682–6691 (2005).

[194] T. D. Fansler, M. C. Drake, and B. Boehm, High-speed Mie-scattering diagnostics for spray-guided gasoline engine development, persented at the 8th International Symposium on Internal Combustion Diagnostics, Baden-Baden, Germany, 2008.

[195] C. M. Fajardo and V. Sick, Flow field assessment in a fired spray-guided spark-ignition direct-injection engine based on UV particle image velocimetry with sub crank angle resolution, *Proc. Combust. Inst.* **31**, 3023–3031 (2007).

[196] N. Graf, J. Gronki, C. Schulz, T. Baritaud, J. Cherel, P. Duret, and J. Lavy, In-cylinder combustion visualization in an auto-igniting gasoline engine using fuel tracer- and formaldehyde-LIF imaging, SAE paper 2001-01-1924 (2001).

[197] R. Collin, J. Nygren, M. Richter, M. Aldén, L. Hildingsson, and B. Johansson, Simultaneous OH- and formaldehyde-LIF measurements in an HCCI engine, SAE paper 2003-01-3218 (2003).

[198] J. E. Dec, W. Hwang, and M. Sjöberg, An investigation of thermal stratification in HCCI engines using chemiluminescence imaging, SAE paper 2006-01-1518 (2006).

[199] Y. Hardalupas, C. S. Panoutsos, and A. M. K. P. Taylor, Spatial resolution of a chemiluminescence sensor for local heat release rate and air-fuel ratio measurements in a model gas turbine combustor, *Exp. Fluids* **49**, 888–909, (2010).

[200] N. Kawahara, E. Tomita, K. Hayashi, M. Tabata, K. Iwai, and R. Kagawa, Cycle-resolved measurements of the fuel concentration near a spark plug in a rotary engine using an in situ laser absorption method, *Proc. Combust. Inst.* **31**, 3033–3040 (2007).

[201] M. Cundy, T. Schucht, O. Thiele, and V. Sick, High-speed laser-induced fluorescence and spark plug absorption sensor diagnostics for mixing and combustion studies in engines, Appl. Opt. **48**, B94–B104 (2009).

[202] Editorial, *New York Tribune* (May 11, 1899).

[203] E. Iglauer, The Ambient Air, *New Yorker* (April 13, 1968).

[204] Advisory Council for Aeronautics Research in Europe. Strategic research agenda. A series of documents, sra-1 October 2002, sra-2 October 2004, the latest of which is the 2008 addendum, available from acare4europe.org.

[205] ICAO Annex 16 – International standards and recommended practices, environmental protection – Vol. II, aircraft engine emissions, 2nd ed. (1993) plus amendments: Amendment 3, 20 March 1997; and amendment 4, 4 November 1999.

[206] J. E. Penner, D. H. Lister, D. J. Griggs, D. J. Dokken, M. M. (eds.), Aviation and the global atmosphere ipcc, prepared in collaboration with the Scientific Assessment Panel to the Montreal Protocol on Substances that Deplete the Ozone Layer Part I (Cambridge University Press, Cambridge, 1999), p. 373.

[207] R. G. McKinney, D. Sepulveda, W. Sowa, A. K. Cheung, The Pratt & Whitney talon × low emissions combustor: Revolutionary results with evolutionary technology, persented at the 45th AIAA Aerospace Sciences Meeting and Exhibit, Reno, NV, 8 – 11 January 2007.

[208] M. P. Spooner, Development of low emissions turbojet combustor, in *Proceedings of the 2nd International Symposium on Japanese National Project for Super/Hyper-sonic Transport Propulsion System* (HYPR Association, Tokyo, Japan, 1995) pp. 91–98.

[209] R. Tacina, C. Wey, P. Laing, and A. Mansour, Sector tests of a low nox, lean direct injection multi point integrated module concept, ASME paper GT2002-30089 (2002).

[210] R. Tacina, A. Mansour, P. Laing, and C. Wey, Experimental sector and flame tube evaluations of a multi point integrated module concept for low nox combustors, IGTI-2004-53263 (2004).

[211] L. J. Spadaccini and J. A. TeVelde, Autoignition characteristics of aircraft-type fuels, NASA paper CR-159888 (1980).

[212] S. Hayashi, H. Yamada, K. Shimodaira, T. Saitoh, S. Horiuchi, Spontaneous ignition of atomised kerosene/air mixtures in a high pressure flowing system. [in Japanese], National Aerospace Laboratory Report TR966, Tokyo, Japan (1988).

[213] C. M. Coates, Experiments on a high pressure premixing/prevaporising combustor, Technical Memorandum P1099 RAE UK (1987).

[214] M. Rachner, J. Becker, C. Hassa, T. Doerr, Modeling of the atomisation of a plain liquid jet in crossflow at gas turbine conditions, Aerospace Sc. Technol. **6**, 495–506 (2002).

[215] J. C. Mankins, Technology readiness levels, A White Paper. Advanced Concepts Office, NASA (April 6, 1995).

[216] A. H. Lefebvre, *Atomisation and Sprays* (Hemisphere, Washington, D.C., 1989).

[217] C. Cha, J. Zhu, N. K. Rizk, and M. S. Anand, A comprehensive liquid fuel injection model for cfd simulation of gas turbine combustors, AIAA paper 2005-0349, presented at the 43rd AIAA Aerospace Sciences Meeting and Exhibit, Reno, NV, (2005).

[218] E. De Villier, A. D. Gosman, and H. G. Weller, Large eddy simulation of primary diesel spray atomisation, SAE paper 2004-01-0100 SAE 2004 World Congress & Exhibition, Detroit, MI, March 2004.

[219] K. Döbbeling J. Hellat, and H. Koch, 25 Years of BBC/ABB/Alstom lean premix combustion technologies, presented at the ASME Turbo Expocsition 2005: Power for Land, Sea and Air, Reno, Tahoe, NV, June 6-9, 2005.

[220] Integrated pollution, prevention and control. Reference document on best available techniques for large combustion plant. European Commission, July 2006.

[221] Standards of performance for stationary combustion turbines: Final Rule. 40CFR, Part 60. Part 3, Environmental Protection Agency, 6 July 2006.

[222] Rule 413, Stationary gas turbines. Proposed Amendments. Sacramento Metropolitan District Staff Report. Williams KJ. February 2005.

[223] San Joaquin Valley pollution control directive, Rule 4703, Stationary gas turbines last amended September 20, 2007. Available at http://www.valleyair.org/rules/currntrules/r4703.pdf.

[224] Directive 2001/80/EC of the European Parliament and of the Council of 23 October 2001 on the limitation of emissions of certain pollutants into the air from large combustion plants.

[225] Tuning approaches for DLN combustor performance and reliability, EPRI, Palo Alto, CA, 2005, 1005037. Available at http://mydocs.epri.com/docs/public/000000000001005037.pdf.

[226] S. Aoki, K. Matsumoto, Y. Douura, T. Oda, M. Ogata, and Y. Kinoshita, Up-
graded lineup of Kawasaki Green Gas Turbine combustion systems Proceedings of the Conference on Power Engineering-09 (ICOPE-09), Kobe, Japan,
November 16-20, 2009.

[227] T. Scarinci, I. J. Day and F. Freeman, U.S. patent 6,698,206B2, March, 2004.

[228] T. Schweiger, M. Underhill, and D. W. Livingston, Rolls-Royce RB211-DLE
combustor Diffuser Design Optimisation, paper 99-GT-237, presented at the
International Gas Turbine and Aero Engine Congress & Exhibition, Indi-
anapolis IN, June, 1999.

5 Future Directions

5.1 Utilization of Hot Burnt Gas for Better Control of Combustion and Emissions
By S. Hayashi and Y. Mizobuchi

Recirculated hot burned gas has long been used to stabilize flames in continuous flow combustion systems, such as furnaces, boilers, gas turbines, and afterburners in jet engines [1]. The recirculation is created by employing swirl vanes, bluff bodies, or baffle plates. The heat and radicals in the hot combustion products act as the ignition source for fresh mixtures of fuel and air flowing into the flame-stabilization region.

In the late 20th Century, air pollution by combustion-generated smoke, partially burned hydrocarbons, carbon monoxide, and nitric oxides, NO_x, became a public concern. With the increasing stringency of the regulations for NO_x emissions, recirculation of combustion products or exhaust-gas came to be used as one of the measures for controlling NO_x formation. Exhaust-gas recirculation (EGR), for example, is extensively used in reciprocating engines for automobiles [2], where the engine exhaust gas is recirculated into the intake manifold in order to mix with fresh air. Another example is recirculation of combustion products in furnaces and boilers, which is sometimes called *internal* EGR. Most of the reduction of NO_x emissions achieved by EGR is attributed to reduced peak flame temperatures that are due to increased specific heat capacity of the diluted mixtures, though some of the reduction is attributed to resulting slower NO formation that are due to reduced oxygen concentrations and increased carbon dioxide and water vapor concentrations.

More recently, in response to the increasing demands for improving fuel efficiency while reducing emissions of pollutants, mainly NO_x, a new area of combustion, which is known as highly preheated combustion [3], has evolved. Auto-ignition is the mechanism for supporting combustion in this new mode of combustion, though flame propagation is the mechanism in the conventional mode of combustion. Major examples are homogeneous charge-compression ignition (HCCI) engines [4], catalytically supported gas-phase reaction for gas turbine engines [5], and mild or flameless combustion for boilers and furnaces [6–8]. A feature common to these new combustion strategies is that the mixture or the air is highly preheated by compression, heat generated by reactions on catalysts, waste heat recovery by heat exchangers or recuperators, or recirculating hot combustion products in the combustion chamber, so that the mixture can be continuously auto-ignited. Some waste heat recovery recuperators made of heat-resistant ceramics can preheat combustion air to temperatures higher than 1300 K, and therefore mixtures of most fuels and air can be ignited even under highly diluted conditions.

As for HCCI engines, a prototype demonstration is in progress, but several technical issues, involving the control of ignition timing by EGR for example, are still to be resolved before commercial production can begin.

Mild or flameless combustion has already been used in commercial material-processing furnaces, boilers, and other combustion systems and successfully reduces NO_x, CO, and soot emissions to very low levels. Additional merits of this new combustion strategy are low combustion oscillations or noise and uniform wall temperature distributions.

For gas turbines, lean premixed combustion is potentially the most promising approach for ultra-low-NO_x emissions because the turbine inlet gas temperatures are a few hundred degrees lower than the peak flame temperature because of the thermal limitation of the materials presently available for turbine blades. This approach has already been extensively used worldwide in large-scale natural gas-fired gas turbines for combined cycle power generation with ultra-low-NO_x emissions, meeting the most stringent emissions regulations for this pollutant. In aviation, the regulations for NO_x emissions are increasingly stringent, and therefore premixed lean combustion technologies for kerosene are being actively developed for the next-generation aeroengines. Some technical issues, such as combustion oscillation, trade-off between CO and NO_x emissions, and, for high-pressure-ratio engines, auto-ignition and flashback in mixture preparation, need to be resolved. However, it is worth mentioning that, in high-pressure-ratio gas turbines, shorter ignition lags resulting from higher air temperature and pressure conditions could make the application of auto-ignition-supported combustion much easier.

Investigation of the application of mild or flameless combustion to gas turbine combustors has continued. Model combustion chambers proposed by some investigators [9, 10] are designed so that large strong vortex flows of internally recirculating burned gas can be produced to establish the conditions required for mild combustion. The ratio of recirculating burned gas to fresh air in the previously mentioned combustion chambers is much larger than that in the recirculation zone of conventional swirl or bluff-body stabilized flames, and therefore the mixture in the vortex is highly preheated and diluted with burned gas.

The pressure to minimize CO_2 emissions from combustion systems for the mitigation of global warming has been increasing worldwide. Capture and sequestration of CO_2 can be one of the major options for reducing CO_2 emissions from fossil fuel combustion. For large combined cycle power generation plants with CO_2 sequestration [11], combustion product recirculation is proposed to increase the CO_2 concentration in the working fluid so that the efficiency for CO_2 separation can be improved. As both combustion and emissions characteristics will be altered significantly by high CO_2 and H_2O concentrations, the use of the combustion products to optimally control the formation of emissions should be more actively investigated.

In gas turbines, the combustor operating conditions, such as inlet air temperature and pressure, equivalence ratio in the combustion region, and overall fuel–air ratio of the mixture, vary widely with power or rotational speed. Therefore, an ideal

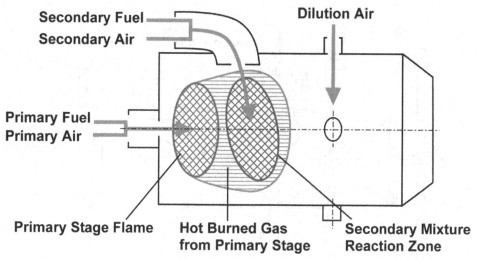

Figure 5.1. A schematic illustration of flow of axially staged lean-mixture injection concept.

mild combustion mode can be established over only a limited range of engine operations. The constraints imposed by the need for recirculating a large amount of combustion products in the previously mentioned approaches would make the flow in the combustor complicated and therefore may result in an increase pressure loss and a complicated combustor wall-cooling structure. Considering that staging of fuel would be required for a combustor or to function satisfactorily over a wide range of operation from engine start to full load, a multi-axially staged lean-mixture injection concept [12–14] was proposed from our laboratory.

This section is devoted to axially staged lean mixture injection. Experimental observations with model combustors are presented and discussed based on the predictions by simple numerical simulation of chemical reactions. A successful application of the concept to a commercial gas turbine is presented, followed by comments on more realistic simulations of the process to be utilized for the design of staged lean-mixture injections.

5.1.1 Axially Staged Lean-Mixture Injection

In this combustion concept, hot combustion products prepared in the upstream stage are used as the diluent for the downstream stage where lean to ultra-lean mixtures, injected into and mixed with the hot burned gas, are reacted in the mild combustion mode. Staged mixture injection in this simple straight-flow system can easily establish the conditions required for mild combustion, even in a gas turbine whose combustor inlet air temperature is lower than the auto-ignition temperature for the fuel. It should be noted that, even though they look similar, this concept is different from the axially staged flame combustion in that no flame of the injected mixture is stabilised. In this simple staging strategy, only the fuel flow rates to each stage are varied to modulate engine power or thrust while the air split between stages is fixed.

The combustion concept of a two-stage configuration [14] is schematically illustrated in Fig. 5.1 in which the secondary mixture is injected into the hot burned

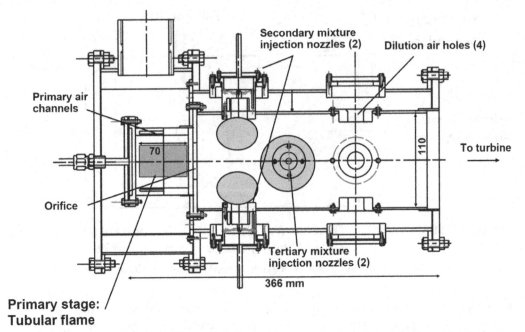

Primary stage:
Tubular flame

Figure 5.2. A schematic illustration of gas turbine combustor with axially staged lean-mixture injection.

gas produced by the conventional flame combustion in the primary stage. Ideally, the injected mixture is instantaneously and uniformly mixed with hot burned gas before ignition so that reaction proceeds in a volumetric manner rather than in a flame sheet in the conventional combustion mode. The temperature and oxygen concentration of the hot burned gas in the secondary stage are modified simply by varying the fuel–air ratio in the first stage.

The capability of this concept was experimentally verified by use of a three-stage, model gas turbine combustor, shown schematically in Fig. 5.2 [15]. Tubular flame combustion of lean mixtures was employed for the first stage because of its inherently better stability at lean conditions and thermal reaction of lean to ultra-lean mixtures was used for the second and third stages. In the first stage the primary mixture enters the cylindrical chamber from the four tangential slots opened on the cylindrical wall to prepare a swirling flow where a tubular flame is stabilized. A pair of mixture injection nozzles, opposing each other, was provided in the second and third stages, respectively. Swirl was imparted to the mixture jets from the nozzles to enhance the mixing of injected fresh mixtures and burned gas. Experiments were conducted at an air flow rate of 72 g/s and at an inlet air temperature of 700 K and atmospheric pressure. Equivalence ratios of the secondary and tertiary mixtures were varied from zero (just air) to about 0.7.

Figure 5.3 shows typical pictures of the combustion phenomena recorded for ultra-lean to lean secondary mixtures injected into the hot burned gas of a temperature of 1230 K. The hot burned gas was prepared by burning a premixed mixture of an equivalence ratio of 0.46 in a tubular flame. The blue ring in the photo (see coloured plate) corresponds to the cross section of the tubular flame in the first stage, and a pair of mushroom-shaped plumes is formed by the secondary mixture

$W_a = 72$ g/s $T_{in} = 700$ K $\phi_1 = 0.46$ (1230 K)

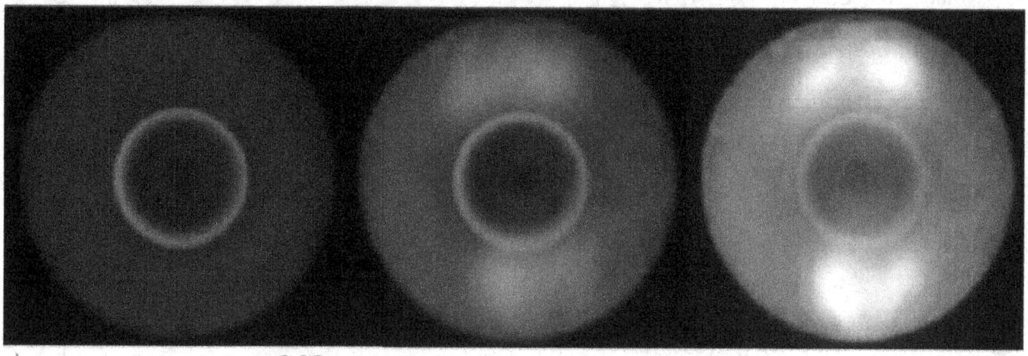

$\phi_2 =$	0.25	0.4	0.6
Combustion efficiencey (%)			
=	99.9	99.9	99.9
NO_x ppm (15%O_2) =	4.9	7.1	35.6

Figure 5.3. Photograph of flames in primary and secondary stages. (See colour plate.)

jets injected into the primary burned gas. The end wall is red-hot. In the third stage, not mixture but air was injected from the tertiary mixture nozzles into the hot combustion product. It was found that the secondary stage combustion was supported by the hot burned gas from the primary stage. No flame was seen with ultra-lean secondary mixtures ($\phi_2 \leq 0.25$), and a pair of mushroom-shaped reaction volumes gradually becomes visible with increasing fuel supply to the secondary stage. Combustion efficiencies greater than 99.9 were achieved over a wide range of ϕ_2 with low-NO_x emissions for ultra-lean mixtures.

The NO_x emissions, corrected to 15% O_2, and combustion efficiencies for two- and three-stage modes are shown in Fig. 5.4 as functions of overall equivalence ratio, ϕ_t, determined by gas analysis. The data with single-stage flame combustion are also shown for comparison. Auxiliary scales for equivalence ratios of the secondary and tertiary mixtures are attached to the abscissa for convenience. The levels of NO_x emissions in the staged mixture injection mode are maintained at very low levels, provided that the equivalence ratios of secondary or tertiary mixtures are smaller than about 0.5. It is found that most of the NO formed in the upstream stages is converted into NO_2 probably by the fuel in the mixtures injected. Small combustion inefficiencies seen when shifting from single- to two-stage modes and from two- to three-stage modes can be avoided by slightly increasing ϕ_1 and ϕ_2 or by increasing the residence times in these stages.

Visualization of the reaction zones of the mixtures injected into hot burned gas was conducted by use of planer laser-induced fluorescence (PLIF) imaging. A rectangular combustor with quartz windows was used for ease of optical access. A nozzle was attached to one of the side walls of the combustor to inject the secondary mixture horizontally into the vertical flow of combustion products of a temperature of about 2000 K. Swirling motion was imparted to the secondary mixture jet to enhance the mixing of the secondary mixture and the combustion products. The equivalence ratio of the secondary mixture was varied from 0.1 to 0.8 and that of the primary mixture was 0.6.

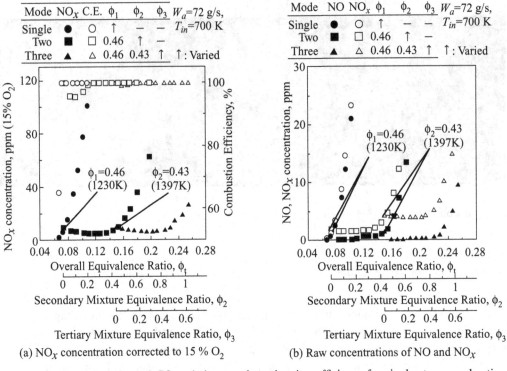

Figure 5.4. NO$_x$ and CO emissions, and combustion efficiency for single-stage combustion and lean-mixture injection in the second and third stages.

Figure 5.5 shows typical instantaneous OH images obtained by PLIF for the reaction of an ultra-lean mixture in a hot burned gas. The mixture was injected into the cross-flowing hot burned gas from a nozzle fitted on the wall of the rectangular combustor. No visible flames are seen to be stabilised on the nozzle exit. Distributed unburned mixture volumes are seen in the bottom-right corner of the photo. A close-up of this region is presented in Fig. 5.5(b). Reacting volumes of small scales are seen to be spatially distributed in the unburned mixture, suggesting that the reaction occurs in a volumetric manner rather than in a flame sheet. For mixtures of equivalence ratios greater than about 0.4, a conventional premixed jet flame is always stabilised on the nozzle exit and no distributed unburned mixture volumes are observed in the hot burned gas.

The potential of staged lean-mixture injection can also be demonstrated by a simple 1D simulation. The schematic of an example is presented in Fig. 5.6. A methane–air mixture of equivalence ratio of $\phi = 0.8$, temperature of 300 K, and pressure of 1 atm flows into the computational domain from the left side. The mixture burns through a premixed laminar flame at the first stage just after the inlet. The secondary and tertiary mixtures are injected at 2-cm and 4-cm downstream locations, respectively. In this 1D simulation, the flow speed is chosen to be very low so that the first-stage premixed flame, which is normal to the flow direction, can be stabilised in the combustor. Therefore the combustor size is correspondingly small to adjust the time scales of the axial flow to be those in 3D combustors.

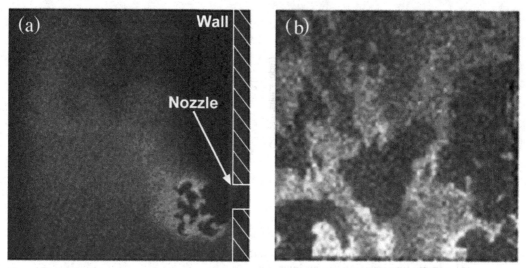

Figure 5.5. PLIF images of OH in the reaction of ultra-lean mixture (0.3 equivalence ratio) injected into flowing hot combustion product: (a) overall view, (b) close-up of the region in (a) where unburned mixture volumes are spatially distributed. The grey scale is arbitrary: light patches represent high OH concentration whereas dark ones represent low concentrations.

In the following discussion, results for staged injection of a very lean mixture at equivalence ratio $\phi = 0.4$, which is leaner than the lower flammability limit are compared with those for staged air injection, which is currently used in practical low-emission engine systems.

In Fig. 5.7(a), a very lean mixture at $\phi = 0.4$ is injected at the secondary and tertiary injections. The temperature and pressure of injected mixtures are 300 K and 1 atm, respectively. At the secondary injection, the same mass of the primary stage mixture is injected and the overall equivalence ratio is reduced to about 0.6. One and

Figure 5.6. Schematic of 1D analysis of staged lean-mixture injection.

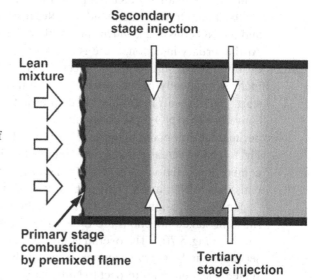

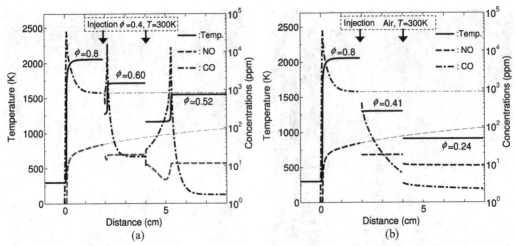

Figure 5.7. NO$_x$ reduction and operation range extension by staged lean-mixture injection: (a) lean-mixture injection, (b) air injection.

a half times the primary stage mass is injected at the tertiary stage, and the overall equivalence ratio becomes about 0.52.

Rapid mixing of burnt gas and injected lean mixture is assumed at each staged injection. The burnt gas of the primary stage combustion is obtained by PREMIX [16]. Then a zero-dimensional (0D) analysis at constant pressure is applied downstream of the secondary injection, where the reacting mixture is advected with the flow velocity. The detailed chemical kinetics is quoted from GRI-MECH 3.0 [17]. In the figure, the solid lines, dashed lines, and dot–dashed lines are the profiles of temperature and NO and CO concentrations in ppm, respectively. The NO and CO profiles for the single-stage combustion are represented by thin lines.

The temperature of the mixture rises to about 2050 K at the primary stage of combustion, and drops down to about 1250 K by the secondary mixture injection. The injected very lean mixture can burn without leaving a significant amount of CO, thanks to the relatively high temperature and radicals produced by the primary stage combustion. On the other hand, the NO formation is suppressed or almost frozen, and the NO emission is much smaller than that of the single-stage combustion case. At the tertiary injection, there is an entry region upstream of the substantial combustion region. Just after the injection, H, OH, and O decrease rapidly and then increase gradually in the entry region until the substantial combustion reactions start. The concentration of NO decreases in the entry region, where the injected O$_2$ converts NO to NO$_2$. The conversion is analogous to the NO-to-NO$_2$ conversion experimentally observed in the very lean staged injection cases, as previously shown in Fig. 5.4. NO$_2$ increases in the entry region and decreases rapidly once the substantial combustion reactions start. Finally, the injected very lean mixture burns, and lower emissions of NO$_x$ and CO are achieved than in the single-stage combustion.

In Fig. 5.7(b), the results for air injection are shown for comparison. The mass ratios of staged injections are the same as those in the lean-mixture injection case shown in Fig. 5.7(a). The overall equivalence ratio goes down to 0.41 and 0.24 at the secondary and tertiary injections, respectively. When one compares the two types of staged injections with respect to NO formation, the lean-mixture injection produces

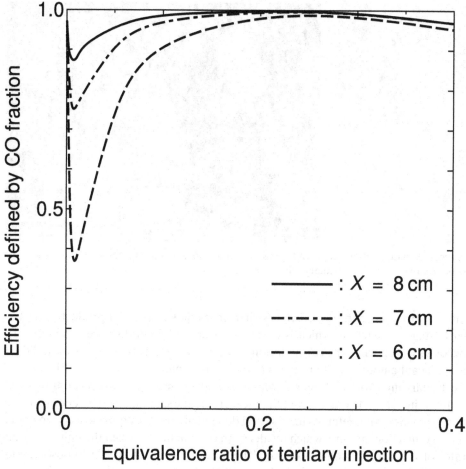

Figure 5.8. Local efficiency response to the equivalence ratio of the tertiary injection in the 1D virtual combustor.

only a little larger but almost equivalent amount. The lean-mixture injection provides a higher working gas temperature that is, higher energy, whereas it achieves an equivalent performance regarding the NO emission compared with the air injection.

The preceding simple simulations indicate the possible advantages of staged lean-mixture injection. The virtual 1D combustor achieves low-NO_x combustion with the operating range from 900 K (air injection case) to 1600 K (lean-mixture injection case) in the burnt gas temperature. The combustion is stabilized by the primary stage rather than rich combustion, and the succeeding combustion by staged injection of a very lean mixture increases the total energy of the system without significant increase of NO_x or CO. Figure 5.8 shows the variation of the efficiency at the three axial locations in the virtual 1D combustor with the change of the equivalence ratio of the tertiary injection mixture. The efficiency is defined by the combustion completeness in terms of the CO mole fraction as, $1 - x_{CO}/x_{CO_2}^{\infty}$, where x_{CO} is the CO mole fraction at the axial measurement location and $x_{CO_2}^{\infty}$ is the CO_2 mole fraction when all the fuel completely reacts to products. When the tertiary injection mixture is very lean, the CO formation is very small and then the efficiency is high. As the tertiary injection

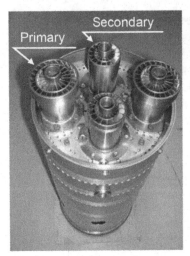
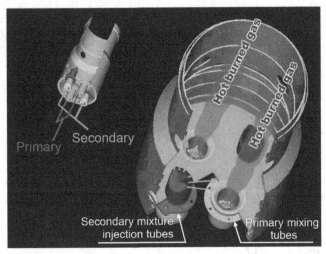

Figure 5.9. (Left) Photograph and (right) schematic illustration of staged tubular combustor with premixing tubes for primary mixture and mixture injection tubes for secondary mixtures.

mixture becomes rich, the efficiency first drops down. The CO production at the early stage of reaction, which is observed at around 5-cm distance in Fig. 5.7(a), becomes larger, while the consumption by the succeeding reactions is slow, and then a significant amount of CO remains at the measurement location. If the richness of the tertiary injection mixture is increased then the CO consuming reactions become faster to increase the efficiency at the measurement location. The efficiency in terms of CO reaction completeness has a strong dependency on the equivalence ratio of the tertiary injection mixture, which results from the balance between the finite reaction rates of CO and the combustor size. The local efficiency drop is also observed in the experiments, as presented in Fig. 5.4(a), and may be attributed to the same mechanism. This example indicates the possibility of optimal design and control of staged injection of lean mixtures. There will be some parameters to be optimised that control the performance of the combustor. The impact of these parameters on the performance should be estimated at an early stage of the system design.

5.1.2 Application of the Concept to Gas Turbine Combustors

This new low-emissions combustion concept was successfully applied to a unique lean–lean axially staged combustor installed in commercial liquid-fueled, regenerative 300-kW gas turbines [18, 19]. A direct photograph and a schematic illustration of the combustor are presented in Fig. 5.9, showing the arrangement of the premixing tubes for the primary mixture and the mixture injection tubes for the secondary mixture on the combustor liner dome. Premixed mixtures are prepared by atomizing liquid fuels at the inlets of the premixing and mixture injection tubes. Premixed flames of lean mixtures are stabilized at the exits of the premixing tubes. The mixture injection tubes extend about half the full length into the primary combustion zone. The jets of the secondary mixture, lean to ultra-lean in fuel composition, are directed towards the two volumes of the primary mixture burned gas by the V-shaped flow deflectors attached to the exits of the mixture injection tubes. Figure 5.10 is

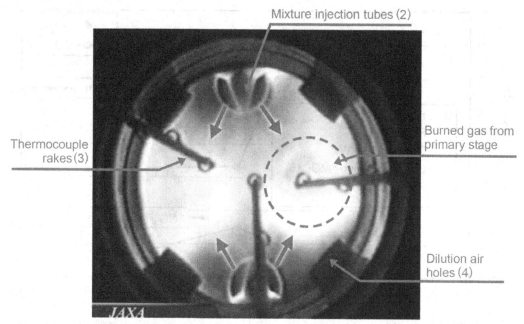

Figure 5.10. Photo showing phenomena in combustion chamber with secondary mixtures being injected into volumes of hot burned gas produced by flame combustion of primary mixtures.

a photograph showing the phenomena in the combustion chamber, installed in the combustor test rig, operating with kerosene at simulated engine full power conditions of 920 K inlet air temperature and 0.4-MPa pressure [20]. The primary burners on the end wall were not visible because of the radiation from the volumes of the primary combustion products (dotted circle). The mixture injection tubes are seen on the top and bottom and the details of the exit of these injection tubes are clearly seen. Four dilution air holes on the combustor liner and three thermocouple rakes for the measurement of the exit gas temperature profiles are also seen. The arrows starting from the exit of the mixture injection tubes show the directions of the secondary mixture jet. No flames are seen to be stabilised at the exit of the secondary mixture injection tubes, as intended while the combustor was designed.

The NO_x and CO emissions levels of the gas turbine are well below 15 and 35 ppm (15% O_2), respectively, for kerosene over a range of loads from 50% to 100%, as shown in Fig. 5.11. Other liquid fuels such as gasoline and solvents disposed from factories are also used maintaining low levels of NO_x and CO emissions. For more than 10 years of operation, neither auto-ignition nor flashback in the tubes has been experienced. Combustion is smooth and free from oscillatory combustion over the full range of power.

5.1.3 Numerical Simulation towards Design Optimization

Thanks to the development of computer power, parallel computing techniques, and numerical methods, numerical simulation is becoming available as a tool for engineering design in some fields. It is expected to assist the optimal design and control

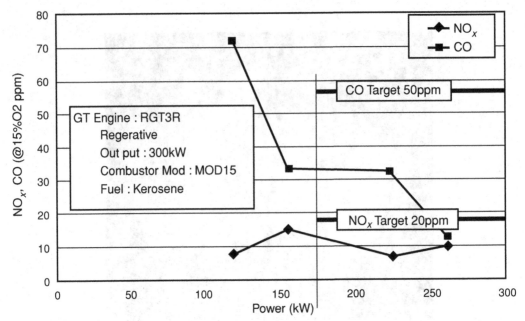

Figure 5.11. Measured NO$_x$ and CO emissions when power is increased by increasing secondary fuel flow while primary fuel flow is maintained in lean–lean two-stage combustion.

of staged lean-mixture injection. The major issues in this combustion configuration are rapid mixing at each staged injection and the combustion reactions of the mixture composed of hot burnt gas and injected fuel-lean mixture. The requirement of a numerical simulation is to estimate the previously mentioned mixing process and chemical reactions that take place in real-size combustors at an acceptable level and throughput.

The mixing process contains two parts. One is the atomization and vaporisation process of a liquid fuel jet mixing into air to make the lean mixture for injection. The other is the mixing of the lean mixture and the hot burnt gas. The two may occur sequentially and simultaneously. The former process is not yet clearly understood or well modeled. Some detailed and large-scale simulations have been conducted to understand the mechanism and to model the phenomena into CFD (computational fluid dynamics) friendly formulations [21–23]. Figure 5.12, for example, shows a snapshot of an atomising liquid jet reproduced by two-phase flow simulation using about 6 billion cells, which includes the ligament production process from the jet and the droplet production process from the ligament. The latter mixing process is strongly affected and controlled by turbulence caused by the interaction between burnt gas flow and the injection jet. As with other mixing problems in the absence of combustion, the improvement of the turbulent flow model is persistently the key for the estimation, whichever approach, either RANS (Reynolds-averaged Navier–Stokes) or LES (large-eddy simulation), one uses.

Concerning the combustion reaction simulation, one of the critical problems is the use of practical fuels that contain heavy hydrocarbons; for example, the carbon number of kerosene is 10–15. It is not easy to understand the rigorous chemical

140.0

-50.0
[m/s]

Figure 5.12. Detailed simulation of atomization process of liquid jet [23]. The suface colour shows the axial velocity. (See colour plate.)

kinetics and to include the huge number of species and elementary reactions in a numerical simulation for the optimal design and the further control process of staged-injection combustion, although these elementary reactions can be important. Given the proper chemical kinetics, the combustion modelling will be the most important issue, as is now the case in many combustion simulations. The combustion we have to deal with may be categorised into a different type from that described by the concept of flamelets [24]. The reactants are not pure unburnt gas components at low temperature but the mixture of the hot products and radicals produced by the previous stage and the injected lean mixture. The combustion occurring in the staged-injection location may be so-called 'mild combustion' or 'flameless combustion' [6, 7, 25, 26], including, auto-ignition processes, to which the conventional concept of a flamelet may not be applied rigorously. The combustion may take place in rather thick layers, and propagative flame may not exist. Direct numerical simulation (DNS) with detailed chemistry and transport properties, which is becoming possible for real-size combustion phenomena [27–29], is expected to be the most promising approach to estimate combustion configurations such as staged lean-mixture injection. But even today's Peta FLOPS computer power is not enough to simulate the previously mentioned phenomena in practical scale combustors with detailed chemistry of a practical fuel. The situation will not drastically change even in the next Exa FLOPS era. Hence, for the near term, the modelling of the combustion of the hot burnt gas and injected lean mixture is an important and indispensable research subject, as well as the development of improved numerical techniques such as discretization methods. The chemistry reduction and tabulation approaches [30, 31] may be of help in the construction of models. DNS applied to rather fundamental configurations will support the construction and validation of these models.

5.2 Future Directions and Applications of Lean Premixed Combustion
By J. F. Driscoll

Combustion technology in the future is expected to progress more and more in the direction toward lean premixed turbulent combustion for several reasons. Premixing creates short, compact flames that allow the size of an engine to be reduced. Smaller engines are needed for small business jets, for mobile gas turbine power generators, and for small automobile engines. Premixing also provides inherently low levels of nitric oxide, CO, and hydrocarbon pollutants. Government regulations of pollutant levels are expected to tighten for automobiles, aircraft, and ground-based power generators. Another motivating factor is the attempt to reduce the 'carbon footprint' and thus the grams of CO_2 emitted per kiloJoule of energy delivered. There is evidence [32] that significant reductions in CO_2 emissions are possible if a lean premixed recuperative gas turbine power plant is operated using a hydrogen-CO synfuel, as subsequently discussed. Commercial aircraft engines also are moving to premixed technology. A lean premixed prevapourised (LPP) combustor in the General Electric GEnx engine will be operated in the next few years in the Boeing 787 aircraft [32]. In the automotive engine area, the HCCI engine represents another new advance in lean premixed combustion technology.

However, this trend towards increased levels of premixing also leads to several new research challenges. Partially premixed combustion often occurs in LPP and HCCI devices, and it presents modelling challenges because any instantaneous reaction interface can contain regions, of premixed flames, which propagate as waves, and non-premixed regions, which do not propagate, so the two regions are governed by different physics. Premixed flames are inherently more difficult to stabilize in a high-speed flow than non-premixed flames, so the trend towards premixed devices makes it especially important to develop models that can realistically and reliably predict flame lift-off, blowout limits, and flashback, and such a model is not available today. This issue is discussed in Subsection 5.2.2. Flame-stability complications arise because engines usually employ complex bluff-body geometries, swirl, and recirculation zones to assist in flame stabilisation. Another future driving force is the need for a model that can reliably predict the onset of combustion instabilities that are coupled with the acoustic pressure waves, which often arise when premixing is employed. Subsection 5.2.3 lists some emerging experimental technologies; the future direction of measurements is to provide high-speed imaging of unsteady combustion processes with improved spatial and temporal resolution. Subsection 5.2.4 contains a discussion of some fundamental issues that still require further work to improve our ability to model and understand premixed turbulent combustion.

5.2.1 LPP Combustors

Several types of gas turbine combustors have been developed that use premixing to reduce both the NO_x emissions as well as the size of the reaction zone. Three examples are called the dry low-NO_x (DLN) injector [32–35], the low-swirl combustor,

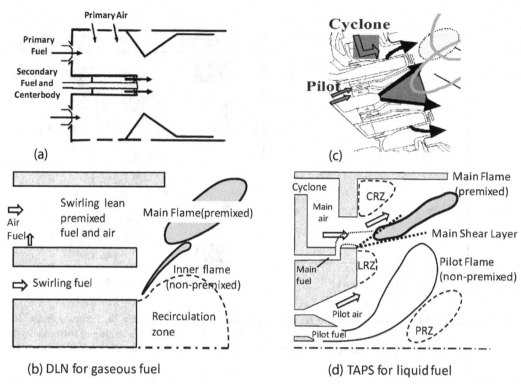

Figure 5.13. Lean premixed technology within two commercially available gas turbine combustors: (a,b) DLN device operated on gaseous fuel (natural gas) for ground-based power
generation (schematic of Maughan et al. [33]). (c,d) GE-TAPS operated on liquid Jet-A fuel
for aviation applications (schematics of Mongia [37] and Dhanuka et al. [38]). PRZ, CRZ,
and LRZ respectively mean primary, corner, and lip recirculation zones.

which was developed by R.K. Cheng [36], and the TAPS (twin annular premixed
swirler) mixer that is described by Mongia [37] (see Section 4.2 also). DLN and
low-swirl devices operate on gaseous fuel (including H_2/CO syngas mixtures) for
ground-based power generation, and TAPS is certified for use in commercial aircraft
that are run on liquid Jet-A fuel.

The DLN fuel injector [32–35] is shown in Fig. 5.13(a). It contains two annular
ducts that are shown in Fig. 5.13(b). The inner duct contains a small flow rate of
pure fuel that is gaseous and swirling. This pilot fuel burns downstream as a non-
premixed pilot flame. Most of the fuel is injected into the outer duct as a fuel jet
into a cross-flow of air. This main fuel and air premix a few millimeters upstream
of the premixed main flame. The primary zone equivalence ratio is designed to vary
from 0.5 to 0.8, and the attempt is to keep the maximum values of the instantaneous
gas temperatures below 1800 K to reduce thermal NO_x formation. Another goal is
to keep the post-flame temperatures between 1500 and 1800 K to oxidise the CO
without producing additional NO_x. Using natural gas fuel, very low NO_x levels of
5 ppm were reported [33].

The DLN device also has been used to burn a synfuel mixture of H_2 and CO
in a gas turbine power generator [32, 34]. DLN was reported [34] to provide a
significant reduction in the cost of capturing CO_2. Using a regenerative cycle, natural

FSD (mm^{-1})

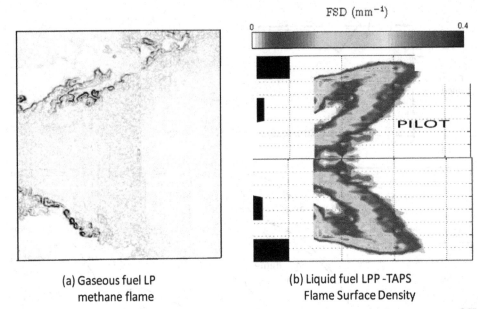

(a) Gaseous fuel LP   (b) Liquid fuel LPP -TAPS
    methane flame   Flame Surface Density

Figure 5.14. Fluorescence images of flames in lean premixed devices: (a) Instantaneous OH PLIF in gaseous fueled LP combustor of Stopper et al. [39]; (b) time-averaged formaldehyde fluorescence of main and pilot flames in a TAPS LPP combustor fueled with liquid Jet-A, by Dhanuka et al. [40].

gas was converted to synfuel and the exhaust gases were recirculated. The DLN injector provides the flexibility to optimize the pressures, temperatures, and CO_2 concentrations within the regenerative cycle so that CO_2 could be more efficiently removed by both absorbing solvents and a CO_2 membrane. The use of synfuel creates two challenges: Flashback occurs for large H_2/CO ratios, and for small H_2/CO ratios the challenge is to consume all of the CO. Synfuel can contain different concentrations of H_2 and CO so its flame speed can vary; it important to have models that can predict the effects of varying flame speed on engine performance.

The low-swirl combustor of R.K. Cheng and co-workers [36] has similarities with the DLN device in that it operates on natural gas or gasesous synfuels and it achieves low NO_x by employing unique methods to stabilise ultra-lean combustion. Diverging walls and swirl force the axial velocity to decrease as the reactants approach a relatively flat premixed flame; this avoids flashback because the flame must propagate back against progressively larger reactant velocities. It was selected for modelling studies because of its simple geometry and because a fairly complete data set of flow properties was measured [36]. Another fairly complete database was measured for a lean premixed design by Stopper et al. [39] and one OH image is seen in Fig. 5.14(a).

For aircraft applications, TAPS is a type of LPP device that is operated on liquid Jet-A fuel [37]. TAPS provides a 50% reduction in NO_x emissions compared with the previous non-premixed 'rich dome' technology, and it is currently employed in the GEnx engine. Figure 5.13(d) shows that the TAPS injector creates an inner pilot flame that surrounds the cone of Jet-A fuel drops that are created by an atomiser. Most of the fuel enters in the main fuel stream. This main fuel jet is atomised when

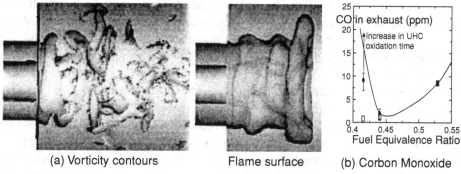

(a) Vorticity contours   Flame surface   (b) Corbon Monoxide

Figure 5.15. LES results for Gasesous-Fueled LP gas turbine combustors, similar to the DLN design of the previous figure. (a) Instantaneous vorticity and flame surface from the G equation solutions of Yang and co-workers [42]; (b) exhaust CO computed using the G equation coupled with a flamelet library, by Eggenspieler and Menon [41].

it interacts with a high-speed cross-flow of heated air. The cross-flow velocity is large enough to efficiently atomise and prevapourise the fuel, and it creates a sufficiently large lift-off distance between the main flame and the fuel injection location. This distance provides the premixing that is necessary to achieve the proper fuel-lean conditions before reactions occur. As with the DLN device, the goal is to keep temperatures in the premixed main flame and post-flame gases between 1500 and 1800 K to minimize NO_x while maximising the oxidation of CO. Residence times of post-flame gases in the three recirculation zones shown in Fig. 5.13(d) are important. A complete database of velocity and flame properties in a TAPS LPP combustor was measured by use of particle image velocimetry (PIV) and PLIF diagnostics [38, 40]. The dotted lines in Fig. 5.13(d) represent the main shear layer that provides the necessary premixing; the main flame is anchored near the low-velocity edge of the shear layer. Unfortunately the shear layer also contains a periodic stream of strong vortices that tend to create oscillations in the flame anchoring position and may lead to combustion instabilities at off-design conditions.

The modeling of LPP devices requires proper simulation of several highly unsteady features, including the acoustic pressure field, the precesion of the recirculation zone, the shedding of vortices in the shear layer, and the unsteady locations of flame anchoring. The primary recirculation zone (PRZ) has a toroidal shape and has a fluid dynamic centreline that precesses about the geometric centreline. LPP premixed flames can flash back, but only to the locations where the main fuel jet is injected into to the main air flow, so this distance is purposely minimised. The main air velocity is designed to be larger than any possible turbulent burning velocity in order to prevent flashback. One research issue is that these devices are difficult to model because partially premixed combustion often occurs. More research is needed to optimise the uniformity of the temperature field, to optimise the flame holding, and to avoid a combustion instability known as engine growl [37].

Figure 5.15 shows two examples of LES results from Yang, Menon, and co-workers [41, 42], who simulated a lean premixed gas turbine combustor that is operated on gaseous fuels. The G equation (see Section 2.2) was used to track the flame surface at the resolved scale in order to ensure that this surface has the proper wave physics; the surface must propagate normal to itself and form cusps that are due to Huygen's Principle. The chemistry was included in a premixed flamelet library that depends on a variable (G) that indicates the distance from any point to the flame surface. Partial premixing is included by defining a conserved scalar and by adding a library for nonpremixed flamelets. The idea is that a flamelet solution provides the number of kilograms per second of CO that is produced per unit flamelet area. This number is then multiplied by the predicted flame surface area per unit volume, which is deduced from the mean gradient of the G field. This yields the computed volumetric CO reaction rate. The post-flame reaction rates and residence times also are modeled to account for NO_x that is created and CO that is oxidised in the post-flame gases. This approach inherently contains assumptions about several values of means and variances that result in the prediction of how much of the fuel burns in the premixed mode and how much burns in the non-premixed mode.

The DLN and TAPS devices previously described contain 'partially premixed combustion' (PPC). Normally the term "premixed flame" implies that fuel and air are fully mixed with no stratification in the reactants. Therefore the term PPC is used to refer to reactants that are not fully mixed. That is, PPC consists of stratified premixed or stratified non-premixed flames that often are close enough to each other to interact. One example of PPC occurs when the fuel equivalence ratio varies along a reaction layer and there are regions where the reactant mixture lies within the flammability limits, so a stratified premixed flame exists. In other regions the reactants lie outside of the flammability limits so a stratified non-premixed flame exists. Where local extinction occurs along the reaction layer, two edge flames may propagate towards each other to reconnect the layer. The similar case of edge (triple) flames [43] near the base of a lifted non-premixed jet flame represents another example of PPC. In addition, PPC has been used to describe a rich premixed Bunsen flame that interacts with a downstream non-premixed flame; the latter exists at the boundary between the rich combustion products and the surrounding air.

A number of new technologies are needed to address the research challenges of modelling partially premixed combustion. Experimental methods are needed to quantify how much of the fuel burns in the premixed mode and in the non-premixed mode. To do so, measurements of the flame index would be helpful. The flame index is the product of the local gradients of the fuel and the oxidiser mass fractions. A positive index identifies a premixed flamelet whereas a negative index identifies a non-premixed flamelet. The DNS of Takeno and co-workers [44] provided computed values of the flame index. To assess such simulations, fuel gradients can be measured by use of line Raman [45] or PLIF of fuels (or fuel markers) that fluoresce. Measurements of the oxidiser gradients are still a challenge. Reaction-rate intensity can be identified by employing simultaneous OH–formaldehyde PLIF diagnostics [46], and this may provide another way to differentiate between premixed and non-premixed combustion. Other experimental challenges are the application of PIV and PLIF to the high-pressure, dense spray environment of LPP devices and the measurement of scalar gradients. At high pressures the background radiation from soot precursors presents a formidable interference for both PIV and PLIF signals. In dense sprays, PIV may incorrectly track the drop velocity instead of the gas velocity.

Models of the chemistry in LP devices [41, 42] considered two regions: the primary reaction layers (flamelets) within the flame brush and the post-flame gases, where distributed secondary reactions continue to form NO_x and to oxidise CO.

A challenge is to correctly model and measure the residence times in post-flame gases. RANS predictions of residence times often are unrealistically small because the actual path of a fluid element in a recirculation zone, for example, is longer than that predicted from the time-averaged flow field. One motivation for applying LES is to achieve realistic predictions of residence times. Another challenge is to develop methods to account for strain rate on the turbulent premixed flamelets. The leading edge of the premixed main flame in an LPP device is lifted to a position within a strong shear layer, where strain rates are large. Images of the flamelet thicknesses in LPP devices would help to determine if a thickened flamelet model or a strained laminar flamelet model is more appropriate.

Another issue is that models are needed to predict flame-stability limits. Better models would assist in achieving operation of LPP devices closer to the lean limit. To maintain safety margins, a non-premixed pilot flame often is used, so additional reductions in pollutant levels should be possible if there is better understanding of how to minimise the pilot fuel yet still stabilize the main flame. An acoustic combustion instability that is called growl can occur if LPP devices are operated outside of their operating margins. This instability has been linked to several factors: the shedding of shear layer vortices, the precessing of recirculation zones, the movement of the premixed flame anchoring position, and the time lag introduced by droplet evaporation. Current industry practice relies heavily on trial-and-error methods to avoid growl, such as varying the location and number of fuel injector ports, the gas velocities, and the fuel-drop diameter.

5.2.2 Reliable Models that Can Predict Lift-Off and Blowout Limits of Flames in Co-Flows or Cross-Flows

A major challenge is to develop models that can correctly predict flame-blowout limits. These limits determine the operating envelope of many practical combustion devices. For safety reasons, fuel and air often are introduced in separate streams in gas turbines, scramjets, rockets, direct-injection internal combustion engines, and industrial burners. Mixing often occurs in the lift-off distance that exists between the fuel injector and the downstream flame, parts of which consist of partially premixed combustion. The fundamental physics of lift-off was studied extensively for the case of a lifted simple jet flame. The flame-base region was observed to consist of a 'triple flame' by Bray and co-workers [43]. Both a lean premixed reaction layer and a rich premixed layer exist, and they create products that come into contact to form a central diffusion flame. When the flow remains laminar the lift-off heights of several laminar jet flames have been predicted accurately by the highly resolved CFD solutions of Smooke and co-workers [47], which include complex chemistry and buoyancy forces. However, practical devices introduce complexities associated with turbulence, swirl, and co-flowing or cross-flowing air. An extensive set of measurements was reported that quantifies the blowout and lift-off limits for all of these cases [48–53], but reliable computations of blowout limits are still in the developmental stage.

To identify the information that is still needed to properly model blowout limits, some scaling arguments can be of help. For a lifted, initially non-premixed jet flame, the basic criterion that controls lift-off height is that the gas velocity incident upon the lifted base region (U_b) must equal the propagation speed of the flame base (S_b) [54]. The propagation speed of the base is the product of the strained laminar flame speed (S_L) and an enhancement factor that is due to turbulence (S_T/S_L) so

$$U_g = S_L(S_T/S_L), \tag{5.1}$$

where S_T is the turbulent burning velocity of an edge flame. To proceed, estimates of the quantities in Eq. (5.1) are needed. When a simple jet is considered, the gas velocity U_g everywhere along the stoichiometric contour is argued to be nearly constant and approximately equal to ($U_F Z_{st}$), where U_F is the jet exit velocity and Z_{st} is the stoichiometric mixture fraction. Thus at any location along the stoichiometric contour the mean velocity U_g/U_F is nearly equal to the stoichiometric value (Z_{st}) of the conserved scalar. The next step is to estimate the strained burning velocity S_L. One proposed idea [55] is to use

$$S_L = S_L^0 \left(1 + \frac{c_1 K \hat{\alpha}}{S_L^{0^2}}\right)^{-1}, \tag{5.2}$$

where K is the strain rate, $\hat{\alpha}$ is the thermal diffusivity, S_L^0 is the unstretched laminar flame speed, and c_1 is a constant. This relation states that S_L decreases from S_L^0 to zero as the flame is strained. It is not the conventional relation among burning velocity, Markstein number, and Karlovitz number that often is applied in the limit of zero strain. Instead, only the limit of a large strain rate is considered; at this limit S_L approaches zero for any value of Markstein number. In some models the strain rate (K) is replaced with the scalar dissipation-rate (SDR) [55], but strain-rate and dissipation-rate concepts are similar. For a simple jet, strain rate K is proportional to the jet centreline velocity divided by the jet width; the former scales as $U_F (d_F/x)$, whereas the latter is proportional to x. Combining these relations with Eq. (5.1), while neglecting the term that is equal to unity in Eq. (5.2), and setting x equal to the lift-off height h, leads to

$$\frac{U_F/h}{S_L^{0^2}/\hat{\alpha}} = Z_{st}^{-1/2} B, \tag{5.3}$$

where

$$B = c_2 \left\{ \left(\frac{S_T}{S_L}\right)^{1/2} \left(\frac{S_L^0 d_f}{\hat{\alpha}}\right)^{-1/2} \right\}. \tag{5.4}$$

Equation (5.3) is in agreement with measured lift-off heights [48], provided that the quantity B defined by Eq. (5.4) is a constant. More research is needed to evaluate B as well as the turbulent burning velocity S_T of an edge flame. It would be attractive to replace S_T/S_L with a formula that was measured in a Bunsen flame, for example, but this may be erroneous because the triple flame may not undergo the same wrinkling and straining process as does a Bunsen flame. The previous arguments that lead to Eq. (5.3) are somewhat speculative, but they do identify the idea that the straining of the flame base can explain the measured values of lift-off heights. Note that the left-hand side of Eq. (5.3) is an inverse Damköhler number, which often has been found to correlate flame lift-off and blowout data. Blowout must occur when the lift-off height (h) exceeds the length of the region within which

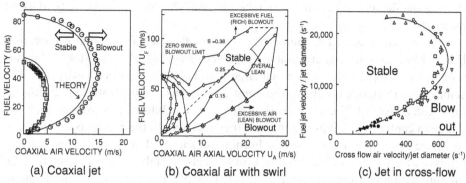

Figure 5.16. Measured flame blowout limits for the (a) coaxial jet of Dahm and Mayman [51]; circles = 5-mm inner-tube diameter, squares = 3-mm diameter. (b) Swirling coaxial jet of Driscoll and co-workers [52]; swirl number is largest for the curve on the right. (c) Jet in a cross-flow of Kalghatgi [50].

the gases are flammable. For a simple jet this length is proportional to (d_F/Z_{st}). Thus if h in Eq. (5.3) is replaced with (d_F/Z_{st}) and if U_F is replaced by the jet blowout velocity U_{BO}, then

$$\frac{U_{BO}/d_F}{S_L^{02}/\hat{\alpha}} = c_3 Z_{st}^{-3/2} B, \qquad (5.5)$$

which is also in reasonable agreement with experimental data [49] if the quantities B and c_3 remain constant. For example, for a typical methane–air flame, the product of the constants c_3 and B in the preceding relation, is 0.024.

A more complex case is that of a coaxial jet flame. Consider a central fuel tube that is surrounded by a larger-diameter tube that contains a coaxial stream of air. This geometry is of practical importance because if coaxial air is forced across the diverging boundaries of the central fuel stream, then the fuel–air mixing process will be more intense and a shorter combustion zone will result. For the coaxial geometry some measurements of flame blowout limits were correlated in [51] by

$$\frac{U_{BO}^*/d^*}{S_L^{02}/\hat{\alpha}} = \frac{c_4}{Z_{st}^*}, \qquad (5.6)$$

where the parameters d^*, U_{BO}^*, and Z_{st}^* are the effective diameter, jet exit velocity at blowout, and stoichiometric mixture fraction of jet fluid for an equivalent single jet that has the same momentum flux and mass flux as the coaxial jet. Simple relations for $d^* U_{BO}^*$ and Z_{st}^* are given in [51] in terms of the diameters and gas velocities of the inner and outer coaxial jets; for example,

$$d^* = d_1 \left(\frac{\rho_2}{\rho_1}\right)^{-1/2} \frac{1 + \rho_2 U_2(d_2^2 - d_1^2)/(\rho_1 U_1 d_1^2)}{[1 + \rho_2 U_2^2(d_2^2 - d_1^2)/(\rho_1 U_1^2 d_1^2)]^{1/2}}. \qquad (5.7)$$

The quantity c_4 is the only empirical constant that was required to achieve good agreement between Eq. (5.6) and the measured blowout limits that are seen in Fig. 5.16(a). This figure shows that coaxial air (with no swirl) always is destabilising. The blowout curve always is below the location where the curve intersects the y axis, which corresponds to a jet with no coaxial air. The coaxial air does significantly shorten the flame, but care must be taken to limit the coaxial air velocity to avoid blowout.

When swirl is added to the outer coaxial air flow, a stable region is measured [52] to have the same general shape as the zero-swirl case, but the stable region is seen in Fig. 5.16(b) to become much larger and taller than the zero-swirl case. The stable region extends far above the point on the y axis that identifies the blowout limit with zero swirl. Therefore swirl usually is very stabilising for two reasons. Swirl creates a central recirculation zone that transfers heat to the reactants upstream of the flame base, and this increases the burning velocity S_L^0. It also creates low-velocity regions, which reduces U_g in Eq. (5.1), and this also is stabilising, providing that the mixture within the low-velocity region is flammable. For the case of a jet flame in a cross-flow, the measured blowout curves [50] are shown in Fig. 5.16(c); the stable region has a shape that is similar to those of the last two examples. However, the cross-flow of air is very destabilising because the blowout limits are lower than the case of zero-cross-flow velocity, which is the extrapolated intersection of the upper part of the curve with the y axis. A fuel jet in a cross-flow occurs in lean premixed gas turbine combustors as well as in scramjets. Scramjets typically inject fuel from a port in the wall jet to avoid the melting of fuel injector struts if struts were placed in the hot airstream.

Several models can correctly predict the blowout limits of a simple turbulent jet flame, but more research is required to make such models robust enough to correctly predict blowout limits when coaxial air, swirl, or cross-flow is added. The RANS model of Bradley and Gu simulates the strained premixed turbulent flame that occurs at the base of a lifted jet flame [56]; their approach is consistent with experimental observations. Their model does correctly predict a linear increase in the lift-off height as the jet velocity is increased. Devaud and Bray [57] simulate lifted jet flames by employing the conditional moment closure (CMC) method, whereas in other studies DNS [58] and LES [59] were applied. However, there are few systematic studies that predict the measured values of lift-off height as the fuel velocity or injector diameter is systematically varied, which is needed for model validation. The preceding scaling analysis suggests that the flame base will lift to a downstream location where the strain rate is sufficiently reduced; a realistic model should be expected to predict that the strain rate decreases in the downstream direction in a manner that is consistent with experiments. Proper validation will also require measurements of strain rates on the lifted base such as those of [60] and [61]. Some other theories are reviewed in [61] that do not assume that the strain rate governs lift-off. Instead they assume that the flame lifts off in order to find a location where the integral scale is larger [49] or where the flame is less likely to be subjected to coherent vortices in the shear layer.

5.2.3 New Technology in Measurement Techniques

New measurement technology is now making it possible to visualise the individual premixed flamelets, the individual turbulent eddies, and the interactions that occur between the two. The eddy–flamelet interaction represents the fundamental building block of premixed turbulent combustion, and it is important to measure it for several reasons. This interaction determines the rate of stretching of the flame surface, and

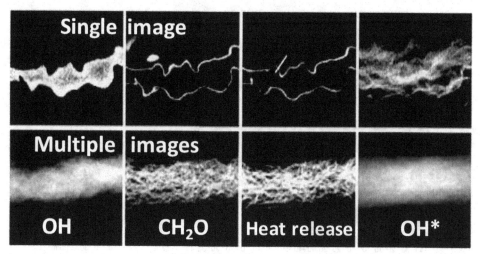

Figure 5.17. Reaction layers and HRRs within a turbulent counterflow flame imaged by Ayoola et al. [46] using simultaneous formaldehyde–OH PLIF diagnostics.

thus it controls the amount of the surface area of the reaction layers as well as the amount of local extinction that may occur. Because eddy–flamlet interactions control the rate of strain that is exerted, they affect the NO_x and CO chemistry. In addition, acoustic instabilities (see Chapter 3 for flame instabilities) often are associated with premixed combustion processes; these instabilities can be triggered when a periodic stream of shed vortices causes periodic oscillations in both the flame surface area and the heat release rate (HRR). Therefore there is a strong motivation to improve the spatial and time resolution of current imaging techniques in order to better image and quantify the eddy–flamelet interactions.

A quantitative measure of the strength of each reaction layer is the HRR, which now can be recorded using simultaneous formaldehyde–OH PLIF diagnostics [46]. This is an important advance because the local HHR quantifies the local flamelet chemistry as well as the degree of local extinction. The HRR was demonstrated to be proportional to the product of the concentrations of formaldehyde and OH, as these two species react to form HCO in the major chemical pathway that releases heat. The HRR can be measured by recording PLIF signal intensities (I) for formaldehyde (CH_2O) and OH and applying

$$HRR = I_{CH_2O} I_{OH} f(T). \tag{5.8}$$

In general the function $f(T)$ is a complicated function of gas temperature; however, this function was shown in [46] to be a constant when the proper fluorescence excitation and emission wavelengths were selected. Figure 5.17 illustrates some images of the OH, formaldehyde (CH_2O), HRR, and OH* chemiluminescence in a turbulent premixed counterflow burner. A different way to image flamelets is to record CH fluorescence [62, 63] because CH is a short-lived radical that exists only within the thin reaction layers.

Another advance is the use of phase-locked diagnostics to simultaneously image the 2D velocity field and flame structure in combustors that undergo periodic oscillations. A combustion instability occurs in the partially premixed gas turbine

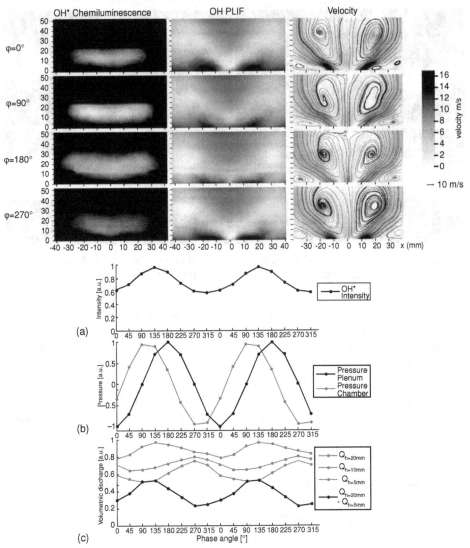

Figure 5.18. Simultaneous phase-locked measurements of velocity field, flame properties, and gas pressure by Stohr et al. [64] to understand acoustic instabilities in a partially premixed gas turbine model combustor.

model combustor of [64]. Lasers are repeatedly pulsed at one particular phase angle φ of the sinusoidal pressure signal, and the fluorescence signal is phase averaged. Some results are plotted in Fig. 5.18. In this figure a gaseous fuel jet issues upwards and is surrounded by coaxial swirling air. The velocity field contours show that a strong central recirculation zone is formed. Simultaneous images were recorded of OH PLIF (to quantify the HRR), the velocity field using PIV, and the gas pressure. The sinusoidal curves in Fig. 5.18 provide the phase-lag and time-response information that is needed to model the combustion instability. Another recent study [65] demonstrated that simultaneous PLIF and PIV images can be recorded at a exceptionally high framing rate of 12 000 frames/s using a diode-pumped YAG laser. Biacetyl vapour was seeded into the fuel to act as a fuel marker and was excited at a wavelength of 355 nm. It also was shown that OH PLIF could be imaged at 1000 frames/s using an Nd:YLF laser [66].

Improved diagnostics also can provide answers to important questions that initially were raised by Damköhler, Markstein, and other pioneering researchers many decades ago. For conditions that fall within what is now called the laminar flamelet regime, they proposed the idea that the primary role of turbulence is to increase the turbulent burning rate by increasing the surface area of the reaction surface. They assumed that the wrinkled surface continues to convert reactants to products locally at a rate (per unit surface area) that is similar to that of a laminar flame. Markstein proposed a relation between the curvature of the reaction surface and its local propagation speed. Several of these ideas were incorporated into the balance equation for the flame surface density. However, this equation requires a model for the source term that describes the basic way by which eddies stretch the flame. A relation for this source term was proposed, but it is based on theoretical speculation and not on hard experimental evidence. The eddy–flamelet interaction is described by the stretch efficiency function (Γ_K) that relates the mean stretch rate (K) to the local turbulence intensity (u_{rms}) and the integral scale of turbulence (Λ) in the following way:

$$K = \Gamma_K \frac{u_{\mathrm{rms}}}{\Lambda} \tag{5.9}$$

Computations were performed [67] to estimate Γ_K by first assuming that turbulence can be represented by an ensemble of vortex pairs that have different sizes. The axis of each canonical vortex pair is assumed to be always perpendicular to the flame surface, and all vortices are counter-rotating such that the vortex pair exerts only a positive stretch rate. Numerical simulations of many laminar canonical vortex-pair–flame interactions were ensemble averaged to estimate the stretch efficiency function, and these computed values of Γ_K are used in several LES and RANS models.

Only recently has it been possible to observe eddy–flame interactions within fully turbulent premixed combustion. Steinberg et al. [68–70] were able to image eddy–flamelet interactions; their goal was to provide measured values of Γ_K that do not require any assumptions about the eddy structure. It was shown that the geometry of the actual eddies often did not correspond to that of the canonical vortex pair that was assumed to occur in previous calculations. Cinema-stereo PIV diagnostics provided high-speed movies of the eddy–flamelet interactions that were recorded at 1100 frames/s. One interaction is shown in Fig. 5.19. Eddies are identified by contours of vorticity; these eddies move to the right and upwards across the flame, which is the black line. Exceptionally good time resolution (0.66 ms/frame) and a spatial resolution of 280 μm were achieved, which resolves the Taylor scale, which was 0.9 mm.

3D movies of the interactions also were reported in [69], but for reduced spatial and temporal resolution. To obtain 3D images of the vorticity field, two simultaneous cinema–PIV systems were employed that required horizontal and vertical laser sheets, respectively. Taylor's hypothesis was applied as eddies moved vertically upwards throught the horizontal sheets. Measurements from the vertical sheets indicated that the frozen-flow Taylor hypothesis was an accurate assumption. Typical

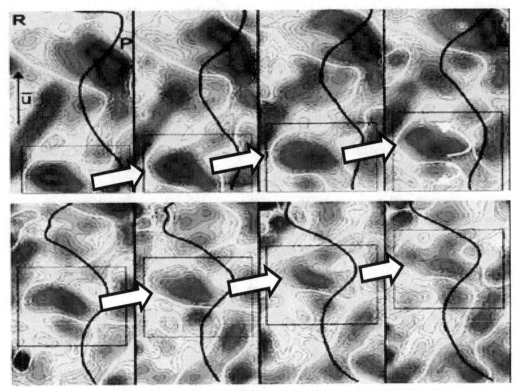

Figure 5.19. High-speed movies of eddy–flamelet interactions in the fully turbulent premixed Bunsen flame of Steinberg et al. [69]. Time between images is 0.66 ms. Field of view is 4 mm × 8 mm. The Black line is the flame surface. Contours of vorticity vary from -1000 s^{-1} (blue) to 1000 s^{-1} (red). Flow is from bottom to top. (See colour plate.)

3D results are seen in Fig. 5.20, which indicate that horseshoe-shaped vortical structures are common. The toroidal structure shown on the right also occurs, but is less common.

Several efficiency functions now have been measured [70] that are useful inputs for LES subgrid models that include the flame surface density balance equation. The mean stretch rate (K) is separated into two components: the subgrid strain rate on the flame $(a_{t,\text{sg}})$ and the component associated with flame curvature $(K_{c,\text{sg}})$. The measured relations that represent these subgrid quantities are given in Table 5.1. The subgrid strain rate on the flame was found to correlate best with the resolved scale fluid dynamic strain rate of the reactants (S'_Δ).

5.2.4 Unresolved Fundamental Issues

It is clear that some fraction of future research in the field of turbulent premixed combustion will be devoted to resolving a number of lingering questions that were studied extensively in the past, but have not yet been answered. The reason why additional research now might be expected to be fruitful is because of recent advances in DNS and time-resolved laser diagnostics and a new understanding of complex chemistry. If we were to list some fundamental questions that still need answers,

Table 5.1. *Measured relations [70] for subgrid stretch rate (K_{sg}), which is the sum of subgrid strain rate ($a_{t,sg}$) and the component of stretch that is associated with flame curvature ($K_{c,sg}$)*

Equations for the subgrid strain rate	Equations for the subgrid curvature stretch rate
$$a_{t,sg} = \overline{\Pi} S'_\Delta$$	$$\kappa_{c,sg} = \overline{\Upsilon} \Xi$$
$$S'_\Delta = \langle \widehat{S'_{ij} S'_{ij}} \rangle^{1/2}$$	$$\Xi = \frac{A_{t,sg}}{A_\Delta}$$
$$\overline{\Pi} = \overline{\Pi}_p \overline{P}_p^a - \overline{\Pi}_n \left(1 - \overline{P}_p^a\right)$$	$$\overline{\Upsilon} = \overline{\Upsilon}_p \overline{P}_p^D - \overline{\Upsilon}_n \left(1 - \overline{P}_p^c\right)$$
$$\overline{\Pi}_p = \left\{ b_1 \exp\left[-\frac{b_2}{\left(S'_\Delta \frac{\delta_l^0}{s_l^0}\right)^{b_3}} \right] \left(\frac{\Delta}{\delta_l^0}\right)^{b_4} \right\} \left(\frac{\Delta}{\delta_l^0}\right)^{-1/3}$$	$$\overline{\Upsilon}_p = e_1 \exp\left[-\frac{e_2}{(\Xi_{sg}-1)^{e_3}} \right] \left(\frac{\Delta}{\delta_l^0}\right)^{e_4}$$
$$\overline{\Pi}_n = \left(\left\{ c_1 \exp\left[c_2 \left(S'_\Delta \frac{\delta_l^0}{s_l^0}\right)^{c_3} \right] + c_4 \right\} \left(\frac{\Delta}{\delta_l^0}\right)^{c_5} \right) \left(\frac{\Delta}{\delta_l^0}\right)^{-1/3}$$	$$\overline{\Upsilon}_n = f_1 \exp\left[-\frac{f_2}{(\Xi_{sg}-1)^{f_3}} \right] \left(\frac{\Delta}{\delta_l^0}\right)^{f_4}$$
$$\overline{P}_p^a = d_1 \exp\left(-\frac{d_2}{\left(\Delta/\delta_l^0\right)^{d_3}} \right) + d_4$$	$$\overline{P}_p^c = g_1 \exp\left(-\frac{g_2}{(\Xi_{sg}-1)^{g_3}} \right) - g_4$$

Notes: The constants in the relations are given in [70]. S' is the subgrid fluid dynamic strain rate, A is the flame area, and Δ is the filter scale. P is the measure probability, Π is the strain-rate transfer function, Υ is the flame curvature transfer function, δ_L^0 is the unstretched laminar flame thermal thickness, and Ξ is the wrinkling factor.

the list might include these: How well can we predict the turbulent burning velocity of fully premixed (see Section 2.3 for recent developments) and partially premixed flames with DNS, LES, or CMC models? What are the measured limits to the flamelet regime, that is, over what range of conditions can flamelet-based LES submodels be expected to remain valid? Does the turbulent burning velocity vary as the square root of the integral scale of the turbulence, as has been postulated in the past? Can we observe and understand the physical process by which turbulent eddies strain and wrinkle the reaction layers in order to apply concepts that are sometimes called the theory of flame stretch [71] to turbulent flames? Can we correctly simulate

3D vortex structure from cinema PIV

Figure 5.20. High-speed cinema PIV movies that record 3D vorticity contours during an eddy–flamelet interaction in the fully turbulent premixed Bunsen flame of Steinberg et al. [69]. Two cinema PIV systems are used that employ vertical and horizontal laser sheets, respectively. Field of view is 4 mm × 8 mm. The contours of vorticity shown corresponds to $700\ \mathrm{s}^{-1}$.

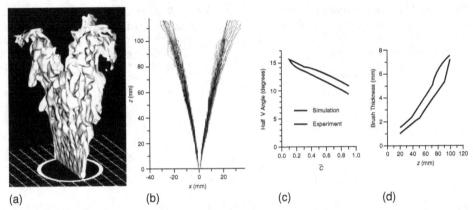

Figure 5.21. DNS of Bell et al. [76] of the V-flame experiment of Cheng and Shepherd: (a)
Computed instantaneous reactedness, (b) computed mean reactedness contours, (c) computed
and measured half-angle of the mean contour, where reactedness is 0.5, which is proportional
to the turbulent burning velocity, (d) computed and measured brush thicknesses.

these eddy–flame interactions using LES subgrid models that employ, for example,
a stretch efficiency function [72]. How do turbulent burning velocities depend on
elevated gas pressures and very large Reynolds number conditions? A review of
some of these questions is contained in [73–75].

To begin such a discussion, it is first noted that DNS results of Bell et al. [76, 77]
demonstrated that, using no modeling constants, it is now possible to predict accurate
values of the turbulent burning velocity of the V-flame experiment of Cheng and
Shepherd [76] and the 2D slot Bunsen experiment of Filatyev et al. [63] for realistic
(but moderate) Reynolds numbers. In [63] the reactants were methane and air, and
the Bunsen burner exit dimensions were 50 mm × 25 mm; the mean velocity of
the reactants was 3 m/s. The conditions in [63] and [76] represent fully turbulent
flames, because the burning velocities (S_T/S_L^0) varied between 2.5 and 4.0, but the
turbulence level was only of moderate intensity, because values of u_{rms}/S_L^0 were
1.1 and 0.8. The spatial resolution of 60 μms of the adaptive mesh in the DNS
was sufficient to resolve the Kolmogorov scale of both experiments, and the DNS
included a 20-species, 84-reaction chemistry mechanism.

Results of the DNS for the V-flame are seen in Fig. 5.21. The computed half-angle
$(\theta/2)$ of the brush agrees with the experiment, as does the turbulent burning velocity
(S_T/S_L^0), which is $(U/S_L^0)\sin(\theta/2)$, where U is the velocity of the reactants. The
normalised burning velocity of the V-flame is 3.95 for the DNS and 3.62 cm/s for the
experiment. Figure 5.22 displays some DNS results for the 2D slot Bunsen flame. The
thick line represents the $\bar{c} = 0.5$ contour, from which the turbulent burning velocities
(S_T/S_L^0) were determined to be 2.45 and 2.55 for the DNS and the experiment,
respectively. The brush thicknesses of the Bunsen experiment and the DNS were
also in agreement.

It is now generally understood that turbulent premixed combustion is geometry
dependent, because turbulent burning velocity depends on the amount of flame
surface wrinkling, and this wrinkling process differs for each burner geometry. It is
useful to define the following categories of turbulent premixed combustion: Bunsen
flames, V-flames, counterflow flames, and spherical flames. The best practice is to

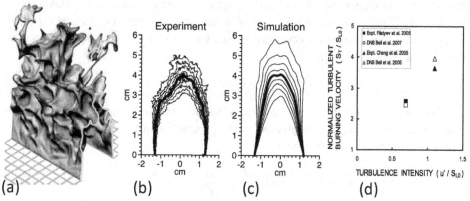

Figure 5.22. DNS of Bell et al. [77] of the slot Bunsen experiment of Filatyev et al. [63]: (a) Computed 1684 K isotherm; (b, c) measured and computed mean reactedness contours (black line: reactedness = 0.5); (d) measured and computed turbulent burning velocities of the experiments of Filatyev et al. [63].

compare computations with experiments only if the burner geometry is the same for both. Similarly, it is best to correlate turbulent burning velocity data only from burners of similar geometry. This fairly obvious rule has not been followed in the past. For example, a spherical flame becomes more wrinkled in time; Bunsen and V-flames become more wrinkled as the distance from the anchoring location increases. Bunsen and V-flames have an imposed mean velocity of the reactants, whereas spherical flames do not. Therefore it does not seem logical to expect that burning velocities that are measured for different types of burners will correlate, even if the turbulence level and integral scales of the reactants are matched (see Section 2.3 for recent developments.)

LES and RANS simulations also have accurately predicted values of the turbulent burning velocity, but unlike DNS, some modelling constants must be specified. LES predictions of Charlette et al. [78] appear in Fig. 5.19(a), while the RANS result of Peters [79] and Duclos et al. [80] appear in Figs. 5.23(b) and 5.23(c). It is noted that all three sets of curves display a 'bending' that is seen in many experiments,

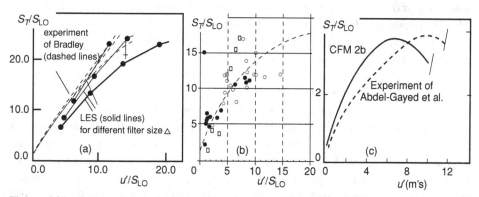

Figure 5.23. Comparisons of turbulent burning velocities predicted by (a) the LES of Charlette et al., [77] (b) the RANS/G-Equation approach of Peters [78] (dashed line), and the RANS–coherent flamelet model of Duclos et al. [79] with measurements of Abdel-Gayed et al. [80].

and all three models yield values that are similar to experiments of Bradley and co-workers [81]. The LES of Charlette et al. assumed that subgrid turbulent diffusion is represented by a Smagorinsky relation (see Chapter 1), and the mean volumetric reaction rate $\overline{\dot{\omega}}$ was set equal to

$$\overline{\dot{\omega}} = \rho_u S_L^0 I_{0,\text{SG}} \Sigma_{\text{SG}}. \qquad (5.10)$$

To model Σ_{SG} the flame surface density balance was considered and the diffusion and convection of Σ_{SG} were neglected, so the production and destruction terms were equated. Equation (5.10) becomes:

$$\overline{\dot{\omega}} = \overline{\rho}(1 - \overline{c}) \exp\left(\frac{-E}{R\overline{T}}\right)\left(\frac{A}{F}\right)\left(1 + \frac{\Gamma_K u'_\Delta}{S_L}\right)^\beta, \qquad (5.11)$$

where $\overline{c}$ and $\overline{T}$ are the resolved progress variable and temperature, respectively. Another relation was included to relate subgrid turbulent velocity u'_Δ to the resolved vorticity field, and the exponent β was set to 0.5 to achieve best agreement with measurements. A is the Arrhenius pre-exponential constant, F is a thickening factor, and Γ_K is the stretch efficiency function [67, 78]. A different LES subgrid model was developed by Hawkes and Cant [82], who included Eq. (5.10) but used the surface density balance equation instead of Eq. (5.11). Their computed values of S_T/S_L^0 were as large as 2.6 for turbulent flame in a box.

Another LES subgrid model involves the G equation, as described by Pitsch et al. [72]. The height of a Bunsen flame was computed, and it differed by only 6% from a corresponding experiment. Because flame height is inversely proportional to turbulent burning velocity, and because the LES does predict the correct flame height, this implies that the LES does accurately predict the experimental burning velocity. The resolved-scale wrinkled interface is thicker and less wrinkled than the true interface, which cannot be resolved. The resolved interface is forced to propagate at a speed of $S_{T,\text{SG}}$, assumed to be equal to

$$\frac{S_{T,\text{SG}}}{S_L^0} = 1 + b_3\left[\frac{\mathcal{D}_T}{\mathcal{D}}\left(1 + \frac{b_4 S_L^0 \Delta}{u'_\Delta \delta_L^0}\right)^{-1}\right]^{1/2}, \qquad (5.12)$$

where $\mathcal{D}_T$ and $\mathcal{D}$ are the turbulent and molecular diffusivities, respectively, Δ is the LES filter scale, and b_3 and b_4 are constants. It was argued that this function behaves properly at the thin-flamelet limit where the laminar flame thermal thickness, δ_L^0, is much smaller than Δ and $S_{T,\text{SG}}$ should be proportional to (u'_Δ/S_L^0). It was also argued that it behaves properly at the thick-flamelet limit, where δ_L^0 is larger than Δ and $S_{T,\text{SG}}$ should be proportional to $(\mathcal{D}_T/\mathcal{D})^{1/2}$.

Another unresolved question is related to the limits of the flamelet regime; these limits have not been measured accurately, and current information is based only on theoretical scaling arguments. It would be useful to know the range of conditions over which flamelet models are applicable. Peters [83] defines a laminar flamelet to be a layer that contains a preheat and reaction zone, has gradients in the normal direction that are significantly larger than gradients in the tangential direction, and has a molecular diffusivity within it that is much larger than its turbulent diffusivity. Peters also defines another regime that occurs when the preheat zone is broadened but there is no broadening of the reaction zone. In this regime small eddies fit in the preheat zone but do not fit into the reaction zone. This regime has been called the regime of the 'broadened preheat zone/thin reaction zones' or 'thickened layers'.

There is experimental evidence that preheat layers of turbulent flamelets vary from about 0.5 to about 4 times the thickness of the unstretched laminar preheat layer; a number of studies are reviewed in [72]. Unfortunately there has been no experimental verification that turbulence is the cause of the few cases of layer broadening that have been recorded. This would require the imaging of vortices that are smaller than the layer thickness, which is not yet possible. Measurements that simply show that the layer is thicker than the unstretched flame thickness are not conclusive evidence that turbulence is the cause of the broadening. A laminar flamelet may be 3–4 times thicker than an unstrained laminar flame, not because of turbulence, but because of the strain rate that is applied, because of the orientation of the layer with respect to a laser light sheet, or because several layers have merged together to give the appearance of a thick reaction zone. There is almost no hard evidence that turbulence alone can cause distributed reaction zones. Documented examples of distributed reactions usually involved the rapid heating of reactants in rapid-compression devices such as HCCI engines and shock tubes. In these cases it is the rapid compression (and not the large turbulence level) that raises the temperature of the reactants above the ignition temperature and thus is the cause of the distributed reactions. However, improved spatial resolution of measurement techniques is expected to provide measurements of the limits of the flamelet regime in the future. This can be achieved by applying PLIF and Rayleigh scattering imaging diagnostics that have thinner laser sheets and have optics that zoom in to resolve smaller scales.

5.2.5 Summary

New designs of practical combustion devices are taking advantage of the benefits of partial premixing, but there remains a gap in our ability to model these types of flames. In real devices the fuel and air often are introduced in separate streams; examples are fuel jets introduced into cross-flowing or co-flowing airstreams. A mixing region is followed by a downstream partially premixed combustion region. This avoids flashback problems and results in low levels of NO_x, CO, and soot. However, partial premixing increases the probability of flame blowout and the occurrence of acoustic instabilities compared with conventional non-premixed combustion. LESs and DNSs have made impressive progress in simulating some important basic concepts, including the wrinkling of flame sheets, the local propagation of the interface where the premixed flames exists, and the unsteady flamelet modelling of the interface region where the non-premixed flames exist. However, proper assessment of models still requires new experimental information. Measurements are needed of the instantaneous structure of the reaction layers, the flame index (which indicates whether the local combustion is premixed or non-premixed), and the stretch efficiency function (which is in the source term of the flame surface density equation in LES subgrid models). One new research direction is the effort to achieve improved experimental spatial resolution in order to zoom in and image the fine-scale flame structure including scalar gradient, and the eddy–flame interactions. Another new direction is the development of kilohertz PLIF and PIV imaging methods to record high-speed movies of combustion instabilities and eddy–flame interactions. Some

DNS results were presented to show that DNS has successfully simulated the experimental data that was obtained in laboratory-scale turbulent premixed V-flames and Bunsen flames.

5.3 Future Directions in Modelling
By K. N. C. Bray and N. Swaminathan

The balance between the various different energy sources that will be in use globally in the years ahead cannot be predicted with confidence. We do not know what fraction will be produced from biofuels or from tar sands, for example, or the nature of the fuels to be derived from these sources, or the extent to which electric vehicles will take over road transport, or the means by which the resulting increased demand for electrical power will be met. What *is* apparent is that, although combustion processes will continue to provide a major source of heat and power for the foreseeable future, the specifications of future fuels are likely to be more variable than at present, creating a range of issues [84], such as flame position, stabilisation, flashback, combustion oscillations, etc., for gas turbines and engine knock and misfires for reciprocating engines, to deal with while optimising combustors or engine designs. At the same time, growing legislative and financial pressures must be anticipated, which will encourage more efficient use of fuel and will further constrain emissions of pollutants and greenhouse gases. In these uncertain circumstances, designers and operators of combustion systems will require the most accurate and reliable ways to predict the influence of changes in fuel specification, operating procedure or plant design on the performance of their equipment, and such predictions must be provided by models and computer codes that are based on a fuller understanding of the physics and chemistry of combustion and the amount of empiricism involved in the modelling must be minimised. Furthermore, the choices made by designers of combustion systems are strongly influenced by policy on energy and environment for which sound scientific advice is imperative [85].

In this closing section, we first identify some developments of turbulent combustion modelling capabilities that appear to be desirable. Strengths and potential weaknesses of existing models are then discussed, in order to assess the ability of each model to meet the challenge of these future requirements and, finally, some unsolved problems are identified and indications of future directions are proposed. Inevitably these conclusions are based on the personal judgement of the present authors; other recent reviews of the present and future status of the subject may be found in [86–90].

5.3.1 Modelling Requirements

1. **Chemical Kinetics:** A detailed and accurate chemical kinetic mechanism, appropriate for the fuel–oxidiser system under consideration, is an essential feature of any model that aims to predict pollutant emissions, auto-ignition, extinction, or flashback phenomena. Practical fuels may contain heavy hydrocarbon components or biofuel fractions for which appropriate data is not always available, and validated mechanisms are often too large to be directly included in turbulent combustion simulations. On the other hand, numerically stiff algebraic

rate expressions of reduced mechanisms available for simple fuels, for example, despite involving fewer chemical species, present their own issues for simulations.

The accurate prediction of NO_x concentrations is a continuing challenge, partly because of the strong sensitivity to temperature of relevant reaction rates, so it may sometimes be necessary to take account of heat losses that are due to radiation, as well as temperature and composition fluctuations that are due to incomplete mixing. If pockets of fuel-rich combustion can occur, which is inevitable in practical engines, then the kinetics of soot formation and oxidation is relevant. Processes of auto-ignition and the quenching of combustion that is due to heat losses must also be included in a complete chemical kinetic specification. An accurate and reliable prediction of CO also remains challenging, as noted in Subsection 4.2.9. The formation of nanoparticles [91] and polyaromatic cyclic hydrocarbons [92, 93] must also be included in view of the impact of combustion on human health [94, 95] and environment [85, 96]; see Section 4.1 for the importance of particulate matters in automobile engine emissions legislation.

2. **Flows with Non-Uniform Composition:** As discussed in earlier sections of this Chapter and Chapter 4, most practical applications of lean premixed combustion are expected to involve incomplete fuel–air mixing, and partial mixing with cooled combustion products is also common. An important special case is the preparation of a combustible mixture through liquid-droplet evaporation, mixing, and the possibility of its simultaneous autoignition.

3. **Influence of Liquid Spray:** Although a liquid fuel spray can be expected to be fully evaporated by the time it reaches a combustion zone, its atomisation, evaporation, and mixing can have a strong influence on auto-ignition and flashback (see Subsections 4.2.9, 4.3.8, 4.3.9, and 5.1.3), and also on the degree of inhomogeneity of combustion.

4. **Flows with Heat Losses:** It may become necessary to take account of radiation heat losses in some systems in order to improve the accuracy of NO_x predictions, particularly if regions of fuel-rich combustion can occur in considerable portions of the combusting flow. Modelling of turbulence–radiation interaction is also of interest, and recent developments on this topic can be found in [97–101]. The quenching of combustion at cold walls and resulting in emissions of unburned hydrocarbon species are also of interest, as explained in Subsections 4.1.3 and 4.1.4.

5. **Unsteady Effects:** Difficulties in simulating the development of certain combustion instabilities were referred to in Section 3.1. The possible occurrence of auto-ignition, flashback, or blowout of combustion is of obvious concern in gas turbine design, as described in Subsection 4.2.8. Auto-ignition is also central to the modelling of HCCI reciprocating engines (see Section 5.1) and mild or flameless combustion (see Subsection 5.1.3). Predictions of flame acoustics–combustion dynamics require knowledge of the *fluctuating* heat release rate, $\dot{\Omega}'$, as noted in Subsection 3.3.1.

6. **Non-Conventional Processes:** Given a high probability for variability in future fuels, non-conventional methods such as catalytic combustion may become more widely used in order to improve combustion characteristics and stability over a range of operating conditions. The application of catalytic combustion for gas

turbines running on syngas/hydrogen is explored in [102]. Catalytic combustion is an important topic although its treatment does not belong within the scope of this book as it involves heterogeneous reactions and catalysis. Interested readers can find a detailed treatment of the subject in other references, for example [103–105].

5.3.2 Assessment of Models

It seems safe to assume that the speed and power of computing systems will continue to increase, while the unit cost of calculations will go on falling, as Moore's law predicts [106]; in these circumstances, the attractions of LES are likely to grow as time passes. Nevertheless, many issues [107] in LES remain unanswered, for example the suitability of LES to predict the law of the wall in wall-bounded flows [108, 109] and the non-trivial influence of grid [110], all of which will be resolved eventually. However, we must acknowledge that RANS calculations will remain substantially less expensive than LES, so a continuing significant demand for both types of simulation is anticipated. The high cost of DNS, particularly with realistic reaction-rate mechanisms, will continue to rule out its use for practical problems for the foreseeable future, although its value for model development is widely recognised. We discuss RANS models first, and then LES.

A. RANS Models
1. **Laminar Flamelet Models:** Here we include the flame surface density models and G-equation formulations, introduced in Section 2.2, as well as simple presumed PDF models, introduced in Sections 1.7.2 and 2.1, all of which can take account of detailed chemical reaction-rate mechanisms in tabulated form; see Subsection 2.2.3. An advantage of these models is that they are relatively inexpensive to run, although cost and complexity are increased if non-gradient turbulent transport processes are to be taken into account, as described in Section 2.1. Predictions of these models are sensitive to the SDR, $\tilde{\epsilon}_c$. Their extension to describe partial premixing is not straightforward, particularly if regions of fuel-rich burning can occur in diffusion flamelets, and auto-ignition and quenching do not belong within this framework. The prediction of CO using these methods is not as good as one would wish. However, these methods work well for species with fast time scales. Additional modelling strategies must be sought for these methods to improve the prediction of species with slow time scales.
2. **SDR Models:** These models, described in Section 2.3, can be applied in several different ways. At sufficiently large values of the Damköhler number, the mean reaction rate $\bar{\omega}_c$ is proportional to the scalar dissipation $\tilde{\epsilon}_c$ [see Eqs. (1.36) and (2.9)], and the algebraic approximations for $\tilde{\epsilon}_c$, introduced in Subsection 2.3.5, then lead directly to algebraic models for $\bar{\omega}_c$. Alternatively, the $\tilde{\epsilon}_c$ transport equation derived and modelled in Subsection 2.3.4 can be interpreted as an equation for $\bar{\omega}_c$ or $\overline{\Sigma}$ [111]. Used in these ways, SDR models can be implemented in conjunction with chemical-reaction-rate data in tabulated form, and have the same advantages and limitations as the laminar flamelet models previously discussed. Some success of this approach is shown

in [112] and [113]. More generally, SDR models provide a promising route to a more accurate and physically based description of scalar dissipation phenomena, central physical processes in turbulent combustion, as an input to other combustion models (see [114]).

3. **Transported PDF Models:** A great advantage of this approach, described in Section 2.4, is its flexibility. Chemical reaction terms are included in closed form, and detailed mechanisms can be included; auto-ignition and quenching processes can be simulated, as can the effects of incomplete mixing and radiation. If velocity components are represented as additional stochastic variables, both gradient and counter-gradient scalar transport can be predicted. Models of this type are more expensive to run in comparison with flamelet models, and cost constraints usually lead to the imposition of simplified reaction rate mechanisms. Transient processes would also be expensive to follow. A more fundamental modelling challenge is met at sufficiently large values of the Damköhler number, when reaction zones resemble laminar flames, as the scalar dissipation and pressure gradient fluctuation terms in Eq. (2.145) are then linked to chemical reaction rates.

4. **Other RANS Models:** The CMC approach is well advanced for non-premixed flames. Its development for premixed flames is hindered mainly by issues related to the modelling of the conditional SDR of the progress variable, as noted in Chapter 1. A preliminary calculation using CMC with a classical SDR model [115] seems encouraging, especially for slow time-scale species. Similarly to transported PDF models, the CMC method can be used to predict auto-ignition and flame quenching with additional second-order modelling. However, this method can be computationally expensive as it involves $(N+1)$ CMC equations, each with four independent variables for a 3D turbulent flame involving $(N+1)$ scalars. However, the computational expense can perhaps be reduced by using cross-stream averaged CMC equations, as has been done for non-premixed flames [116], but the applicability of this strategy for premixed combustion has yet to be studied.

B. LES Models

1. As noted in Subsection 1.7.3, RANS modelling can be extended to LES and thus the preceding comments apply equally to the LES approach. If sufficiently fine numerical grids with small filter widths are used, then LES solutions are found to be relatively insensitive to subgrid scale models, which is clearly attractive. Also, transient flame processes such as auto-ignition, flashback, and lift-off height can be predicted, and details of combustion oscillations can be obtained, as discussed in Section 3.3. However, the numerical discretisation scheme and the nature of the solver can affect the details of the simulated flow and flame, and the high computational cost of LES limits its regular use in industry. This implies that there can be a strong incentive to improve subgrid closures, so that larger filter scales and coarser meshes can be used. Significant advances have yet to be made on this front. A review [87] of industry-standard approaches to the numerical simulation of gas turbine combustion has identified that the unsteady RANS (URANS) methodology has the potential to predict combustion-induced oscillations at a moderate computational cost. It is also unclear how well dynamic LES models perform

in predicting the SDR in premixed flames. Issues relating to the recovery of law-of-the-wall behaviour [108, 109] in LES also raise uncertainties in circumstances in which the heat and momentum fluxes through the combustor walls need to be evaluated as design variables.

5.3.3 Future Directions

Turbulent combustion is a socioeconomically important topic, and its modelling will continue to be an active area of research as long as we go on burning fuel to meet society's energy needs in a controlled manner. All established combustion models will continue to be developed to improve their accuracy, reliability, and capabilities. The following are some of the topics that are believed to need immediate attention in order to assist the development of lean premixed combustion equipment.

1. Detailed chemical kinetic mechanisms applicable to present and future fuels are required. These mechanisms must be able to accommodate variability in fuel composition.
2. There is an immediate requirement to identify ways to improve CO predictions because flamelets and transported PDF approaches give similar results whose agreement with experimental data is only marginal [113].
3. Practical lean-burn systems often involve flow regions with both high and low Damköhler numbers, and thus models with a wider range of applicability are required. Specifically, the combustion submodel and the chemical kinetics should be able to accommodate both fuel-lean and fuel-rich combustion because practical lean premixed combustion almost always include both lean and rich regions involving different physics.
4. Reliable and accurate models capable of simulating auto-ignition, flashback, blowout, and quenching processes are needed.
5. Radiative heat loss and its interaction with turbulence need to be addressed. The formation rates of oxides of nitrogen are very sensitive to temperature, and thus their accurate prediction can sometimes require reliable models for radiation and its interaction with turbulence.
6. In situations in which liquid fuels are burned, models are needed to simulate atomisation and droplet evaporation and their interaction with turbulence and turbulent mixing of the fuel and air, in order to correctly predict the composition of the mixture entering the combustion zone. Can the modelling of atomisation and evaporation be decoupled from thermochemistry in spray combustion? If so, under what conditions would this decoupling be acceptable?

There is a real need to capture the complex geometries of practical combustors in CFD simulations, and the most computationally expensive combustion models may not be attractive for regular use in these circumstances. There will be a continuing need for both LES and RANS models. Also, the level of confidence to be placed in model predictions must be established. One way to achieve this is through a posteriori validation of the models using as much experimental data as possible. There is an urgent need for carefully designed experiments for model validation, both fully and partially premixed, with comprehensive scalar and velocity field information and covering the widest possible range of burning regimes. In particular, experiments are needed that identify conditions under which flamelet assumptions fail, and the nature of the failure. A clear demonstration of the circumstances under which laminar flamelet assumptions fail would be of great value. Although DNS provides very valuable insights, together with a tool to test model hypothesis and data for model validation, Reynolds numbers remain unrealistically low. A few test cases at the highest possible Reynolds numbers would enable the influence of this parameter to be explored.

To conclude, we would like to suggest this: *Nature is usually kind to us. Simple modelling methods based on a very good understanding of the underlying physics is generally adequate and it is seldom necessary to complicate our approach.*

REFERENCES

[1] J. M. Beer and N. A. Shiger, Swirling flows, in *Combustion Aerodynamics* (Applied Science Publisher, London, 1972), Chap. 5.

[2] *Bosch Automotive Handbook*, 3rd ed. (Robert Bosch GmbH, Germany, 1993).

[3] A. Cavliere, M. deJoannon, and R. Ragucci, Highly preheated lean combusiton, in D. Dunn-Rankin (ed.) *Lean Combustion: Technology and Control* (Academic, Burlington, VT, 2007), pp. 55–94.

[4] R. H. Stanglmaier and C. E. Roberts, Homogeneous charge compression ignition (HCCI): Benefits, compromises and future engine applications, SAE, paper 1999-01-3682 (1999).

[5] W. J. Kuper, M. Blaauw, F. van der Berg, and G. H. Graaf, Catalytic combustion concept for gas turbines, *Catalysis Today* **47**, 377–389 (1999).

[6] J. Wünning, Flammenlose Oxdation von Brennstoff mit hochvorgewärmter Luft, *Chem. Ing. Tech.* **63**, 1243–1245 (1991).

[7] A. Cavaliere and M. de Joannon, Mild combustion, *Prog. Energy Combust. Sci.* **30**, 329–366 (2004).

[8] M. Katsuki and T. Hasegawa, The science and technology of combustion in highly preheated air, *Proc. Combust. Inst.* **27**, 3135–3146 (1998).

[9] Y. Levy, V. Sherbaum, and V. Erenburg, The role of the recirculating gases in the MILD combustion regime formation, presented at the ASME Turbo Exposition, ASME paper GT2007-27766 (2007).

[10] L. Marco, B. Claudio, C. Giorgio, G. Eugenio, and R. Marco, Turbulent combustion in trapped vortex combustors, in *Proceedings of the XXV Event of the Italian Section of the Combustion Institute* (The Combustion Institute, Pittsburg, 2002), paper II, Vol. 13, pp. 3–5.

[11] A. D. Rao, G. S. Samuelsen, F. L. Robson, and R. A. Geisbrecht, Power plant system configurations for the 21st century, presented at the ASME Turbo Exposition, ASME paper GT2002-30671 (2002).

[12] S. Hayashi and H. Yamada, NO_x emissions in combustion of lean premixed mixtures injected into hot burned gas, *Proc. Combust. Inst.* **28**, 2443–2449 (2000).

[13] H. Yamada, H. Takagi, and S. Hayashi, Low-NO_x emissions over an enlarged range of overall equivalence ratios by staged lean premixed tubular flame combustion, presented at the ASME Turbo Exposition, ASME paper GT2005-68849 (2005).

[14] N. Aida, T. Nishijima, S. Hayashi, H. Yamada, and T. Kawakami, Combustion of lean prevaporized fuel–air mixtures mixed with hot burned gas for low NO_x emissions over an extended range of fuel–air ratios, *Proc. Combust. Inst.* **30**, 2885–2892 (2005).

[15] S. Adachi, A. Iwamoto, S. Hayashi, H. Yamada, and S. Kaneko, Emissions in combustion of lean methane–air and biomass–air mixtures supported by

primary hot burned gas in a multi-stage gas turbine combustor, *Proc. Combust. Inst.* **31**, 3131–3138 (2007).

[16] R. J. Kee, J. F. Grcar, M. D. Smooke, and J. A. Miller, A FORTRAN program for modeling steady laminar one-dimensional premixed flames, Sandia Rep. SAND85-8240 (1985).

[17] G. P. Smith, D. M. Golden, M. Frenklach, N. W. Moriarty, B. Eiteneer, M. Goldenberg, C. T. Bowman, R. K. Hanson, S. Song, W. C. Gardiner, V. V. Lissianski, and Z. Qin, GRI-Mech 3.0, available online at http://www.me.berkeley.edu/gri_mech/

[18] H. Fujiwara, M. Koyama, S. Hayashi, and H. Yamada, Development of a liquid-fueled dry low emissions combustor for 300 kW class recuperated cycle gas turbine engines, presented at the ASME Turbo Exposition, ASME paper GT2005-68645 (2005).

[19] R. Shibata, Y. Nakayama, K. Kobayashi, M. Koyama, and H. Fujiwara, The development of 300 kW class high efficiency, ultra low emission micro gas turbine RGT3R, presented at the CIMAC Congress, paper 52 (2004).

[20] S. Hayashi, H. Yamada, and M. Makida, Extending low-NO_x operating range of a lean premixed-prevaporized gas turbine combustor by reaction of secondary mixtures injected into primary stage burned gas, *Proc. Combust. Inst.* **30**, 2903–2911 (2004).

[21] R. Lebas, T. Menard, P. A. Beau, A. Berlemont, and F. X. Demoulin, Numerical simulation of primary break-up, and atomization: DNS and modelling study, *Int. J. Multiphase Flow* **35**, 247–260 (2009).

[22] M. G. Pai, O. Desjardins, and H. Pitsch, Detailed simulations of primary breakup of turbulent liquid jets in crossflow, in *CTR Annual Research Briefs* (Center for Turbulence Research, Stanford University, Stanford, CA, 2008).

[23] J. Shinjo and A. Umemura, Simulation of liquid jet primary breakup: Dynamics of ligament and droplet formation, *Int. J. Multiphase Flow* **36**, 513–532 (2010).

[24] F. A. Williams, Recent advances in theoretical descriptions of turbulent diffusion flames, in S. N. B. Murthy (ed.), *Turbulent Mixing in Nonreactive and Reactive Flows* (Plenum, New York, 1975), pp. 189–208.

[25] J. A. Wünning, J. G. Wünning, Flameless oxidation to reduce thermal NO formation, *Prog. Energy Combust. Sci.* **23**, 81–94 (1995).

[26] D. Tabacco, C. Innarella, and C. Bruno, Theoretical and numerical investigation on flameless combustion, *Combust. Sci. Technol.* **174**, 1–35 (2002).

[27] C. K. Westbrook, Y. Mizobuchi, T. J. Poinsot, P. J. Smith, and J. Warnatz, Computational combustion, *Proc. Combust. Inst.* **30**, 125–157 (2004).

[28] J. B. Bell, M. S. Sheperd, M. R. Johnson, R. K. Cheng, J. F. Grcar, V. E. Beckner, and M. J. Lijewski, Numerical simulation of a laboratory-scale turbulent V-flame, *Proc. Natl. Acad. Sci. USA* **102**, 10006–10011 (2005).

[29] C. S. Yoo, R. Sankaran, and J. H. Chen, Three-dimensional direct numerical simulation of a turbulent lifted hydrogen jet flame in heated coflow: Flame stabilization and structure, *J. Fluid Mech.* **640**, 453–481 (2009).

[30] U. Maas and S. B. Pope, Simplifying chemical kinetics: Intrinsic low-dimensional manifolds in composition space, *Combust. Flame* **88**, 239–264 (1992).

[31] O. Gicquel, N. Darabiha, and D. Thévenin, Laminar premixed hydrogen/air counterflow flame simulation using flame prolongation of ILDM with differential diffusion, *Proc. Combust. Inst.* **28**, 1901–1908 (2000).

[32] A. M. El Kady, A. Evulet, T. Ursin, and A. Lynghjem, Application of exhaust gas recirculation in a DLN F-class combustion system for post-combustion carbon capture, *J. Eng. Gas Turbines Power* **131**, 034505-1 (2009).

[33] J. R. Maughan, K. M. Elward, S. M. De Pietro, and P. J. Bautista, Field test results of a dry low NO_x combustion system for the M3002J regenerative cylce gas turbine, *J. Eng. Gas Turbines Power* **119**, 50 (1997).

[34] S. Hoffman, S. Bartlett, M. Finkerath, A. Evulet, and T. Ursin, Performance and cost analysis of advanced gas turbine cycles with postcombustion CO2 capture, *J. Eng. Gas Turbines Power* **131**, 021701-1 (2009).

[35] B. Andreini and B. Facchini, Gas turbines design and off-design performance analysis with emissions – Evaluation, *J. Eng. Gas Turbines Power* **126**, 83–91 (2004).

[36] M. Johnson, D. Littlejohn, W. Nazeer, K. Smith, and R. K. Cheng, A comparison of the flowfields and emissions of high-swirl injectors and low-swirl injectors for lean premixed gas turbines, *Proc. Combust. Inst.* **30**, 2867–2874 (2005).

[37] H. C. Mongia, TAPS – A 4th generation propulsion combustor technology for low emissions, AIAA paper 2003–2657 (2003).

[38] S. Dhanuka, J. Temme, and J. F. Driscoll, Vortex shedding and mixing layer effects on periodic flashback in a lean premixed prevaporized gas turbine combustor, *Proc. Combust. Inst.* **32**, 2901–2908 (2009).

[39] U. Stopper, M. Aigner, W. Meier, R. Sandanandan, M. Stohr, and I. S. Kim, Flow field and combustion characterization of premixed gas turbine flames by planar laser techniques, *J. Eng. Gas Turbines Power* **131**, 021504-1 (2009).

[40] S. Dhanuka, J. Temme, and J. F. Driscoll, Unsteady aspects of lean premixed-prevaporized gas turbine combustors: Flame-flame and flame-shear layer Interactions, submitted to *J. Propul. Power*.

[41] G. Eggenspieler and S. Menon, Large eddy simulation of pollutant emission in a DOE-HAT combustor, *J. Propul. Power* **20**, 1076–1085 (2004).

[42] Y. Huang, H. G. Sung, S.-Y. Hsieh, and V. Yang, Large eddy simulation of combustion dynamics of lean premixed swirl stabilized combustor, *J. Propul. Power* **19**, 782–794 (2003).

[43] P. N. Kioni, K. N. C. Bray, D. A. Greenhalgh, and B. Rogg, Experimental and numerical studies of a triple flame, *Combust. Flame* **116**, 192–206 (1999).

[44] Y. Mizobuchi, S. Tachibana, J. Shinio, S. Ogawa, and T. Takeno, A numerical analysis of the structure of a turbulent hydrogen jet lifted flame, *Proc. Combust. Inst.* **29**, 2009–2015 (2002).

[45] R. S. Barlow, G. H. Wang, P. Anselmo, M. S. Sweeney, and S. Hochgreb, Application of Raman/Rayleigh/LIF diagnostics in turbulent stratified flame, *Proc. Combustion Inst.* **32**, 945–953 (2009).

[46] B. O. Ayoola, R. Balachandran J. H. Frank, E. Mastorakos, and C. F. Kaminski, Spatially resolved heat release rate measurements in turbulent premixed flames, *Combust. Flame* **144**, 16 (2006).

[47] B. A. V. Bennett, Z. Cheng, R. W. Pitz, and M. D. Smooke, Computational and experimental study of oxygen-enhanced axisymmetric laminar methane flames, *Combust. Theory Model.* **12**, 497–527 (2008).

[48] G. T. Kalghatgi, Liftoff heights and visible lengths of vertical turbulent jet diffusion flames in still air, *Combust. Sci. Technol.* **41**, 17–29 (1984).

[49] G. T. Kalghatgi, Blow out stability of gaseous jet diffusion flames. Part 1: In still air, *Combust. Sci. Technol.* **26**, 233–239 (1981).

[50] G. T. Kalghatgi, Blow out stability of gaseous jet diffusion flames. Part 2: Effect of cross wind, *Combust. Sci. Technol.* **26**, 241–244 (1981).

[51] W. J. A. Dahm and A. G. Mayman, Blowout limits of turbulent jet diffusion flames for arbitrary source conditions, *AIAA J.* **28**, 1157–1162 (1990).

[52] D. Feikema, R. H. Chen, and J. F. Driscoll, Enhancement of flame blowout limit by the use of swirl, *Combust. Flame* **80**, 183–195 (1990).

[53] J. F. Driscoll and C. C. Rasmussen, Correlation and analysis of blowout limits of flames in high-speed airflows, *J. Propul. Power* **21**, 1035–1044 (2005).

[54] L. Vanquickenborne and A. van Tiggelen, The stabilization mechanism of lifted diffusion flames, *Combust. Flame* **10**, 59–69 (1966).

[55] Y. C. Chen and R. W. Bilger, Stabilization mechanisms of lifted laminar flames in axisymmetric jet flows, *Combust. Flame* **123**, 23–45 (2000).

[56] D. Bradley, P. H. Gaskell, and X. J. Gu, The mathematical modeling of liftoff and blowoff of turbulent nonpremixed methane jet flames at high strain rates, *Proc. Combust. Inst.* **27**, 1199–1206 (1998).

[57] C. B. Devaud and K. N. C. Bray, Assessment of the applicability of conditional moment closure to a lifted turbulent flame: First order model, *Combust. Flame* **132**, 102–114 (2003).

[58] P. Domingo, L. Vervisch, and J. Reveillon, DNS analysis of partially premixed combustion in spray and gaseous turbulent flame-bases stabilized in hot air, *Combust. Flame* **140**, 172–195 (2005).

[59] S. A. Ferraris and J. X. Wen, Large eddy simulation of a lifted turbulent jet flame, *Combust. Flame* **150**, 320–339 (2007).

[60] L. Muniz and M. F. Mungal, Instantaneous flame-stabilization velocities in lifted-jet diffusion flames, *Combust. Flame* **111**, 16–30 (1997).

[61] A. Upatnieks, J. F. Driscoll, C. C. Rasmussen, and S. L. Ceccio, Liftoff of turbulent jet flames – assessment of edge flame and other concepts using cinema-PIV, *Combust. Flame* **138**, 259–272 (2004).

[62] W. Meier, X. R. Duan, and P. Weigand, Reaction zone structures and mixing characteristics of partially premixed swirling CH_4–air flames in a gas turbine model combustor. *Proc. Combust. Inst.* **30**, 835–842 (2005).

[63] S. A. Filatyev, J. F. Driscoll, C. D. Carter, and J. M. Donbar, Measured properties of turbulent premixed flames for model assessment, including burning velocities, stretch rates and surface densities. *Combust. Flame* **141**, 1–21 (2005).

[64] M. Stohr, R. Sadanandan, and W. Meier, Experimental study of unsteady flame structures of an oscillating swirl flame in a gas turbine model combustor, *Proc. Combust. Inst.* **32**, 2925–2932 (2009).

[65] C. M. Fajardo, J. D. Smith, and V. Sick, Sustained simultaneous high-speed imaging of scalar and velocity fields using a single laser, *Appl. Phys. B* **85**, 25–31 (2006).

[66] M. E. Cundy and V. Sick, Hydroxyl radical imaging at kHz rates using a frequency-quadrupled Nd:YLF laser, *Appl. Phys. B* **96**, 241–245 (2009).

[67] C. Meneveau and T. J. Poinsot. Stretching and quenching of flamelets in premixed turbulent combustion, *Combust. Flame* **86**, 311–332 (1991).

[68] A. M. Steinberg, J. F. Driscoll, and S. L. Ceccio, Temporal evolution of flame stretch due to turbulence and the hydrodynamic instability, *Proc. Combust. Inst.* **32**, 1713–1721 (2009).

[69] A. M. Steinberg, J. F. Driscoll, and S. L. Ceccio, Three-dimensional temporally resolved measurements of turbulence–flame interactions using orthogonal-plane cinema-stereoscopic PIV, *Exp. Fluids* **47**, 527–547 (2009).

[70] A. M. Steinberg and J. F. Driscoll, Stretch-rate relationships for LES subgrid models measured using temporally resolved diagnostics, *Combust. Flame* **156**, 2285–2306 (2010).

[71] C. K. Law and C. J. Sung, Structure, aerodynamics, and geometry of premixed flamelets, *Prog. Energy Combust. Sci.* **26**, 459–505 (2000).

[72] H. Pitsch and D. L. Duchamp, Large-eddy simulation of premixed turbulent combustion using a level-set approach. *Proc. Combust. Inst.* **29**, 2001–2008 (2002).

[73] J. F. Driscoll, Turbulent premixed combustion: Flamelet structure and its effect on turbulent burning velocities, *Prog. Energy Combust. Sci.* **34**, 91–134 (2008).

[74] A. N. Lipatnikov and J. Chomiak, Turbulent flame speed and thickness: Phenomenology, evaluation, and application in multi-dimensional simulations, *Prog. Energy Combust. Sci.* **28**, 1–74 (2002).

[75] A. N. Lipatnikov and J. Chomiak, Molecular transport effects on turbulent flame propagation and structure, *Prog. Energy Combust. Sci.* **31**, 1–73 (2005).

[76] J. B. Bell, M. S. Day, I. G. Shepherd, M. R. Johnson, R. K. Cheng, and J. F. Grcar, Numerical simulation of a laboratory scale turbulent V-flame, *Proc. Natl. Acad. Sci. USA* **102**, 10006–11 (2005).

[77] J. B. Bell, M. S. Day, J. F. Grcar, M. J. Lijewski, J. F. Driscoll, and S. A. Filatyev, Numerical simulation of a laboratory-scale turbulent slot flame, *Proc. Combust. Inst.* **31**, 1299–1307 (2006).

[78] F. Charlette, C. Meneveau, and D. Veynante, A power law flame wrinkling model for LES of premixed turbulent combustion Part I – Non-dynamic formulation and initial tests, *Combust. Flame* **131**, 159–180 (2002).

[79] N. Peters, The turbulent burning velocity for large-scale and small-scale turbulence. *J. Fluid Mech.* **384**, 107–132 (1999).

[80] J. M. Duclos, D. Veynante, and T. J. Poinsot, A comparison of flamelet models for premixed turbulent combustion. *Combust. Flame* **95**, 1–16 (1993).

[81] R. G. Abdel-Gayed, D. Bradley, M. N. Hamid, and M. Lawes, Lewis number effects on turbulent burning velocity, *Proc. Combust. Inst.* **20**, 505–512 (1984).

[82] E. R. Hawkes and R. S. Cant, Implications of a flame surface density approach to large eddy simulation of premixed turbulent combustion, *Combust. Flame* **126**, 1617–1629 (2001).

[83] N. Peters, *Turbulent Combustion* (Cambridge University Press, Cambridge, 2000).

[84] G. A. Richards, M. M. McMillian, R. S. Gemmen, W. A. Rogers, and S. R. Cully, Issues for low-emission, fuel-flexible power systems, *Prog. Energy Combust. Sci.* **27**, 141–169 (2001).

[85] R. F. Sawyer, Science based policy for addressing energy and environmental problems, *Proc. Combust. Inst.* **32**, 45–56 (2009).

[86] J. B. Moss, Combustion research – 25 years on, *Combust. Sci. Technol.* **98**, 337–340 (1994).

[87] J. J. McGuirk and A. M. K. P. Taylor, Numerical simulation of gas turbine combustion processes-present status and future trends, *J. Gas Turbulence Soc. Jpn.* **30**, 347–361 (2002).

[88] R. W. Bilger, S. B. Pope, K. N. C. Bray, and J. F. Driscoll, Paradigms in turbulent combustion research. *Proc. Combust. Inst.* **30**, 21–42 (2005).

[89] C. K. Law, Combustion at a crossroads: Status and prospects, *Proc. Combust. Inst.* **31**, 1–29 (2007).

[90] N. Peters, Multiscale combustion and turbulence, *Proc. Combust. Inst.* **32**, 1–25 (2009).

[91] A. D'Anna, Combustion-formed nanoparticles, *Proc. Combust. Inst.* **32**, 593–613 (2009).

[92] J. H. Miller, Aromatic excimers: Evidence for polynuclear aromatic hydrocarbon condensation in flames, *Proc. Combust. Inst.* **30**, 1381–1388 (2005).

[93] S. Honnet, K. Seshadri, U. Niemann, and N. Peters, A surrogate fuel for kerosene, *Proc. Combust. Inst.* **32**, 485–492 (2009).

[94] G. E. Hatch, E. Boykin, J. A. Graham, J. Lewtas, F. Pott, K. Loud, and J. L. Mumford, Inhalable particles and pulmonary host defense: *In vivo* and *in vitro* effects of ambient air and combustion particles, *Environ. Res.* **36**, 67–80 (1985).

[95] I. M. Kennedy, The health effects of combustion-generated aerosols, *Proc. Combust. Inst.* **31**, 2757–2770 (2007).

[96] K. Pikoň, Environmental impact of combustion, *Appl. Energy* **75**, 213–220 (2003).

[97] P. J. Coelho, Numerical simulation of the interaction between turbulence and radiation in reactive flows, *Prog. Energy Combust. Sci.* **33**, 311–383 (2007).

[98] M. Roger, C. B. Da Silva, and P. J. Coelho, Analysis of the turbulence–radiation interactions for large eddy simulations of turbulent flows, *J. Heat Mass Transfer* **52**, 2243–2254 (2009).

[99] A. Wang, M. F. Modest, D. C. Haworth, and L. Wang, Monte Carlo simulation of radiative heat transfer and turbulence interactions in methane/air jet flames, *J. Quant. Spectrosc. Radiat. Transfer* **109**, 269–279 (2008).

[100] Y. Wu, D. C. Haworth, M. F. Modest, and B. Cuenot, Direct numerical simulation of turbulence/radiation interaction in premixed combustion systems, *Proc. Combust. Inst.* **30**, 639–646 (2005).

[101] L. H. Liu, X. Xu, and Y. L. Chen, On the shapes of the presumed probability density function for the modeling of turbulence–radiation interactions, *J. Quant. Spectrosc. Radiat. Transfer* **87**, 311–323 (2004).

[102] J. Wu, P. Brown, I. Diakunchak, and A. Gulati, Advanced gas turbine combustion system development for high hydrogen fuels, presented at the ASME Turbo Exposition: Power for Land, Sea and Air, ASME paper, GT2007-28337 (2007).

[103] R. Prasad, L. A. Kennedy, and E. Ruckenstein, Catalytic combustion, *Catal. Rev. Sci. Eng.* **26**, 1–58 (1984).

[104] W. C. Pfefferle and L. D. Pfefferle, Catalytically stabilized combustion, *Prog. Energy Combust. Sci.* **12**, 25–41 (1986).

[105] R. E. Hayes and S. T. Kolaczkowski, *Introduction to Catalytic Combustion* (Gordon & Breach, Amsterdam, 1997).

[106] G. E. Moore, Cramming more components onto integrated circuits, *Electronics*, **38** (8) (1965).

[107] S. B. Pope, Ten questions concerning the large-eddy simulation of turbulent flows, *New J. Phys.* **6**, 1–24 (2004).

[108] N. V. Nikitin, F. Nicouda, B. Wasistho, K. D. Squires, and P. R. Spalart, An approach to wall modeling in large-eddy simulations, *Phys. Fluids* **12**, 1629–1632 (2000).

[109] J. G. Brasseur and T. Wei, Designing large-eddy simulation of the turbulent boundary layer to capture law-of-the-wall scaling, *Phys. Fluids* **22**, 021303-1-21 (2010).

[110] S. E. Gant, Reliability issues of LES-related approaches in an industrial context, *Flow Turbulence Combust.* **84**, 325–335 (2010).

[111] K. N. C. Bray and N. Swaminathan, Scalar dissipation and flame surface density in premixed turbulent combustion, *C. R. Mec.* **334**, 466–473 (2006).

[112] H. Kolla and N. Swaminathan, Strained flamelets for turbulent premixed flames, I: Formulation and planar flame results, *Combust. Flame*, **157**, 943–954 (2010).

[113] H. Kolla and N. Swaminathan, Strained flamelets for turbulent premixed flames, II: Laboratory flame results, *Combust. Flame* **157**, 1274–1289 (2010).

[114] M. Stöllinger and S. Heinz, Evaluation of scalar mixing and time scale models in PDF simulations of a turbulent premixed flame, *Combust. Flame* **157**, 1671–1685 (2010).

[115] S. M. Martin, J. C. Kramlich, G. Kosaly, and J. J. Riley, The premixed conditional moment closure method applied to idealised lean premixed gas turbine combustors, *J. Eng. Gas Turbine Power* **125**, 895–900 (2003).

[116] A. Y. Klimenko and R. W. Bilger, Conditional moment closure for turbulent combustion, *Prog. Energy Combust. Sci.* **25**, 595–687 (1999).

Nomenclature

A		localised area of flame front
$A(s), F(s)$		acoustic transfer functions
$\overline{A}(n)$		logarithmic growth rate of perturbation
A'		empirical constant $= 16$, in relationship between λ and Λ
$A_{\bar{c}}$		constant in Eq. (3.25)
$\mathcal{A}_\varphi$		area of iso-φ surface
a		perturbation of flame radius, amplitude of sin component
a_{turb}		turbulent strain rate $1/\tau_T$
$a_{t,\text{sg}}$		subgrid tangential strain rate
B		numerical constant for a mixture; see Eq. (3.13)
b		amplitude of cos component
b_{ij}		anisotropy tensor
$C_1,\quad C_1^*$		
$C_2,\quad C_3,\quad C_3^*$		Reynolds stress closure constants
$C_4,\quad C_5,\quad C_S$		
$C_{S\epsilon},\quad C_{\epsilon 1},\quad C_{\epsilon 2}$		dissipation equation constants
$C_\phi,\quad C_\phi^*$		time-scale ratio constants
C_μ		eddy viscosity model parameter $= 0.09$
c		reaction progress variable
$\bar{c}$		mean reaction progress variable
$\hat{c}$		acoustic speed
$\widetilde{c''^2}$		Favre variance of c
c_p		specific heat capacity at constant pressure for the mixture
$c_{p,i}$		specific heat capacity at constant pressure for species i
$\mathcal{D}$		molecular diffusivity
$\hat{D}$		fractal dimension
$\mathcal{D}_T$		turbulent diffusivity, subgrid scale turbulent diffusivity
$\mathcal{D}_i$		molecular diffusivity for species i
Da		Damköhler number ($\equiv \Lambda\, S_L^0/(u_{rms}\delta_L^0)$)
Da$_\Delta$		Damköhler number at filter scale ($\equiv \Delta\, S_L^0/(u'\delta_L^0)$)
Da$_L$		local Damköhler number

Da_L^*	density-weighted local Damköhler number ($\rho_u Da_L / \overline{\rho}$)
d	jet exit diameter; distance to pressure measurement
$\boldsymbol{d}$	data vector, $[p_{\text{ref}} \dfrac{V_c}{s + z_c}]^T$
$\overline{d}$	d/δ_ℓ
E_r^a	activation energy for reaction r
$e_{\ell k}$	symmetric part of strain tensor $\partial u_\ell / \partial x_k$
$e_\alpha, e_\beta, e_\gamma$	eigenvalues of $e_{\ell k}$ with $e_\alpha > e_\beta > e_\gamma$
F	ratio of flame speeds, S_n / S_s
f	lag factor for n_s; frequency in Hz
f_i	body force per unit volume for species i
$G(s)$	overall open-loop transfer function
g	variance parameter ($\widetilde{c''^2} / \widetilde{c}(1 - \widetilde{c})$)
g_0	sign of the high-frequency gain
$H(s)$	flame transfer function
h	specific enthalpy of the mixture
h_i	specific enthalpy of species i
i	$\sqrt{-1}$
$J_{i,k}$	molecular diffusion flux vector
K	strain rate; mean stretch rate; flame stretch parameter ($= 0.157\,\text{Ka}_L$); turbulent Karlovitz stretch factor, $(u_{\text{rms}}/\lambda)(\delta_\ell/u_\ell)$
$K(s)$	controller transfer function
$K_{0.2}$	turbulent Karlovitz stretch factor (K) for probability of flame propagation of 0.2
$K_{0.8}$	turbulent Karlovitz stretch factor (K) for probability of flame propagation of 0.8
K_ℓ	laminar Karlovitz stretch factor, $\alpha^* \delta_\ell / u_\ell$
K_c	laminar Karlovitz stretch factor, $\alpha^* \delta_\ell / u_\ell$, for flame curvature
K_s	laminar Karlovitz stretch factor, $\alpha_s^* \delta_\ell / u_\ell$, for aerodynamic strain
K_{cl}	laminar critical Karlovitz stretch factor at the onset of instability
$K_{c,sg}$	stretch rate due to subgrid flame curvature
Ka	Karlovitz number $[(\delta/\eta_K)^2 = \tau_c/\tau_K]$
Ka_L	local Karlovitz number
k	wave number $\omega/\hat{c}$; turbulent kinetic energy; control parameter
$\widetilde{k}$	Favre-averaged turbulent kinetic energy
k_1, k_2, z_c, p_c	controller parameters
$\boldsymbol{k}$	controller vector $[k_1\, k_2]^T$
L	length of Rijke tube
L_a	characteristic length scale of the acoustic perturbations
L_b	burned-gas Markstein length
L_f	characteristic length scale of the turbulent flame
Le_i	Lewis number for species i ($\hat{\alpha}/\mathcal{D}_i$)
l_{eff}	effective neck length of Helmholtz resonator

$M, \mathcal{M}$	Mach number
Ma	general, undifferentiated, Markstein number
Ma_b	flame-speed Markstein number
Ma_{cr}	Markstein number for flame curvature
Ma_{sr}	Markstein number for aerodynamic strain rate
M_ℓ	$-\partial \tilde{c}/\partial x_\ell / |\nabla \tilde{c}|$
m_a	mass flow rate of air
m_f	mass flow rate of fuel
N	size of the nonlinear eigenvalue problem
	number of particles
N_c	instantaneous SDR $(\hat{\alpha} \partial c/\partial x_k \partial c/\partial x_k)$
n	wave number and pressure exponent
n_k	unit normal in direction k for an iso-scalar
	surface pointing towards smaller scalar values
n_l	smallest unstable wave number
n_s	largest unstable wave number
n^*	relative degree of transfer function
$n_{\mathbf{u}}$	field of interaction index
$(n_{\text{ref},1}, n_{\text{ref},1}, n_{\text{ref},1})$	reference unit vector
$(n_{\text{BC},1}, n_{\text{BC},1}, n_{\text{BC},1})$	outward unit vector normal to the boundary
n_{cl}	wave number at Pe_{cl}
$n_{s,cl}$	largest unstable wave number at Pe_{cl}
$\overline{P}$	RANS or LES PDF
$\widetilde{P}$	Favre (density-weigted) PDF
P_{kk}	Reynolds stress production term
Pe	Peclet number
Pe_{cl}	critical Peclet number
Pe_0	Pe at $\bar{t} = 0$
Pr	Prandtl number $(\mu\, c_p / \hat{\lambda})$
p	pressure
p'	fluctuating pressure
p_f	probability of flame propagation $(f = 0.2 \text{ and } 0.8)$
p_{ref}	measured sensor pressure
Q	heat of reaction
	heating value of the fuel
$\dot{Q}_{\text{rad}}$	radiative energy exchange per unit volume
q	heat release rate per unit valume
q_{tot}	volume averaged heat release rate used for scaling purpose
q'	fluctuating volumetric heat release rate
$\mathcal{R}$	universal gas constant
Re_T	turbulent Reynolds number $(u_{\text{rms}} \Lambda / \nu_u)$
r	flame or hot-spot radius, difference in heat capacities $(c_p - c_v)$
r_0	initial hot-spot radius
S	stretched flame speed; cross-section area; neck area of
	Helmholtz resonator
S_n	unstable flame speed

S_s	flame speed at zero stretch
S_L	laminar flame speed
S_d	dispalcement speed
S_T	turbulent flame speed
$S_{T,\mathrm{SG}}$	propagation speed of resolved interface
S_L^0	unstrained planar laminar flame speed
$S_{\kappa,L}$	curvatured-induced laminar flame speed
$S_{\ell k}$	mean fluid dynamic strain tensor
Sc_i	Schmidt number for species i
s	entropy per mass unit; Laplace transform variable
T	absolute temperature
T_a	burning velocity activation temperature
T_b	burnt side temperaure
T_p	pilot flame temperature
T_u	unburnt side (reactant) temperature
t	time
t_a	time for sound wave to propagate
t_{r0}	acoustic residence time $(= r_0/\hat{c})$
$\bar{t}$	dimensionless time (tu_ℓ^2/ν)
$\bar{t}_a$	dimensionless time for acoustic wave
U_{BO}	blowout velocity
U_{bulk}	bulk velocity of the burner (scaling purpose)
U_{ref}	a reference velocity
u_a	auto-ignition front propagation velocity
u_c	unstable burning velocity
u_k	velocity component in direction x_k
u_ℓ, u_L	unstretched laminar burning velocity
u_{nr}	stretched laminar burning velocity
u_{rms}	turbulence rms velocity
u_t	turbulent burning velocity
$u_{\bar{c}}$	turbulent burning velocity at flame surface defined by $\bar{c}$
u'	rms velocity at filter scale Δ
V	net volume generated at reaction front; cavity volume of Helmholtz resonator
v	specific volume
V_c	actuator signal used to drive fuel valve
v_K	Kolmogorov velocity
W_i	molecular weight of species i
W_{ac}	actuator transfer function
W_{cl}^*	closed-loop transfer function resulting from stabilising controller
$W_{\ell k}$	mean vorticity tensor
X_i	mole fraction of species i
$\mathbf{x}(x_1, x_2, x_3)$	position vector
Y_i	mass fraction of species i
Z	mixture fraction; reduced boundary impedance

Z_{st}	stoichiometric mixture fraction
$\langle \cdots \rangle_s$	surface average; filtered surface average; gradient weighted average
$\tilde{\cdots}$	Favre average

Greek Symbols

α	relaxation coefficient; efficiency factor; constant in Eqs. (3.25) and (3.26); efficiency function
$\hat{\alpha}$	thermal diffusivity ($\hat{\lambda}/\rho c_p$); diffusivity of c
$\hat{\alpha}_T$	eddy mass diffusivity
α_c^*	contribution of flame curvature to flame stretch rate
α_{cl}^*	total stretch rate at $\mathrm{Pe_{cl}}$
α_s^*	contribution of aerodynamic strain to flame stretch rate
α^*	total flame stretch rate $= \alpha_s^* + \alpha_c^*$
α, β, γ	weights in the BML PDF for c
β	constant in Eqs. (3.25) and (3.26)
$\hat{\beta}$	Zeldovich number
Γ_K	stretch efficiency function; diagonal matrix containing convergence coefficients
γ	ratio of specific heats (c_p/c_v)
γ^*	sample space variable for scalar gradient magnitude
Δ	filter scale
Δp	increase in pressure above ambient
δ	Zeldovich thickness ($\hat{\alpha}/S_L^0$)
δ_f	effective 1D premixed flame thickness
δ_ℓ	laminar flame thickness, given by ν/u_ℓ
$\delta_{\ell k}$	Kronecker delta
δ_L^0	unstrained planar laminar flame thermal thickness
ϵ	small parameter of the order of p_1/p_0
$\tilde{\epsilon}_c$	Favre-averaged scalar dissipation rate of progress variable ($\overline{\rho\hat{\alpha}\partial c''/\partial x_k \partial c''/\partial x_k}/\overline{\rho}$)
$\tilde{\epsilon}_Z$	Favre-averaged scalar dissipation rate of mixture fraction ($\overline{\rho\hat{\alpha}\partial Z''/\partial x_k \partial Z''/\partial x_k}/\overline{\rho}$)
$\tilde{\epsilon}_{cz}$	Favre-averaged cross dissipation rate ($\overline{\rho\hat{\alpha}\partial c''/\partial x_k \partial Z''/\partial x_k}/\overline{\rho}$)
ε	acoustic residence/excitation time (t_{r0}/τ_e)
$\tilde{\varepsilon}$	dissipation rate of $\tilde{k}$
ζ	sample space (stochastic) variable for c
η_K	Kolmogorov length scale
η_c	inner cut-off scale
Λ	turbulence integral length scale
$\hat{\Lambda}$	normalised wavelength
$\hat{\Lambda}_l$	largest normalised unstable wavelength
$\hat{\Lambda}_s$	smallest normalised unstable wavelength
λ	Taylor length scale of turbulence, excess air factor $= 1/\phi$

$\hat{\lambda}$	thermal conductivity
μ	molecular dynamic viscosity
μ_t	dynamic eddy viscosity
ν	kinematic viscosity; stoichiometric coefficient
$\hat{\nu}$	subgrid-scale viscosity
ν_T	kinematic eddy viscosity
Ξ	flame wrinkling factor
ξ	stochastic variable for Z, $= \hat{c}/u_a$
ξ_l	lower detonation limit of ξ
ξ_u	upper detonation limit of ξ
ρ	fluid mixture density
ρ_u	reactant density
Σ	FSD
Σ_{sg}	subgrid FSD
$\overline{\Sigma}$	generalised FSD
σ	unburned-, burned-gas density
σ_i	turbulent Schmidt number for quantity i
τ	heat release parameter $[(T_b - T_u)/T_u]$; time delay; wave period
τ_c	chemical time scale
τ_e	excitation time
τ_i	ignition delay time
τ_K	Kolmogorov time scale
τ_s	scalar integral time scale ($\widetilde{c''^2}/\widetilde{\epsilon_c}$)
τ_T	turbulence time scale
$\tau_{\mathbf{u}}$	field of time delays
τ_{lk}	shear stress tensor
τ_K	Kolmogorov time scale
ϕ	equivalence ratio
φ_v	subgrid variance for φ
$\overline{\varphi}$	Reynolds mean of φ (resolved in LES)
$\widetilde{\varphi}$	Favre mean of φ (resolved in LES)
χ_{ij}	instantaneous cross dissipation rate ($\mathcal{D}\nabla Y_i \cdot \nabla Y_j$)
ψ	sample-space variable for stretch rate K
Ω	instability parameter; computational domain
$\dot{\Omega}$	heat release rate per unit valume
ω	growth rate parameter; angular frequency
$\dot{\omega}_i$	mass reaction rate of species i
ϖ_i	molar reaction rate of species i
$\partial\Omega$	boundary of the computational domain
$\partial\Omega_N$	subset of the boundary where $\nabla\hat{p} \cdot \mathbf{n}_{BC} = 0$
$\partial\Omega_D$	subset of the boundary where $\hat{p} = 0$
$\partial\Omega_Z$	subset of the boundary where the reduced impedance Z is imposed

Acronyms

ACARE	Advisory Council for Aeronautical Research in Europe
ATDC	after top dead centre
AFR	air fuel ratio
BACT	best available control technology
BARCT	best available retrofit control technology
BDC	bottom dead centre
BFG	blast furnace gas
BMEP	brake mean effective pressure
BSFC	brake specific fuel consumption
BTDC	before top dead centre
CAA	clean air act
CAEP	Committee on Aviation and Environmental Protection
CFD	computational fluid dynamics
CI	compression igrition
COG	coke oven gas
CSP	computational singular peturbation
DI	direct injection
DLE	dry low emissions
DLN	dry low NO_x
DLTD	Darrieus–Landau thermodiffusive instability
DNS	direct numerical simulation
DPF	diesel particulate filter.
EGR	exhaust-gas recirculation
EI	emission index: grams of pollutant per kilogram of fuel burnt
EPA	Environmental Protection Agency
FGM	flame-generated manifold
FLOPS	floating-point operations per second
FPI	flame prolongation ILDM
FSD	flame surface density
GHG	green house gas
GLM	generalised langevian model
HCCI	homogeneous charge compression ignition
ICAO	International Civil Aviation Organisation
ILDM	intrinsic low-dimensional manifold
IMEP	indicated mean effective pressure
ISA	international standard atmosphere
ISAT	in-situ adaptive tabulation
ISFC	indicated specific fuel consumption
LBDI	learn burn direct injection
LES	large-eddy simulation
LDG	Linz–Donawitz gas (a by-product of steel making)
LMS	least mean squares
LNG	liquefied natural gas
LPP	lean premixed prevapourised
LTO	landing and take off (cycle)

LTC	low temperature combustion
MFB	mass fraction burnt
MFM	multidimensional flamelet-generated manifolds
NAAQS	national ambient air quality standards.
NGV	nozzle guide vane
OLTF	open-loop transfer function
OPR	overall pressure ratio at 100% SLS thrust
PDF	probability density function
PFI	port fuel injection
PIV	particle image velocimetry
PLIF	planar laser-induced florescence
PM	particulate matter
PRF	primary reference fuel.
RANS	Reynolds-averaged Navier Stokes
RHP	right half-plane
RQL	rich–quench–lean (NO_x reduction strategy)
SCR	selective catalytic reduction
SDR	scalar dissipation rate
SFC	specific fuel consumption
SGS	subgrid scale
SLS	sea-level static
SMD	Sauter mean diameter.
STR	self-tunning regulator
TAPS	twin annular premixing swirler
TDC	top dead centre
TET	turbine entry temperature
TRL	technology readiness level
UHC	unburnt hydro carbon.
ULBDI	ultra lean burn direct injection
URANS	unsteady RANS
VOC	volatile organic compound
VTEC	variable value timing and lift control.
YAG	Yttrium aluminium garnet
YLF	Yttrium lithium flouoride

Index

Printed in the United States
By Bookmasters